CONSUMER

AUTOMOBILE BOOK

THE COMPLETE NEW CAR BUYING GUIDE

ALL NEW 1989 EDITION

Contents

Acknowledgments

Contributing Writer
James M. Flammang
(Safety, Insurance and Consumer Complaints sections)

CONSUMER GUIDE®

INTRODUCTION

Once upon a time, when cars were much cheaper, buying a new car was mostly a matter of choosing which color and body style you liked at the nearest showroom for your favorite brand name. Now, for most people, it's a complicated decision, starting with, should they buy a car, or should it be a compact van or a 4-wheel-drive off-road vehicle?

Buyers also face several other decisions, some of which produce unavoidable conflicts. Sporty, high-performance cars are coveted mainly by younger buyers, but they can least afford the high insurance premiums for such cars. Small cars are generally cheaper to buy and more economical to operate, yet big cars generally offer better occupant protection. Domestic cars are now less expensive than most of their Japanese competitors, yet the promise of better reliability makes the Japanese models more attractive to many buyers despite their higher prices. Twenty years ago nearly all cars sold in the U.S. had rear-wheel drive. Now it's a choice among rear drive, front drive and the increasingly popular 4WD.

After wrestling with those decisions, then the buyer faces the challenge of dealing with dealers, and that can be challenging indeed. If you're looking at a car that's in short supply and high demand, you're lucky if you can get the time of day from a salesman, let alone a price quote unless you put down a $500 deposit. If you're looking at a slow-selling model that's been collecting dust, you might be getting a good price on the new car, but watch out for low trade-in value for your old car. And, then you have to decide if you need the $700 "protection package" and the $600 extended warranty that the salesman is pushing, and whether the dealer's financing is cheaper than the bank's.

Buying a car isn't easy or fun for most people, especially when the average price of a new car today is over $13,000. Owning a car isn't always fun either, especially if it turns out to be a lemon or a car plagued by safety problems. The 1989 edition of the *Automobile Book* addresses the major concerns of buying and owning a car, from judging how safe a car is to finding the best insurance at the lowest cost, and from shopping for financing to taking a car company to court over unresolved problems.

As in previous years, most of the *Automobile Book* is devoted to the Buying Guide, which this year covers 165 passenger models of all sizes and shapes. The information under "What's New" for each model is supplied by the manufacturers, as are the specifications. We caution you that every year we find that some of what is promised by the car companies never materializes; engines, certain options even models are delayed until the next model year or scrapped as the market changes. If you have trouble finding what's listed in this issue at a dealer, then contact us and we'll try to give you current information. The address is listed below.

Our judgments of the vehicles in the Buying Guide under the "Summary" for each model are based on road tests, driving impressions of new models and ownership of previous models. We rely on our readers to keep us informed of how well these cars perform after the new wears off. Since we have limited space in which to judge whole model lines, we concentrate on recommending the engines, transmissions and suspensions that provide the best all-around performance, economy and comfort. In addition, we highlight the availability of anti-lock brakes and air bags since those are two safety features we highly recommend.

The Best Buys section lists the vehicles that the editors have selected as the top choices in 10 different classes. Vehicles are grouped by size, type and price so an $8000 subcompact competes with other low-priced small cars, not a $20,000 car that appeals to a different audience. The most expensive models are often—but not always—the best vehicles in their class. However, since money is a limited resource for most people, we make price an important criterion in choosing Best Buys. Because consumers have to fit car expenses into a budget that includes housing, food, clothing, medical expenses and other necessities, we recommend the cars that offer the most value for the least money.

Since price is usually a major factor when consumers choose a new car, we have made every effort to see that the price lists in this issue are accurate and the latest available at time of publication. In some cases, the latest available prices are for 1988 models; some manufacturers hadn't announced 1989 prices when this issue went to press. No prices are listed for some new models that hadn't gone on sale yet, such as the Ford Thunderbird and Mercury Cougar. We're sorry about that, but we can't guess at the car companies' price strategies. We have provided only manufacturer's suggested retail prices for some models because the dealer invoice and low price weren't available to us. In other cases, the prices are labeled "preliminary," meaning that the manufacturer released this information to dealers with a caution that they may be changed.

The invoice prices in this issue are what the dealer pays to buy the car from the factory. On most domestic cars, the invoice price includes the cost of preparing the car for delivery, advertising fees and other costs charged to the dealer. That isn't true for most imported cars. The "Low Price" listed for most cars is an estimate of what you should expect to pay for a particular model based on national market conditions. The actual selling price for any vehicle depends on market conditions for the car you're buying in your immediate area. Supply, demand and competition rule the market, so the "Low Price" should be used only as a guide. Market conditions vary greatly for different cars in different parts of the country. The amount of competition among dealers in your area and local economic conditions also affect price.

Car companies are free to change their prices at any time. As the value of the U.S. dollar has fallen against foreign currencies the last three years, Japanese and European car companies have raised their prices as many as four times a year. This could continue, so bear in mind that all prices in this book are subject to manufacturer's change.

Buying a new car is a trying experience for many, traumatic for some. We hope the 1989 edition of the *Automobile Book* eases the pain of new-car shopping and helps you make a more informed buying decision. The editors invite your comments. Address them to:

CONSUMER GUIDE®
7373 N. Cicero Ave.
Lincolnwood, IL 60646

SHOPPING TIPS

The telephone rings and the reader on the other end says, "I've been looking at the Wombat Z-70 4-door and a dealer said he could sell me one for $15,000. Is that a good deal?"

Occasionally we can give a clear "yes" or "no" answer to that kind of question; most of the time we can't. First, it helps if the caller has more information, such as a list of all the factory equipment on the car, the total manufacturer's suggested retail price for the car and options, and whether any dealer-installed accessories are included such as rustproofing, pinstripes or "fabric protector." Once we know where the dealer started from, we can talk about how far he's gone with his price.

Second, and of greater importance, it also helps if the caller has actually been "shopping." That means searching for the best deal by going to three or more dealers to compare prices on the same car with the same equipment, getting written price quotes that spell out what's on the car and what's not, and assessing supply and demand for that particular model in that area. Once a reader sizes up the market for the car he wants, it's much easier to determine whether he's getting a "good deal."

Even then, it's not cut-and-dried. A good deal on a slow-selling car such as the Dodge Lancer might be $200 above dealer invoice, while a good deal on a Honda Accord 4-door with automatic transmission might be full suggested retail price. It all depends on market conditions for that car in that city and how much competition there is among dealers. Imports are more popular in the Northeast and the West Coast than in the Midwest, so it figures that you should expect to pay more for a Japanese car in California or Connecticut than in Illinois or Michigan. However, there are more dealers selling Japanese cars in Los Angeles than in Chicago, so there should be more competition, which tends to lower selling prices. We can't tell you what the best price is in your city for the particular model you're interested in, though we can estimate what you should expect to pay based on national market conditions. You will have to do the legwork in your locale, shopping at dealers in your area to gauge supply and demand and determine the lowest selling price.

No matter what others might tell you, there are no magic formulas for calcu-

lating the best price. You can't just knock 10 percent off suggested retail to determine dealer invoice and add $200 as a "fair profit." The difference between retail and invoice prices varies from manufacturer to manufacturer, and from car to car. Supply and demand, not dealer invoice prices, determine the actual selling price. Nor can you count on dealers lowering their prices at the end of the month. The theory is that at the end of each month dealers want to clear out inventory to make room for the next shipment and salesmen want to earn more commission. What if they sold most of their stock last week and the salesmen have already earned fat commissions for the month? Neither can you walk in 20 minutes before closing time and expect to negotiate a great deal because they're in a hurry to get home. A smart salesman will keep you there long after the lights are turned off, or tell you to come back tomorrow.

You simply can't rely on gimmicks or try to outwit car dealers to get a good deal on a new car. The proper approach involves planning what you want to buy, carefully looking at all aspects of the deal to make sure you're not spending

Shopping Tips

more than you should and diligently shopping for the best price. While price should certainly be important, don't make it your only consideration. How much have you gained if you spend another full day shopping and save $50 on a $12,000 car? You should buy the car that you really want or need, and you should buy it from a dealer who will give you the best service in the long run. A dealer with a reputation for good service and giving its customers the benefit of the doubt deserves to charge a higher price. Remember also, if you're buying a car that many other people really want, you're going to pay a high price for it.

Here are several more suggestions for your new-car shopping.

Sell, Don't Trade

Most new-car buyers trade in their old car because it's faster and easier than trying to sell it, plus they use the trade-in value as all or most of their down payment. As with most conveniences, there's a cost. Because the new-car market is so competitive, dealers often make more money on the trade-in than on the new car. If they plan to sell your old car on their used-car lot, they'll probably give you wholesale value for it. If they plan to wholesale your old car to another dealer or to a used-car auction, then they'll probably give you less than wholesale value. Either way, they'll make a profit, so you'll get less than the real value of the car.

The difference between wholesale and retail value is usually at least $500 on a late-model car. For example, one used-car price guide lists the average retail value of a 1984 Ford Escort L hatchback in good condition at $3400 and the average wholesale value at $2800. The higher the value of the car, the bigger the difference between retail and wholesale.

You should try to sell your old car yourself since you might get close to full retail value from a private buyer. Ask friends and relatives if they're interested and advertise the car in local papers. It's worth the effort to find out what you can get on the open market versus trade-in value. The worst thing to do is trade in your old car before it's paid for. The dealer will gladly pay off your old loan, use any value left from your trade-in as a down payment, and write you another loan for the new car. Be assured the dealer will make out on that kind of arrangement, not you.

Consult used car guides and local newspapers to get an idea of what your old car is worth at retail and wholesale.

Be realistic about the value of your car; while mileage and condition are major factors in used-car prices, supply and demand play roles as well.

Be Flexible

Narrow your choices down to one or two models that fit your needs and budget to make it easier to compare prices among dealers. You should be comparing vehicles of similar size, type and price to find the best value for the money. If your shopping list includes a sporty hatchback, a luxury sedan and a 4-wheel-drive off-road vehicle, how can you make meaningful comparisons? It's a good idea to do some preliminary shopping at several dealers: Pick up brochures, talk to salesmen and test drive several brands of cars to see what you like within your price range. However, don't try to compare prices on six different cars and expect to come up with a good deal. Before you start serious price negotiations you should have a good idea of what you want to buy.

While you should narrow your choices down to one or two cars, don't become rigid about what you'll actually buy. If a salesman knows your sights are fixed on a bright red 2-door with a sunroof and aluminum wheels, and he has exactly what you want, he's going to hold out for a higher price. Be willing to take a different paint color, equipment or even body style to save money. For example, the 4-door sedans with automatic transmission are the hottest sellers in the Honda Accord line. You can probably buy an Accord 3-door hatchback with a 5-speed manual transmission for considerably less. If you're trying to save money, look for slow-selling models the dealer is anxious to sell. Watch for television commercials and print advertisements that tout rebates and factory discounts, tip-offs that the manufacturers are trying to unload excess cars.

Most domestic cars and some imports have option packages that carry hefty discounts below normal retail prices or include some equipment at no charge. While these packages may include features you don't want, they can still be cheaper than buying fewer items as individual options. Few domestic dealers are anxious to order a car from the factory exactly the way you want it, and practically no import dealers will do that. It takes at least six weeks to special-order a domestic car, often longer; if you get tired of waiting and buy a different car elsewhere, the dealer gets stuck with another unsold car, so he doesn't want to mess with it.

That's why they'd much rather sell you a car from their inventory. If you aren't willing to compromise about the car or equipment you'll buy, you'll limit your own options and your chances of saving money.

Shop for a Loan

Early in your shopping, get a quote on a loan from at least one bank, credit union or other lending institution. This will help you decide how expensive a car you can afford. You may have champagne tastes, but you might have to live on a beer budget. Just as you should shop for the best price on your new car, you should also shop for the best price on a loan. The "price" on a loan is the interest rate, expressed as an "annual percentage rate" or APR. Besides the APR, you should also know how much your monthly payments will be and the total amount you will have to pay over the life of the loan. When you compare quotes on loans, get all three figures in writing so you know what you're getting into. Someone may tell you he's giving you a 48-month loan at 8-percent interest, but the document you actually sign might turn out to be for 54 months at 12-percent interest. You'll pay for whatever you sign your name to, not what someone promises you.

Having a loan quote in your hand when you go car shopping means that when a car dealer offers to finance your car, you'll have something to compare it with, so you'll have a better idea of how good a deal it really is. Also, don't let a salesman or banker convince you that you have to take credit life insurance, which insures the loan is paid off if you die. It's an option, like an extended service contract for your car, and you don't have to buy it. If you die before the car is paid for, the car can be sold to cover the balance of the loan.

The best loan is the shortest one you can afford. The shorter the loan (say, 36 months instead of 48), the higher the monthly payments, but the lower the interest rate. The longer the loan (54 or 60 months instead of 48) the lower the monthly payments, but the higher the interest rate—and the more you'll pay in the end. That's why it's bad practice to shop for a monthly payment. If you tell a dealer you can afford $225 a month, he'll write a loan to fit that budget—with a higher interest rate, more monthly payments and more money in total.

Examples: If you borrow $10,000 for 48 months at 10 percent interest, you'll pay about $254 a month, or $12,192 total. If you borrow $10,000 for 60 months at, say, 11 percent, you'll pay only $217 a month, but $13,020 total. That's $828 more that you'll pay over five years instead of four. Also, with a 60-month loan you'll probably be replacing normal wear items such as the battery, brake linings, tires and muffler while you're still making monthly payments on the car. Can you afford both?

Car manufacturers frequently offer you a choice of low-rate financing or a cash rebate. You might save more money by taking the rebate and arranging your own loan through a bank. Ask for prices on a new car with and without dealer financing to see which is the cheapest in the long run. People with spotty credit records may have a better chance of getting a loan through a dealer. Banks tend to be conservative about lending money because they want to be sure you can pay it back. Dealers tend to be more liberal because they want to sell you a car. You should also investigate other possibilities. You might be able to get a loan from your insurance company. Homeowners may be able to arrange a home equity loan, though that is usually a more involved transaction.

Keep Price Separate

Settle on the price of the new car—and get it in writing—before you talk to a dealer about trading in your old car. This way you have a firm price on the new car, regardless of what you do with your old one. If the dealer asks early on if you're trading in, tell him, "Maybe." Insist on a price without your old car figured in; otherwise, you may be committing yourself to a deal that is based on low trade-in value.

We suggested earlier you get a quote on a new-car loan before you go to a dealer. Don't talk financing with a dealer until you've done that, settled the price of the new car and set a trade-in value for your old one. Here's why: Most low-rate loans offered by car dealers require that they contribute part of the money to lower the interest rate (called "buying down" the rate). If a dealer has to kick in $300 to offer you a lower rate, he'll surely try to make it up elsewhere in the deal. If you've already settled on the price of the new car and trade-in value, you've limited his opportunities to do this. What can the dealer do then? He probably won't offer you his lowest interest rate, since he would have to eat the cost of his contribution, but he may still offer you a fairly low rate. Compare it to the quotes from your bank and other lending institutions.

Dealer Price Stickers

Federal law requires that all cars sold in the U.S. have a manufacturer's price sticker posted in a window showing the suggested retail price for the vehicle and all factory-installed options, plus the destination charge for shipping the vehicle to the dealer and the EPA fuel economy estimates. This law does not apply to light trucks; some passenger vans and most 4-wheel-drive vehicles are classified as trucks, not cars. Most dealers add a second window sticker to cars in their showrooms showing accessories they have installed (pinstripes, rustproofing, "protection packages," burglar alarms, etc.), plus a variety of other charges, few of which are legitimate.

For example, many dealers add a "prep charge" for preparing domestic cars for delivery, even when the manufacturer's price sticker clearly states that those charges are included in the suggested retail price. Another favorite is charging the buyer an "advertising fee." Some dealers pay an advertising fee as part of the dealer invoice price for the car, some are reimbursed for it later and some don't pay any such fee because the advertising program is voluntary and they don't participate. In any case, it's their cost of doing business, not yours, so don't pay it. After all, are they paying for your costs of shopping for a car?

Other examples of dealer charges reported by our readers: A $900 "import tariff" on a Honda; no such tariff exists, so this is merely added-dealer profit. A charge of $319 for "ocean freight" and $225 for "inland destination" on a Hyundai Excel (on top of the $255 for destination listed on the manufacturer's price sticker), plus $239 for "dealer prep." None of these charges are legitimate; the dealer was merely trying to take advantage of high demand for the Excel, currently the most popular imported car.

You should refuse to pay for anything on a dealer's price sticker that you don't want or don't feel is justified. Stick with the manufacturer's price sticker and bargain on that, not the inflated price posted on the dealer's sticker. Once you've settled on the price of the car, keep your guard up because the salesman is far from finished. He's likely to unveil all kinds of extras that sound good to the consumer, who's now in a much more relaxed mood. Even with factory corrosion warranties of five years or longer on most cars these days, dealers are still pushing rustproofing, "paint shields" and "pro-

tection packages" that can add $800 to the price of a car. If you're buying a car that's backed by a 6-year/100,000-mile factory corrosion warranty, you don't need extra-cost rustproofing, so you should refuse to pay for it. Most manufacturers who provide long factory corrosion warranties advise in the owner's manual, or even on the price sticker, that additional rustproofing isn't necessary. But wait, there's still more. Now the salesman wants to sell you an extra-cost service contract or extended warranty. It's $600, but it covers "everything on the car." Sounds great doesn't it? Before you sign up for any of this, see the Warranties and Service Contracts section for additional advice.

Read First, Sign Second

The best defense against getting ripped off is to be an informed shopper who reads what he signs—before he signs it. If a dealer sees that you're in a hurry to complete the deal and probably won't read what you're signing, you're inviting him to take advantage of you. Before you sign on the dotted line to buy your new car, you should read the entire contract and be sure you understand what you are actually paying. The dealership is not the best place to do this. The salesman is going to pressure you to sign because that way he has a contract, a legally binding document that sets the terms of your purchase. You will probably be in a hurry because you're anxious to drive away in your new car. Once you sign a contract, it's difficult, if not impossible, to get it changed. Take the contract home, where you can go over at your own pace and call the dealer if you have questions. If the dealer doesn't want you to take the contract home, get a written purchase agreement that spells out all the details. Once you're satisfied with the purchase agreement, then it can be written into a contract.

Here's what to look for:

Sale price: This is the amount you've agreed to pay for the car and optional equipment, plus any dealer-installed accessories.

Dealer prep charges: Dealers are supposed to clean the car, in some cases install some options and make sure everything works properly. On all domestic cars dealer prep is supposed to be included in the retail price; don't let them charge you for it. If a preparation charge is listed for an imported car, try to negotiate it out of the deal. In every case, inspect the car to be sure it

Shopping Tips

was actually "prepped" by the dealer and not just washed.

Down payment: How much do you have to pay in cash or with your trade-in immediately?

Trade-in value: This is the amount you're getting for your old car if you're trading it in. This often covers most or all of the down payment. Trade-ins give dealers plenty of room for manipulation, so be sure you know exactly how much you're getting and that it's an equitable amount.

Destination charge: The cost of shipping the car to the dealer is listed as a separate item on the window sticker. There is no discount to dealers; they pay the same amount listed on the factory price sticker. Don't pay this charge twice; some dealers hide the destination charge in the sale price and then add it again as a separate item, pocketing the money.

Sales tax: Sales tax rates vary among states and localities. In some states they are calculated on a net price after your trade-in value has been deducted; in most states, you pay sales tax on the full purchase price of the new car. Check with your state or local governments to determine how sales tax is assessed in your area.

Total cost: Be sure the bottom line is filled in, so you know the total price including options, accessories, destination charge, dealer prep and taxes. Don't leave it blank because you can end up paying more than you bargained for.

Loans: Federal regulations require lenders to disclose all charges to borrowers. Be sure you know how much you are actually borrowing, the interest rate (expressed as an Annual Percentage Rate, or APR), what your monthly payment will be, the length of the loan (48 months, for example) and the total amount of money you will pay over the course of the loan.

Buy or Lease?

Leasing has become more popular in recent years, but generally it is cheaper in the long run only if you can deduct car expenses as a business expense on your income tax return. We recommend that before you make a decision, consult an accountant, tax adviser or the Internal Revenue Service for advice on how federal tax laws will apply to you and whether leasing will be cheaper than buying. Leasing usually benefits those who drive less than 15,000 miles a year, trade for a new model every three years or so, and

keep their cars clean and well maintained.

One of the traditional advantages of buying a car has been that loan interest could be deducted on federal income tax returns. However, that deduction is being phased out. Consumers can deduct 40 percent of interest paid in 1988, 20 percent in 1989 and 10 percent in 1990. After 1990, the interest deduction disappears. One of the main advantages of leasing is that you don't have to make a large down payment (though some leases do require a substantial amount down). Also, monthly payments on a lease are often lower than a monthly loan payment for similarly equipped cars. Here are some things to look for in leases:

Types of leases: The two basic types are "closed-end" and "open-end." With a closed-end lease, also called a "walk-away" lease, you simply walk away from the deal at the end of the lease with no additional payments, unless the vehicle has serious damage or excessive wear. That reduces the risks to the consumer.

With an open-end lease, you and the lessor estimate what the car will be worth at the end of the lease. If the car is worth less than was estimated, you are responsible for making up the difference, though in most cases the consumer does not have to pay more than the equivalent of three monthly payments. However, the value of a leased car can drop unexpectedly due to changes in fuel prices, safety problems in particular models or other matters beyond the consumer's control. Thus, monthly payments are usually lower in an open-end lease, but the risks are greater for the consumer at the end.

Monthly payments: Make sure you know how much the monthly payment is, plus the total amount you will pay over the term of the lease (usually 36, 48 or 60 months), including the maximum amount for which you could be liable.

Mileage limits: Most leases specify a mileage limit of around 15,000 miles per year. If you exceed the total number of miles at the end of the lease, you'll have to pay extra (usually 10 or more cents per mile).

Security deposit: Find out how much you have to put down at the start of the lease, and whether you have to pay the first and/or last monthly payment in advance. Also, find out whether you or the lessor pays for the state sales tax, license and title fees.

Insurance and maintenance: The lease must specify whether you are

providing insurance coverage and paying for maintenance of the car. Generally, it is cheaper for the consumer to pay for his own insurance and maintenance.

Early termination and purchase options: Find out before you sign the lease whether you can terminate it early (and what the penalty is) or buy the car (the purchase price or method of calculating the price should be specified). This could be important if you have to move to another state before a lease expires.

End-of-lease costs: Find out now if you could owe money at the end based on the value of the car (open-end leases). Also, know if you will have to pay for excessive wear (make sure "excessive" is explained) or have the car prepped for resale.

Shop for a lease the same way you shop for a car, comparing price quotes from different leasing agents for the same car and same leasing terms. The Federal Trade Commission publishes a 16-page booklet called "A Consumer Guide to Vehicle Leasing" that provides additional information. The booklet is available for 50 cents from:

Consumer Information Center
Department 458P
Pueblo, CO 81009

The "Today Buyer"

Eventually, you're bound to run into a pitch that goes like this: "If you buy this car from me today, here's what I can do for you . . . " Or, it could be, "What do I have to do to get you to buy a car from me today?" The operative word is "today," with the salesman usually warning you that this great deal is good for one day only. The goal is to get you to buy before you have a chance to comparison shop or think about it for more than a few minutes. The fear is that if they let you walk out the door, you will either find a better price elsewhere, or get nailed by another salesman with a more convincing "today buyer" pitch.

Consumers seldom win in this situation, so it's best avoided. Tell the salesman you're not a "today buyer," but you will be buying soon, and not necessarily from him. If he can't give you a price that's good tomorrow and next week, then take your business elsewhere. He's not treating you with the respect you deserve as a prospective customer about to spend a great deal of money. Don't buy until you're ready.

WARRANTIES AND SERVICE CONTRACTS

The warranty wars are heating up. Automakers have adorned their 1989 cars with unprecedented guarantees against defects in materials and workmanship, but not every change is a clear improvement.

Manufacturers obviously believe a warranty—their pledge to absorb certain repair or replacement costs over a specific period—is a big selling tool. It allows them to tout engineering advances that make their cars more reliable and to boast about assembly quality. We don't recommend choosing a car simply for its warranty, but it is a factor that should weigh in the buying decision.

Nine of the roughly two dozen major car companies doing business in the U.S. have made significant changes in their warranty coverage for 1989. The trend generally is to longer "bumper-to-bumper" basic warranties, but some automakers have gone a step further. Audi, for instance, has broadened its basic coverage to include routine maintenance items that are usually excluded from other manufacturers' warranties. And Pontiac, Oldsmobile and Chrysler are experimenting with a totally new concept: pilot programs in which a buyer can exchange a new car for another model.

Good as they may sound, none of these warranty packages comes without its share of loopholes, catches and fine print. And in some cases, 1989's "upgraded" coverage may not really be that much better than coverage it replaces. The rule is to look at warranties with a critical eye.

The most dramatic change for '89 comes at General Motors. The nation's No. 1 automaker has increased the basic bumper-to-bumper coverage on most of its cars from 12 months or 12,000 miles, to 36 months or 50,000 miles. But while more parts are now covered for a longer period, GM has eliminated an extended warranty of 72 months or 60,000 miles that had applied to the powertrain—the engine, transmission and associated parts. GM retained its requirement that owners pay a deductible of $100 per warranty repair after the first 12 months or 12,000 miles.

Alfa Romeo, Honda, Hyundai, Subaru, Isuzu, Mitsubishi, Subaru and Suzuki are other manufacturers who have lengthened their basic warranties, usually from 12 months/12,000 miles to 36 months/36,000 miles. Most dropped extended powertrain warranties in the process.

Perhaps 1989's most innovative twist on manufacturers' warranties is a new-car return policy. Chrysler tested its program in two markets during the fall of 1988. In Washington, buyers could return a new 1988 or '89 Chrysler Motors car or truck within 30 days or 1000 miles and receive the full purchase price, less any accrued finance and insurance charges. In Denver, buyers could exchange their vehicle and get credit toward the purchase of a new Chrysler Motors product. In the other temporary pilot program, buyers of 1989 Oldsmobile Cutlass Supremes and Pontiac Grand Prixs could return their cars within 30 days or 3000 miles and receive full credit toward the purchase of another Olds or Pontiac. The automakers are evaluating these test programs to see how they affect sales.

It's clear that car shoppers are faced with a more complex array of warranty packages than ever before, including expansive guarantees against corrosion, and tricky extra-cost service contracts. Detailed warranty information should be readily available through your auto dealer or the manufacturer. To help, here is an overview of how factory warranties and service contracts work and what they generally cover. The accompanying chart compares the major features of manufacturers' 1989 factory warranties.

Warranties

Factory warranties are divided into specific areas of coverage.

Basic warranty: Usually covers the entire car, except for the tires and battery, which are warranted by their manufacturers. Other important exclusions typical to basic factory warranties are normal wear and maintenance items (oil, air filters, brake linings); damage from the environment (hail, floods,

other "acts of God"); damage due to improper maintenance (incorrect fuel or lubricants); and damage caused by the owner or vehicle occupants.

Extended warranty: Coverage for a specific period beyond the basic warranty. Extended warranties usually apply to the car's powertrain, which consists of such major components as the engine, transmission or transaxle, fuel system, turbocharger, drive shafts and related parts, and the transfer case on 4-wheel-drive vehicles. Powertrain warranties usually exclude the same sorts of damage and maintenance items as the basic warranties.

While Ford and Chrysler have retained extended warranties, most import makers—and now GM—include powertrain coverage under their longer basic warranties. A few extended warranties—Lincoln's, Merkur's and Chrysler Fifth Avenue's, for example—also cover major systems, such as air conditioners.

Deductible: Domestic manufacturers and Yugo charge owners for warranty work performed after the basic coverage has expired. The deductible is generally $100, meaning the owner pays the first $100 of each repair made under the extended warranty and the car manufacturer picks up the rest. Though GM has dropped its extended warranty, it continues to charge a deductible after 12 months/12,000 miles.

Corrosion: This is warranty protection against perforation rust, in which sheet-metal is eaten through from the inside, and against surface rust, which starts on the outside and works inward. Coverage is generally longer for perforation rust than for surface rust. Most corrosion warranties now last at least three years and some have no mileage limits. Typical exclusions include surface corrosion caused by hail, stone chips and industrial pollution.

Transfers: Most manufacturers allow factory warranties to be transferred from owner to owner at no charge for as long as the warranty is in effect. Chrysler charges $100 to transfer warranties

Warranties and Service Contracts

to subsequent buyers of most of its cars, while Subaru charges $25. There also may be specific second-owner exclusions attached to leased vehicles and cars purchased for fleet use.

Rustproofing

Improved anti-corrosion techniques allow automakers to provide rust warranties that often run five years or longer. Extensive use of 2-sided galvanized steel, sealed body seams and joints, high-tech primer application and anti-chip paint are among steps taken by manufacturers to combat rust. Despite this, many new-car dealers persist in selling expensive after-market rustproofing that is often redundant and in some cases potentially damaging.

Dealers do this, of course, because the rustproofing they sell is tremendously profitable. An aftermarket rustproofing wholesaler recently advertised in a publication aimed at dealers, "Rustproofing only $4.95 per car! Paint shield and fabric protection at similar savings." These prices don't take into account the equipment or manpower a dealer must devote to applying the protectant. But when retail buyers are charged $300 or more for rustproofing, you get an idea of the profits involved. The situation is compounded by dealers who routinely rustproof new vehicles and add the cost to the selling price. This denies shoppers a choice, and forces customers to bargain down from a sticker price that already includes rustproofing.

Most automakers recommend against aftermarket rustproofing, despite its potential to generate profits for their dealers. General Motors warranty guides, for instance, state, "application of additional rust inhibiting materials is not necessary or required . . . " Depending on the application technique, some aftermarket rustproofing may actually reduce the car's corrosion protection, GM warns. And GM notes that repairs to correct damage or malfunctions caused by aftermarket rustproofing are not covered by its warranties.

Contracts

Once your salesman has sealed the deal, he is likely to bring up the subject of an extra-cost service contract or extended warranty. Sure, you're eager to protect your investment, but proceed cautiously. The manufacturers and other companies who write these service contracts expect to make a profit from them: They're betting they'll pay out less in repairs than you did to buy the contract. The odds are in their favor.

Extra-cost warranties cover items not protected by factory warranties, or provide longer coverage. And most of what they do cover probably won't need repair or replacement during the duration of the contract. If you pay $500 to $1,000 or more for a service contract, you'll need at least that amount in repairs to make it worthwhile.

About the only use for a service contract during the time your car is under the manufacturer's basic warranty is if it covers towing expenses. If your car has an extended factory warranty, say Chrysler's 84-month/70,000-mile powertrain coverage, chances are you'll never use an extra-cost service contract to cover engine or transmission repairs. Powertrain components are certainly among the most costly to repair, but they are also among the most durable.

Your decision to buy a service contract may of course be influenced by its price and provisions. Be aware that several car manufacturers offer their own contracts. GM, for instance, sells the General Motors Protection Plan and Ford the Extended Service Plan. Dealers may offer you the factory-sponsored plan, or one from another source, such as an independent underwriter who offers the dealer a higher profit margin. Compare plans from more than one source. After you buy a car, you might receive information in the mail on service contracts offered by the manufacturer or an underwriter at less cost than the dealer's plan.

All but the most expensive service contracts carry deductibles for each repair, usually $25 to $75. Some specify where you can get your car repaired, limiting your choices and perhaps making repairs more costly or inconvenient. Extra-cost contracts issued by automakers require that repairs be done at an authorized dealership, which may also be costly or inconvenient. And some service contracts require that you pay for repairs when they're made and then file for reimbursement.

In their favor, service contracts can cover your car well beyond factory warranties, so you may be able to avoid major repair bills down the road. This is especially useful on cars with sophisticated mechanical and electronic components that are expensive to replace.

Our advice is to give any service contract long and careful scrutiny. And think about it at home, not in the dealer's showroom. Get a copy of the contract, not just a brochure summarizing the coverage. Don't be in a hurry; you don't always have to buy a service contract at the time you buy your car. You can back out of some of them for a small fee. You may decide to put your money into a savings account instead.

1989 Manufacturers' Warranties

Make/Model	Basic Warranty (Months/Miles)	Extended Warranty (Months/Miles)	Deductible	Corrosion Warranty (Months/Miles)	Transferable?
Acura	36/36,000	none	none	36/unltd	Yes/no cost
Alfa Romeo	36/36,000	none	none	72/60,000	Yes/no cost
Audi	36/50,000	none	none	120/unltd	Yes/no cost
BMW	36/36,000	none	none	72/unltd	Yes/no cost
Buick	36/50,000	none	$100 after 12/12,000	72/100,000	Yes/no cost
Cadillac (exc Allante)	48/50,000	none	$100 after 12/12,000	72/100,000	Yes/no cost
Cadillac Allante	48/50,000	none	$25 after 12/12,000	84/100,000	Yes/no cost
Chevrolet	36/50,000	none	$100 after 12/12,000	72/100,000	Yes/no cost

Make/Model	Basic Warranty (Months/Miles)	Extended Warranty (Months/Miles)	Deductible	Corrosion Warranty (Months/Miles)	Transferable?
Chrysler (exc Fifth Avenue, New Yorker & Imports)	12/12,000	84/70,000 powertrain	$100 after 12/12,000	84/100,000	Yes/$100
Chrysler Fifth Avenue	12/12,000	84/70,000 powertrain; 60/50,000 major systems	$100 after 12/12,000	84/100,000	Yes/$100
Chrysler New Yorker	60/50,000	84/70,000 powertrain	$100 after 60/50,000	84/100,000	Yes/$100
Chrysler Imports	36/36,000	60/50,000 powertrain	none	72/unltd[1]	Yes/no cost
Daihatsu	12/12,500	36/36,000 powertrain	none	36/unltd	Yes/no cost
Dodge	12/12,000	84/70,000 powertrain	$100 after 12/12,000	84/100,000	Yes/$100
Eagle (exc Summit)	12/12,000	84/70,000 powertrain	$100 after 12/12,000	84/100,000	Yes/$100
Eagle Summit	36/36,000	84/70,000 powertrain	$100 after 36/36,000	84/100,000	Yes/$100
Ford	12/12,000	72/60,000 powertrain	$100 after 12/12,000	72/100,000	Yes/no cost
Geo	36/50,000	none	$100 after 12/12,000	72/100,000	Yes/no cost
Honda	36/36,000	none	none	36/unltd	Yes/no cost
Hyundai	36/36,000	none	none	36/unltd	Yes/no cost
Isuzu	36/36,000	none	none	36/unltd	Yes/no cost
Jaguar	36/36,000	none	none	72/unltd	Yes/no cost
Jeep	12/12,000	84/70,000 powertrain	$150 after 12/12,000[2]	84/100,000	Yes/$100
Lincoln	12/12,000	72/60,000 powertrain & major systems	$100 after 12/12,000	72/100,000	Yes/no cost
Mazda	36/50,000	none	none	60/unltd	Yes/no cost
Mercedes-Benz	48/50,000	none	none	48/50,000	Yes/no cost
Mercury	12/12,000	72/60,000 powertrain	$100 after 12/12,000	72/100,000	Yes/no cost
Merkur	12/12,000	72/60,000 powertrain & major systems	$100 after 12/12,000	72/100,000	Yes/no cost
Mitsubishi	36/36,000	36/50,000 powertrain[3]	none	60/unltd outer panels; 36/unltd all panels[4]	Yes/no cost
Nissan	36/36,000	none	none	60/unltd	Yes/no cost
Oldsmobile	36/50,000	none	$100 after 12/12,000	72/100,000	Yes/no cost
Peugeot	36/36,000	60/50,000 powertrain	none	36/36,000	Yes/no cost
Plymouth	12/12,000	84/70,000 powertrain	$100 after 12/12,000	84/100,000	Yes/$100
Pontiac	36/50,000	none	$100 after 12/12,000	72/100,000	Yes/no cost
Porsche	24/unltd	none	none	120/unltd	Yes/no cost
Range Rover	36/36,000	none	none	72/unltd	Yes/no cost
Saab	36/36,000	none	none	72/unltd	Yes/no cost
Sterling	36/36,000	none	none	72/unltd	Yes/no cost
Subaru	36/36,000	none	none	60/60,000	Yes/$25
Suzuki	24/24,000	none	none	36/unltd	Yes/no cost
Toyota	36/36,000	none	none	60/unltd	Yes/no cost
Volkswagen	24/24,000	none	none	72/unltd	Yes/no cost
Volvo	12/unltd	36/unltd powertrain & major systems	none	36/unltd surface; 60/unltd body; 96/unltd structure	Yes/no cost
Yugo	12/12,000	48/40,000 powertrain	$25 after 12/12,000	12/12,000	Yes/no cost

1. 84/100,000 on Dodge/Plymouth Colt.   2. $100 after 12/12,000 on 2-wheel drive models.   3. 36/36,000 on Precis.   4. 84/100,000 outer panels on Galant and Mirage.

CONSUMER COMPLAINTS

If dealing with your new-car dealer is less than pleasant at purchase time, chances are it'll be no fun at all later on, when repairs are needed. All too often that friendly salesman gets replaced by an indifferent service manager. Keep coming back with troubles and his indifference can turn into hostility.

By far the best warranty is a good car sold by a good dealer. No multi-month, unlimited-mileage guarantee is as pleasing. Cars are indeed better than they used to be, but they're hardly perfect.

Still, by selecting a make and model known for reliability, and buying from a dealer with a good reputation, you reduce the risk immensely. A reputation for tempting prices isn't the same as a reputation for good service after the sale. The dealer who comes close to your price isn't always the one who's eager to see you later on or who's able to fix your car without a fuss.

A little observation before you buy can reveal a lot. Look around the service department. Is the shop well-equipped? Do you notice any obviously disgruntled customers? And if the person making a ruckus is an employee— or, worse yet, a manager—think seriously about showing the place your heels. You could be the next victim of his rage.

What you want is a service department staffed by people who will listen carefully to customer complaints and lean toward giving customers the benefit of the doubt in case of a dispute.

Probably the best way to find one is to as for recommendations from friends and neighbors. Did a particular dealer treat them fairly and courteously? Was the car ready when promised, with the repair made correctly? If they had to come back with a recurring problem, did the staff grow nasty, or display concern? Were they ever overcharged, or given a part or service they didn't ask for? Most important, would they buy from this dealer again?

One general rule is that smaller dealers, especially in rural areas and small towns, are the best bets for good service—though not always for low prices. They're more likely to rely on repeat customers than big-city mega-dealerships.

Dealers that advertise the heaviest, especially the frantic screamers on late-night TV commercials, may be the worst choice. Constantly on the lookout for new customers, they aren't necessarily trying to hang onto former ones. Big isn't always bad, of course, but it is something to consider when thinking about service.

Your antenna for good service ought to be up while you're negotiating your new-car purchase, also. Good treatment on the sales floor isn't a perfect predictor of good service later, but there is a correlation. And don't forget to ask about service hours, whether appointments must be made, if loaners are available or whether they're willing to shuttle you to your job. Ask about the hourly wage rate for repairs: $50 an hour isn't unusual, but remember that only a small portion of that goes to the mechanic. Give the local Better Business Bureau a call. They can tell you whether this dealer has had an unusually large number of complaints, and how they were resolved.

Dealers are learning that handling complaints courteously and efficiently is in their best interest. Keep one customer happy, and he or she will bring in new ones. The automakers are even encouraging dealers to solicit customer beefs and reactions about service visits. Dealers are also being advised to listen closely to complaints. A quick and concerned response can be just as important as what's eventually done to resolve the problem. Unfortunately, a few dealers have not yet gotten these messages.

Inspect Before You Accept

Problem prevention also means refusing to sign the sales contract until you're satisfied that everything you've paid for on the car is present and operating properly. The dealer-prep process, for which you probably paid extra, is supposed to take care of this, but don't trust it implicitly.

Occasionally, a dealer-installed accessory isn't available right away, and you have to come back later. If that happens, be sure each detail about the accessory and its installation is spelled out in writing.

Accessories, like cars as a whole, are more complex nowadays, so don't drive away without asking how the stereo works, how to set the air conditioner and cruise control, even how to start the car properly. Much of this, of course, is explained in the owner's manual, so be sure there's one supplied with your car. Read it, and study the manufacturer's recommended maintenance schedule.

Cars are often damaged in transit, so inspect carefully before taking delivery. Dealers have been known to rack up several hundred miles on a car, then palm it off as a demonstrator. Demos can be demons. Proceed with utmost caution. They've also been know to sell as like-new a car that's been repossessed from a defaulting customer. Suspicion is valuable. Be sure your new car really is new. Careful inspection before accepting the car avoids a lot of trouble later.

Look for evidence of body repair, such as color mismatch between a door and fender, or within a panel; for glass fragments on the floor; for loose or missing pieces. Any significant damage means, at the very least, that you deserve a discount off the price. Details belong in the contract, in case the initial minor flaw leads to anything serious. Better yet, skip that car and turn to another one.

Have the scheduled maintenance performed. There's no question that well-maintained cars last longer and have fewer troubles. Get the oil changed no later than the schedule advises; a little more often wouldn't hurt. Keep accurate records of what you've done, and what the service department has handled. This last point is extremely important, as we'll see when discussing recurring repair problems.

Minor Complaints

A lot of problems seem to turn up in the first few days of ownership. With today's computerized engine controls and with many electronic gadgets inside the dash, it's a wonder more things don't go wrong. Many failures have trivial causes—no more than a loose wire or bolt. How a dealer handles a small problem is perhaps the best predictor of how the biggies will be dealt with.

But turning up at the service department every other day with one piddling problem after another isn't the way to endear yourself to the staff. No need to dash off to the dealership because the radio sounds a little fuzzy or a remote-control mirror moves stiffly. Just make a note and bring it up on your next scheduled service visit. Of course, turning up on the last day of the warranty period with a three-page list of troubles isn't the wise course either.

When making notes, write down every detail you can think of. When you visit the service department, the more information the service advisor gets, the better the chance for a successful fix. Vagueness is asking for failure. Be sure the service writer listens to every detail and writes down a complete and correct description of your complaint. Most of the time, the written service order is all the mechanic has to go by, so it must include all the relevant information.

Don't become angry if you're told parts have to be ordered. That's an unfortunate fact of modern automotive life. Dealers can't stock every component that's likely to need replacement. Many of these parts can be obtained overnight, however, especially in urban areas, where access to warehouses is good. Parts for low-volume import makes often take longer to obtain. If it's a rare component, or a rare car, parts can take a week—or a month—to arrive from some distant source. This is something to consider when you're new-car shopping.

Once your car is finished, don't leave the premises until you're satisfied with the repair. Check your list of requested work. If anything has been overlooked, ask the manager to explain. And if a "remedied" noise or malfunction recurs on the way home, turn right around and go back to the service writer. It won't help to get excited or demanding, but make your displeasure known.

Even if your warranty permits oil changes or other minor tasks to be done outside the dealership, it often pays to bring the car back to the dealer for this work. That way, they grow used to seeing you and your car in a non-problem, non-threatening situation. Regular customers tend to get closer attention when a real problem comes up later. They may even warrant a slight bending of the rules, performing a repair gratis, say, when just beyond the warranty period. Your car is also in the dealer's records, which, when combined with your own, could make a real difference in case of serious warranty-related flaws.

Intermittent Problems

Something in the back end may have been grinding away like mad all morning, sounding like it wouldn't go another 10 feet. But as you glide onto the dealer's driveway, it makes a miraculous recovery. Maybe the engine idles crazily upon each morning's start-up, or stalls every day after you've gone two miles. Unfortunately, the dealer may be three miles away, so by the time you arrive, everything's fine.

Most of us, including the dealer's mechanics, have had these things happen, and can sympathize. Don't be embarrassed or apologetic if you can't make the problem appear during a brief test drive with the service writer. Just describe the symptoms clearly and explain the exact circumstances under which they occur. Does the engine stall only when cold? Only when accelerating? Does the brake pedal feel squishy in every light rain? Does that clanking noise appear only when turning left? Ask that they try the car again later, when it's perhaps more likely to falter.

The electronic brains in some newer cars have the ability to store information about these intermittent problems, for retrieval later by a service department's diagnostic computer. Up-to-date service departments will also have access to a manufacturer's data bank, which sometimes tracks tough complaints and contains details on possible remedies. Computers can help as well as hinder.

Long-Term and Recurrent Problems

Sad to say, some cars just don't stay fixed, and some problems have no solution that ordinary experts can unearth. We've had letters from people who've returned to the dealer three, four or more times with the same complaint.

At some point, you'll have to suggest that the service manager contact the manufacturer's service representative. If he's on his toes, he'll make the suggestion himself. You might not be alone. Others may have suffered the same malady, and the company computer may stand ready to disgorge details on an appropriate fix. It might even be published in the latest service bulletin, not yet sent to the dealerships.

After three or four fruitless sessions in the service department, speak to the owner or manager of the dealership, if you haven't already done so. Explain everything clearly, and show all your records. Assume he's eager to help. He may well be. He can then prod the staff into making one more stab at the problem, perhaps trying a different tack.

Consumer Complaints

Don't be shy about trying another dealership. A fresh approach at another location can be exactly what's needed.

What If Nothing Helps?

Some problems simply have no evident solution. After the dealer's mechanics run out of tricks, gather up your notes and contact the manufacturer's zone office. Check the owner's manual for the address and phone number (often a toll-free hotline). Ask for the customer service or technical-support representative. Don't expect miracles, though. Often as not, the factory rep just sends you back to the dealer to start the whole routine again. Even if someone higher up takes a look at your car, he just might declare it "fixed" or normal.

Mediation and Arbitration

The next step is a third-party dispute resolution program. Except for Mercedes-Benz, Hyundai, Subaru, Suzuki and Range Rover, every major automaker participates in a mediation or arbitration program.

Generally, the programs allow the consumer to present his or her complaint, along with any documentation, to a mediation or arbitration panel. The manufacturer's representative does the same, and the panel renders a decision or a recommendation. Some findings are binding on the manufacturers, others are not. The good news is that disputes handled in this way often result in the consumer finally getting some satisfaction, be it something as "minor" as a new paint job or, in some cases, as major as a new vehicle. When the panel sides with the manufacturer, however, it can be very difficult for the consumer to pursue the complaint further.

The Council of Better Business Bureaus operates an Auto Line program that combines mediation and arbitration. Mediation alone is provided by the National Automotive Consumer Action Program (AUTOCAP), a service directed by the National Automobile Dealers Association (NADA) but performed by state or local dealer associations. Some automakers work only with Auto Line, others can be dealt with through either program. Several manufacturers operate their own programs.

With either Auto Line or AUTOCAP, the organization's staff will first try to settle the dispute informally. Their recommendation is not binding on either party, but the vast majority of cases are

Consumer Complaints

resolved in some way without going beyond this mediation stage into arbitration or legal action. Only 14 percent of AUTOCAP's cases in 1987 progressed to the consumer/dealer mediation panel.

AUTOCAP panels are made up of ordinary consumers and industry-connected members. They attempt to settle warranty and product-reliability disputes between customers and participating import manufacturers (see chart). Deborah Hopkins, AUTOCAP's manager of consumer affairs, describes it as an attempt "to get both parties to agree . . . with AUTOCAP as the conduit." The panel's settlement recommendation is not legally binding, though Hopkins says "peer pressure" convinces most dealers to abide by an AUTOCAP determination. Arbitration, by contrast, is usually binding on the manufacturer, but not the customer.

If you're dissatisfied with the recommendation of the mediators, your next, more serious option is arbitration. Details differ, but the procedure is basically the same for all. Arbitration is the last step before legal action. It's similar to a court hearing, but less formal. A total of 173,000 consumers used Auto Line in 1987, and 29,000 went as far as arbitration.

Ford and Chrysler run their own arbitration programs; Toyota works through the American Automobile Association Complaint Arbitration Services. General Motors and the other automakers that accept arbitration deal with BBB'S Auto Line. Your owner's manual or other booklet should tell you which service may be used by the company that built your car.

Chrysler's arbitration panel accepts only written arguments, and covers only cars under warranty. Dealer and zone office representatives will be present, but cannot vote in the outcome. Ford's Consumer Appeals Board takes oral arguments pertaining to any Ford product less than four years old with fewer than 50,000 miles.

Whether you're participating in one of these programs or in that of the Better Business Bureau, you'll present your case orally or in writing. You'll need complete records, including work orders, receipts, letters and notices. You must prove the car doesn't work properly, that both dealer and manufacturer have failed in several attempts to fix it, and that you've exhausted other avenues of complaint. The manufacturer and/or dealer will then present its case, assisted by a representative familiar with repair problems.

Finally, a decision is handed down. That decision is binding on the company, but not on you. You can accept the verdict, or go on to the next step: litigation. You usually get 60 days to decide, but if you accept the arbitrator's decision, you can't change your mind and take legal action.

Think twice if the manufacturer wants you to settle before the case goes into mediation or arbitration. It is a good way to avoid the hassle of arguing your case, but is the company just worried that it will lose in arbitration?

Don't expect miracles. Statistics show that about one-third of the consumers who complain have the dispute settled their way; another third lose to the automaker or dealer, and a final third gain only a part of what they wanted.

Lemon Laws

Motorists in about 40 states have one other source of assistance: some form of "lemon law" that requires a manufacturer to refund the purchase price or to replace a car that's proven critically flawed. Instances of consumers actually gaining such dramatic results under lemon laws are quiet rare, though, so here again, don't expect miracles.

In most states, you must exhaust all other possible remedies before qualifying for consideration under a lemon law. That means you must make a specified number of tries at the dealership and pass through arbitration without successful resolution. A car might have to be inoperable for at least 30 days (not necessarily consecutive) during your first 12 months or 12,000 miles of ownership. Details vary, so obtain guidance from the state attorney general's office, a consumer protection agency, or the Center for Auto Safety. And while arbitration is a process you can handle yourself, fighting under a lemon law requires you to hire a lawyer.

Lemon laws typically stipulate that a manufacturer be given one last chance to remedy the complaint. Even if you "win," the manufacturer nearly always is allowed to deduct a portion from your award to compensate for mileage you've put on the vehicle. If your state has no lemon law, you'll need a lawyer anyway to enter the final step of this treacherous process: the all-out lawsuit.

Step-by-Step

Here's a rough list of steps to follow when your new car develops a serious problem:

1. Speak to someone in authority at the dealer, going up one step at a time until you reach the owner or top manager. Keep a detailed record of your service visits, including all receipts.

Third-Party Dispute Programs

Manufacturer(s)	Arbitration Program
General Motors	Better Business Bureau Auto Line
Ford/Lincoln/Mercury	Ford Consumer Appeals Board (contact district office or call 1-800-241-8450)
Chrysler/Dodge/Plymouth/Jeep/Eagle	Chrysler Consumer Arbitration Board (contact any zone office or call 313-956-5970)
AMC/Jeep/Renault (before 1988 models)	Auto Line (BBB)
Audi, Porsche, Volkswagen	Auto Line (BBB)
Honda/Acura, Nissan, Peugeot, Saab	AUTOCAP (consumer can also use BBB Auto Line)
Alfa Romeo, BMW, Isuzu, Jaguar, Mazda, Mitsubishi, Rolls-Royce, Sterling, Volvo, Yugo	AUTOCAP (consumers may be able to use other programs as well)
Toyota	Complaint Arbitration Services of the American Automobile Association 1-800-331-4331)

---

2. Contact a customer relations person at the manufacturer's zone or national office. Zone offices and their phone numbers are listed in your new car ownership materials. Keep a detailed record of your contact with the manufacturer. Jot down the names and numbers of the service representatives with whom you speak.

3. Contact a third-party dispute-resolution program for mediation and, if necessary, arbitration.

4. Consult a lawyer about filing a lawsuit.

At each step of the process, be prepared to provide full vehicle data, including mileage, date of purchase and vehicle identification number; describe the problem; describe what's been done to correct it and explain why you're dissatisfied. Above all, be clear about exactly what action or solution you're seeking.

Government Agencies and Consumer Groups

The following list includes telephone numbers and addresses for auto arbitration programs, government agencies and consumer groups that may be of help to consumers with car problems. In addition, many states and some local governments have their own consumer protection agencies that may be able to act faster than a federal agency.

American Automobile Association
8111 Gatehouse Road
Falls Church, VA 22047-0001
(703) 222-6000

AAA services Toyota's mediation/arbitration program.

Better Business Bureau

Check your local phone book for the nearest office.

Center for Auto Safety
2001 S Street, NW
Washington, DC 20009
(202) 328-7700

This is a non-profit consumer group that lobbies on behalf of consumer interests in vehicle safety and quality, highway safety and mobile home safety. It also provides information on lemon laws and provides a lawyer referral service.

Chrysler Motors Owner Relations Department
P.O. Box 1718
Detroit, MI 48228-1718
(313) 956-5970

This department handles complaints not resolved at the dealer or zone-office level and provides information about Chrysler's regional customer arbitration boards.

Environmental Protection Agency
401 M Street, SW
Washington, DC 20460
(202) 382-2090

EPA publishes the EPA Gas Mileage Guide and enforces emissions control regulations.

Federal Trade Commission
6th & Pennsylvania Avenue, NW
Washington, DC 20580
(202) 326-2222

The FTC provides information on arbitration and other consumer complaints. The FTC also has regional offices in major cities.

Ford Consumer Appeals Board
1-800-241-8450

Call this number for information on Ford's arbitration program.

General Motors Arbitration Handbook
1-800-824-5109

Your name and address will be taken on a recording and you will receive a handbook about GM's arbitration program.

General Motors Technical Bulletins
1-800-551-4123

Call this number for information on technical bulletins that might apply to your car.

Highway Loss Data Institute Insurance Institute for Highway Safety
Watergate 600
Washington, DC 20037
(202) 333-0770

This is an insurance industry lobbying group that compiles information on vehicle and highway safety.

National Automobile Dealers Association
AUTOCAP
8400 Westpark Drive
McLean, VA 22102
(703) 821-7144

NADA headquarters will refer you to a local office to answer questions on the AUTOCAP arbitration system and tell you if your dealer is a member.

National Highway Traffic Safety Administration
400 7th Street, SW
Washington, DC 20590
(202) 366-5972

NHTSA investigates safety defects and enforces safety regulations. NHTSA operates a consumer hotline for information on safety recalls and on how to report a safety problem: 1-800-424-9393 (In Washington, D.C., 366-0123)

National Insurance Consumers Organization
121 N. Payne Street
Alexandria, VA 22314
(703) 549-8050

This consumer-advocate group works to educate consumers on insurance issues, researches public policy and rate making, and does some lobbying.

U.S. Department of Justice Office of Consumer Litigation
10th Street, NW
Washington, DC 20530
(202) 724-6786

This office enforces federal laws covering price labeling of new cars.

SAFETY

Does safety sell? Five years ago, most automakers would have responded with a resounding "No!" Mercedes-Benz was the first company to offer air bags in recent years when it introduced them as an option in 1984. Ford took a tentative lead among domestic manufacturers in 1986 by making air bags available to Tempo/Topaz buyers, insisting that customers were ready to pay for safety. After a slow start, that prediction appears to be gradually coming true.

For years, manufacturers were a big part of the problem, resisting many proposals to make cars safer. The prevailing industry view, according to Jim Mooney of the Insurance Institute for Highway Safety, was that "you don't remind people that they can be in an auto crash." Not until after Mercedes and then BMW made strong moves in the direction of safety with anti-lock brakes and air bags did real change begin. Even Chrysler Chairman Lee Iacocca admitted earlier this year that automakers had been wrong not to promote safety. Of course, this change in attitude may never have happened were it not for the efforts of safety-minded groups who campaigned for safety devices to be made mandatory.

A fair amount of blame must also rest on the public. Automakers were largely correct in insisting that motorists weren't prepared to pay the price for safety. Early attempts to provide safety devices were not greeted favorably. Worse yet, many drivers displayed little evidence that they even wanted safer cars. The efforts of governmental agencies, and of activist groups such as the Center for Auto Safety, were derided by millions. Safety, many seemed to be saying, was for wimps. Real Americans knew how to drive, and didn't need any Nader-inspired do-gooders to tell them what kind of cars they could have.

As the 1989 models roll out, that sort of attitude, while hardly disappeared, is losing favor. "No question about it," says IIHS's Mooney, "consumers think in terms of auto safety." For one thing, more women are closely involved in buying decisions, and women tend to be more safety-conscious. Drivers might still be unwilling to pay a lot of money for extra safety measures, but can at least acknowledge their desirability. And that's a strong step in the right direction. As the population ages, safety concerns probably will gain even greater attention.

How Safe Is Safe?

Definitions of auto safety are loaded with controversy. For example, many consumer safety organizations campaign for laws that mandate cars be made safer in crashes. Yet some of these groups pay scant attention to technical features that might help cars avoid crashes, such as anti-lock brakes. Beyond basic driver's education in high schools, little effort is made to improve the abilities of drivers on the road. The only promising campaign is the one aimed at removing the drunks from behind the steering wheel. And as we all know, that one has quite a way to go, since 40 percent of fatalities are still connected to alcohol.

"No one can guarantee safety," Lee Iacocca warned in an ad for Chrysler cars with newly standard air bags. "But that doesn't mean the car industry shouldn't keep trying to do more." Finally, the industry seems to be making a serious attempt, even if such moves would never have gotten off the ground without outside instigation.

We can't tell you which cars are safest. Even the most knowledgeable experts, who devote their lives to the subject, can't do that. Too many technical factors are involved, which can't be measured neatly. We can discuss the various aspects of safety, and describe some features that you might want to look for, and ask about, at the dealers this year. But remember: No car will ever be safe if driven recklessly, or occupied by unbelted passengers. Nobody likes to admit it, but driving is a risky activity. Each of us should have safety in mind every time we get into a car—or search for one to buy.

Buying a Safe Car

Until recently, safety has hardly been mentioned except in negative terms. We heard about the perils (real or alleged) of Corvairs, Pintos and, more recently, the Audi 5000. Seldom did a manufacturer promote safety, even if its cars had an enviable record or carried safety-related features. Marketing people were convinced that safety talk wouldn't help sales.

Not many of us would make safety Number One in choosing a vehicle to buy. Yet more than ever, we're inclined to make it one of the influencing factors, especially when picking between otherwise equivalent models. We should be thinking about:

1. A car's basic size and weight.
2. Standard safety equipment.
3. Optional safety features.

Getting down to basics, safety takes two forms: (1) how well the vehicle can avoid an accident (provided that the driver can recognize dangers and is capable of taking appropriate evasive action); and (2) how well the driver and passengers will be protected if the worst happens and a collision occurs. Unfortunately, while plenty of statistics are available on actual crashes, it's far tougher to measure a car's ability to evade trouble. Even if it were possible to find out how many close calls were avoided, we could never be sure how much of that laudable result was due to the car's design, and how much to driver skill (or to dumb luck).

Available statistics aren't always wholly reliable, either. Many are gathered by organizations supported by the insurance industry, such as the Highway Loss Data Institute, likely to ignore evidence that doesn't back the insurers' viewpoint. The federal government issues data obtained from cars that are crash tested in a carefully controlled environment, not actual conditions. Rating "scores" are publicized, but reasons for differences among cars are largely ignored. Within these limits, though, we can learn how different cars fare in both real-world driving and when intentionally crashed to help us decide which ones offer better passenger protection.

The opinion of most experts is that a larger, heavier car is safer in a collision. It's basic physics: a big car has more momentum and can more easily absorb the energy of hitting (or being hit by) another car, especially a smaller one. On the other hand, big wagons and sedans that rate best are normally owned by older, more experienced drivers who have children. Smaller, sportier (and cheaper) cars appeal to younger, less experienced drivers. So is an injury or damage rating the fault of the vehicle or its driver? Besides, not every serious accident in a small or sporty car leads to tragedy, and occupants of full-size wagon are hardly invulnerable.

Since safety has become almost a trendy topic, beware of advertising puffery. Assertions that accidents involving a certain manufacturer's cars result in smaller claim costs are often based solely on the performance of their big wagons, conveniently ignoring the record of their subcompact and compact cars.

Setting safety standards remains a battle among interest groups and the government, with consumers in the middle. Consumer groups like the Center for Auto Safety petition the federal government for rule changes or recalls. The National Highway Traffic Safety Administration (NHTSA) investigates these petitions, acts on some and dismisses others, prompting harsh criticism either way. Automakers often fight each step. Conflicting study results are published, amid charges that statistics are biased or distorted to favor one viewpoint over the other. In the end, nobody is entirely sure what's safe and what's not. Up to now, though, no one has come up with a better system.

Government Crash Tests

For years, NHTSA has been wrecking cars, using a pair of wired-up dummies to bear the impact that would otherwise be felt by flesh-and-blood humans. Cars are chosen that either haven't been tested before, or have new designs. A total of 37 vehicles were tested for the 1988 model year under the agency's New Car Assessment Program. Each is crashed head-on into a fixed barrier at 35 miles per hour. That's five mph faster than the minimum standard that all passenger cars must meet, and produces actual forces one-third higher.

The dummies are intended to represent the average adult male. The head, chest and upper legs are fitted with instruments to monitor impact force and acceleration. All available safety restraints are secured in place (an unrealistic beginning, since more than half of motorists still don't buckle up). Crashes produce three ratings: head injury criterion (HIC); chest G's (force against the chest); and potential injury to the femur (thigh bones). Separate figures are released for the driver and passenger. In fact, some vehicles show a dramatic difference between the two, offering acceptable (even excellent) protection on one side of the front seat, yet failing in the other.

Minimum acceptable ratings in a 30-mph crash are 1000 (HIC), 60 G's (chest), and 2250 pounds (femur). Naturally, higher levels should be expected from the 35-mph test crash. In this past year's testing, only three passenger vehicles scored above 1000 for both driver and passenger in the HIC category. Chevrolet's Astro van rated 1003 and 1424 respectively; the Peugeot 505 GLS rated 1701 and 1457; and the Volkswagen Fox rated 1114 for the driver, 1424 for passenger. Vehicles rating higher than 1300 for either the driver or passenger (not both) included the Dodge Colt wagon, Mitsubishi Montero 4x4, Renault Medallion 4-door, Buick Park Avenue, Chrysler New Yorker, and Volkswagen Vanagon. Another five vehicles rated between 1000 and 1300 for either driver or passenger. Worst of all was Chevrolet's G20 van, with ratings of 3665 and 1452.

Even NHTSA doesn't advise using these crash figures as a basis for a buying decision. They're intended only to compare vehicles of similar size and weight. Critics have outlined a host of flaws in the test procedure, including the fact that the "average size" dummies obviously represent only a fraction of the total population. Nothing is shown about how people taller or shorter, lighter or heavier, might fare. Crashing into a solid barrier is also meant to simulate hitting a car of equal weight, at equal speed. In the real world, the opposing vehicle might be bigger or smaller. A car that rates highly in the test won't necessarily survive well if it hits one that weighs half a ton more, while a heavy car that scores poorly in the crash test is likely to come out of a real encounter with a subcompact relatively unscathed.

NHTSA used to select all the vehicles to be tested. More recently, automakers have been allowed a retest at their own expense. Many have taken advantage of that when the original result looked bad. Critics charge this can distort the results. The Ford Taurus scored a dramatic improvement in its retest after poor initial scores. Ford said it had made significant structural changes. The VW Fox also improved, but still failed in one (HIC) rating. Taking note of improvement, in fact, may be more useful than paying attention only to the raw scores.

Testing more than one example of a model might produce different results. NHTSA once tested a dozen precisely identical Chevrolet Citations, built on the same day at the same assembly line. Variance up to 100 in the HIC score was recorded.

Another criticism is that test crashes are head-on only, while most collisions in the real world come from other directions. Cars slice into trees, or hit other vehicles at an angle, or are hit from the rear, or roll over. There is a federal standard for side-impact protection, but it hasn't been updated since 1973. It also exempts vans. A new proposal would use an "angled moving barrier" to simulate a car traveling 30 mph hitting a perpendicular vehicle going 15 mph. If this proposal becomes effective, it would be phased in beginning in 1991. The twin dummies, named "Vince and Larry" when participating in NHTSA's traveling promotion on auto safety, might also get a long-sought upgrade. For the moment, however, the old standard prevails.

In addition, some cars that have scored well in the crash tests have done poorly in injury ratings compiled by insurance companies. Other cars have earned mediocre scores in crash tests, yet have compiled impressive injury ratings from the insurance industry. NHTSA testing is far from perfect, but the results aren't as useless as some foes suggest. Marginal "pass" or "fail" ratings shouldn't be taken as gospel, but we ignore notably bad results at our peril. A manufacturer that barely meets the minimum standard is perhaps not taking safety quite seriously enough. On the other hand, critics charge automakers spend too much effort on beating the competition and improving crash-test results, as if NHTSA ratings were a game.

Injury and Collision Ratings

The other major source of information is the Highway Loss Data Institute (HLDI), which gathers data for the insurance industry and issues charts that show bodily injury and collision damage experience for specific cars. The best 10 in an HLDI survey of injury ratings for 1985-87 models were: Mercedes-Benz SDL/SEL, Pontiac Safari wagon, Buick Electra Estate wagon, BMW 735i, Saab 9000, Buick LeSabre Estate wagon, Jaguar XJ6, Oldsmobile Custom Cruiser wagon, Ford LTD Crown Victoria wagon and Mercury Grand Marquis 2-door.

All of these are either full-size, rear-drive domestic cars or expensive European luxury cars. Drivers of these cars tend to be older and have fewer accidents than average, which of course reduces the chances of injury. Of these 1985-87 cars, only the Mercedes SDL/SEL was equipped with a driver's-side air bag. The BMW 735i and Jaguar XJ6 were redesigned for the 1988 model

Safety

year and statistics from the previous models may not be applicable to current models.

Not surprisingly, the worst injury ratings were earned by smaller, less expensive cars that are often driven by younger, less-experienced motorists, who tend to have more accidents than average. The worst injury ratings were for the Chevrolet Spectrum and Sprint (both from Japan), Hyundai Excel, Ford Escort EXP, Isuzu I-Mark (similar to Spectrum), Yugo GV, Pontiac 1000 and Nissan Pulsar. The Ford EXP and Pontiac 1000 are no longer produced, while the Sprint has been restyled and renamed the Geo Metro for 1989.

Ratings for collision claims traditionally have favored big cars, but compact passenger vans and even subcompacts have been among the best for 1987 and 1988 models, according to HLDI statistics. High-performance, sporty and expensive models dominate the list of cars with the highest collision claims. The latest HLDI injury and collision ratings are in the Insurance section, where there is additional discussion of the differences among cars. Note that the lists for best and worst injury and collision ratings are by no means identical.

Given that all cars, regardless of vehicle size or type, or who's doing the driving, are covered by these statistics, there are good reasons for avoiding models that regularly make the bottom of the injury claims list. However, as with the government's crash tests, it's best to compare similar models only. While small cars generally don't score as well as mid-size and large cars, some small cars clearly do better than others, year after year.

Is that a product of the design of these cars, or of who's driving them? Insurance statistics don't answer that question, but if a certain car, or a certain manufacturer's cars, continue to show up among the worst, then we have to think that at least part of that is due to the design. However, the charts shown in the Insurance section reflect only claims experience by insurance companies and don't "prove" any particular flaw in a car's design. In using these statistics, be sure the model that interests you hasn't changed significantly since the results were published.

Copies of the HLDI brochure, "Injury and Collision Loss Experience," are available from:

Highway Loss Data Institute
600 Watergate
Washington, DC 20037

Restraint Systems

If all of us had buckled up from the beginning, we might not have to face the annoyance of automatic seat belts or the expense of air bags. Seat belt usage has increased dramatically, from just 11 percent in 1982 to 43 percent today, using estimates from the U.S. Department of Transportation, but some skeptics think find that estimate optimistic. NHTSA Administrator Diane Steed claims that this increase has saved at least 8000 lives. A recent consumer survey by the Insurance Division of SRI Research Center found that 74 percent of Americans claimed to "usually or always" wear safety belts—a figure we find hard to believe—and 89 percent agreed that using belts reduces the risk of injury.

Voluntarily buckling up won't do any longer. At least 32 states (and Washington, D.C.) have laws requiring seat belt use. In addition, 40 percent of this year's new cars must have some form of passive restraint—a device that protects driver and passenger without the need to take any action. For 1990, every car sold has to have such a system, at least for the front seat. The rule doesn't state which type of restraint must be installed, but only two are currently in use: automatic seat belts and air bags. At the instigation of Ford, NHTSA agreed to an extension of current rules (until 1994) allowing cars with a driver's side air bag to have a manual front passenger belt rather than automatic protection.

Air Bags

Finally here to stay, bags go back a long way. General Motors installed some 10,000 bags in the mid-1970s, but stopped when they didn't sell (though they weren't promoted strongly either). Mercedes made driver's bags optional on its North American cars in 1984, standard in 1986. Ford offered them as an option on Tempo/Topaz, mainly on fleet cars at first. In 1986, fewer than 1000 buyers paid $815 for the Ford bag. On the other hand, critics charge that Ford initially failed to push their sale.

With ample fanfare, Chrysler made a driver's side bag standard on half a dozen models late in the 1988 model year. They've also been standard or optional on some Acura, BMW, Volvo, Porsche and Oldsmobile models, and others. More are standard or optional for 1989. Japanese automakers have been slower to offer air bags than European and domestic companies, but

will have to increase their offerings to remain competitive.

Until this year, Porsche was the only make to offer a passenger's-side air bag. They've presented an installation problem because of the distance between the front-seat passenger and the instrument panel, and the fact that passengers tend to move around more. Driver's-side bags are mounted in the steering wheel hub, but there's no equivalent spot on the right side of the dashboard. The 1989 Lincoln Continental has both driver's-and passenger's-side bags as standard equipment this year. GM expects to phase them in as standard on some models by 1992.

When an air bag-equipped car collides head-on with a solid object at about 12 mph or more (or another car at equivalent speed), bumper-or seat-mounted sensors signal an igniter. A solid chemical (sodium azide) fills the fabric bag with nitrogen. It happens within 50 milliseconds (quicker than an eyeblink), before your head can hit the windshield or your chest the steering wheel. Then, the bag quickly deflates. Potential customers have expressed two main worries:

1. Will the bag inflate when it's needed?
2. Will it ever inflate accidentally and obscure vision through the windshield?

Cases of inadvertent inflation are rare, says Jim Mooney of the Insurance Institute for Highway Safety. Those few cases typically resulted from a car "bottoming out" when running into a ditch or otherwise suffering a severe jolt other than a collision. The fact that air bags in GM cars from the 1970s have properly inflated in accidents should alleviate other fears that the bag mechanism "wears out" after a time. That probably was a pleasant surprise for some used-car buyers who had no idea that their old GM vehicle was equipped with an air bag.

An air bag has the advantage of being "unobtrusive," Mooney says, yet it's "there when you need it." Bags alone are not enough for real protection. They're called supplemental restraints for good reason, designed to open only in frontal collisions. Side or rear impacts have no effect. As Fred Ranck of the National Safety Council has said, "Forty percent of collisions are frontal. Only a seat belt will be effective in the other 60 percent."

Unfortunately, air bags remain expensive. They haven't come down much from the $815 price asked on Tempo/Topaz a couple of years back; they're even more expensive as options on other cars. Mass production is expected to cut the price down to less than $300, but that hasn't happened yet. The SRI Research Center survey found that 72 percent of respondents said they would consider buying a car with air bags; 22 percent would not. Oddly, people of modest income were willing to pay a bit more ($550 average) for air bags than more affluent families. Ford in the past has given $500 rebates to air-bag buyers, bringing the price down to the $300 level that researchers feel will attract motorists in larger numbers.

What many people don't realize is that the cost of reinstalling a bag after it has deployed comes to considerably more than its original price. Of course, that's just one of the repairs needed after a collision, and should be covered by an insurance policy. Dealers have also been slow to push bags, some even discouraging customers who show interest.

One minor drawback: a tilt steering wheel hasn't been available on most cars with air bags, because the air bag must be precisely located in the steering wheel. This is unfortunate because tilt wheels have grown so popular, and are standard in most of the luxury cars that are the best candidates for air bags.

Except in a handful of states, a number of insurance carriers give discounts for bag-equipped cars, ranging up to 30-40 percent. However, the discount only applies to the personal injury or medical payment portion of a premium, so we're not talking about big savings. The lion's share of insurance premiums is for liability coverage. Many insurance companies offer a similar discount for automatic belts.

Automatic Seat Belts

Effective as air bags have been proven to be in front-end collisions, protecting the head as no other device can, they're not enough. Belts are essential. Since Americans haven't been willing to buckle themselves, even with state laws mandating belt use, many of today's cars do so automatically. Most automakers have found automatic seat belts the most expedient and cost-effective way of meeting the government mandate for passive restraints.

Three systems for the outboard front seats are currently available. A motorized arrangement, found on some Toyotas, Fords and others, uses a 2-point shoulder harness that slides on a track above the door. The belt locks into place when the key is switched on.

Since a separate lap belt has to be buckled manually, it's not 100 percent automatic. A good many motorists—including those who have used seat belts for several years—forget to buckle the lap belt, so they get only partial protection. Even though some studies suggest that a 2-point system is almost as effective as a 3-point, why be two-thirds safe?

Chrysler's is also a 2-point motorized system with separate lap belt, but its automatic belts are attached to the door itself, which interferes with entry and exit. Furthermore, the manual belt is buckled at the outside, and in hard to reach along the center console. Many owners find the system cumbersome and uncomfortable, and try to defeat it by detaching the emergency release.

Another fault we have noticed with motorized shoulder belts is that they don't lie over the shoulder and across the chest for all people, where they do the most good. In addition, some of the manual lap belts that come with motorized shoulder belts actually lie across the top of the thighs, not the pelvis, where they are supposed to be.

General Motors took a different tack, developing a system more similar to conventional seat belts. One study even found that most owners didn't know GM's belts were automatic. This 3-point system is door-mounted, and can be left permanently buckled. The setup is attached to two points on the door, a third at your inside hip. When the belts are left buckled, you just slide into the seat under them, and they shift into position as you shut the door. Some people don't mind, but others find entry and exit more difficult. It's also easy to use this system manually—or not at all. John Fobian of AAA says the GM belts "meet the letters of the NHTSA regulation, but certainly not the spirit." In response, engineer Richard A. (Dick) Wilson of GM says they wanted "a system that meets the law but is not a great departure for the habits of people who normally wear belts." Perhaps for that reason, many drivers seem to find them less annoying than the competition. Honda uses a similar system.

Obviously, a setup that people find irritating enough to try to disconnect (and often succeed) isn't what safety experts had in mind. None of the three automatic belt systems is perfect. NHTSA's goal of creating as many options as possible allowed manufacturers to choose a system least likely to meet buyer resistance. As AAA's Fobian points out, "It's not a choice for consumers; it's a choice for manufac-

turers." No automaker makes an alternate system optional, so you take (and live with) what's there. Some people worry that they'll somehow get tangled in automatic belts, have eyeglasses knocked off, or a cigarette clipped in half. Most of the reaction, though, is more irritation than actual mishap.

Seat Belt Slack

For proper effect, NHTSA advises that belts should fit snugly at all times. Shoulder belts should pass over the shoulder and across the chest, with little or no slack. Lap belts are supposed to run across the pelvis (not the abdomen or thighs). Peek into any passing car, though, and that's not what you're likely to see. Too many people are running around with belts so loose with slack as to be virtually useless in the event of collision.

Americans, it seems, are unwilling to pay much of a price, whether in dollars or discomfort, for the benefits of safety. Imported cars are designed to keep their occupants snugly belted. An inertia reel automatically takes up any slack in the belt. Most domestic cars have belts designed with comfort uppermost in mind. If you lean forward, the shoulder harness follows along. Lean back, and the harness doesn't return to its starting point, often leaving several inches of slack. That much looseness could allow the belted person to hit the steering wheel, dashboard—even the windshield.

Domestic automakers say removing the slack would discourage motorists from wearing them at all. Statistics suggest otherwise. Imports with their snug belts have a higher rate of usage.

Safety

Of course, that could merely mean that import buyers are more willing to use seat belts.

No matter what kind of restraint system your car has, you can keep the belts properly snug. A quick tug, or pulling it out then letting go, should take up any excess slack.

Rear Seat Belts

Seat belts for back-seat passengers is another source of controversy. In 1986, the National Transportation Safety Board cited a study of 26 frontal crashes in which rear seat lap belts caused as many injuries as they prevented. That study has been harshly criticized on two grounds: (1) the number was too small to be statistically valid; and (2) it didn't study other types of accidents, where lap belts are more valuable. In any case, unless rear passengers are belted, they can shoot forward in a frontal crash, seriously injuring the front-seat occupants as well as themselves.

A year later, a NHTSA study estimated that about 100 lives were saved and 1500 serious injuries prevented annually by rear belts—but only 16 percent of passengers use them. Both studies agree on one point: a combination shoulder harness and lap belt is far safer than a lap belt alone. The NHTSA report recommended 3-point harnesses for both front and rear occupants. This was advocated by safety lobbyists back in the early 1970s, but opposed by automakers, who convinced the government not to require such a system. Anchor mountings for shoulder belts have been installed in most cars for years, but few have had rear shoulder belts until recently.

By the 1987 model year most automakers had already changed or were changing to combined lap and shoulder belts for rear passengers. Thomas H. Hanna, president of the Motor Vehicle Manufacturers Association, testified to a Congressional subcommittee that "by 1990, virtually every (domestic car) will be equipped with lap and shoulder belts in outboard rear seating positions." Hanna added that a requirement to make retrofit kits available for earlier models may be acceptable to automakers.

Many 1989 models (including all GM cars) have them as standard. Older models, and those lacking rear lap/shoulder belts, can be retrofitted. A recent survey of 30 dealers, though, found that not one had such a kit in stock, and most didn't think one existed. The kit was expensive, too,

priced from $99 to $400 at GM dealers. Our recommendation is simple: We advise that you refuse to accept a new car of any make unless rear shoulder belts are installed.

Child Restraints

Transportation Secretary Jim Burnley has stated that "The risk to small children can be reduced by two-thirds through proper use of a child safety seat." Better yet, those children who are belted in for every trip might develop a buckle-up habit that many of their elders never attain. Every state requires a baby carrier or child safety seat, though laws vary as to the ages of children who must use them. Ask at your state's motor vehicle department or state police headquarters.

Ordinary seat belts are definitely not the way to protect children. A child's head is more likely than an adult's to jerk forward during a sudden stop. Babies and children under four (or weighing under 40 pounds) belong in an approved child's car seat, not in the laps of adults. That seat belongs in the back, not the front of the car. Otherwise, a child could be injured by hitting the dashboard.

Child restraint systems vary in design and how they're secured. As a rule, the child is held to the carrier or seat by straps for each shoulder, and another between the legs. The seat itself is secured by the car's seat belts. Some car seat manufacturers offer a tether strap that bolts to the car. That prevents the seat from falling forward or sideways in a serious crash. Some cars come with a bolt anchor or bracket; others must be installed.

Why not have this done before you buy the car? If the salesman isn't sure whether a tether strap can be used, details should be in the owner's manual. Make sure any child restraint system was built after January 1, 1981, so it conforms to Federal Motor Vehicle Safety Standard 213.

Seat Belt Laws

At least 32 states (and Washington, D.C.) have laws requiring seat belt use. Only five state laws cover all occupants of passenger vehicles (including vans and pickups). Rear passengers are often exempted.

A University of Michigan study in 1987 of the first eight states that adopted mandatory seat belt laws found that front-seat fatalities declined an average of 8.7 percent. In Texas and New York, where strict laws allow ticket-

ing drivers and passengers for failing to wear belts, the decline was greater yet. States that permitted ticketing for noncompliance only when a car was stopped for other reasons experienced a more modest improvement (or no change). Researchers say the study clearly shows that seat belt laws are saving lives. Fatality rates for rear-seat passengers, motorcyclists and pedestrians were unchanged over the test period, showing that the drop was not due to seasonal or other variables.

NHTSA Administrator Diane Steed estimates that 1350 lives were saved in 1987 because of seat belt laws. NHTSA further estimates that 8035 lives have been saved and 95,500 injuries prevented since 1983 by voluntary compliance. Whether required by state law or not, buckling up does save lives.

Anti-lock Brakes

Those who have tried it nearly all agree: an anti-lock braking system (ABS) is one option that's well worth the hefty price. This is the one that could save your life on a slippery road—a safety device potentially ranking close to 4-wheel brakes in benefit to motorists. Rather than prevent injury during an accident, ABS can help you avoid collision in the first place by bringing your car to a halt in a straight line on just about any kind of surface.

ABS prevents the wheels from "locking," which can cause a car to skid and lose steering control. If the wheels stop turning, simple laws of physics take over. The car keeps on going in the same direction as before, no matter how you twist and tug at the steering wheel. With ABS, sensors at each wheel detect when a wheel is about to lock up under hard braking. You can use full braking force without risking an uncontrolled slide, stopping in the quickest possible time but still able to steer around any obstacle.

In dry weather, ABS lets you hit the brakes as hard as you want in an emergency, without worrying about skidding sideways. Actual stopping distances aren't necessarily much shorter, but the knowledge that you can maintain control in a panic situation adds to driver confidence. Where ABS really excels, though, is on roads slick with rain, ice or snow. Can you imagine stomping on the brake as hard as you can on icy pavement? That's exactly what we are taught not to do. In effect, ABS "pumps" the pedal as experienced drivers have always done, but much faster and more precisely than any human could. Stopping distances on highly slippery surfaces are much shorter with ABS. An average driver in an ABS-equipped car might even be better able to avoid a collision than an expert driver in a car without it.

No, it won't prevent every possible accident, or make everyone an expert. Anti-lock braking has limits. On super-slick glare ice, nothing can deliver a quick stop. Be assured, though, that it will be quicker and shorter with ABS.

Why doesn't every car have it? ABS has two drawbacks: price and availability. So far, it's standard or optional only on a relative handful of cars, most of which are expensive, and costs a fair amount itself—usually over $900. Mercedes-Benz brought ABS to the U.S. on some 1985 models and BMW was first to make it standard on all models. For 1989, it will be available on even more models, including several under $20,000.

More people might want ABS if the price fell, or dealers offered incentives; but neither course seems likely at the moment. In fact, not all dealers have been pushing the system, claiming it's hard to demonstrate and to sell. Customers, they fear, begin to ask what's wrong with the ordinary brakes that have been there all along. Drivers who try it, though, are likely to concur that even $1000 is a small price to pay for what you get, even if (or especially if) you only have to use its full capability once.

If you're considering a car with ABS, ask the dealer to let you give it a try, preferably on a rainy or snowy day. Get away from traffic, onto an empty, spacious parking lot. Start slowly until you get the feel of it, and don't worry about any odd pulsations in the pedal when you hit it hard; they're normal. The

Unintended Acceleration

For more than 20 years, a succession of makes and models have been charged with sudden, unwanted acceleration. Most allegations claim that as the car was shifted from park into drive or reverse, the engine would suddenly race, shooting the vehicle forward (or backward), out of control.

Audi's 5000 was by no means the only vehicle so charged, but it suffered the worst publicity, culminating in a name change for current models. Audi suffered most because such a large portion of the cars it sold in the U.S. were affected by a recall, television coverage was intense, and government

studies revealed more related deaths and injuries than for the other makes.

Similar problems have been investigated in models from Chrysler, Ford, GM, Honda and Acura, Mercedes, Austin Rover (Sterling), Nissan, Toyota, Volvo and Volkswagen. Not many manufacturers are missing from that list. The Center for Auto Safety claims the Acura Legend, to take one example, has an accident rate due to sudden acceleration that's second only to Audi's. The Center further claims that at least 50 occurrences of unintended acceleration were noted by GM employees and dealers on H-body cars (Buick Le Sabre, Oldsmobile 88 Royale and Pontiac Bonneville), but the corporation rejected an engineering recommendation that the cars be recalled.

At least 1650 accidents, according to the Center, have been linked to acceleration surges. It's a problem that never seems to go away. After years of study by NHTSA and consumer groups, no universal cause has been unearthed to the experts' satisfaction. What all cases have in common is an automatic transmission. Audi has continued to claim that the cause is simply "the errant right foot of the driver," hitting the gas instead of the brake. No one has been able to duplicate a case where stepping firmly on the brake failed to stop a car, even with the gas pedal pressed to the floor and the engine revving at full speed.

Critics insist that the large number of incidents reported could not logically all result from driver error, and that so many involve Audis cannot be written off as coincidental. Some experts believe the fault lies in the computerized complexities of the fuel and ignition systems. On many cars, engine speed can increase, or surge, without warning, though such increases are usually brief and minor. Whether this could cause a driver to panic and hit the gas instead of the brake is a matter of dispute. Many Audi owners further claim that their cars accelerated out of control and could not be stopped, even though the drivers maintain they were pressing as hard as they could on the brake pedal. Since there's no physical connection between the throttle and brake systems, it seems unlikely that both systems would suddenly fail, yet the brakes would work normally afterward.

A few cases have been traced to floor mats that kept a gas pedal from fully releasing, and there have been several recalls to fix flaws in cruise control units. Audi 5000 models and Nissan 280ZX and 300ZX models were recalled to install a shift-lock device that

prevents the vehicle from being shifted into drive or reverse without first stepping on the brake. Many manufacturers also have repositioned the two pedals, to minimize the possibility of driver error. For 1989, a shift lock is standard on Audi 100/200, Nissan 300ZX and Toyota Cressida and Supra models with automatic transmission.

Lawsuits against Audi remain in litigation. The company's offer to provide discounts of up to $2000 on new Audis to owners of the affected 1978-86 models was rejected by the court this past summer. The judge ruled that nobody should have to buy another car to take advantage of the offer. Clarence Ditlow of the Center of Auto Safety pointed out, "If any individual has had a sudden acceleration problem with an Audi car, probably the last thing they want to do is get another Audi." Audi's response was that it would consider lawsuits against the Center, "60 Minutes" (which twice aired a program critical of the Audi) and the Audi Victims Network, for causing the loss of resale value without anything having been proven wrong. In fact, no defect has been proven—and perhaps it never will.

At this writing, NHTSA still has the Audi case under formal investigation. Other models are in the engineering analysis stage of investigation: 1985-88 C-body GM cars (Oldsmobile Ninety-Eight, Buick Electra/Park Avenue, Cadillac De Ville/Fleetwood); 1979-87 Nissan 280ZX and 300ZX; 1986-88 Acura Legend and Sterling 825; and the most recent, all 1986-88 gasoline-powered Mercedes models. NHTSA assigned an independent agency to investigate the sudden acceleration question, but no conclusion has been reached.

If your car ever accelerates suddenly, make sure you step on the brake—even if it takes a fraction of a second longer to be certain. Why not practice this maneuver, so you'll be most familiar with the pedal positions if it has to be done in a hurry? Keep your foot on the brake whenever you shift into a drive gear. Even if the engine surges, it shouldn't move if the brake is applied firmly. Properly maintained brakes should be able to stop a car even with the engine running at full speed. Be ready to shut off the ignition quickly.

Rolling Over

Not long after the furor over Audi's alleged problem with unwanted acceleration began to fade, Consumers Union issued a scathing denunciation of the Suzuki Samurai for its purported propensity for rolling over. Several months earlier, the Center for Auto Safety had petitioned NHTSA to start proceedings to recall 140,000 of the 4-wheel-drive sport-utility vehicles.

Just as the story broke, Ted Koppel on ABC-TV's "Nightline" show carried a program on the Samurai. Many viewers must have been startled to see a series of Samurai commercials during the breaks. Clearly, the company was prepared to fight back. At first glance, the facts appear clearer than in the sudden acceleration cases. Samurai, like some other sport-utility vehicles, has a short wheelbase (distance between front and rear wheels), narrow track (distance between left and right wheels), and high center of gravity. The result, following the basic laws of physics, is less stability. As Jim Mooney of the Insurance Institute for Highway Safety warns, such vehicles "definitely have more potential for rollovers." By federal regulation, most 4WD vehicles carry warning labels cautioning that they are more likely to tip over than passenger cars.

Critics charged that the Samurai could become unstable not just in off-road driving, but when trying to veer away from a pothole or obstacle (or another car) on an ordinary road. Similar vehicles, including earlier Jeeps, have faced like charges (and lawsuits). More than a year earlier, the Insurance Institute for Highway Safety had found that small utility vehicles had a fatality rate five times that of large passenger cars, and that almost half the deaths were caused by rollovers.

Other current 4WD vehicles have wider tracks than Samurai, which measures 51.2 inches between the front wheels and 51.6 inches between the rear wheels, and longer wheelbases than the Suzuki's short 80 inches. A Colorado inventor even came up with a bolt-on device to widen the Samurai's track by three inches. Suzuki would have none of that, insisting that Samurais were as safe and stable as any 4WD vehicle. Sales fell sharply after the charges surfaced, but soon recovered—partly as a result of a $2000 dealer rebate. In a letter to a trade publication, Doug Mazza, general manager of Suzuki of America, cited a "nearly 200 percent increase in Samurai sales" the next month as evidence that customers were confident of the vehicle's "outstanding safety record."

Early in September 1988, NHTSA rejected a petition from two consumer groups to investigate the Samurai for a potential recall. After studying fatal accidents, the agency found "overwhelming support for the human factor" as a cause, rather than structural problems. "Every single one of the fatalities involved ejections," said NHTSA spokesman Ron DeFore, suggesting that seat belts weren't being used—a risky omission in vehicles less stable than the norm. Half the accidents involved alcohol, and most drivers were under 25, having owned their Samurais an average of eight months, according to NHTSA. In sum, the record for Samurai fell right in the middle of similar vehicles.

At the same time, however, NHTSA granted a petition to initiate work on a rollover standard for such vehicles. This followed rejection of an earlier petition to base standards on their "stability factor" (half the track size, divided by center of gravity height), even though "a high correlation exists" between that factor and danger, according to NHTSA. So the question is not yet resolved.

Speed Limits and Safety

Highway fatality rates dropped significantly after the 55-mph limit became law in 1974. The rate fell 17 percent, down from 4.24 deaths per 100 million miles a year earlier. Death rates continued to fall over the next decade, even as traffic increased. Experts have hotly debated the contribution of the 55-mph limit to that decline. Their argument continues, now that the limit has changed to 65 mph on rural interstates in most states.

Within six months of the new law's enactment in April 1987, all but a dozen states had made the switch. The basis for the change to 65 mph was that rural interstates appeared to be our safest highways, accounting for only nine percent of traffic fatalities in 1984. On the other hand, federal figures showed that the fatality rate in that year was actually higher on rural interstates than urban ones. Either way, the National Highway Safety Council soon estimated that at least 450 more people would die each year because of the higher rural limit.

A NHTSA report from 22 states that changed to 65 mph found a 52 percent increase in fatalities over the following 3-month period, compared to the corresponding period in 1986. In 13 of those 22 states, the rate went up on affected interstates, but declined on other roads. The Insurance Institute for Highway Safety found an overall 21 percent increase in deaths on those interstates for 1987, compared to less than five percent the year before. Further analysis comparing 1987 to the 1975-86 period revealed a 20 percent increase in the first 28 states that raised the limit (but death rates in the other states also went up). Death rates on rural roads in general declined slightly. The fatality rate on all U.S. roads dropped slightly in 1987, to a record low 2.4 deaths per 100 million miles traveled (from 2.5 the year before and 3.3 in 1980).

Whew! Plenty of statistics. While they certainly suggest that 65 has brought more deaths, the extent has yet to be learned. Any increase reverses a trend toward fewer highway deaths that began decades ago, a result of improved auto and road safety, driver education, drunken-driving laws and emergency medical techniques. Many advocates of lower speed limits consider the 55-mph limit a critical factor in the decline in traffic deaths.

As anyone who's traveled the interstates recently has observed, though, few drivers actually obey the posted limit. Just as large numbers of motorists viewed the 55-mph limit as a license to travel 60, 65 and beyond, today's drivers in 65-mph areas soon edge up to 70 and 75. Lack of compliance with seat belt laws is another example of attitude problems. In Britain, at least 90 percent of motorists abide by the law; in Canada, at least 60 percent comply. American rates, while improving, still fall short of where they should be. Widespread use of radar detectors contributes to speeding. Brian O'Neill of the Highway Loss Data Institute has equated ownership of a radar detector with the possession of burglar tools. Other critics may not go quite that far, but insist detectors have no purpose other than to enable speeders to break the law—and do so with impunity.

The relationship of detector use to accident rates is a matter of dispute. Critics claim drivers who use radar detectors are involved in more accidents, though that's partly because they tend to drive more miles. Even if radar detectors have no direct correlation to increased accident rates, their existence produces an aura of defiance—a twisting of the law to suit one's personal convenience—that can pose as great a danger as exceeding the speed limits. Currently, at least five groups are trying to ban the devices, including the AAA, International Association of Chiefs of Police and the National Safety Council. Only two states, Connecticut and Virginia, and the District of Columbia, have done so, at this writing.

INSURANCE

Consumers are fighting back. Especially in California, motorists unhappy with skyrocketing insurance premiums have banded together to try to call a halt. By the filing deadline in spring 1988, five separate initiatives dealing with insurance rates were proposed for the November ballot, including one backed by Ralph Nader from the Voter Revolt to Cut Insurance Rates. Insurance companies sought to block several proposals, which range from a one-time rate cut to massive restructuring of the rating system, and issued their own initiative.

California's movement has been compared to the Proposition 13 battle a decade ago, which cut property taxes dramatically. At least $43 million has been spent on advertising by the insurers alone. A pro-consumer result against what is perceived as a ruthless, profit-hungry insurance industry could provide fuel for other protests. Already, activity has taken place in states such as New Jersey, which holds the dubious record for the highest auto insurance premiums in the country. California is third. Pennsylvania and Massachusetts also rank among the states with the biggest insurance problems.

High insurance costs have even blamed for the demise of some sporty models, including Pontiac's Fiero. One magazine reports that urban Fiero owners pay an average of $2000 to $3000 a year for coverage. A Mazda spokesman notes there are some "owners whose insurance payments came out to about 75 percent of their monthly car payments." Performance-oriented cars appeal most to the young, of course, who are charged the most for insurance.

Can't Do Without It

Unlike most commodities we buy, insurance is required (in about 40 states). Even in a non-mandatory state, logic dictates that only a fool or an irresponsible person drives without proper coverage. This is a litigious country, and all your property is put at risk for even a modest accident, unless you have adequate coverage. *Best's Review*, an insurance trade magazine, claims that one in five drivers have an accident each year. Yet a psychological study found that individuals believed their

own risk to be half that high. Clearly, many of us persist in the belief that accidents "happen to the other guy." In Illinois alone, an estimated 2 million drivers (out of 7 million total) have no insurance at all.

Insurers have held the reins for decades, setting rates according to their own standards. They even have "rate bureaus" to exchange information. That's possible, says Robert Hunter of the National Insurance Consumer Organization (NICO), "because insurance companies are exempted from anti-trust laws." They exist in a "never-never land of weak regulation and competition," where the Federal Trade Commission is "not even allowed to study insurance." Nearly all regulation is at the state level.

Insurers thus occupy a unique position. While you may be required to have insurance, no company is compelled to sell it to you, much less at what you consider a fair price. Only if you're placed in a high-risk pool are you "guaranteed" that an insurer will cover you. All companies operating in a state share the risk of "pool" policies, and charge plenty for their service.

It would be a lot easier if we could determine our own rates by consulting a chart. The California proposition backed by NICO includes a system whereby a customer could type facts into a computer, and get direct quotes from the cheapest companies. As it stands, the complexity is overpowering.

Why So High?

Premiums grew by an average 9-13 percent in 1986 and 1987, according to Hunter of NICO. He predicts a 20 percent rise this year, which would raise the average premium to more than $550 (versus just $276 back in 1981).

That's not to say that increases cannot be justified. Cars cost more than they used to, and more to fix. People drive more (though not as much as is generally believed). Lawsuits cost money, and jury awards are growing. Medical costs have skyrocketed. The Insurance Information Institute says that over the 1977-86 decade, average payments for bodily injury claims rose 156 percent and damage claims 139 percent, while premiums only went up 93 percent. Average payments edged up four percent in 1988. Insurers further

insist that they lose money on car insurance, especially in states like California.

Insurers also are less than thrilled about mandatory car insurance for the 10 or so states that still have no such law on the books. They insist that such laws are too difficult (and costly) to enforce, haven't put uninsured drivers off the road, and wind up costing insured drivers more in premiums. As a matter of pure economics, insurers have to oppose compulsory coverage because many of the drivers they will have to cover are viewed as high risk. And the last thing insurers want is to have to provide coverage for any more risky motorists than they're now compelled to deal with under "pool" plans.

"Financial responsibility" laws prevail in non-mandatory states. These require a motorist involved in an accident to prove ability to pay for damages, whether by insurance or personal assets. State Farm maintains that such laws are more fair because they're "aimed only at drivers who cause accidents, not at every driver in the state." Massachusetts adopted mandatory insurance in 1927, but most others were enacted in the 1970s as part of a switch to no-fault programs.

Types of Coverage

Six forms of coverage are included in most policies:

Bodily injury liability—Pays injury/death claims against you, and legal costs, if your car injures or kills someone.

Property damage liability—Pays claims for property that you damage in an accident.

Medical payments—Pays for injuries to yourself and occupants of your car.

Uninsured motorist protection—Covers you for injuries caused by an uninsured or hit-and-run driver; in some states, property damage is included.

Collision coverage—Pays for damage to your own car in an accident, up to its book value, unless covered by a liability claim against the other party.

Comprehensive (physical damage)—For damage to your car from fire, wind, flood, theft, vandalism and other causes unrelated to accidents.

Liability coverage is expressed by three figures: "$25/50/15" means

Insurance

$25,000 bodily injury coverage for one person, $50,000 total for a single accident, and $15,000 for property damage. It's the coverage required by law in most states, since it protects the other party.

Collision coverage has a deductible: a specified amount that you have to pay (in cash) before the coverage takes effect. Deductibles run as low as $50, or as high as $1000 ($2000 in a few states). Comprehensive may have a similar deductible but many people, especially in non-urban areas, have full coverage, which costs little more. Coverage for medical payments ranges from $1000 on up. In today's world, you might even have to consider "underinsured motorist protection," in case you're hit by a driver who has inadequate coverage.

No-fault programs, operating in half the states, were supposed to pull down premium rates by reducing the number of lawsuits. In 14 states, lawsuits are indeed restricted. Lower premiums haven't necessarily appeared, however, prompting at least two states to rescind their no-fault law. Regulations vary by state, but regardless of who "caused" an accident, each party may (or must) make all claims for collision and medical reimbursement to his or her own insurance company. Separate personal injury coverage, similar to medical payments, may be available (or required).

Setting Insurance Rates

Insurers consider a lot more than a person's record of accidents and traffic violations. Three basic factors are involved:

1. Where you live.
2. Who you are (age, gender, driving record, etc.).
3. Type of car insured (mainly for collision and comprehensive coverage).

Most of us quickly discovered as teenagers that young men (or their parents, for use of the family car) paid a lot more than young women, or older males. That's still true, and the dividing line is sometimes set at age 30 rather than 25, as it used to be. Getting married usually (not always) changes the young person's status. Warren L. MacKenzie, an insurance agent who argues against denial of coverage (or higher rates) for under-25s, has written that "basic insurance principles dictate spread of risk over all ages."

Others criticize reliance on sex. Proponents of unisex insurance laws claim that gender-based rates are inherently unfair, whether they truly predict risk or not. Five states have agreed. Hawaii and Massachusetts prohibit both sex and age as a rate-setting factor. Unfortunately for women, the change has jacked up some of their rates considerably. *Insurance Review*, a trade publication, reports that young single women in Montana now pay $91 to $274 more than they would have before that state's unisex law. A survey of "household representatives" found that 43 percent agreed with the law that prevented consideration of gender. But when told that such a law would boost their rates up to $100, 68 percent decided that would be unfair.

Similar criticism targets geographic standards (rating territories). Urban dwellers tend to pay far more—as much as triple—for equivalent coverage than rural or small-town drivers. Insurers can easily pull out statistics that show higher accident rates in crowded urban areas. A major reason for New Jersey's "top" ranking in premium rates is said to be the fact that 85 percent of its drivers (half falling into the high-risk pool) live in densely populated areas.

Some companies boost your premium significantly for having a single

---

Injury Ratings for 1985-87 Models
Highway Loss Data Institute

Best Injury Ratings	Worst Injury Ratings
42 Mercedes-Benz SDL/SEL	178 Chevrolet Spectrum 4-dr
44 Pontiac Safari wgn	167 Chevrolet Sprint 4-dr
48 Buick Electra Estate wgn	167 Hyundai Excel 4-dr
50 BMW 735i	164 Ford EXP
51 Saab 9000	163 Isuzu I-Mark 4-dr
52 Buick LeSabre Estate wgn	162 Chevrolet Sprint 2-dr
54 Jaguar XJ6	161 Chevrolet Spectrum 2-dr
55 Olds Custom Cruiser wgn	159 Yugo GV
57 Ford LTD Crown Vic wgn	158 Pontiac 1000 4-dr
57 Mercury Gr Marquis 2-dr	157 Nissan Pulsar NX

Note: HLDI injury ratings are based on frequency of injury claims filed under personal injury coverage for 1985-87 models. Ratings are expressed in relative terms: 100 represents the average of all cars. A rating of 42 is 58 percent better than average; a rating of 178 is 78 percent worse than average. Some of the models listed have been discontinued or redesigned since 1987, so these ratings may not apply to current models.

Collision Ratings for 1987 Models
Highway Loss Data Institute

Best Collision Ratings	Worst Collision Ratings
35 GMC Safari van	329 Chevrolet Corvette conv.
45 Chevrolet Astro van	329 Chrysler Conquest
53 Dodge Caravan van	235 Porsche 924
54 Ford Crown Victoria 4-dr	224 Chevrolet Corvette
54 Cadillac Fleetwood 4-dr	205 Porsche 944
54 Plymouth Voyager van	187 Mazda RX-7
57 Dodge Diplomat 4-dr	186 Audi 5000 4-dr
57 Chevrolet Celebrity wgn	183 Mitsubishi Mirage 2-dr
58 Honda Civic wgn	166 Pontiac Fiero
59 Ford Aerostar van	165 Volkswagen GTI

Note: HLDI collision ratings are based on average loss payments for collision damage for 1987 models. Ratings are expressed in relative terms: 100 represents the average of all cars ($194). A rating of 35 is 65 percent better than average; a rating of 329 is 229 percent worse than average. Some models in this list have been discontinued or structurally modified since the 1987 model year, so these ratings may not apply to 1988-89 models.

---

traffic violation or minor accident. Some won't take you at all with any kind of blemish on your record. Others are more lenient. Ask about the company's standard for changing rates later, but don't expect a dependable response.

Insurers insist they can't rely wholly on driving records as a factor, because information available from states varies too much in content and quantity. Nor can they be sure a policy-seeker's claim of driving under a set number of miles is true. So they rely on factors that are easy to determine, but may or may not be the best predictors. The National Insurance Consumers Organization advocates a system whereby "people with good driving records be entitled to coverage with the company of their choice."

Shopping for Insurance

Many of us keep the same insurance carrier year after year, and have no intention of shopping. The fact is, though, you can pay a vastly different amount for the same coverage—even when buying from agents of the same company. In a California study of price quotes from 20 agents, some were ready to charge more than double the rate of others, for the same coverage. Differences between agents of the same insurance company varied almost as vastly. Other surveys have found people paying up to three times as much as others in the same area.

"People are beginning to look at the total picture of car costs," says Jim Mooney of the Insurance Institute for Highway Safety; and that includes insurance. Unless you're satisfied that your current premium is in line with those offered by other agencies, it's wise to investigate at least one or two others when buying a new car.

Of course, shopping takes some effort. You actually have to read some of that fine print in your policy—the chore you've probably been putting off for years. No way can you compare what's offered by someone else unless you know exactly what you have already, and exactly how much it costs. Call or visit at least one agent for each company you have in mind. May as well start with State Farm or one of the other biggies. Explain exactly what you want. If quoted rates are far apart, call back to confirm that each agent is talking about the same coverage. Ask to see an itemized list if in doubt. Don't be surprised if some won't quote a price on the phone, or show anything in writing. Focus on those who cooperate.

Price is only one factor. You also want a company that's ready to provide quick, efficient claims service when needed. Robert Hunter of the National Insurance Consumers Organization feels that "companies with the best service records tend to have the lowest prices." Before making up your mind, contact your state insurance department or a consumer agency to see if the company you're considering has any serious complaints against it.

Basic rates are just the beginning. Ask about discounts that might apply. In addition to discounts and surcharges tacked onto certain car models (see below), there might be a discount for:

1. Students with a B or better average in school (as much as 25 percent off).
2. More than one car insured by the company.
3. Anti-theft alarms.
4. Safety devices installed (air bags, automatic seat belts, ABS).
5. Insuring car and home with the same company.
6. Flawless driving record.
7. Driving less than a set number of miles per year.
8. Driving only for pleasure (not for business or commuting).
9. Taking a defensive-driving class.
10. Age over 50-55 (or retired).
11. Nonsmokers.
12. No male drivers in household.
13. Drivers in specified professions.

Most discounts apply only to one portion of the policy, so don't expect a dramatic overall price cut. And even if you qualified for 10 cuts of 10 percent each, that doesn't take your premium down to zero. Ironically, the one factor that seems most logically related to risk, your driving and accident record, doesn't always warrant as great a discount as factors that have little to do with driving skill. A good student in the family often saves more dollars than a spotless motoring record.

How Much Do You Need?

Years ago, taking the minimum coverage was good enough for most drivers. With the massive rise in claims and jury awards, those minimum levels barely begin to cover the costs involved in anything beyond the slightest accident. A state may require a minimum as low as 15/30/10, or even less. Only the foolhardy should accept those amounts, especially since increased liability coverage doesn't usually cost a whole lot more than the minimum. A typical $10,000 minimum for property damage won't help if you collide with a new Mercedes or BMW.

"The more you have in assets and income at risk," warns Bob Sasser of State Farm, "the more you need." He adds that 100/300/50 ($100,000 per person, $300,000 per accident, $50,000 property damage) is an appropriate amount "for a typical middle-class family." Better to save money on other parts of the policy than to cut corners on liability, since a lawsuit could destroy most of us financially.

If your assets are substantial, a million-dollar liability limit may not be out of line. That could be obtained for as little as $100-150 under a "personal liability umbrella" in conjunction with homeowner's coverage.

Probably the best way to cut costs is to increase the deductible amount for collision and/or comprehensive. Switching from a $100 deductible to $500 could slash premium cost by a third, says Hunter of NICO, who recommends $500 as the minimum for most families. Going up to $1000 might even cut it in half. Increasing the deductible maintains coverage for catastrophic incidents, but bypasses the fender-benders. Experienced motorists think twice about filing small claims anyway. For a few dollars' gain, you risk earning a sharp rise in later premiums, or even cancellation. Why pay for something you don't intend to use? This fear of filing small claims is why many people involved in minor accidents try to settle "on the street."

If your car is more than five or six years old, there's little point in maintaining collision coverage at all. Claim payments are limited to book value, so unless the car is worth markedly more than the deductible amount, you're throwing money away. Unless a creditor demands collision coverage, you're free to do without it even for a valuable car, but most of us don't wish to take on that kind of risk.

Shipping a teenager off to college in another city should bring the family premium down quite a bit. Making him or her share the family car helps too. If the child simply must have a personal car, a cheap one may lack pizzazz but costs less to insure than an expensive model.

Medical payments might be eliminated too, if your regular group health insurance policy is a good one. The difference isn't always great, though. Does your policy include any extras such as towing costs, or the cost of a rental car if yours is laid up? If you'd be

Insurance

reluctant to make a claim for that feature, there's no point paying for it. A policy may also contain exclusions, such as exempting a CB radio from theft coverage. Fine-print items can change during the course of coverage, so be sure to check any updates before renewing. Review your policy at least every couple of years. Another company might have a better offer waiting for you. Then again, it might not.

How Cars Compare

The rule is simple: an expensive car demands expensive insurance. With few exceptions, even a minor fender-bender costs less to repair on a budget-priced vehicle. Rates that depend on the car rather than the driver make some sense, since at least half the typical premium cost is for property damage, not personal injury protection.

Collision coverage rates, then, vary with the type of vehicle. Each insurer offers a set of surcharges that apply if you own a car they've found to be involved in many accidents or is costly to repair. Discounts come when their experience reveals fewer or less expensive claims. Note that it's the insurer's experience that matters, not any measure of the car's real safety factors. Surcharges and discounts are also based upon the likelihood that a certain car will be stolen, or broken into.

Surcharges don't always amount to a large sum, since they apply only to collision and comprehensive. State Farm, for instance, adds 10 to 30 percent to the basic rate for cars with a "bad" claims record, and lops off like amounts for those on its discount list. Each insurer has a similar (but not identical) list—another reason to comparison shop.

Exactly which cars get discounts and which are hit with a surcharge sometimes defies logic. Sporty and performance models are likely to make the extra-cost list, but a few surprises pop up too, such as the Plymouth Colt and Toyota Corolla on one such roster. Larger, plainer family vehicles are likely to warrant discounts, but so might a Volkswagen Vanagon, Jaguar XJ6 or even a Ferrari.

Statistics from the Highway Loss Data Institute, also discussed in the Safety section of this book, show injury and collision loss experience of various car models. Insurers use these as a guide in setting rates, but don't consider any related factors. For instance, the HLDI ratings ignore the fact that small cars (with highest injury claims) tend to be driven by younger drivers,

viewed as more accident-prone.

On the whole, big cars have fewer injury claims, while high-performance, sporty and luxury models show higher collision claims. However, the collision ratings for 1987 models show that compact vans and the subcompact Honda Civic wagon rank among the best.

For the 1985-87 period, all large cars rated better than average in HLDI injury claim frequency, except for the Cadillac De Ville coupe and Oldsmobile Delta 88, which rated average. The Mercedes-Benz SDL/SEL sedan, which comes with anti-lock brakes and a driver's-side air bag standard, headed the list, but full-size, rear-drive station wagons accounted for five of the top 10. Highest injury claim experience came mainly from small, imported front-drive models, including Chevrolet Sprint and Spectrum (both built in Japan), Isuzu I-Mark, Hyundai Excel, Nissan Pulsar and Yugo GV. Domestic models among those with the highest injury claims were the Ford EXP and Pontiac 1000, neither of which are still produced.

Fighting Car Thieves

Auto theft is big business. Frederick Cripe of Allstate testified to Congress that the "total value of property stolen as a result of auto theft countrywide in 1986 was nearly $7 billion." Lt. Robert F. Morgan of the New York City Police Department estimated that 55 percent of the cars stolen annually were taken by pros, for the purpose of dismantling and selling parts. Joy riders are definitely in the minority. Both Allstate and State Farm paid out more than $400 million in theft claims in 1987.

What's the best way to hang onto your car? Move out of the city. More cars are stolen in big cities like New York, Boston, Chicago and Los Angeles than in the rest of the country combined. Short of hiring an armed guard to watch over the car day and night, the best that urban dwellers can do is take steps to reduce the risk to a tolerable level.

If moving to the country isn't practical, you can sell that Chevrolet Corvette or Pontiac Firebird and replace it with a humdrum station wagon or nondescript, stripped-down little sedan. Invariably, the cars that top the NHTSA list of cars favored by thieves are sporty and sport-luxury models. That shouldn't be surprising. Thieves just follow the law of supply and demand. Their "customers" want popular cars (and the parts that come off those

cars), so why take a risk grabbing a vehicle that nobody wants? Cars that thieves avoid include the likes of Ford Escort and the Dodge/Plymouth Colt Vista.

Near the bottom of the 1987 list for fewest thefts (see chart for the "top 10") were the Saab 9000, Toyota Tercel, and Mercedes-Benz 300DT. This list rates cars in proportion to the number produced. While more Chevrolet Camaros are stolen than any other single model, Firebird actually fared a worse theft rate in 1987 because fewer were built. Ranking close to the Chevrolet Monte Carlo in number of total thefts was the Oldsmobile Cutlass Supreme, but Olds barely made the top ten because more of them were put on the road. Yugo ranked No. 51, with 193 stolen out of a total population of 35,000. Corvette, one of the traditional favorites among thieves, rated only No. 17, with 278 being stolen; and 537 thefts put the Lincoln Town Car into the 31st position. Not a single Rolls-Royce Corniche was reported stolen.

Another type of list is published annually by the Highway Loss Data Institute. Year after year, Volkswagen models head the list of claim frequencies. The reason is that the HLDI ranking includes any kind of claim, whether for the loss of a radio or for the whole

Most Frequent Theft Claims 1987 Passenger Cars	
Model	**Claim Frequency**
Volkswagen GTI	1030
Volkswagen Cabriolet	817
Volkswagen Golf 2-dr	746
Volkswagen Jetta 4-dr	740
Volkswagen Golf 4-dr	545
Saab 900 2-dr	522
Cadillac De Ville 2-dr	338
Cadillac Brougham	304
Hyundai Excel	293
Cadillac De Ville 4-dr	276

Source: Highway Loss Data Institute. Theft claim frequency is expressed in relative terms, with 100 representing the average for all 1987 cars. A claim frequency of 1030 is 10.3 times the average of all 1987 cars, and a claim frequency of 276 is 2.76 times the average of all 1987 cars. Theft claims include those for stolen parts and accessories, such as stereos, as well as vehicle thefts. This list covers only frequency of claims, not dollar value.

car. VW's reason for prominence stems from the fact that its "premium" stereo system (and what thief would want anything less than premium merchandise?) is mounted high on the dash, in a slide-out bracket. It's easy to pry out. So are radios in Audi, BMW, Jaguar, Mercedes, Peugeot and Saab.

Since 1986, most of these European cars have contained a pre-wired alarm, with a disabling circuit in the stereo. When pried out of the dash, or disconnected from the battery, the stereo won't work until a certain code is punched into the system. Volkswagen (except for the Fox) added such a system in 1987. Has that stopped the thieves? Not quite, it seems, since VW is still up there at the top of 1987's list (though with a lower ranking than before). Saab's ranking, while lower than in 1986, hasn't fallen dramatically either.

The top three in average payouts per claim in the 1985-87 period, according to HLDI, were the '87 Buick Regal, Monte Carlo, and Camaro, each averaging over $6200. That indicates that many of these cars are never recovered, the victims of "chop shops." On the other hand, payouts for Volkswagens averaged between $941 for the GTI to $1522 for the Cabriolet, indicating that these cars were broken into but not cut up and sold in pieces.

What You Can Do

Rather than being successful crime-preventers, anti-theft alarms are more often annoying devices that wake neighbors in the middle of the night, wailing away because they've been triggered by a gust of wind, a passing car, or stray radio signal. We're just glad that many of them shut themselves off within a minute or two. While a good alarm can help, a low-budget version is likely to become more of an irritant than a helper.

A good alarm that won't go off without cause, and takes time to disable or override, can easily cost $500. While a thief isn't likely to give up because a car has an alarm, he can at least be encouraged to pick an easier target. One available alarm can differentiate between significant sounds (a hammer breaking glass, for example) and ordinary street noises. Others cut off the ignition as they sound an alarm. Some can be turned on/off from outside the car by a remote control unit. Those with coded numbers that must be entered before a car can be started offer possible combinations that number in the billions.

Corvette was the first to offer a standard computerized disabling device, on 1986 models. A microchip embedded in its ignition key contains an electronically-set code, which must be "read" by a hidden control module before the car can be started. Theft claims dropped 45 percent in the first year of the device's availability. For 1989, this "Pass-Key" system is also standard on Camaro and Firebird.

Complacency is foolish, no matter how sophisticated the alarm. Criminals follow the news of the latest technology, too, and soon learn to deal with each new obstacle. Since 1980, luxury and sports cars have had standard theft-deterrent systems of some sort, or had to have major body and powertrain components marked with a serial number. Despite these efforts, theft of sporty cars certainly hasn't ground to a halt.

Are alarms worth the price? Think about the amount of your insurance policy's deductible, and about the inconvenience if your car is stolen, and they might seem a fairly modest "investment." Probably the best source of information is your insurance company. They have to pay out the claims on auto thefts, and can recommend what's most effective in the protection market, and may be able to tell you which ones aren't worth the bother.

Some insurance companies offer a discount (5 to 25 percent) if a car contains an acceptable alarm. In a few states (currently Illinois, Michigan, Massachusetts, Kentucky and New York), such a discount is mandatory. It only applies to comprehensive coverage, so it doesn't amount to a whole lot.

A much different use of high-tech electronics might help cut down thefts after the fact. The LoJack MicroMaster, an electronic vehicle recovery system, is a tiny radio transceiver that hides under the hood. If that car is stolen, a police car equipped with a tracking device can spot its location (within a 2-3 mile radius). Priced at $595, the unit has become popular with Corvette and Camaro owners. Some dealers have even offered them as sales premiums, or at a reduced price.

On a more elementary level, we've all heard the warnings over and over to take the keys out of the ignition. Yet surprising numbers are still left there, demonstrated when recovered stolen cars show no evidence of damage to the lock. Other simple rules also play a role:

1. Leave only the ignition key with parking attendants (and make sure it has no identifying numbers on it).
2. Park in busy, well-lit areas, and keep valuables hidden (better yet, not there at all).
3. Always lock the car (and garage), even if you're only leaving for a minute.
4. Engage the parking brake and turn wheels sharply toward the curb, to make the car harder to tow away.

Most Frequently Stolen Cars
1987 Models

Model	Number Stolen	Theft Rate (Per 1000)
Pontiac Firebird	2424	30.1
Chevrolet Camaro	3333	26.0
Chevrolet Monte Carlo	1516	20.3
Toyota MR2	381	19.3
Buick Regal	910	14.8
Mitsubishi Starion	100	14.6
Ferrari Mondial	2	13.6
Mitsubishi Mirage	357	12.8
Pontiac Fiero	563	12.7
Oldsmobile Cutlass Supreme	1338	11.7

Source: National Highway Traffic Safety Administration, based on data supplied by the National Crime Information Center. Number Stolen is the total number of thefts per car line during calendar 1987. Theft Rate is the number of cars stolen for each 1000 cars manufactured for that car line during 1987. For example, 2424 Pontiac Firebirds were stolen and 80,414 were produced in 1987. Only two Ferrari Mondials were stolen, but only 147 were imported.

CAR BUYER'S GLOSSARY

If you're about to venture into a new-car showroom for the first time in several years, you may discover that your automotive vocabulary is out of date. Such terms as anti-lock brakes, electronically adjustable suspension, 4-wheel steering, fuel injection and multi-valve engines await the new-car shopper. The unprepared face an avalanche of high-tech double-talk. To reduce your risk, we've compiled a glossary of some recent technical innovations. The list will also help define many of the terms used to describe vehicles in the *Automobile Book*.

Aerodynamic body parts: Body panels shaped to reduce wind resistance or lift at high speeds. Usually a front air dam, mounted beneath the front bumper, and a spoiler, mounted at the trailing edge of the rear deck. Some cars also have side skirts, usually nonfunctional lower-body extensions designed to achieve a race-car-like appearance.

Aerodynamic headlamps: Also called composite headlamps, these often have separate high-and low-beam bulbs housed in a single unit covered by a flush-fitting lens to reduce wind resistance. They also frequently use halogen gas. American manufacturers have recently adopted this European design for many of their newer models. Aerodynamic headlamps are usually more expensive to replace than traditional American sealed-beam units. (See "Halogen headlamps")

Air bags: A fabric bag that inflates in frontal collisions of around 12 mph or higher to prevent front-seat occupants from hitting the steering wheel, dashboard or windshield. Air bags are designed to inflate within fractions of a second, then immediately deflate so as not to impede the driver's ability to control the car. Most cars equipped with air bags have only one, for the driver, and house it in the steering wheel hub. The Porsche 944 Turbo and Lincoln Continental also have passenger's-side air bags mounted in the dashboard. (See "Automatic seat belts" and "Passive restraints")

All-season tires: Designed with tread that allows water and snow to escape from under the tire for better traction.

Ordinary tread designs tend to trap water, causing "hydroplaning," in which the tire rides on a thin film of water. Most all-season tires are designed for a soft ride rather than good handling, though some tire companies now offer all-season radials that have a high-performance tread. (See "Ride quality" and "Handling")

Anti-lock braking systems (ABS): A computer controls braking force to prevent the wheels from locking in a panic stop. This critically acclaimed safety feature is designed to dramatically reduce stopping distances and to improve steering control during braking in rain or snow or on icy pavement. Locked brakes result in skids. When ABS brakes are applied, the computer senses when a wheel is about to lock up and then "pumps" the brakes many times per second. The driver simply continues to apply steady force to the brake pedal, but the computer allows the wheels to rotate while slowing the car.

Automatic seat belts: Front seat belts that engage automatically upon entering a car or turning on the ignition. There are two basic types. Motorized shoulder belts anchored to the door frame pivot around front-seat occupants when the front doors are closed and the ignition is on (separate lap belts usually have to be buckled manually). Non-motorized automatic belts can be left buckled, but will automatically extend or retract when the front doors are opened or closed. Automatic seat belts are part of a passive-restraint system. (see "Air bags" and "Passive restraints")

Balance shafts: Shafts mounted inside an engine and designed to rotate in a direction opposite the crankshaft so as to balance and reduce the shaking forces, making the engine quieter and smoother. Most commonly found on larger displacement 4-cylinder engines and some V-6s.

Body lean/body sway: In cornering, the transfer of weight causes a car's body to lean toward the outside of the turn. Balance and traction are distorted because the outside wheels are squashed down, while the inside wheels want to lift. Generally, the less a car leans in turns, the better it "han-

dles." Cars with soft suspensions have more body lean than cars with firm suspensions. (See "Handling," "Neutral Cornering," "Oversteer," "Stabilizer bar," "Suspension" and "Understeer")

Child-proof door lock: Allows a rear door to be opened from the outside only. There usually is a switch in the doorjamb to set the locks in the child-proof mode.

Clutch: Mechanical device that couples and uncouples a drive shaft from its power source. An automobile's clutch is placed between the engine and transmission.

Curb weight: The weight of the vehicle when ready for the road, including fuel and other fluids. Curb weights listed in this issue are for base models without optional equipment. Options such as a larger engine, automatic transmission, air conditioning, and 4-wheel-drive can add several hundred pounds.

Direct ignition: Ignition coils, which supply the spark that ignites the air/fuel mixture inside the engine, are mounted directly atop the spark plugs, eliminating the ignition distributor and ignition wires to the plugs. (See "Multi-coil ignition systems")

Disc brakes: Brake design in which a caliper squeezes two friction pads against a disc that's attached to the wheel. Considered more efficient than a drum brake, in which an open-ended drum-shaped iron casting is attached to the wheel. Curved brake shoes press against the inside of the drum to provide braking action.

Double overhead camshafts: Also called twin overhead cams. Two camshafts per cylinder bank, with one operating the intake valves and the other the exhaust valves. Double-overhead cam engines usually have four valves per cylinder, two intake, two exhaust. A V-type engine with one overhead camshaft per cylinder bank is still said to be a single-cam engine. (See "Multi-valve engines" and "Overhead camshaft and overhead-valve engines")

Driveability: Term used to describe a combination of mechanical and tactile impressions, often subjective. Includes how smoothly an engine runs, especially at idle and at high speed. It also describes the ability of an automatic transmission to shift smoothly and to downshift promptly for passing, and the smoothness of a manual transmission's shift action and clutch movement. (See "Clutch")

Electronically adjustable suspension: A suspension system that changes the ride quality to suit road or driving conditions, usually by altering shock-absorber damping or air-spring rates. Some electronically adjustable suspensions are designed to enhance ride comfort by softening settings for rough roads. Others stiffen settings to improve handling, at the expense of ride quality. Some systems are automatic, with microprocessors sensing inputs from steering, braking, throttle and other sources to determine the suspension setting. Others are adjustable by the driver from the cockpit. (See "Handling," "Ride quality" and "Suspension")

Entry/exit: Subjective rating of the ease with which the driver and passengers can get into and out of a vehicle.

Final-drive ratio: Describes the number of turns the wheels make compared to the number of turns of the power-output shaft. Thus, if the wheels turn four times per one turn of the output shaft, the final-drive ratio would be 4 to 1, or 4:1. A final-drive ratio of 4:1 would thus result in quicker acceleration than a final-drive ratio of 2:1, but it would also force the engine to run faster per turn of the wheel, meaning higher fuel consumption, more engine wear and more noise.

Four-wheel drive: Engine power is delivered to all four wheels. The chief advantage is added traction on slippery surfaces. Disadvantages include a higher purchase price, extra weight, reduced fuel economy and higher repair costs. The most common system is "part-time, on-demand" 4WD, which allows the driver to shift from 2WD into 4WD via a transfer-case lever or an electronic switch inside the cabin. This system is meant for use only on slippery pavement or off road. "Full-time" 4WD can be used on any surface and is of two types: on-demand (you have to shift a transfer case lever or push a button to engage it) and permanent (it's always engaged). The most sophisticated full-time system is automatic 4WD, which uses a viscous, or fluid, coupling to sense which wheels need

traction and then delivers engine power accordingly. The driver does nothing to engage or disengage the system. Some of the least sophisticated 4WD systems lack "shift on-the-fly" capability. That is, they require that the vehicle be stopped and that its front-wheel hubs be locked from outside before shifting into 4WD.

Four-wheel steering: Allows rear wheels to steer the vehicle in tandem with the front wheels. Introduced on mass-produced cars in 1988, it's offered for '89 on the Honda Prelude Si 4WS and on the Mazda MX-6 GT 4WS. At slow speeds, the rear wheels steer opposite the fronts, turning a few degrees to decrease turning radius and increase maneuverability. At higher speeds, the rear wheels turn a few degrees in the same direction as the fronts to enhance corning and stability. Honda steers its rear wheels with a mechanical linkage governed by steering-wheel angle. Mazda uses an electronic system geared to vehicle speed.

Front-wheel drive: Engine power is transmitted to the front wheels only. Effect is to pull the car by the front wheels, rather than push it by the rears. FWD eliminates the drive shaft extending from the transmission to the rear-mounted differential and hence, does away with the drive-shaft tunnel running along the floor of the interior.

Fuel injection: Replaces the carburetor as a system to supply fuel for combustion. Fuel injection is of two basic designs, both of which meter fuel more efficiently than a carburetor and result in smoother operation and better fuel economy. One design is called port (or multi-point) fuel injection, in which fuel is squirted directly into each cylinder at its intake port. The other is throttle-body fuel injection (or single-point fuel injection), in which fuel is squirted from one or two injectors into an intake manifold on top of the engine and then distributed to the cylinders. Multi-point injection is the more expensive, but because it is also the more precise, it results in more efficient combustion and thus more power, better mileage and lower emissions. Today's fuel-injection systems are controlled by a microprocessor or electronic module. The controls turn the fuel injectors on and off, based on signals received from sensors around the engine that measure oxygen in the exhaust, coolant temperature, engine speed and other operating conditions.

Halogen headlamps: Headlamps that use a bulb with halogen gas. They can be either the sealed-beam or composite type. They produce a stronger, whiter light than ordinary incandescent bulbs. (See "Aerodynamic headlamps")

Handling: Refers to the way in which a car responds to the driver's steering input. A good handling car remains stable and holds the road well in turns and over rough surfaces. This is not the same as "ride," which is a judgment of a car's over-the-road comfort. (See "Body lean/body sway," "Neutral Cornering," "Oversteer," "Stabilizer bar," "Suspension" and "Understeer")

Hatchback: Sedan or coupe body with a rear liftgate that opens to a cargo area. Cars with two passenger doors and a liftgate are called 3-door hatchbacks; four passenger doors and a liftgate makes it a 5-door hatchback. (See "Notchback")

Horsepower: A measurement of an engine's ability to perform work. Usually measured by connecting crankshaft to a dynamometer and expressed as brake horsepower (bhp). Engines that produce a majority of their horsepower at low rpm (revolutions-per-minute of the crankshaft), say 3000 rpm, often provide more low-speed pickup than engines that develop more horsepower at higher rpm, 5800 rpm, for example. Horsepower, however, is not the only measurement of an engine's overall strength. Its ability to produce torque—a twisting or turning motion—is also a key factor. Torque, expressed in pound-feet, is the engine's ability to twist the crankshaft and to therefore transmit power to the transmission and eventually, to the wheels. (See "Torque," "Redline," "Tachometer" and "RPM")

Intercooler: Found on some turbocharged and supercharged cars. Acts as a radiator to cool air before it enters the combustion chamber. Cooling makes the air denser and increases the power of the ignited fuel-air mixture. (See "Naturally aspirated engine," "Turbocharger" and "Supercharger")

Liftover: The height of the rear sill of the trunk or cargo area.

Lockup torque converter: Fuel-saving device found on some automatic transmissions. Designed to overcome efficiency lost in the process of delivering engine power through the automatic transmission's viscous, or fluid, couplings. A lockup torque converter "locks" the crankshaft to the transmission input shaft so the two turn as one, eliminating slippage caused by the fluid coupling. Usually activated at highway speeds. In most cars, the lockup torque converter can be felt engaging and disengaging, much as when the transmission shifts gears. Lockup torque converters save fuel, but can prove an annoying interruption to the smooth flow of power.

Multi-coil ignition systems: Ignition coils grouped as a unit or module and connected directly to the spark plugs. Usually one coil serves two cylinders. Eliminates the need for an ignition distributor, one of the major service points in ignition-system maintenance. Electronic engine controls signal the coils when to send voltage to the right spark plug to ignite the air-fuel mixture. If a multi-coil system fails, you usually have to replace the entire unit, which can be costly. (See "Direct ignition")

Multi-valve engines: Engines with three or four valves per cylinder instead of the customary two. Multiple-valve cylinder heads move more air and fuel through the engine quickly, letting the engine "breathe" better. In all engines, separate valves open in sequence to allow air-fuel mixture into the combustion chambers and then out as exhaust gases. Instead of simply enlarging one intake valve and one exhaust valve, using three or four valves of smaller size and lighter weight allows higher engine speeds, more power and more efficient combustion. (See "Double overhead camshafts" and "Overhead camshaft and Overhead-valve engines")

Naturally aspirated engine: Engines that draw air/fuel mixture into their cylinders without aid of a supercharger or turbocharger. (See "Intercooler," "Turbocharger" and "Supercharger")

Neutral cornering: A car's ability to change direction with little body lean and a balance between forces exerted on front and rear tires. A car that can corner neutrally remains relatively level through the turn, neither understeering nor oversteering drastically. (See "Body lean/body sway," "Handling," "Oversteer," "Stabilizer bar," "Suspension" and "Understeer")

Nosedive: Front end of the car dips under hard braking, caused by weight shifting forward. If nosedive is pronounced, the front brakes will carry excessive share of braking load, while a lack of weight in the tail will cause the back wheels to lift, increasing chances that the rear brakes will lock.

Notchback: A sedan or a coupe body with a separate trunk. (See "Hatchback")

Optical horn: Flashes headlights or high-beams when the turn-signal stalk is pulled or pushed. Usually used to signal an intention to pass.

Overdrive: Fuel-saving transmission gear that lowers engine speed to save fuel and reduce noise. Usually the fourth gear of an automatic transmission and the fifth gear of a manual transmission. "Overdrive" transmissions allow the driveshafts to spin at a faster speed than the engine; those without overdrive have a less-efficient "direct drive" top gear, in which the engine and driveshafts turn at the same speed.

Overhead-camshaft and overhead-valve engines: Distinguishing feature is placement of the camshaft. The camshaft in either design activates the valves. In an overhead-valve engine (ohv), the camshaft is located in the engine block, below the valves. It activates the valves by means of pushrods and associated components. In most foreign-made engines and an increasing number of domestic-made engines, the camshaft is located in the cylinder head and acts directly on the valves mounted below. This overhead-cam design (ohc) is more expensive to manufacture, but has fewer parts and acts more directly on the valves, thus allowing the engine to run more efficiently and at higher speeds. Some engines use double-overhead cams (dohc) for still more efficiency. (See "Double overhead camshafts" and "Multi-valve engines")

Oversteer: Tendency for the rear wheels of a car to lose traction and skid sideways in a turn. Most common in high-performance, rear-wheel drive cars or in tail-heavy vehicles. (See "Body lean/body sway," "Handling," "Neutral Cornering," "Stabilizer bar," "Suspension" and "Understeer")

Passive restraints: Safety features designed to limit injuries to occupants in crashes. Distinguishing feature is that the occupants do not have to actively engage the restraints. Typical examples are automatic shoulder belts and air bags. Associated components may include under-dashboard knee bolsters. Federal regulations require passive restraints on 40 percent of 1989 model cars and on all cars sold in the U.S. starting with the 1990 model year. (See "Air bags" and "Automatic seat belts")

Performance: An overall evaluation of how a car accelerates, holds the road, corners and brakes. Some vehicles have a balance of all these elements, but it's more common to find a car that does well in one or in several aspects of performance, falling short in others. Good performance is usable daily. It enables a driver to merge easily with expressway traffic, pass quickly on a 2-lane road, negotiate turns and bumpy roads with good control, and stop safely. A car can have good performance without being a high-performance machine.

Powertrain: All items necessary to transmit power to the wheels: engine, transmission or transaxle, clutch (on manual transmission cars), torque converter (on automatic transmission cars), and drive shafts.

Rear-wheel drive: Only the rear wheels are supplied with power from the engine. The most common configuration has the engine mounted in front. Advantages are simplicity and better front-rear weight balance than front-drive cars. Disadvantages include reduced wet-weather traction compared to front-wheel drive and less efficient use of space. Passenger room is reduced because the drive shaft that links the transmission with the rear-mounted differential runs the length of the interior. The differential and its axle housings also take up passenger and cargo space.

Redline: An engine's maximum recommended speed in revolutions per minute (rpm). Tachometers mark this maximum engine speed with a red line at the appropriate rpm. (See "Horsepower," "Torque," "Tachometer" and "RPM")

Ride quality: A subjective judgment of a car's over-the-road comfort. Unfortunately, ride quality is often the antithesis of handling and automotive suspensions are a study in design compromise. Often, a car with a suspension taut enough to remain stable at high speed and through turns is too stiff to ride comfortably on a rough road. If the ride is soft and absorbent, the suspension may be so compliant that maneuverability, steering response and overall control are impaired. (See "Handling" and "Suspension")

RPM (Revolutions per minute): A measurement of engine speed, based on how fast the crankshaft is rotating. Rpm is displayed on a tachometer. (See "Horsepower," "Torque," "Redline" and "Tachometer")

Stabilizer (anti-roll) bar: A bar linking the left and right suspension systems. Can be used at the front, rear or both ends of a car. It reduces body roll by helping the suspension system resist the lifting forces that come to play when a car changes direction. (See "Body lean/body sway" "Handling," "Neutral Cornering," "Oversteer," "Suspension" and "Understeer")

Steering response/feedback: Prompt, precise steering response means an immediate, precise change in direction when you turn the steering wheel and easily gauged steering inputs. Steering with "road feel" transmits some feedback from the tires. With poor steering response, there's little sensory connection between the position of the steering wheel and the tires. It's difficult to tell whether you're turning the wheel too much or too little. Over-assisted power steering usually has vague steering response and poor feedback, giving little clue about whether the front tires are gripping the road or are slipping. (See "Handling")

Supercharger: A supercharger is similar to a turbocharger in that it forces additional air/fuel mixture into the combustion chamber to produce more power. An important difference is that a turbocharger is driven by spent exhaust gases; a supercharger is driven by the engine's crankshaft, and thus reacts more directly to throttle pressure and with less lag than with many turbochargers. (See "Intercooler," "Naturally aspirated engine" and "Turbocharger")

Suspension: The components that support the weight of the car and its occupants, keep the tread of the tires on the road, absorb bumps and road shocks, and control force produced during acceleration, braking and cornering. Suspensions vary in design and elements, but parts typical to most include springs, shock absorbers, control arms and stabilizer bars. An independent suspension means that each wheel operates independently of the others, so that when one tire hits a pot hole, the

spring and shock absorber at that corner of the car absorb the bump while the rest of the suspension is unaffected. This improves ride and control. (See "Body lean/body sway," "Handling," "Neutral Cornering," "Oversteer," "Ride quality," and "Understeer")

Tachometer: Dashboard gauge that displays engine speed in revolutions per minute (rpm) of the crankshaft. (See "Horsepower," "Torque," "Redline" and "RPM")

Tire sizes: A series of numerals and letters that denote the tire's intended use, tread width and sidewall height, construction and the size of the wheel for which it is intended. The code is stamped into the sidewall. An example is a P185/70R14 tire. "P" stands for passenger car tire; "185" is the width of the tread in millimeters; "70" is the tire's "profile," meaning the height of the sidewall is 70 percent of the width of the tread; "R" stands for radial; and "14" is the wheel diameter in inches. Tire widths generally increase and decrease in increments of 10 millimeters (185, 195, 205). Tires with a 70-series profile or less (60, 50, 45) are called "low-profile" tires and are designed for better handling. Tires with a 75-series profile or taller (80, 85) are designed for ride comfort. Some tires have an "H," "V" or some other letter as part of the size description, such as "P185/70HR14." Those letters denote maximum speed ratings: "S" for up to 113 mph; "H" for up to 130 mph; "V" for up to 150 mph; and "Z" for over 150 mph. Some tires now have the speed-rating letter imprinted separately on the sidewall.

Torque: The amount of twisting force generated by an engine. An engine's torque rating, expressed in pound-feet, is its ability to transmit power to the transmission and eventually, to the wheels. This is different from horsepower. The amount of torque an engine produces is an important measure of its ability to accelerate and to move heavy loads. As with horsepower, engine speed is a factor with torque. An engine that develops ample torque at low rpm, say, 250 lb/ft at 2000 rpm, can accelerate well from low speeds. It also tends to work well with automatic transmission. An engine that produces 150 lb/ft of torque at 4000 rpm, will have to be revved much higher to accelerate well from low speeds and will have a harder time hauling heavy loads. An engine with poor low-end torque is probably better suited for use with a manual

transmission, which gives the driver more control over engine speed. Engines that produce maximum horsepower at high rpm often have little torque at low rpm. (See "Horsepower," "Tachometer" and "RPM")

Track: The distance between the center of the wheels on the same axle; sometimes referred to as "tread."

Turbocharger: An air compressor that delivers more air to an engine than could be drawn in naturally, thus increasing engine power. Exhaust gas is recirculated through a turbine, which in turn drives the compressor to force more air into the engine. The amount of pressure generated is described as "boost" and measured in terms of pounds per square inch (psi). Turbos enable engineers to increase horsepower without increasing the displacement of an engine. In addition to the additional underhood heat and complexity associated with turbochargers, a prime disadvantage is the lag between the time the throttle pedal is depressed and the time sufficient exhaust gasses are generated to spin the turbo and create boost. (See "Intercooler," "Naturally aspirated engine" and "Supercharger")

Understeer: The tendency for a car to resist turning and for the front tires to lose grip in a turn before the rear tires. The result is that the car continues largely in a straight line, its front tires "plowing" rather than rolling. Many cars are designed to understeer when they enter a turn too quickly. Understeer is thought to be safer and easier to manage for the average driver than "oversteer," in which the rear wheels skid before the front ones do. Understeer is prevalent on most front-wheel drive cars because of their severe forward weight bias. (See "Body lean/body sway," "Handling," "Neutral Cornering," "Oversteer," "Suspension" and "Stabilizer bar")

VIN (vehicle identification number): A series of letters and numbers stamped into a metal plate attached to the left front top side of the dashboard and visible from the outside of a car through the windshield. The VIN contains such coded information as the vehicle's model year, production series and place of manufacture.

Wheelbase: The distance between the center of a car's front wheels and the center of its rear wheels.

BEST BUYS

The Auto Editors of CONSUMER GUIDE® have selected Best Buys among passenger cars, compact vans and 4-wheel-drive vehicles. Vehicles are grouped by size, price and market position, so an $8000 subcompact competes with other low-priced small cars, not against a $50,000 luxury car, for example. Road-test results play a major role in the selections, and the editors consider only cars they have driven. Other factors include cost of ownership, warranty coverage, reputation for reliability and durability, and safety record. At least one Best Buy has been selected in each group, though some groups also include recommended choices that are worthy of attention.

For comparison purposes, the vehicles are priced as if they were equipped with the most popular options, including automatic transmission and air conditioning. We quote suggested retail prices, which are subject to change by the manufacturers. Price is not the only factor in our Best Buys selection. Some of the lowest-priced models aren't the best values in the long run because they may not be as durable or reliable as more expensive competitors.

Subcompacts

Base prices for the **Dodge Omni America** and identical **Plymouth Horizon America** are up by $600, to $6595, for 1989. But these front-drive subcompacts still can be equipped with air conditioning, automatic transmission, power steering, rear window defogger and a stereo radio for under $9000, making them Best Buys. Few other comparably equipped subcompacts match that price, and none match the performance of the Omni/Horizon's standard 2.2-liter 4-cylinder engine. Unlike the smaller engines of most competitors, its 93 horsepower provides brisk acceleration with automatic transmission and doesn't weaken significantly with the air conditioner on. The Omni/Horizon twins debuted for the 1978 model year, which is ancient history in the auto industry, and they come only as 5-door hatchbacks with cramped interiors. They lack contemporary styling and the latest technology, but offer more value than any rivals.

The **Mazda 323** is a competent, well-made and economical car that also is a Best Buy. It comes in two body styles: a 3-door hatchback and a 4-door sedan. Except for a 4-wheel-drive GTX hatchback, all models have front-wheel drive. The base and SE models offer the most value: their suggested retail prices start at $6299 and run to $8299. Air conditioning, automatic transmission, power steering and a stereo radio add about $2000 to those prices, so figure on paying from $8000 to almost $10,500 for a 323 with those features. The 323's 82-horsepower 1.6-liter 4-cylinder engine delivers adequate acceleration and returns good gas mileage. The hatchback has a rear seat that folds down for extra cargo room, while the sedan boasts a deep, roomy trunk. Mazda's generous warranties cover the car for 3 years/50,000 miles and rust for 5 years/unlimited miles.

The **Honda Civic**, traditionally one of our favorite small cars because of its sprightly performance, fuel economy, durability and high resale value, is recommended. The current generation, which arrived for 1988, is no exception.

Plymouth Horizon America

Mazda 323 SE 3-door

Honda Accord LX 4-door

Toyota Camry V6 4-door

Unfortunately, Honda pricing keeps us from naming it a Best Buy. DX models go from nearly $8500 for a 3-door hatchback with manual transmission to more than $10,000 for a 4-door sedan with automatic transmission. An LX sedan, which includes power windows and door locks, starts at over $10,000. You can count on paying at least another $1000 for air conditioning and a stereo, both of which are dealer-installed options that don't carry manufacturer's suggested retail prices. If you want power steering, you have to buy one of the more expensive models: an LX, a DX with automatic, or a 5-door station wagon. High demand for Civics allows most Honda dealers to charge full retail price or more.

Compacts

Despite higher prices, the **Honda Accord** remains a Best Buy among compacts because of its quality, durability, commendable performance and excellent resale value. Accord comes as a 4-door sedan, 3-door hatchback and 2-door coupe, with the 4-door by far the most popular. Our favorite remains the Accord LX 4-door, which has a 98-horsepower, carbureted 2.0-liter 4-cylinder engine, air conditioning, power windows and door locks, and a stereo radio and cassette player. Prices range from around $14,200 with a 5-speed manual transmission to $14,800 for automatic, and there aren't any major options to increase the cost. An Accord DX 4-door starts out much cheaper, but by the time the dealer adds air, a stereo, dual outside mirrors and other options, the price approaches that of an LX. LXi and SEi models have a 120-horsepower, fuel-injected engine, plus a firmer suspension, larger tires, and additional convenience features. To us, the LXi just doesn't have enough additional equipment to justify its roughly $1700 premium over the LX.

Toyota Camry has all that the Honda Accord offers: high quality, durability, strong resale value, so it too is a Best Buy. But it has several things the Accord doesn't: a 4-wheel-drive All-Trac model, the availability of anti-lock brakes and a V-6 engine, and a roomy 5-door station-wagon body style. As with other Japanese cars, Camry is no longer a bargain: a top-line LE sedan with front-wheel drive, the V-6, anti-lock brakes, leather upholstery and other options can cost more than $20,000. Drop down to a Camry Deluxe 4-door with front-drive, the standard

115-horsepower 4-cylinder engine, automatic transmission, air conditioning, power windows and door locks, stereo with cassette player and cruise control and you're looking at about $14,800, or the same as an Accord LX sedan, which has similar equipment. For about $1300 more, Camry can be equipped with a 153-horsepower V-6 engine that's potent, quiet and commendably smooth, giving this car refined performance that some premium sedans can't match. Anti-lock brakes, an important safety feature, are available only on the LE sedan with the V-6 and on the 4WD All-Trac sedan, two of the more expensive models. The Deluxe models offer the most value.

Load a **Chevrolet Corsica** with the optional V-6 engine, automatic transmission, air conditioning, power locks and windows, stereo and other features and its suggested retail price will still be under $14,000, much less than an Accord LX or Camry Deluxe, so this car is recommended. Better still, most Corsicas are sold for well below suggested retail, unlike the Japanese rivals. Corsica is not the equal of Accord or Camry in quality. But most Corsica buyers have generally been satisfied owners, according to early surveys. A 5-door hatchback body style joins the 4-door sedan for '89. General Motors' new warranty covers the car for 3 years/50,000 miles, though the owner has to pay a $100 deductible for each repair visit after 1 year/12,000 miles. Choose the optional V-6 engine over the weak, noisy standard 4-cylinder. We also recommend the optional F41 sport suspension and 14-inch tires. Ride quality suffers, but the combination improves handling over the standard suspension and 13-inch tires.

The **Dodge Aries America** and identical **Plymouth Reliant America**, which are recommended, have a $7595 base price and even loaded models won't run much more than $11,000, which make the American-made compacts less expensive than many Japanese subcompacts. Aries and Reliant come in 2- and 4-door sedan styling with a 93-horsepower 2.2-liter 4-cylinder engine and a 5-speed manual transmission standard. We suggest you opt for the 100-horsepower 2.5-liter engine, which is quieter and smoother than the 2.2, and 3-speed automatic transmission. Add air conditioning and one of the discount option packages and you'll have a fully equipped front-drive sedan with decent performance, acceptable gas mileage, room for five adults and a

fairly spacious trunk. These cars debuted for 1981, so they lack contemporary design and technical features. They're not exciting to drive or to look at, and their overall quality falls well short of Japanese standards. But their low prices and long warranties should warm the hearts of budget-conscious shoppers who can't afford $14,000 for a new car.

Mid-size Cars

The front-drive **Ford Taurus** and nearly identical **Mercury Sable** are easily recognized because of their trend-setting styling, but they're our Best Buys among mid-size cars because they also have good performance, capable handling, roomy interiors and reasonable prices. The only way to go in these cars is with a V-6 engine, either the 3.0-liter V-6 that's standard in all Sables and some Tauruses, or the 3.8-liter V-6 that's optional in both lines. The 2.5-liter 4-cylinder engine standard in lower-priced Taurus models doesn't have enough power for these intermediates. The front-drive Taurus and Sable are functionally the same and both are available as a 4-door sedan or 5-door wagon. Since they're similarly priced, look for a Taurus GL or Sable GS with one of Ford's "preferred equipment" packages. A nicely optioned Taurus GL or Sable GS runs $14,000 to $16,000, while a loaded Taurus LX or Sable LS goes for $16,000 to $19,000. Despite the popularity of these cars, competition has forced Ford to offer rebates on both. You might be able to get one of these at a surprisingly hefty discount.

Recommended are the **Chevrolet Celebrity** and its General Motors sibling, the **Pontiac 6000**. They are older and more conservatively styled than Taurus/Sable, but they're just as spacious inside. Both Chevrolet and Pontiac sell 4-door sedan and 5-door wagon versions of this front-drive car. Taurus/Sable has more capable handling in base form, but road manners of the GM cars improve considerably when you order the Eurosport option package for the Celebrity or the S/E model of the 6000. The firmer suspensions and larger tires on these models improve cornering ability, with some compromise in ride comfort. Figure on about $15,000 for a fully equipped Eurosport with the optional and recommended 2.8-liter V-6, and around $16,000 for a 6000 S/E with the same engine. With discounts on option packages, and the possibilities of cash rebates, selling prices can be lower.

Ford Taurus LX 4-door

Buick LeSabre Custom 2-door

Acura Legend Coupe L

Acura Legend Sedan LS

Buick offers the similar Century, and Oldsmobile the Cutlass Ciera. Both are more expensive than the Celebrity or 6000, but unlike the Chevy or Pontiac, they offer 2-door coupe body styles and an optional 3.3-liter V-6 that produces 160 horsepower, 35 more than the 2.8-liter V-6.

Full-size Cars

The **Buick LeSabre, Oldsmobile 88 Royale** and **Pontiac Bonneville**, family cars built from the same design, earn our Best Buy endorsement for their overall competence and because they are among the lowest priced cars available with anti-lock brakes. The 88 Royale also is available with an optional driver-side air bag, making it the least expensive car available with both anti-lock brakes, which can help you avoid an accident, and an air bag, which reduces the chances of injury in case of an accident. Neither of these safety features is cheap: Anti-lock brakes cost $925 and the air bag, available only on the 88 Royale 4-door, is $850. However, they will more than pay for themselves if just once they prevent a serious accident or injury. All three have roomy interiors and sufficiently large trunks. Their 165-horsepower 3.8-liter V-6 engine delivers satisfying acceleration. Front-wheel drive also gives them better traction in rain and snow than rear-drive rivals. While styling is different for each brand, all three are

built with corrosion-resistant 2-sided galvanized steel. GM bases its luxurious Buick Electra and Oldsmobile Ninety-Eight on the LeSabre/88/Bonneville platform. Since all these cars use the same engines, transmissions, brakes and suspensions, its the plusher furnishings in the Electra and Ninety-Eight that you pay a premium for. Not that the LeSabre, 88, or Bonneville are inexpensive. Figure on a sticker of $18,000 to $20,000 for a fully equipped one, though most dealers should be offering substantial discounts.

Our recommended choice, the rear-drive **Chevrolet Caprice**, comes as a 4-door sedan or 5-door wagon. A well-equipped Caprice Classic runs from about $17,000, while a luxurious LS version can be nearly $20,000. Sure, their aging design dates to the 1977 model year, but a base Caprice 4-door with a handful of options can be purchased for around $15,000, less than some mid-size and compact cars. You get a standard V-8 engine (170 horsepower in sedans; 140 horsepower in wagons), plus traditional size and styling, commendable refinement and trailer-towing ability.

For those who want the traditional full-size, rear-drive American family car, two we recommend are the **Ford LTD Crown Victoria** and its plusher and costlier sibling, the **Mercury Grand Marquis.** Each is available as a 4-door

sedan or a 5-door wagon with body-on-frame construction. Their 150-horsepower 5.0-liter V-8 delivers plenty of refined power. Properly equipped, they can tow a 5000-pound trailer, something most front-drive rivals can't. Expect to pay $16,000 to $18,000 for either brand, depending on model and equipment, more if you want additional comfort and convenience.

Premium Coupes

The Best Buy **Acura Legend Coupe** challenges more expensive rivals with its impressive performance and standard equipment. Plus, the stylish front-drive Honda-made 2-door promises to be reliable and a relative bargain to maintain and insure. Its smooth, quiet 2.7-liter V-6 engine gives prompt around-town response and ample passing power on the open road. All Legend Coupes come with a standard driver-side air bag, while the mid-level L model and top-line LS add anti-lock brakes. There's plenty of room for four people and their luggage. Base models start around $25,000 and the top-line LS gets over $30,000. These are solid, high-quality cars that represent fine value in this league.

The **Cadillac Eldorado** is recommended for its satisfying acceleration from its 155-horsepower V-8 engine, generous standard equipment, luxurious accommodations and optional anti-lock brakes ($925). This year's base price is $20,738, nearly $2000 higher than last year's, and you can easily top $30,000 by adding options. Some of this year's price increase is due to additional standard equipment, such as power front seats, a premium stereo system and a new theft-deterrent system. These cars have been selling at large discounts the past few years, so you should be able to buy one for less than suggested retail.

Also recommended is the **Lincoln Mark VII LSC**, a coupe with fine road manners, power, luxurious equipment and good quality. The LSC is priced the same as the Mark VII Bill Blass Designer Series, but has a sportier, tauter-riding suspension, wider, high-performance tires and firmer power steering. The standard 225-horsepower 5.0-liter V-8 delivers satisfying acceleration, and is impressively quiet and refined. Anti-lock brakes are standard. Prices start around $27,000 and $29,000 is just a few options away.

Premium Sedans

Built by Honda, the **Acura Legend Sedan** was the first $20,000 Japanese car sold in America when it debuted in 1986. Our Best Buy pick is the same basic car for 1989, though the engine is larger and more powerful now, and other improvements have been made in the past three years. We remain highly impressed by Legend's polished, smooth engine, the competent handling from its front-drive chassis and the compliant ride. Then there's the roomy interior that can hold four or five adults and coddle them with loads of comfort and convenience features. It's a luxury sedan with Honda reliability and a sporting personality that makes it a roadworthy alternative to European sedans. Anti-lock brakes are standard on the mid-level L model and top-line LS,

while all Legends now have a driver's-side air bag. Prices start in the $22,000 range, and a top-line LS model gets up to $30,000.

We recommend the **Lincoln Continental.** All-new last year, this front-drive, six-passenger sedan has one major change for 1989: It's the first American car with standard air bags for both the driver and front-seat passenger. The base price is up $1390 for '89 to $27,468, but along with the dual air bags and a lengthy list of luxury equipment, standard features include a computer-controlled suspension system designed to maintain a smooth ride by adjusting in fractions of a second to changes in the road surface; variable-assist power steering that requires more turning effort at high speeds than at low speeds; and 4-wheel disc brakes with anti-lock control. A 140-horsepower 3.8-liter V-6 engine gives adequate performance. The 4-door Continental has a roomy trunk. With anti-lock brakes and this year's new air bags, it's a safe bet among luxury cars.

Also recommended is the rear-drive **Mazda 929**, which went on sale a year ago as a competitor for the Acura Legend. It's a conservatively styled premium sedan that goes about its business quietly and efficiently. The 929's lack of personality may limit its box-office appeal, but this is a substantial, comfortable car with well-balanced performance, a spacious interior and enough standard equipment to justify its $22,000 base price. It uses a 158-horsepower 3.0-liter V-6 engine and a 4-speed automatic transmission. Anti-lock brakes are an expensive ($1400 last year) but recommended option. Even with such options as a power moonroof, power driver's seat, leather upholstery and a compact disc player, a 929 won't be much more than $25,000, which is well below most comparably equipped rivals.

Sporty Coupes

The **Acura Integra** has a blend of power, agility, efficiency and quality unmatched in this hotly competitive market segment, which includes rivals such as the Ford Probe, Honda Prelude, Mazda MX-6 and Toyota Celica. A Best Buy, the Integra delivers brisk acceleration from its smooth and sophisticated 1.6-liter, twin-cam 4 cylinder engine, fine handling and road holding from its front-drive chassis, and ample stopping power from its all-disc brake system. It even has passable rear-seat room for this class. Integra comes in 3-and 5-door hatchback body styles, both of which have fold-down rear seats for extra cargo room. Even with optional air conditioning and a stereo radio, a base RS version is hard to beat at under $13,000; a top-line LS runs about $1500 to $2000 more. Automatic transmission is optional, but Integra is best enjoyed with the 5-speed.

The **Chevrolet Beretta** is recommended as a roadworthy front-drive coupe that starts at under $13,000 with a V-6 engine, 5-speed manual transmission and performance handling package. Air conditioning and the most expensive option package pushes the price to nearly $15,000, but Chevy has been offering cash rebates on Beretta and most dealers discount their prices well below suggested retail. Beretta is a 2-door coupe built from the same design as the Chevrolet Corsica 4-door sedan, but has its own exterior styling and a more sporting personality. A 2.0-liter 4-cylinder is the base engine, but we recommend the 130-horsepower 2.8-liter V-6 that's standard on the Beretta GT and optional on the base Beretta.

Sports & GT Cars

Since most cars in this class cost at least $20,000, the **Ford Mustang GT**'s

Acura Integra LS 5-door

Ford Mustang GT

Best Buys

base price of $13,000 makes it a clear Best Buy. A fully optioned GT hatchback won't list for more than $16,000. A GT convertible is considerably more expensive at $17,099 base, but you'll be hard pressed to get one of those up to $20,000, even with optional leather upholstery. The standard 225-horsepower 5.0-liter V-8 makes the GT a tire-burning sprinter. The well-furnished, comfortable interior has the convenience of fold-down rear seats on the hatchback. Mustang's rear-drive design dates to 1979, it's no Porsche on twisty roads and you can't get anti-lock brakes to improve stopping ability. However, there's nothing stale about the Mustang's appeal as a genuine American muscle car. It's also a prime candidate for collectable-car status. Before you buy, check with your insurance agent to see what a Mustang GT will do to your premiums.

Compact Vans

Our Best Buy is the **Dodge Caravan** and **Plymouth Voyager**, the front-drive compact vans that Chrysler Motors builds from the same design and sells at identical prices. They're available in two sizes: a 176-inch-long body on a 112-inch wheelbase and a Grand Caravan or Grand Voyager 190.5-inch-body on a 119-inch wheelbase. Seating is for a minimum of five or a maximum eight, depending on body size and seating configuration. A turbocharged 150-horsepower version of the standard 100-horsepower 2.5 is newly available on short-wheelbase SE and LE models, but we recommend the Mitsubishi-built 3.0-liter V-6 engine as the best choice. The 141-horsepower V-6 is standard on the Grand LE models and optional on the Grand SE and short-wheelbase LE. This year it

comes with Chrysler's new 4-speed overdrive automatic. Caravan and Voyager are still the only front-wheel-drive minivans, so they have better traction in rain and snow than their rear-drive rivals and are more like cars than trucks to drive. Here's the bad news: Demand remains high for these vehicles, so prices are steep. With the V-6, automatic transmission and air conditioning plan on spending at least $16,000; if you choose a Grand model or an LE in either size, brace yourself for a $20,000 sticker.

4WD Vehicles

Already blessed with the most powerful engine among compact 4-wheel-drive vehicles, our Best Buy **Jeep Cherokee** gains anti-lock brakes as a new option for 1989. The anti-lock system is available on Cherokees and the higher-priced Jeep Wagoneer Limited with the 6-cylinder engine, automatic transmission and the more sophisticated of Jeep's two 4WD systems, Selec-Trac. These are the first light trucks with an anti-lock system that works on all four wheels; other light trucks use anti-lock systems that work only on the rear wheels. Another distinction is that the Jeep system works in both 2WD and 4WD; most anti-lock systems on 4WD cars work in 2WD only. Cherokee and Wagoneer are basically trim and equipment variations on the same 3-and 5-door bodies, but Wagoneer starts out at more than $22,000, so we're only recommending Cherokee, which has a base price of around $13,000 with 4WD. The base Cherokee 4×4 model comes with a 121-horsepower 2.5-liter 4-cylinder engine and Command-Trac, a part-time 4WD system (for use on wet or slippery surfaces

only) with shift-on-the-fly capability. The optional Selec-Trac system, available only with automatic transmission, is a full-time 4WD system that can be left permanently engaged. Though it's more convenient, we think most owners can manage with the less-expensive Command-Trac system. We heartily recommend the 177-horsepower 4.0-liter inline 6-cylinder engine, which delivers ample power without strain. While Cherokee is as rugged as any 4WD vehicle, it also is the most civilized of this breed. Choose a 5-door wagon with the six, air conditioning and a few other options, and expect to pay $17,000 or so.

The **Chevrolet S10 Blazer** and identical **GMC S15 Jimmy** come only in a 3-door body style, but they are the best-selling 4×4s in America and our recommended choice. A 160-horsepower 4.3-liter V-6 engine added as an option last spring nudges them toward the performance of the Jeep Cherokee. Standard rear anti-lock rear brakes are new for '89, though they're not as sophisticated as Jeep's optional 4-wheel ABS system. A 125-horsepower 2.8-liter V-6 and a 5-speed manual are standard on 4WD models. While this combination is sufficient, we prefer the more muscular 4.3-liter V-6, available only with a 4-speed overdrive automatic transmission. GM's Insta-Trac 4WD system, a part-time system not for use on dry pavement, has automatic locking front hubs and full shift-on-the-fly capability. Unlike Jeep, GM doesn't offer a full-time 4WD system. With the 4.3-liter V-6 engine, automatic transmission and air conditioning, an S10 Blazer/S15 Jimmy will cost $16,000 to $17,000 for a basic model, and close to $20,000 for a plush version.

Dodge Grand Caravan LE

Jeep Cherokee Sport

Buying Guide

Acura Integra

Acura Integra LS 3-door

What's New

American Honda's upscale division is set to bring out an all-new Integra next spring as an early 1990 model, so last year's versions are carried over unchanged for 1989. Integra debuted in spring 1986 as the sporty, small Acura, using a front-drive platform derived from the Honda Civic. It comes in 3-door and 5-door hatchback body styles and two trim levels, RS or LS. All versions have a dual-overhead cam, 16-valve 1.6-liter 4-cylinder engine. Transmission choices are a 5-speed manual or a 4-speed overdrive automatic. The 1990 Integra reportedly will have a 1.8-liter engine for more power and be positioned slightly upmarket from today's car.

For

- Performance
- Handling/roadholding
- Fuel economy

Against

- Ride
- Noise
- Air conditioner

Summary

Integra is quick, agile, well built and stylish, with the right sound and feel to enthrall enthusiasts. Its silky engine begs for rushes to the redline, though you'll have to rev it above 3500 rpm or so to draw out its real punch. Cornering is a delight thanks to sharp, responsive power-assisted steering, minimal body roll and good grip from the tires. The strong brakes inspire confidence. On smooth roads, the ride is fine, but the suspension can be jarring over rougher pavement. Road, engine and wind noise can be intrusive if you're not in a sporting mood, and the air conditioner can't seem to cool the glassy interior sufficiently on hot days. Still, the cabin is a paragon of driver-oriented efficiency, and there's reasonable room in the rear seat considering the sporty-coupe design. The 5-door's longer wheelbase gives it slightly more rear-seat room than the 3-door's. Integra is priced about $3000 higher than the gutsy Honda CRX Si and about $3000 below the slick Prelude Si.

Specifications

	3-door hatchback	5-door hatchback
Wheelbase, in.	96.5	99.2
Overall length, in.	168.5	171.3
Overall width, in.	65.6	65.6
Overall height, in.	53.0	53.0
Front track, in.	55.9	55.9
Rear track, in.	56.5	56.5
Turn diameter, ft.	31.5	32.1
Curb weight, lbs.	2326	2390
Cargo vol., cu. ft.	10.0	16.0
Fuel capacity, gal.	13.2	13.2
Seating capacity	5	5
Front headroom, in.	37.5	37.5
Front shoulder room, in.	NA	NA
Front legroom, max., in.	40.7	40.7
Rear headroom, in.	35.8	36.3
Rear shoulder room, in.	NA	NA
Rear legroom, min., in.	33.3	33.9

Powertrain layout: transverse front engine/front-wheel drive.

Engines

	dohc I-4
Size, liters/cu. in.	1.6/97
Fuel delivery	PFI
Horsepower @ rpm	118 @ 6250
Torque (lbs./ft.) @ rpm	99 @ 5500
Availability	S

EPA city/highway mpg

5-speed OD manual	26/30
4-speed OD automatic	25/30

Prices

Acura Integra	Retail Price	Dealer Invoice	Low Price
RS 3-door hatchback, 5-speed	$11260	$9458	$10960
RS 3-door hatchback, automatic	11870	9970	11570
RS 5-door hatchback, 5-speed	12060	10130	11760
RS 5-door hatchback, automatic	12670	10642	12370
LS 3-door hatchback, 5-speed	13070	10978	12770
LS 3-door hatchback, automatic	13680	11491	13380
LS 5-door hatchback, 5-speed	13900	11676	13600
LS 5-door hatchback, automatic	14510	12188	14210
Destination charge	295	295	295

Standard Equipment:

RS: 1.6-liter PFI 16-valve 4-cylinder engine, 5-speed manual or 4-speed automatic transmission, power steering, power 4-wheel disc brakes, cloth reclining front bucket seats, split folding rear seat, tilt steering column, digital clock, tachometer, coolant temperature gauge, intermittent wipers, bodyside molding, remote mirrors, tinted glass, rear defogger and wiper/washer, remote hatch and fuel door releases, center console, front door pockets, right visor mirror, cargo area light, cargo tiedown straps, cargo area cover, Yokohama 195/60R14 tires. **LS 3-door** adds: power sunroof, cassette storage in console, cruise control, AM/FM ST ET cassette, Michelin MXV tires on alloy wheels. **LS 5-door** deletes sunroof and adds power windows and locks.

OPTIONS are available as dealer-installed accessories.

KEY: ohv = overhead valve; **ohc** = overhead cam; **dohc** = double overhead cam; **I** = inline cylinders; **V** = cylinders in V configuration; **flat** = horizontally opposed cylinders; **bbl.** = barrel (carburetor); **PFI** = port (multi-point) fuel injection; **TBI** = throttle-body (single-point) fuel injection; **rpm** = revolutions per minute; **OD** = overdrive transmission; **S** = standard; **O** = optional; **NA** = not available.

Prices are accurate at time of printing; subject to manufacturer's change

Acura Legend

Acura Legend sedan

What's New

Few visible changes have been made to these Honda-made imports, but the Legend Sedan gets the Coupe's double-wishbone rear suspension and all Legends now have a driver's-side air bag as standard. The Sedan's new rear suspension replaces a design that used struts and trailing links. The air bag—a passive safety device designed to inflate in a frontal collision—was previously standard only on the top-rung LS versions of the Sedan and Coupe. Both the 4-door-sedan and 2-door coupe body styles are powered by a 2.7-liter V-6 engine. Base, L and LS versions of each are available. All are front-drive and have 4-wheel disc brakes as standard; L and LS models add anti-lock systems. A 5-speed manual transmission is standard; a 4-speed overdrive automatic is available at extra cost. The Coupe's wheelbase is two inches shorter and its rear track is wider than the Sedan's, plus its suspension is tuned for a firmer ride. Acura reportedly is planning to introduce a larger sedan with a V-8 engine as a flagship companion to the Legend, but the company says it won't be sold during the '89 model year. The present Sedan debuted for 1986 and the Coupe followed a year later.

For

- Performance • Handling/roadholding
- Anti-lock brakes • Air bag

Against

- Driveability (automatic transmission)
- Rear seat room (Coupe)

Summary

Legend sets the standard among affordable premium cars. Of course, prices ranging from roughly $23,000 to $33,000, aren't everyone's idea of "affordable." But to get the Legend's wonderfully smooth and powerful engine, thoughtful interior design, solid construction, sport-sedan road manners and commendable quality in a European car, you'd have to pay lots more. Plus, independent surveys show Acura has an impressive record of owner satisfaction, and that's an important factor in its high resale value. Flaws

include an automatic transmission that shifts abruptly and harshly at times. Torque peaks at a high 4500 rpm and bursts of speed require quicker downshifts than this transmission provides. We haven't driven a Sedan with the new rear suspension, but previous Sedans had a ride that could be a touch jarring over highway expansion joints. Now that all models have a driver's-side air bag standard, look to the mid-level L series, with its standard anti-lock brakes, as probably the best value in a car line packed with it.

Specifications

	2-door notchback	4-door notchback
Wheelbase, in.	106.5	108.6
Overall length, in.	188.0	189.4
Overall width, in.	68.7	68.3
Overall height, in.	53.9	54.7
Front track, in.	59.1	59.0
Rear track, in.	59.1	57.5
Turn diameter, ft.	36.5	36.1
Curb weight, lbs.	3089	3077
Cargo vol., cu. ft.	14.7	16.6
Fuel capacity, gal.	18.0	18.0
Seating capacity	5	5
Front headroom, in.	37.2	37.2
Front shoulder room, in.	55.4	55.8
Front legroom, max., in.	42.9	42.3
Rear headroom, in.	36.3	36.1
Rear shoulder room, in.	54.3	55.4
Rear legroom, min., in.	30.3	34.2

Powertrain layout: transverse front engine front-wheel drive.

Engines

	ohc V-6
Size, liters/cu. in.	2.7/163
Fuel delivery	PFI
Horsepower @ rpm	161 @ 5900
Torque (lbs./ft.) @ rpm	162 @ 4500
Availability	S

EPA city/highway mpg

5-speed OD manual	19/23
4-speed OD automatic	18/22

Prices

Acura Legend	Retail Price	Dealer Invoice	Low Price
Sedan			
4-door notchback, 5-speed	$22600	$18532	$22400
4-door notchback, automatic	23400	19188	23200
4-door notchback w/sunroof, 5-speed	23485	19257	23285
4-door notchback w/sunroof, automatic	24285	19913	24085
L 4-door notchback, 5-speed	25900	21238	25700
L 4-door notchback, automatic	26700	21894	26500
L w/leather trim, 5-speed	26850	22017	26650
L w/leather trim, automatic	27650	22673	27450
LS 4-door notchback, 5-speed	29160	23911	28960
LS 4-door notchback, automatic	29960	24567	29760
Coupe			
2-door notchback, 5-speed	24760	20303	24560
2-door notchback, automatic	25560	20959	25360
L 2-door notchback, 5-speed	27325	22406	27125
L 2-door notchback, automatic	28125	23062	27925
L w/leather trim, 5-speed	28275	23185	28075
L w/leather trim, automatic	29075	23841	28875
LS 2-door notchback, 5-speed	30040	24632	29840
LS 2-door notchback, automatic	30840	25288	30640
Destination charge	295	295	295

KEY: ohv = overhead valve; **ohc** = overhead cam; **dohc** = double overhead cam; **I** = inline cylinders; **V** = cylinders in V configuration; **flat** = horizontally opposed cylinders; **bbl.** = barrel (carburetor); **PFI** = port (multi-point) fuel injection; **TBI** = throttle-body (single-point) fuel injection; **rpm** = revolutions per minute; **OD** = overdrive transmission; **S** = standard; **O** = optional; **NA** = not available.

Prices are accurate at time of printing; subject to manufacturer's change.

CONSUMER GUIDE®

Acura Legend Coupe LS

Standard Equipment.

4-door: 2.7-liter SOHC 24-valve PFI V-6, 5-speed manual or 4-speed automatic transmission, power steering, power 4-wheel disc brakes, driver's side airbag, air conditioning, cruise control, power windows and locks, tilt steering column, remote fuel door and trunk releases, door pockets, lighted visor mirrors, rear defogger, map lights, illuminated entry system, reclining front bucket seats, moquette upholstery, driver's seat lumbar and thigh support adjustments, rear armrest, AM/FM ST ET cassette with EQ and power diversity antenna, maintenance interval reminder, tachometer, coolant temperature gauge, trip odometer, intermittent wipers, Michelin MXV 205/60HR14 tires on alloy wheels; 2-door has power sunroof, V-rated tires. **L 4-door** has: anti-lock braking system, memory power driver's seat, security system; 2-door adds power driver's seat, heated mirrors, driver's seatbelt presenter, driver's information center. **LS** has: automatic climate control, power passenger seat, Bose sound system, driver's information center.

OPTIONS are available as dealer-installed accessories.

Alfa Romeo Milano and Spider

Alfa Romeo Milano

What's New

Alfa carries on for 1989 with two model lines, the venerable Spider 2-seat convertible and the Milano 4-door sedan, biding time until the 1990 model year, when a new sedan is scheduled to arrive. The new sedan is the 164, a front-drive 4-door styled by Pininfarina and built in Italy from the same basic design as the Saab 9000. The 164 will be powered by a 3.0-liter V-6 similar to the one current used in the Milano. Alfa says the 164 will debut in the U.S. next fall. For 1989, Spider returns for its 22nd year in the U.S. unchanged from last year. Three models are again offered: base Graduate, mid-level Veloce and top line Quadrifoglio. All are powered by a 2.0-liter, dual-cam 4-cylinder engine, teamed with a 5-speed manual transmission. Milano also returns unchanged. The rear-drive 4-door sedan comes three ways: base Gold model, midrange Platinum and top-shelf 3.0 Litre (formerly called Verde 3.0). Gold and Platinum models are powered by a 154-horsepower 2.5-liter V-6. The Gold comes with either a 5-speed manual or 3-speed automatic transmission, while the Platinum comes only with the automatic. The 3.0 Litre moniker for the top model describes its larger V-6 engine, which makes 183 horsepower and comes only with a 5-speed manual transmission. Anti-lock brakes are standard on the Platinum and 3.0 Litre.

For

● Handling   ● Anti-lock brakes (Milano)

Against

● Driving position/controls (Spider)   ● Noise (Spider)

Summary

Since we haven't driven the Milano, all our comments are on the Spider, whose timeless styling remains its most endearing feature. In fact, time has passed this car by, so there's really nothing else quite like it on the market today. Except for the base Graduate, Spider prices are hardly cheap, yet you don't get great performance or workmanship for your money. Spider's rear-drive chassis provides good handling, and the Quadrifoglio benefits from larger tires and wider wheels. However, you can get similar handling—even better—on less expensive cars. In addition, you'll probably also get a more comfortable ride and greater refinement on any of several sporty cars in the same price range. The Spider's driving position is hampered by a steering wheel that's too high and pedals that are too close to the seat, so many people can have a hard time getting comfortable behind the wheel. For this kind of money, you expect better assembly quality than you generally get on the Alfa convertible. All ragtops have squeaks and rattles, but most don't have so many fresh from the factory. Despite the attractive styling and charm of this Italian 2-seater, its ancient design, spotty dealer network and mediocre performance are good reasons to put your money into a more modern sports car.

1988 Specifications

	Spider 2-door conv.	Milano 4-door hatchback
Wheelbase, in.	88.6	98.8
Overall length, in.	168.8	170.5
Overall width, in.	64.1	64.2
Overall height, in.	48.8	53.1
Front track, in.	52.1	53.9
Rear track, in.	50.1	53.5
Turn diameter, ft.	34.5	33.1
Curb weight, lbs.	2548	2907
Cargo vol., cu. ft.	NA	10.0
Fuel capacity, gal.	12.2	17.6
Seating capacity	2	5
Front headroom, in.	NA	37.2
Front shoulder room, in.	NA	52.8
Front legroom, max., in.	NA	38.2
Rear headroom, in.	—	35.8
Rear shoulder room, in.	—	51.8
Rear legroom, min., in.	—	35.1

Powertrain layout: longitudinal front engine/rear-wheel drive.

Engines	dohc I 4	ohc V-6	ohc V-6
Size, liters/cu. in.	2.0/120	2.5/152	3.0/182
Fuel delivery	PFI	PFI	PFI
Horsepower @ rpm	115 @ 5500	154 @ 5500	183 @ 5800

Prices are accurate at time of printing; subject to manufacturer's change.

	dohc I-4	ohc V-6	ohc V-6
Torque (lbs./ft.) @ rpm	120 @ 2750	152 @ 3200	181 @ 3000
Availability	S[1]	S[2]	S[3]
EPA city/highway mpg			
5-speed OD manual	21/27	18/24	18/25
3-speed automatic		18/22	

1. Spider 2. Milano 3. Milano 3.0 Litre

Prices

Alfa Romeo (1988 prices)	Retail Price	Dealer Invoice	Low Price
Graduate 2-door convertible	$16429	—	—
Veloce 2-door convertible	19961	—	—
Quadrifoglio 2-door convertible	23113	—	—
Milano Gold 4-door notchback, 5-speed	18077	—	—
Milano Gold, automatic	18798	—	—
Milano Platinum, automatic	22094	—	—
Milano Verde, 5-speed	22300	—	—
Destination charge	350	350	350
Mechanical prep & delivery	160	160	160

Dealer invoice and low price not available at time of publication.

Standard Equipment:

Graduate: 2.0-liter DOHC PFI 4-cylinder engine, 5-speed manual transmission, power 4-wheel disc brakes, limited-slip differential, tachometer, coolant temperature and oil pressure gauges, bronze tinted glass, power mirrors, mahogany steering wheel rim, intermittent wipers, reclining front bucket seats, remote decklid release, 185/70R14 tires. **Veloce** adds: power windows, AM/FM ST cassette, leather upholstery, alloy wheels. **Quadrifoglio** adds: air conditioning, hardtop with rear defogger, power antenna, leather-wrapped steering wheel, 195/60HR15 tires on alloy wheels. **Milano Gold** has: 2.5-liter PFI V-6, 5-speed manual or 3-speed automatic transmission, self-leveling suspension (with automatic transmission only), power steering, power 4-wheel disc brakes, air conditioning, power windows and locks, heated power mirrors, AM/FM ST cassette, power antenna, tilt/telescopic steering column, rear defogger, leather-wrapped steering wheel, tinted glass, fog lamps, tachometer, coolant temperature and oil pressure gauges, headlamp washers, remote decklid release, velour reclining front bucket seats, 195/60VR14 tires on alloy wheels. **Platinum** adds: anti-lock braking system, first aid kit, lighted right visor mirror, toolkit, leather/suede upholstery, 195/55VR15 tires. **Verde** adds: 3.0-liter engine, Recaro cloth seats.

Optional Equipment:

Air conditioning, Graduate & Veloce	995	—	—
Power sunroof, Milano	795	—	—
Cruise control, Milano	200	—	—
Metallic paint, Veloce & Quadrifoglio	275	—	—
Milano	350	—	—
California emissions pkg.	48	48	48

Audi 80/90

What's New

Audi redesigned its 4000 series for 1988 to look like the aerodynamic Audi 100/200 line and renamed it the 80 and 90. The German automaker's smallest sedans are unchanged for '89, though two developments are expected later in the model year: a 20-valve, 160-horsepower version of the 5-cylinder engine is scheduled to debut in the 4-wheel-drive 90 Quattro and a coupe version of the 80/90 may be introduced. Until then, the 80 and 90 continue to share the same notchback body. The more expensive 90 is distinguished by a higher level of standard equipment, in-

Audi 80 Quattro

cluding a leather-and-wood-trimmed interior and power sunroof. Audi 90s equipped with a 5-speed manual transmission get a 2.3-liter 5-cylinder engine; those with a 3-speed automatic get the 2.0-liter four that's standard on the 80. Both models are available as Quattros, Audi's term for its permanent 4-wheel drive system. All Quattros have the 5-cylinder and 5-speed. Four-wheel disc brakes are standard; an anti-lock system is standard on the 90, optional on the 80 Quattro. Fallout from unintended-acceleration allegations leveled at Audi 5000s (renamed the 100/200 series for '89) have hurt sales of all Audis. Audi responds this year with the "Audi Advantage," a warranty program standard on all 1989 models. It covers all scheduled maintenance for three years or 50,000 miles, including oil changes and such wear items as brake pads. In addition, owners who trade a 1989 Audi on a new Audi within four years will be reimbursed by Audi for any difference between the value of their car and that of comparable models from Mercedes-Benz, BMW and Volvo.

For

- Handling/roadholding   • Anti-lock brakes
- 4-wheel drive (Quattro)

Against

- Performance (4-cylinder models)   • Price
- Cargo space

Summary

Audi has positioned the 80/90 to take on such tough rivals as the Acura Legend and the BMW 3-Series. Among its ammunition is a refined, sport-sedan manner, distinctive styling, an ergonomically sound interior, excellent build quality and a fine array of standard equipment. The Quattro option adds a dimension that only the 4WD BMW 325iX can match in a high-performance sport sedan. But even with these attributes, the 80/90 has a problem with perceived value. For the $19,000 base price of an 80, you get a smooth but otherwise ordinary 4-cylinder engine that gives lackluster performance with automatic transmission. You have to spend at least $25,000 on a 90, and even its 5-cylinder engine is short of inspiring. And all 80s and 90s are hampered by a serious shortage of trunk space, which limits their appeal as all-around family cars. If the 80/90 has a leg up on such similarly priced competitors as the V-6 powered Legend Sedan and Nissan Maxima SE, it's a true European character. It's not easy to put a price on that.

Specifications

	4-door notchback
Wheelbase, in.	100.2
Overall length, in.	176.3
Overall width, in.	66.7
Overall height, in.	55.0
Front track, in.	55.6
Rear track, in.	56.3
Turn diameter, ft.	33.8
Curb weight, lbs.	2568[1]
Cargo vol., cu. ft.	10.2
Fuel capacity, gal.	17.9[2]
Seating capacity	5
Front headroom, in.	NA
Front shoulder room, in.	NA
Front legroom, max., in.	NA
Rear headroom, in.	NA
Rear shoulder room, in.	NA
Rear legroom, min., in.	NA

1. 2904 lbs, Quattro   2. 18.5 gals, Quattro

Powertrain layout: longitudinal front engine/front-wheel drive or permanent 4WD.

Engines

	ohc I-4	ohc I-5
Displacement, l/cu. in.	2.0/121	2.3/141
Fuel delivery	PFI	PFI
Horsepower @ rpm	108 @ 5300	130 @ 5700
Torque (lbs./ft.) @ rpm	121 @ 3200	140 @ 4500
Availability	S	S

EPA city/highway mpg

5-speed OD manual	22/30	20/26
3-speed automatic	23/27	

Prices

Audi 80/90 (1988 prices)	Retail Price	Dealer Invoice	Low Price
80 4-door notchback, 5-speed	$19160	—	—
80 4-door notchback, automatic	19685	—	—
80 Quattro 4-door notchback	23380	—	—
90 4-door notchback, 5-speed	25060	—	—
90 4-door notchback, automatic	25060	—	—
90 Quattro 4-door notchback, 5-speed	28550	—	—
Destination charge	NA	NA	NA

Dealer invoice and low price not available at time of publication.

Standard Equipment:

80: 2.0-liter PFI 4-cylinder engine, 5-speed manual or 3-speed automatic transmission, power steering, power 4-wheel disc brakes, air conditioning, tinted glass, rear defogger, cruise control, power windows and locks, AM/FM ST ET with diversity antenna, power mirrors, velour reclining front bucket seats with height adjusters, rear head restraints and lap/shoulder belts, automatic front seatbelt tensioners, tachometer, coolant temperature gauge, trip odometer, digital clock, intermittent wipers, lighted right visor mirror, lockable fuel cap, 175/70SR14 SBR tires. **80 Quattro** adds: 2.3-liter PFI 4-cylinder engine, close-ratio 5-speed, permanent 4-wheel drive, rear spoiler, 195/60VR14 tires on alloy wheels. **90** has: 2.3-liter 5-cylinder engine with 5-speed or 2.0-liter 4-cylinder with 3-speed automatic, anti-lock braking system, fog lights, clearcoat metallic paint, leather interior, wood dashboard trim, power sunroof, individual passenger reading lamps, Auto Check, cassette stereo, 195/60VR14 tires (H-rated on 4-cylinder).

Optional Equipment:

	Retail Price	Dealer Invoice	Low Price
Anti-lock brakes, 80 Quattro	1150	1035	1093
Leather interior, 80	1090	NA	NA

Audi 90

	Retail Price	Dealer Invoice	Low Price
Clearcoat or pearlescent paint	375	300	338
California equipment	00	90	90
Sport seats, exc. 90 Quattro	295	236	266
Power sunroof, 80/80 Quattro	795	668	732
Auto Check System, 80/80 Quattro	100	80	90
Trip computer	245	196	221
Heated front seats	225	180	203
Heated front & rear seats	375	300	338
Cassette stereo, 80/80 Quattro	495	396	446
Ski sack	120	96	108

Audi 100/200

What's New

Audi's biggest car drops its 5000-series designation, gains an updated interior and minor exterior alterations. Models equipped with the naturally aspirated 2.3-liter 5-cylinder engine are now called the 100; those with the turbocharged and intercooled 2.2-liter five are 200s. Audi says it isn't trying to hide the car's association with its Audi 5000 predecessor, whose sales suffered from allegations of unintended acceleration. The name change standardizes the German automaker's worldwide badging. Meanwhile, Audi announced that the 200 will form the basis for the Audi V-8, a $50,000 flagship sedan coming to the U.S. in mid-1989. It will have a 240-horsepower 3.6-liter V-8 with 4 valves per cylinder and Audi's first 4-speed automatic transmission. The car will also come standard with Audi's permanently engaged Quattro 4-wheel drive system. A unique grille and headlights and a new taillamp treatment will distinguish the Audi V-8 from the 100/200. Among the revised 100/200 exterior touches are flush-fitting door handles, new wheels and a dual-diversity radio antenna implanted in the windshield and rear window. The new interior echoes the look begun last year in the 80/90 series. Zebra-wood panels adorn a dashboard that curves into the fronts of the doors, and a new instrument pod includes a complete array of analog gauges. Automatic-transmission shift levers now move through a gated pattern and Audi's Automatic Shift Lock, which requires the driver to depress the brake pedal before shifting out of park, now has an audible signal that sounds if a door is open while the shifter is out of park. Like the 5000 series, 100s and 200s come as 4-door-notchback sedans and 5-door wagons. The line starts with the 100E Automatic, a new entry-level model that adds a standard

> **KEY: ohv** = overhead valve; **ohc** = overhead cam; **dohc** = double overhead cam; **I** = inline cylinders; **V** = cylinders in V configuration; **flat** = horizontally opposed cylinders; **bbl.** = barrel (carburetor); **PFI** = port (multi-point) fuel injection; **TBI** = throttle-body (single-point) fuel injection; **rpm** = revolutions per minute; **OD** = overdrive transmission; **S** = standard; **O** = optional; **NA** = not available.

Prices are accurate at time of printing; subject to manufacturer's change.

Audi

Audi 100

height-adjustable driver's seat and a manual sunroof to last year's base equipment. The 100 sedan and wagon get anti-lock brakes and plusher interior accommodations. Wagons also have a 70/30 split folding rear seatback. Leather-covered power seats are among the features standard on 200s. The Quattro models (available with manual transmission only) have Audi's permanently engaged 4-wheel drive system instead of the front drive on other 100s and 200s. A driver-side air bag is now standard on 200 models, optional on 100s and all models are covered by a new warranty and trade-in guarantee (see description of the "Audi Advantage" in the Audi 80/90 review).

For

●Interior room ●Handling/roadholding ●Ride
●Anti-lock brakes ●Air bag ●4-wheel drive (Quattro)

Against

●Performance (2.3-liter engine) ●Resale value

Summary

Under any name, these Audis are exemplary sport sedans, with a distinctive and functional design, lots of standard equipment, a fine ride, composed road manners and quality workmanship. The '89 editions replace a serviceable but bland interior with one that's sophisticated and luxurious. But shortcomings remain. The naturally aspirated 2.3 has sub-par muscle for this class, while the turbo 2.2 gives good performance when pushed. Most of these cars are sold with automatic, and a $32,000-plus car deserves a 4-speed overdrive automatic rather than the 3-speed Audi offers. Finally, there's the issue of whether Audi can regain its sales appeal. The "Audi Advantage" should help. So should the new interior. But the Japanese, taking a cue from the success of Honda's upscale Acura line, are poised to raid Audi's market segment with sedans from Toyota's new Lexus division and Nissan's new Infiniti line. Audi will fight back by touting its German heritage and its new warranty. New-car buyers will decide the outcome.

> **KEY: ohv** = overhead valve; **ohc** = overhead cam; **dohc** = double overhead cam; **I** = inline cylinders; **V** = cylinders in V configuration; **flat** = horizontally opposed cylinders; **bbl.** = barrel (carburetor); **PFI** = port (multi-point) fuel injection; **TBI** = throttle-body (single-point) fuel injection; **rpm** = revolutions per minute; **OD** = overdrive transmission; **S** = standard; **O** = optional; **NA** = not available.

Specifications

	4-door notchback	5-door wagon
Wheelbase, in.	105.8	105.8
Overall length, in.	192.7	192.7
Overall width, in.	71.4	71.4
Overall height, in.	54.7	55.7
Front track, in.	57.8	57.8
Rear track, in.	57.8	57.8
Turn diameter, ft.	34.2	34.2
Curb weight, lbs.	2932[1]	3042[1]
Cargo vol., cu. ft.	16.8	38.5
Fuel capacity, gal.	21.1	21.1
Seating capacity	5	5
Front headroom, in.	37.5	37.5
Front shoulder room, in.	NA	NA
Front legroom, max., in.	NA	NA
Rear headroom, in.	36.5	36.5
Rear shoulder room, in.	NA	NA
Rear legroom, min., in.	NA	NA

1. 3306 lbs, Quattro   2.3439 lbs, Quattro

Powertrain layout: longitudinal front engine/front-wheel drive or permanent 4WD.

Engines

	ohc I-5	Turbo ohc I-5
Size, liters/cu. in.	2.3/141	2.2/136
Fuel delivery	PFI	PFI
Horsepower @ rpm	130 @ 5600	158 @ 5500
Torque (lbs./ft.) @ rpm	140 @ 4000	166 @ 3000
Availability	S	S

EPA city/highway mpg

5-speed OD manual	18/25	18/26
3-speed automatic	19/22	18/22

Prices

Audi 100/200	Retail Price	Dealer Invoice	Low Price
100E 4-door notchback, automatic	$24980	—	—
100 4-door notchback, 5-speed	27480	—	—
100 4-door notchback, automatic	28030	—	—
100 5-door wagon, 5-speed	28960	—	—
100 5-door wagon, automatic	29510	—	—
100 Quattro 4-door notchback, 5-speed	30805	—	—
200 Turbo 4-door notchback, 5-speed	32455	—	—
200 Turbo 4-door notchback, automatic	33005	—	—
200 Quattro 4-door notchback, 5-speed	36355	—	—
200 Quattro 5-door wagon, 5-speed	37855	—	—
Destination charge	NA	NA	NA

Dealer invoice and low price not available at time of publication.

Standard Equipment:

100E: 2.3-liter PFI 5-cylinder engine, 3-speed automatic transmission, power steering, power 4-wheel disc brakes, air conditioning, power windows and locks, heated power mirrors, velour reclining bucket seats with driver's side height adjustment, folding rear armrest, outboard rear lap/shoulder belts, reading lamps, lighted visor mirrors, anti-theft alarm, tinted glass, rear defogger, cruise control, intermittent wipers, AM/FM ST ET cassette with diversity antenna, manual sunroof, 185/70HR14 SBR tires. **100** adds: 5-speed manual or 3-speed automatic transmission, anti-lock braking system, automatic climate control, Zebrano wood inlays, rear head restraints, front seatback pockets, leather-wrapped steering wheel, shift knob and boot, power sunroof, 205/60VR15 SBR tires on alloy wheels. **Wagon** deletes rear head restraints, rear armrest and diversity antenna and adds: heavy-duty suspension, asymmetrically split folding rear seatback, rear wiper/washer. **100 Quattro** has permanent 4-wheel drive. **200** adds to 100: 2.2-liter turbocharged PFI 5-cylinder engine, trip computer, chenille velour upholstery,

---

Prices are accurate at time of printing; subject to manufacturer's change.

CONSUMER GUIDE®

Audi 200 Quattro wagon

ski sack, driver's side air bag (delayed introduction), Audi/Bose music system (delayed introduction). **200 Quattro** has permanent 4-wheel drive, leather upholstery.

Optional Equipment:

	Retail Price	Dealer Invoice	Low Price
Anti-lock braking system, 100E	1215	—	—
California compliance equipment	95	—	—
Partial leather upholstery, 200 Turbo	900	—	—
Leather upholstery (NA 100E)	1250	—	—
Clearcoat paint, 100s	450	—	—
Pearlescent paint, 200s	450	—	—
Sport seats, 100, 200 Turbo	350	—	—
Cold weather pkg., 100s	350	—	—
200s	500	—	—
Heated front seats, windshield washer nozzles and door lock cylinders.			
Power front seats, 100	825	—	—
Roof rails, wagons	225	—	—
Ski sack, 100s exc. wagon	125	—	—
Trip computer, 100	260	—	—

BMW 3-Series

BMW 325i 2-door

What's New

All 3-Series BMWs save the 325i Convertible get new body-color bumpers that trim overall length 5.4 inches, and all but the super-performance M3 now employ a 168-horse-power 2.5-liter six. The 2.5 ousts the economy-oriented 127-horsepower "eta" engine from the base 325 2-and 4-door, which become 325i models for 1989. Higher-speed V-rated tires (replacing H-rated) is their only other change. They're priced slightly above last year's 325 and well below the 325is. Last year's luxury 325i four-door was supposed to be replaced by a Luxus option package, but BMW was still debating this and a possible 2-door Luxus at press time. If and when they appear, they'll have standard leather seating, rear headrests, premium 8-speaker sound system

and electric sunroof (instead of manual). To these the sporty 325is 2-door adds front air dam, rear spoiler, half-inch-wider wheels, firmer suspension, limited-slip differential (optional elsewhere), and sport seats. The 325i Convertible's only change is a slimmer center rear brake lamp. Introduced last year as a 2-door with manual-shift only, the all-wheel-drive 325iX is now offered with four doors, optional automatic transmission and, as a price-capper, fewer standard features. Except for losing its roof antenna for one in the windshield, the M3 is a carryover. Powered by a 2.3-liter dual-overhead-camshaft, 16-valve four rated at 192 horsepower, it's the only 3-Series BMW limited to a 5-speed manual transmission.

For

- Performance
- Handling
- Anti-lock brakes
- 4WD traction (325iX)
- Instruments/controls

Against

- Passenger room
- Cargo space

Summary

Don't let the name and equipment switching throw you. BMW has simply repositioned its entry-level models by giving them a more exciting engine, and giving you more car for the money in the bargain. Note that price hikes elsewhere are modest: the 325is up $550, the Convertible $1350. Sales of the 3-Series have slid following several price increases the past two years. All the "i" models are great fun to drive. Though a bit noisy, the smooth and free-revving 2.5 six provides excellent performance, while the 3-Series' compact size and well-sorted chassis mean nimble handling, good grip and a firm but comfortable ride. Standard anti-lock braking is a plus for all models. The 3-Series's main drawbacks are limited passenger and cargo space, rather plain interiors for the money, and mediocre economy despite the brisk performance. The M3 is strictly for wealthy would-be racers. Its 4-cylinder engine is potent but throbby and high-strung, its much harsher ride too high a price for extra handling capability that most buyers will seldom use. The 4WD 325iX, on the other hand, sticks to pavement so well it's practically idiot proof.

Specifications

	2-door notchback	4-door notchback	2-door conv.
Wheelbase, in.	101.2	101.2	101.2
Overall length, in.	170.2	170.2	175.2
Overall width, in.	64.8	64.8	64.8
Overall height, in.	54.3	54.3	53.9
Front track, in.	55.4	55.4	55.4
Rear track, in.	55.7	55.7	55.7
Turn diameter, ft.	34.4	34.4	34.4
Curb weight, lbs.	2811[1]	2855	3055
Cargo vol., cu. ft.	14.3	14.3	11.0
Fuel capacity, gal.	16.4	16.4	16.4
Seating capacity	5	5	4
Front headroom, in.	37.7	37.7	NA
Front shoulder room, in.	52.0	52.0	NA
Front legroom, max., in.	NA	NA	NA
Rear headroom, in.	36.4	36.4	NA
Rear shoulder room, in.	52.4	52.4	NA
Rear legroom, min., in.	NA	NA	NA

1. 2865 lbs., M3, 3010 lbs., 325iX

Powertrain layout: longitudinal front engine/rear-wheel drive (permanently engaged 4WD, 325iX).

Prices are accurate at time of printing; subject to manufacturer's change.

BMW

Engines

	ohc I-6	dohc I-4
Size, liters/cu. in.	2.5/152	2.3/140
Fuel delivery	PFI	PFI
Horsepower @ rpm	168 @ 5800	192 @ 6750
Torque (lbs./ft.) @ rpm	164 @ 4300	170 @ 4750
Availability	S	S[1]
EPA city/highway mpg		
5-speed OD manual	18/23	17/29
4-speed OD automatic	18/22	

1. M3

Prices

BMW 3-Series	Retail Price	Dealer Invoice	Low Price
325i 2-door notchback	$24650	—	—
325i 4-door notchback	25450	—	—
325is 2-door notchback	28950	—	—
325i 2-door convertible	33850	—	—
325iX 2-door notchback	29950	—	—
325iX 4-door notchback	30750	—	—
M3 2-door notchback	34950	—	—
Destination charge	325	325	325

Dealer invoice and low price not available at time of publication.

Standard Equipment:

325i: 2.5-liter PFI 6-cylinder engine, 5-speed manual transmission, power steering, power 4-wheel disc brakes, anti-lock braking system, air conditioning, cloth or leatherette reclining bucket seats with height/tilt adjustments, outboard rear lap/shoulder belts, power windows and locks, cruise control, power mirrors, AM/FM ST ET cassette, power antenna, tinted glass, Service Interval Indicator, Active Check Control, digital clock, manual sunroof, toolkit, 195/65VR14 tires on alloy wheels. **Convertible** adds: leather sport seats with adjustable thigh support, map lights, trip computer, premium sound system. **325is** adds: limited-slip differential, front and rear spoilers, power sunroof. **325iX** has: permanent 4-wheel drive, sill extensions, leatherette upholstery, manual sunroof, folding rear armrest with ski sack, 205/55VR15 tires. **M3** has 325is equipment plus 2.3-liter DOHC PFI 4-cylinder engine, sport suspension, 205/55VR15 tires.

Optional Equipment:

4-speed automatic transmission (NA M3) .	645	—	—
Metallic paint	375	—	—
Limited-slip differential (std. 325is, M3) . .	370	—	—
Heated front seats, 325iX & conv.	200	—	—
Hardtop, convertible	3500	—	—

BMW 5-Series

What's New

BMW's midrange 5-Series sedan has been fully redesigned this year, replacing a model introduced to North America in 1975. Except for powertrains, it's all new, but with a general look and many features drawn from the second-generation 7-Series that bowed here in early '87. Compared to its predecessor, the new 5-Series rides a 5.4-inch-longer wheelbase, is two inches wider and fractionally lower, but measures 3.2 in. shorter. The new rear-drive 5-Series 4-door also is 260-300 pounds heavier, reflecting the growing industry trend toward weightier, thirstier but more solid-feeling cars in a time of plentiful, fairly cheap fuel. Two models are offered initially. The 525i, powered by the 2.5-liter, 168-horsepower straight six introduced in last year's 3-Series, takes over for the economy-oriented 528e. Replacing the

BMW 525i

previous 535i and 535is is a new 535i with the 208-horse-power 3.4-liter six also used in the 735i/735iL sedans and 635CSi coupe. Both are available with 5-speed manual or extra-cost 4-speed automatic transmission. A high-performance M5 model with a sport suspension and BMW Motorsport's 3.5-liter, 256-horsepower 24-valve six is iffy, but should appear for 1990. Below are first impressions of the 535i gleaned at BMW's press preview.

For

● Handling   ● Anti-lock brakes   ● Ride   ● Quietness
● Comfort

Against

● Fuel economy   ● Price

Summary

As you'd expect from what amounts to a scaled-down version of the impressive 7-Series, the new 535i makes a great first impression. It's too bad there's so much added weight, as it's taken the edge off standing-start acceleration and midrange passing ability in the 535i. It's especially evident with automatic, where you quickly lose steam going up steep hills unless you select a lower gear—disappointing for a $43,000 sports sedan. We weren't able to try the $37,000 525i, but it weighs only about 100 pounds less and lacks strong low-end torque, so it will doubtless be even more strained. Both models also disappoint in EPA fuel economy estimates, though that's not a big consideration for most buyers in this league. Otherwise, the news is good. Handling is fluid and responsive, grip ample, body roll well checked, the ride taut but comfortable, and noise levels quite low. Stopping ability is stunning, thanks to standard ABS and big 4-wheel disc brakes borrowed from the 7-Series. There's more back-seat space now—not a great deal, but enough to be useful—and the interior is well designed, borrowing much from the 7-Series. Cargo space is unchanged, but loading is easier from a trunk lid that now lifts a full 90 degrees from a bumper-height opening. Cabin decor remains rather "cold," though materials are first-rate. In all, the new 535i is more than a match for Mercedes' comparable and costlier 300E.

Specifications

	4-door notchback
Wheelbase, in.	108.7
Overall length, in.	185.8
Overall width, in.	68.9
Overall height, in.	55.6
Front track, in.	57.7
Rear track, in.	58.5

Prices are accurate at time of printing; subject to manufacturer's change.

CONSUMER GUIDE®

	4-door notchback
Turn diameter, ft.	37.7
Curb weight, lbs.	3305[1]
Cargo vol., cu. ft.	16.2
Fuel capacity, gal.	21.1
Seating capacity	5
Front headroom, in.	NA
Front shoulder room, in.	NA
Front legroom, max, in.	NA
Rear headroom, in.	NA
Rear shoulder room, in.	NA
Rear legroom, min., in.	NA

1. 525i; 3530 lbs., 535i

Powertrain layout: longitudinal front engine/rear-wheel drive.

Engines	ohc I-6	ohc I-6
Size, liters/cu. in.	2.5/152	3.4/209
Fuel delivery	PFI	PFI
Horsepower @ rpm	168 @ 5800	208 @ 5700
Torque (lbs./ft.) @ rpm	164 @ 4300	225 @ 4000
Availability	S[1]	S[2]

EPA city/highway mpg		
5-speed OD manual	18/24	15/23
4-speed OD automatic	18/23	15/19

1. 525i   2. 535i

Prices

BMW 5-Series	Retail Price	Dealer Invoice	Low Price
525i 4-door notchback	$37000	—	—
535i 4-door notchback	43600	—	—
Destination charge	325	325	325
Gas Guzzler Tax, 535i 5-speed	650	650	650
535i automatic	850	850	850

Dealer invoice and low price not available at time of publication.

Standard Equipment:

525i: 2.5-liter PFI 6-cylinder engine, 5-speed manual or 4-speed automatic transmission, power steering, power 4-wheel disc brakes, anti-lock braking system, air conditioning with individual temperature controls, leather power front bucket seats, folding center armrests, rear armrest with storage, outboard rear lap/shoulder belts, power windows and locks, heated power mirrors, fog lights, adjustable steering wheel, tinted glass, tachometer and coolant temperature gauge, rear defogger, seatback pockets, front and rear reading lights, dual LCD trip odometers, Service Interval Indicator, fuel economy indicator, trip computer, power sunroof, toolkit, 205/65VR15 tires on alloy wheels. **535i** adds: 3.5-liter engine, leather-wrapped steering wheel, driver's side airbag, automatic climate control, 225/60VR15 tires.

Optional Equipment:

Limited-slip differential	390	—	—
CD changer	775	—	—
Heated front seats	200	—	—

BMW 6-Series

What's New

BMW's "classic sports-luxury coupes," the low-volume 635CSi and the even lower-volume M6, are mostly reruns. The M6, in fact, is a 1988 leftover. It does get a new option, though: a dealer-installed compact-disc player/changer that handles up to six CDs, a $775 item also available for the 635CSi. The M6's base price is unchanged; the 635CSi's is

BMW 635CSi

up $1000. Last year, the 635 got the higher-compression, 208-horsepower version of the 3.5-liter inline six that powers the 735 sedans. Now it picks up the 7-Series' revised Servotronic power steering. Unlike the previous system, Servotronic varies assist proportionally with car speed rather than engine speed. An electronically controlled 4-speed overdrive automatic with driver-selectable economy, sport and manual shift programs remains standard, and a 5-speed manual is a no-cost alternative. M6, built by the Munich automaker's Motorsport division, comes only with a heavy-duty 5-speed and a 3.5-liter engine with dual overhead camshafts, four valves per cylinder and 256 horsepower. Other differences on the M6 include a firmed-up sports suspension, wider wheels and tires, larger-diameter brakes and a small trunklid spoiler.

For

●Performance ●Anti-lock brakes ●Handling
●Instruments/controls

Against

●Fuel economy ●Rear seat room ●Ride (M6) ●Price

Summary

"Classic" is a good word for the 6-Series, which continues to find a small but steady market even after 14 years and few major changes. This explains why BMW may continue it after its erstwhile 8-Series replacement, a shorter 2-door version of the big 750iL sedan, goes into production in 1990. But if the 6-Series is hardly modern any more (it owes much to the original 5-Series of 1975), it's aged gracefully. But other than a tight back seat, limited trunk space and a driving position that won't suit everyone, the 6-Series offers those who can afford it a lot to like. As before, its strong suits are brisk, smooth, refined performance; agile handling; excellent braking with the safety of standard ABS; and power assists and creature comforts aplenty. The M6 is for connoisseurs: almost $9000 costlier than the 635, faster but harder-riding, noisier and quite thirsty (only 10/19 EPA mpg).

KEY: **ohv** = overhead valve; **ohc** = overhead cam; **dohc** = double overhead cam; **I** = inline cylinders; **V** = cylinders in V configuration; **flat** = horizontally opposed cylinders; **bbl.** = barrel (carburetor); **PFI** = port (multi-point) fuel injection; **TBI** = throttle-body (single-point) fuel injection; **rpm** = revolutions per minute; **OD** = overdrive transmission; **S** = standard; **O** = optional; **NA** = not available.

Prices are accurate at time of printing; subject to manufacturer's change.

Specifications

	2-door notchback
Wheelbase, in.	103.3
Overall length, in.	189.6
Overall width, in.	67.9
Overall height, in.	53.7
Front track, in.	56.3
Rear track, in.	57.5
Turn diameter, ft.	37.4
Curb weight, lbs.	3530
Cargo vol., cu. ft.	14.6
Fuel capacity, gal.	16.6
Seating capacity	4
Front headroom, in.	37.7
Front shoulder room, in.	55.4
Front legroom, max., in.	41.7
Rear headroom, in.	35.7
Rear shoulder room, in.	54.1
Rear legroom, min., in.	29.9

Powertrain layout: longitudinal front engine/rear-wheel drive.

Engines

	ohc I-6	dohc I-6
Size, liters/cu. in.	3.4/209	3.5/211
Fuel delivery	PFI	PFI
Horsepower @ rpm	208 @ 5700	256 @ 6500
Torque (lbs./ft.) @ rpm	225 @ 4000	243 @ 4500
Availability	S	S[1]

EPA city/highway mpg

5-speed OD manual	15/23	10/19
4-speed OD automatic	15/19	

1. M6

Prices

BMW 6-Series	Retail Price	Dealer Invoice	Low Price
635CSi 2-door notchback	$47000	—	—
M6 2-door notchback (1988)	55950	—	—
Destination charge	325	325	325
Gas Guzzler Tax, 635CSi 5-speed	650	650	650
635CSi automatic	850	850	850
M6	2250	2250	2250

Dealer invoice and low price not available at time of publication.

Standard Equipment:

635CSi: 3.4-liter PFI 6-cylinder engine, 4-speed automatic transmission, power steering, power 4-wheel disc brakes, anti-lock braking system, dual air conditioning system, cruise control, fog lamps, intermittent wipers, heated windshield washer jets, power mirrors, heated driver's door lock, power windows, central locking including trunk, telescopic steering column, eight-way power front seats with three-position memory, hand-stitched Nappa leather (on seats, doors, lower dash and console), time-delay courtesy light, tinted glass with shaded upper windshield band, tachometer and coolant temperature gauge, trip odometer, Active Check Control 7-function monitor, Service Interval Indicator, fuel economy indicator, onboard computer, rear defogger, rear window sunshade, power two-way sunroof, AM/FM cassette with power amp, equalizer and anti-theft circuitry, toolkit, 220/

KEY: **ohv** = overhead valve; **ohc** = overhead cam; **dohc** = double overhead cam; **I** = inline cylinders; **V** = cylinders in V configuration; **flat** = horizontally opposed cylinders; **bbl.** = barrel (carburetor); **PFI** = port (multi-point) fuel injection; **TBI** = throttle-body (single-point) fuel injection; **rpm** = revolutions per minute; **OD** = overdrive transmission; **S** = standard; **O** = optional; **NA** = not available.

55VR390 Michelin TRX tires on forged alloy wheels. **M6** adds: 3.5-liter DOHC 24-valve PFI 6-cylinder engine, 5-speed manual transmission, sport suspension, limited-slip differential, 245/45VR415 Michelin TRX tires.

Optional Equipment:	Retail Price	Dealer Invoice	Low Price
Limited-slip differential, 635CSi	390	—	—
Heated front seats	200	—	—

BMW 7-Series

BMW 735i

What's New

The first of BMW's second-generation 7-Series, the 6-cylinder 735i, arrived in the spring of '87 as an early 1988 model. Last winter brought the 750iL, a stretched-wheelbase version powered by a 5.0-liter V-12 with 300 horsepower. Arriving last summer was the 735iL, basically the 735's 208-horsepower six in the long 750 body. For 1989, all three models have infrared remote control for their standard central locking/anti-theft system, plus ZF Servotronic power steering with "revised assist characteristics," according to BMW. Assist now varies with vehicle speed, not engine speed. To improve acceleration on 6-cylinder models with automatic transmission, the final drive ratio is shortened from 3.64 to 3.91:1. As before, only the 735i offers a 5-speed manual transmission as a no-charge option. Otherwise, you get a ZF 4-speed overdrive automatic with electronic control and, new this year, two shift programs instead of three; the previous "S" program is now a "Sport" position on the quadrant to improve mileage in EPA tests. The "Economy" and "Manual" programs are retained. A trunk-mounted compact-disc player/changer that holds six CDs is now standard on the 750iL and a dealer-installed option for the 735iL.

For

- Performance
- Anti-lock brakes
- Air bag
- Handling
- Ride
- Passenger room
- Quietness

Against

- Fuel economy
- Price

Summary

The latest 7-Series is tremendously impressive—a good thing for BMW, as much of its design has gone into the all-new 5-Series and will show up in the 8-Series coupe and, ultimately, the next 3-Series. All are big, heavy, quiet

---

Prices are accurate at time of printing; subject to manufacturer's change.

CONSUMER GUIDE®

and smooth-riding cars with ample room, especially the limousine-like 735iL and 750iL models. Alas, they're also gas guzzlers, as the EPA ratings and our own tests show. Then again, if you can afford a 7-Series in the first place, you probably won't care. More worrisome is its sheer complexity. There's a lot to go wrong with so many electric and electronic gizmos—and one or two have gone wrong on our various test cars. Performance depends on engine. With its mighty V-12, the 750iL does 0-60 mph in 7.6 seconds. The 6-cylinder 735i with automatic needs about two seconds more; the heavier 735iL is a little slower than that. What they all do best is to eat up lots of highway miles in a hurry and with great comfort. Yet they're surprisingly capable on tight, twisty roads, with a nimbleness that belies their bulk. Thanks to ABS, stops are smooth, straight and very short. In all, we prefer the 7-Series to Mercedes' aging S-Class sedans for superior roadability combined with comparable performance, room and luxury. If only they didn't cost so much.

Specifications

	735i 4-door notchback	735iL/750iL 4-door notchback
Wheelbase, in.	111.5	116.0
Overall length, in.	193.3	197.8
Overall width, in.	72.6	72.6
Overall height, in.	55.6	55.1
Front track, in.	60.1	60.2
Rear track, in.	61.0	6.13
Turn diameter, ft.	38.1	39.4
Curb weight, lbs.	3835	4015[1]
Cargo vol., cu. ft.	17.6	17.6
Fuel capacity, gal.	21.5	24.0
Seating capacity	5	5
Front headroom, in.	NA	NA
Front shoulder room, in.	NA	NA
Front legroom, max., in.	NA	NA
Rear headroom, in.	NA	NA
Rear shoulder room, in.	NA	NA
Rear legroom, min., in.	NA	NA

1. 4235 lbs., 750iL

Powertrain layout: longitudinal front engine/rear-wheel drive.

Engines

	ohc I-6	ohc V-12
Size, liters/cu. in.	3.4/209	5.0/304
Fuel delivery	PFI	PFI
Horsepower @ rpm	208 @ 5700	300 @ 5200
Torque (lbs./ft.) @ rpm	225 @ 4000	332 @ 4100
Availability	S	S[1]

EPA city/highway mpg

5-speed OD manual	15/23	
4-speed OD automatic	14/19	12/17

1. 750iL

Prices

BMW 7-Series	Retail Price	Dealer Invoice	Low Price
735i 4-door notchback	$54000	—	—
735iL 4-door notchback	58000	—	—
750iL 4-door notchback	70000	—	—
Destination charge	325	325	325
Gas Guzzler Tax, 735i 5-speed	650	650	650
735i automatic	850	850	850

	Retail Price	Dealer Invoice	Low Price
735iL	850	850	850
750iL	1850	1850	1850

Dealer invoice and low price not available at time of publication.

Standard Equipment:

3.4-liter PFI 6-cylinder engine, 5-speed manual or 4-speed automatic transmission, power steering, power 4-wheel disc brakes, anti-lock braking system, driver's side airbag, automatic climate control system with separate left and right controls, outboard rear lap/shoulder belts, cruise control, fog lamps, speed-sensitive intermittent wipers, heated wiper parking area, heated windshield washer jets, heated power mirrors, heated driver's door lock, central locking including trunk, power windows, leather-wrapped steering wheel, 8-way power front seats with 3-position driver's side memory (including outside mirrors), driver's side airbag, rear head restraints, leather seating, front center armrests, rear armrest with storage compartment, Bubinga wood trim, time-delay courtesy light, map lights, rear reading lights, tinted glass, tachometer and coolant temperature gauge, LCD main and trip odometers, Service Interval Indicator, fuel economy indicator, Active Check Control, trip computer, rear defogger, cruise control, roll-up rear sunshade, power two-way sunroof, AM/FM ST ET cassette, power antenna, lockable glovebox, toolkit, 225/60VR15 tires on cast alloy wheels. **735iL** adds: power rear seat, self-leveling rear suspension. **750iL** adds: 5.0-liter PFI V-12, anti-theft warning device, additional leather trim, Elmwood trim, cellular telephone, forged alloy wheels.

Optional Equipment:

Limited-slip differential	390	—	—
Heated front seats (std. 750iL)	200	—	—
CD changer	775	—	—

Buick Century

Buick Century Custom 2-door

What's New

Fresh exterior styling and a new V-6 engine are in store for the mid-size, front-drive Century, which returns for its eighth season. Main styling changes at the front are a new vertical bar grille and flush composite headlamps. A new rear window and roof pillars give Century a rounder appearance at the back, while full-width taillamps and a new trunk lid add a family resemblance to the Buick Regal. The new engine is a 3.3-liter V-6 derived from Buick's 3.0 V-6, used last year in the compact Skylark. Dubbed the "3300" V-6, it replaces 125-horsepower 2.8-liter and 150-horsepower 3.8-liter V-6s as this year's only optional engine for Century. The 3300 produces 160 horsepower. A 98-horsepower 2.5-liter 4-cylinder engine remains standard, though later

Prices are accurate at time of printing; subject to manufacturer's change.

Buick

in the year a revised version producing 110 horsepower will be introduced. Century also gains Buick's Dynaride suspension, which uses deflected disc shock absorber valving. Buick says the revised shock absorbers allow a boulevard ride on smooth roads and an absorbent ride on washboard surfaces without letting the suspension bottom out. Other changes are that dual outside mirrors and, as with all 1989 GM cars, rear shoulder belts are standard. Three body styles are again offered: 2-door coupe, 4-door sedan and 5-door wagon. Century is part of GM's A-body family, which includes the Chevrolet Celebrity, Oldsmobile Cutlass Ciera and Pontiac 6000.

For

- Performance (V-6) • Passenger room
- Trunk space • Quietness

Against

- Performance (4-cylinder)
- Handling (base suspension)

Summary

While Century's new 3.3-liter V-6 engine has more horsepower than the 3.8, it has less torque (185 pounds/feet versus 200). Torque is what you feel more than horsepower when accelerating from rest or in highway passing, and the more you have the better. However, the 3.3 doesn't suffer a big performance deficit compared to the 3.8, and it has more torque than the 2.8 V-6, which was rated at 160 pounds/feet. In addition, the 3.3 is a smooth and fairly quiet engine that should serve Century owners better than the weaker, noisier 4-cylinder that remains standard. The other major mechanical change is the Dynaride suspension, which might improve ride comfort some, but makes little apparent difference in handling ability. The base suspension lacks sufficient roll resistance to prevent the front end from plowing in hard cornering, and the soft, narrow standard tires aren't much help. In routine, gentle driving, Century does fine; it just isn't set up for spirited driving. Otherwise, this remains a roomy, comfortable intermediate that's competitively priced against the Ford Taurus/Mercury Sable. The similar Chevrolet Celebrity and Pontiac 6000 offer the same basic package at a slightly lower price.

Specifications

	2-door notchback	4-door notchback	5-door wagon
Wheelbase, in.	104.9	104.9	104.9
Overall length, in.	189.1	189.1	190.9
Overall width, in.	69.4	69.4	69.4
Overall height, in.	53.7	54.2	54.2
Front track, in.	58.7	58.7	58.7
Rear track, in.	56.7	56.7	56.7
Turn diameter, ft.	38.5	38.5	38.5
Curb weight, lbs.	2707	2753	2912
Cargo vol., cu. ft.	16.2	16.2	74.4
Fuel capacity, gal.	15.7	15.7	15.7
Seating capacity	6	6	8
Front headroom, in.	38.6	38.6	38.6
Front shoulder room, in.	56.1	56.2	56.2
Front legroom, max., in.	42.1	42.1	42.1
Rear headroom, in.	37.9	38.0	38.9
Rear shoulder room, in.	57.0	56.2	56.2
Rear legroom, min., in.	36.1	35.9	34.8

Powertrain layout: transverse front engine/front-wheel drive.

Powertrains

	ohv I-4	ohv V-6
Size, liters/cu. in.	2.5/151	3.3/204
Fuel delivery	TBI	PFI
Horsepower @ rpm	98 @	160 @
	4800	5200
Torque (lbs./ft.) @ rpm	135 @	185 @
	3200	2000
Availability	S	O

EPA city/highway mpg

3-speed automatic	23/30	20/27
4-speed OD automatic		20/29

Prices

Buick Century

	Retail Price	Dealer Invoice	Low Price
Custom 4-door notchback	$12429	$10726	$11578
Custom 2-door notchback	12199	10528	11364
Custom 5-door wagon	13156	11354	12255
Limited 4-door notchback	13356	11526	12441
Estate 5-door wagon	13956	12044	13000
Destination charge	450	450	450

Standard Equipment:

Custom: 2.5-liter TBI 4-cylinder engine, 3-speed automatic transmission, power steering, power brakes, dual outside mirrors, bumper guards and rub strips, tinted glass, map lights, instrument panel courtesy lights, engine compartment and trunk lights, cloth notchback bench seats, AM/FM ST ET, headlamps-on chime, deluxe wheel covers, P185/75R14 all-season SBR tires. **Wagon** adds: split folding rear seatback, two-way tailgate. **Limited and Estate** add: moldings (windsplit, wide rocker panel), 55/45 notchback cloth seat, armrest with storage, trunk trim (4-door).

Optional Equipment:

3.3-liter PFI V-6	710	604	653
4-speed automatic transmission (V-6 req.)	175	149	161
HD engine & transmission cooling, w/A/C	40	34	37
w/o A/C	70	60	64
Rear defogger	145	123	133
California emissions pkg.	100	85	92
Decklid luggage rack, exc. wagons	115	98	106
AM/FM ST ET cassette	122	104	112
Power antenna	70	60	64
Bodyside stripes	45	38	41
HD suspension, wagon	27	23	25
195/75R14 WSW tires, 2- & 4-doors	61	52	56
Wagons	34	29	31
Cloth 55/45 seat w/storage armrest, Custom	183	156	168
Locking wire wheel covers	199	169	183
Styled steel wheels	99	84	91
Alloy wheels	199	169	183
Rear wiper/washer, wagons	125	106	115
Woodgrain applique, wagons	350	298	322
Gran Touring Pkg. (NA wagons)	500	425	460

Gran Touring Suspension, leather-wrapped steering wheel, 215/60R14 Goodyear Eagle GT tires.

Popular Pkg. SB, Custom 2- & 4-door	700	595	644

Air conditioning, rear defogger, tilt steering column, intermittent wipers.

Premium Pkg. SC, Custom 2-door	1125	956	1035
Custom 4-door	1185	1007	1090

Pkg. SB plus cruise control, 55/45 seat, AM/FM ST ET cassette, carpet savers, seatback recliners, power door locks, door edge guards.

Luxury Pkg. SD, Custom 2-door	1492	1268	1373
Custom 4-door	1627	1383	1497

Pkg. SC plus wire wheel covers, lighted right visor mirror, power windows.

Prestige Pkg. SE, Custom 2-door	1780	1513	1638
Custom 4-door	1915	1628	1762

Pkg. SD plus power mirrors, power driver's seat, power antenna, power trunk release.

Prices are accurate at time of printing; subject to manufacturer's change.

Buick Century Custom 4-door

Buick Electra T Type

	Retail Price	Dealer Invoice	Low Price
Special Pkg. SJ, Custom 2-door	925	786	851
Custom 4-door	1010	859	929
Air conditioning, rear defogger, tilt steering column, intermittent wipers, door edge guards, power windows.			
Special Pkg. SK, Custom 2-door	1065	905	980
Custom 4-door	1115	948	1026
Deletes power windows and door edge guards from Pkg. SJ and adds carpet savers, power door locks, cruise control.			
Premium Pkg. SC, Limited	965	820	888
Air conditioning, rear defogger, tilt steering column, carpet savers, cruise control, intermittent wipers, power door locks.			
Luxury Pkg. SD, Limited	1336	1136	1229
Pkg. SC plus AM/FM ST ET cassette, seatback recliners, wire wheel covers, door edge guards, power windows.			
Prestige Pkg. SE, Limited	1826	1552	1680
Pkg. SD plus power mirrors, lighted right visor mirror, power drivers' seat, power antenna, power trunk release, front reading lights, premium rear speakers.			
Popular Pkg. 1SB, Custom wagon	815	693	750
Air conditioner, rear defogger, roof rack, intermittent wipers, tilt steering column.			
Premium Pkg. SC, Custom wagon	1602	1362	1474
Pkg. SB plus carpet savers, power door locks, power tailgate release, cruise control, AM/FM ST ET cassette, third seat, swing-out rear quarter windows, front seatback recliners, 55/45 front seat, rear window air deflector, door edge guards.			
Luxury Pkg. SD, Custom wagon	2044	1737	1880
Pkg. SC plus wire wheel covers, power windows, lighted right visor mirror.			
Prestige Pkg. SE, Custom wagon	2282	1940	2099
Pkg. SD plus power mirrors, power driver's seat, power antenna.			
Premium Pkg. SC, Estate	1220	1037	1122
Air conditioning, rear defogger, roof rack, intermittent wipers, tilt steering column, carpet savers, power door locks, power tailgate release, cruise control, front seatback recliners.			
Luxury Pkg. SD, Estate	1881	1599	1731
Pkg. SC plus AM/FM ST ET cassette, third seat, swing-out rear quarter windows, rear window air deflector, door edge guards, wire wheel covers, power windows, power antenna, lighted right visor mirror.			
Prestige Pkg. SE, Estate	2158	1834	1985
Pkg. SD plus power driver's seat, power mirrors, front reading lights, premium rear speakers.			

Buick Electra/Park Avenue

What's New

A new top-line luxury model joins the Electra/Park Avenue line at mid-year as Buick's 4-door flagship. Called Park Avenue Ultra, the new model has a leather interior and 20 way power adjustments for the driver and front passenger seats. Conventional 6-way power seat adjustments are controlled by switches on the front door armrests. A separate control panel folds out of the center armrest to reveal buttons for seatback recliners, thigh and lumbar supports, and power headrests. Anti-lock brakes, 2-tone paint and 15-inch aluminum wheels also are standard on Ultra, whose price will be announced later. The rest of Buick's full-size luxury lineup consists of the Electra Limited, plusher Park Avenue and sporty Electra T Type, all available only in 4-door styling. A 165-horsepower 3.8-liter V-6 and a 4-speed overdrive automatic is the only available powertrain. New for '89 are standard front seat belts that can be left buckled for automatic protection, a 100-mph speedometer for all models, blue leather upholstery as a new option for Park Avenue, and a remote keyless entry system available on all models. Unlike keyless entry systems that use infrared signals, GM's radio-frequency system doesn't require that the signal be aimed at a receiver inside the car. It can lock or unlock the doors or trunk from as far away as 30 feet. Electra has more option packages to choose from this year, offering discounts of $100 to $500 compared to ordering the same options individually. The basic design for the front-drive Electra/Park Avenue also is used for the Buick LeSabre, Oldsmobile Ninety-Eight and 88, and Pontiac Bonneville.

For

- Performance
- Room/comfort
- Driveability
- Anti-lock brakes

Against

- Fuel economy
- Ride (T Type)

Summary

The main difference between Electra and the less expensive LeSabre is Electra's mission as a luxury sedan instead of a family sedan. Thus, Electra offers plusher furnishings, more standard equipment and a few options not available on LeSabre. However, Electra offers no advantages in its mechanical components, which are shared with LeSabre. The 3.8-liter V-6 provides brisk acceleration, though it uses too much gas (expect 15-18 mpg around town) and isn't the smoothest engine. The front-drive Electra is a traditional Buick in most respects, providing a soft ride, roomy and posh interior, and quiet highway cruising. The T Type has a more sporting personality, due to its firmer suspension and larger tires, which improve handling and roadholding, but

---

KEY: ohv = overhead valve; **ohc** = overhead cam; **dohc** = double overhead cam; **I** = inline cylinders; **V** = cylinders in V configuration; **flat** = horizontally opposed cylinders; **bbl.** = barrel (carburetor); **PFI** = port (multi-point) fuel injection; **TBI** = throttle-body (single-point) fuel injection; **rpm** = revolutions per minute; **OD** = overdrive transmission; **S** = standard; **O** = optional; **NA** = not available.

Prices are accurate at time of printing; subject to manufacturer's change.

Buick

at a noticeable loss in ride comfort. We highly recommend the anti-lock brakes, standard on the T Type and Ultra, optional on the others, for the extra safety margin they provide. However, since you can also get the anti-lock feature on the cheaper LeSabre, we recommend that car over this one.

Specifications

	4-door notchback
Wheelbase, in.	110.8
Overall length, in.	196.9
Overall width, in.	72.4
Overall height, in.	54.3
Front track, in.	60.3
Rear track, in.	59.8
Turn diameter, ft.	39.4
Curb weight, lbs.	3289
Cargo vol., cu. ft.	16.4
Fuel capacity, gal.	18.0
Seating capacity	6
Front headroom, in.	39.3
Front shoulder room, in.	58.9
Front legroom, max., in.	42.4
Rear headroom, in.	38.1
Rear shoulder room, in.	58.8
Rear legroom, min., in.	41.5

Powertrain layout: transverse front engine/front-wheel drive.

Engines

	ohv V-6
Size, liters/cu. in.	3.8/231
Fuel delivery	PFI
Horsepower @ rpm	165 @ 4800
Torque (lbs./ft.) @ rpm	210 @ 2000
Availability	S

EPA city/highway mpg

4-speed OD automatic	19/28

Prices

Buick Electra

	Retail Price	Dealer Invoice	Low Price
Limited 4-door notchback	$18525	$15987	$17256
T Type 4-door notchback	21325	18403	19864
Park Avenue 4-door notchback	20460	17657	19059
Ultra 4-door notchback	NA	NA	NA
Destination charge	550	550	550

Standard Equipment:

Limited: 3.8-liter PFI V-6 engine, 4-speed automatic transmission, power steering and brakes, air conditioning, 55/45 cloth front seat with storage armrest, power driver's seat, tilt steering column, power windows, left

Buick Park Avenue Ultra

remote and right manual mirrors, headlamps-on tone, trip odometer, remote fuel door release, tinted glass, front reading and courtesy lights, engine compartment and trunk lights, bodyside moldings, AM/FM ST ET, automatic front seat belts, outboard rear lap/shoulder belts, automatic level control, 205/75R14 all-season SBR tires. **T Type** adds: anti-lock braking system, Gran Touring Suspension, sport steering wheel, carpet savers, overhead console, quartz analog gauge cluster (includes tachometer, voltmeter, coolant temperature and oil pressure, low-fuel warning), rear headrests, lighted visor mirrors, power mirrors, 45/45 front seat with console and armrest, leather-wrapped steering wheel, 215/65R15 Goodyear Eagle GT + 4 tires on alloy wheels. **Park Avenue** adds to base: cruise control, power door locks, power mirrors, intermittent wipers, coach lamps, rear reading and courtesy lights, lighted right visor mirror, added sound insulation, carpet savers, upgraded carpet, WSW tires. **Ultra** adds: anti-lock braking system, 55/45 leather front seat with 20-way power adjustments, two-tone paint, rear armrest, 205/70R15 tires on alloy wheels.

Optional Equipment:

	Retail Price	Dealer Invoice	Low Price
Anti-lock braking system, Ltd & Park Ave	925	786	851
HD engine & transmission cooling (std. T Type)	40	34	37
Rear defogger (SA Pkg. req.)	145	123	133
Quartz analog gauges, Ltd & Park Ave	126	107	116
Electronic instrumentation, Park Ave & Ultra	299	254	275
Decklid luggage rack	115	98	106
Firemist paint (exc. T Type)	210	179	193
AM/FM ST ET cassette	132	112	121
AM/FM ST ET cassette w/EQ	352	299	324
Delco/Bose music system	905	769	833
Power antenna, Park Ave & Ultra	70	60	64
Astroroof	1230	1046	1132
Bodyside stripes	45	38	41
Theft deterrent system	159	135	146
Heavily padded vinyl top, Ltd & Park Ave	260	221	239
Leather/vinyl 55/45 seat, Park Ave	425	361	391
T Type	325	276	299
Alloy wheels, Ltd	255	217	235
Park Ave	220	187	202
Gran Touring Pkg., Ltd	548	466	504
Park Ave	447	380	411

Gran Touring Suspension, 215/65R15 Goodyear Eagle GT tires on alloy wheels, 2.97 axle ratio, HD cooling, leather-wrapped steering wheel.

Popular Pkg. SB, Ltd	877	745	807

Rear defogger, 205/75R14 WSW tires, intermittent wipers, cruise control, power door locks, AM/FM ST ET cassette, wire wheel covers.

Premium Pkg. SC, Ltd	1170	995	1076

Pkg. SB plus carpet savers, lighted right visor mirror, power antenna, power trunk release, door edge guards, passenger seatback recliner.

Luxury Pkg. SD, Ltd	1562	1328	1437

Pkg. SC plus power passenger seat, automatic climate control, HD battery, power mirrors, door courtesy and warning lights.

Prestige Pkg. SE, Ltd	2652	2254	2440

Pkg. SD plus AM/FM ST ET cassette w/EQ, cornering lights, lamp monitors, Twilight Sentinel, illuminated entry, automatic power door locks, automatic day/night mirror, lighted left visor mirror, power driver's seatback recliner, self-sealing tires, remote keyless entry system.

Popular Pkg. SB, Park Ave	341	290	314

AM/FM ST ET cassette, rear defogger, wire wheel covers.

Premium Pkg. SC, Park Ave	1042	886	959

Pkg. SB plus power antenna, door edge guards, HD battery, power passenger's seatback recliner, power passenger seat, automatic climate control, Concert Sound speakers, Twilight Sentinel.

Luxury Pkg. SD, Park Ave	1517	1289	1396

Pkg. SC plus illuminated driver's door lock and interior light control, four-note horn, low fuel and washer fluid indicators, cornering lights, light monitors, low fuel indicator, power decklid pulldown, remote keyless entry system, lighted left visor mirror, automatic power door locks.

Prestige Pkg. SE, Park Ave	2223	1890	2045

Pkg. SD plus AM/FM ST ET cassette w/EQ, automatic day/night mirror, power mirrors with heated left, power driver's seatback recliner, memory power driver's seat, self-sealing tires, deluxe trunk trim with mat.

Popular Pkg. SB, T Type	722	614	664

Intermittent wipers, cruise control, rear defogger, power door locks, AM/FM ST ET cassette, power antenna, power trunk release.

Prices are accurate at time of printing; subject to manufacturer's change.

CONSUMER GUIDE®

	Retail Price	Dealer Invoice	Low Price
Premium Pkg. SC, T Type	1293	1099	1190

Pkg. SB plus door edge guards, power passenger seatback recliner, power passenger seat, automatic climate control, HD battery, Concert Sound speakers

Prestige Pkg. SE, T Type	2482	2110	2283

Pkg. SD plus automatic power door locks, AM/FM ST ET cassette w/EU, automatic day/night mirror, power mirrors with heated left, power driver's seatback recliner, memory power driver's seat, deluxe trunk trim with mat.

Buick LeSabre

Buick LeSabre Limited 4-door

What's New

A tilt steering column and trip odometer are now standard on all models in Buick's most popular car line, while a new front-seat armrest with a built-in storage compartment is standard on Limited models and optional on Customs with 55/45 split front seats. New speedometer graphics provide readings up to 100 mph instead of 85, and Custom and Limited models get more fake wood interior trim. The full-size, front-drive LeSabre is built from the same design as the luxury Electra/Park Avenue, but has less standard equipment and a lower price. LeSabre uses the same 165-horsepower 3.8-liter V-6 as Electra/Park Avenue. LeSabre also shares its design with the Oldsmobile 88 Royale and Ninety-Eight and Pontiac Bonneville. This year's lineup includes a base LeSabre 2-door coupe, Custom 4-door sedan, and Limited coupe and sedan. A sporty T Type package is again available for the base coupe, while a Gran Touring suspension package is available for all models. Most options are grouped into packages that are discounted $100 to $600 compared to what they would cost individually.

For

●Performance ●Room/comfort ●Driveability
●Anti-lock brakes

Against

●Fuel economy ●Ride (T Type)

Summary

We favor LeSabre over the similar but more expensive Electra simply because there are no substantive differences between the two. While Electra is clearly the plusher of the two, LeSabre is hardly a stripped econocar by comparison. The interiors are fairly plush even on the base and Custom models, plus air conditioning, tinted glass and a stereo

radio are standard. In addition, all body panels except the roof are made of 2-sided galvanized steel (same as Electra) for long corrosion resistance. LeSabre's 165-horsepower V-6 engine has ample low-end power for quick getaways and brisk passing response, but we think the EPA estimates (19 mpg city and 29 mpg highway) are optimistic; we usually average 16-18 mpg in urban driving and 22-24 mpg on the highway. The T Type and Gran Touring packages offer better handling than LeSabre's base suspension and tires, but at considerable sacrifice in ride comfort. We recommend the optional anti-lock brakes, which are available on all models. Though they're expensive at $925, they can prevent costly collision damage and serious injuries by stopping a car in situations where ordinary brakes can't. In other ways, LeSabre follows a traditional approach to a large family car: Conservative styling; soft, flat bench seats that can hold six people; and adequate trunk space for a full-size car. They're hardly cheap, but they don't cost as much as some smaller imported cars. This is a good choice in a front-drive, full-size car; you should also look at its corporate cousins at Oldsmobile and Pontiac.

Specifications

	2-door notchback	4-door notchback
Wheelbase, in.	110.8	110.8
Overall length, in.	196.5	196.5
Overall width, in.	72.4	72.4
Overall height, in.	54.7	55.4
Front track, in.	60.3	60.3
Rear track, in.	59.8	59.8
Turn diameter, ft.	40.0	40.0
Curb weight, lbs.	3227	3267
Cargo vol., cu. ft.	15.7	16.4
Fuel capacity, gal.	18.0	18.0
Seating capacity	6	6
Front headroom, in.	38.1	38.9
Front shoulder room, in.	59.0	59.5
Front legroom, max., in.	42.4	42.4
Rear headroom, in.	37.6	38.3
Rear shoulder room, in.	57.8	59.5
Rear legroom, min., in.	37.0	38.7

Powertrain layout: transverse front engine/front-wheel drive.

Engines

	ohv V-6
Size, liters/cu. in.	3.8/231
Fuel delivery	PFI
Horsepower @ rpm	165 @ 4800
Torque (lbs./ft.) @ rpm	210 @ 2000
Availability	S

EPA city/highway mpg

4-speed OD automatic	19/28

Prices

Buick Le Sabre	Retail Price	Dealer Invoice	Low Price
2-door notchback	$15425	$13312	$14369
Custom 4-door notchback	15330	13230	14280

KEY: ohv = overhead valve; **ohc** = overhead cam; **dohc** = double overhead cam; **I** = inline cylinders; **V** = cylinders in V configuration; **flat** = horizontally opposed cylinders; **bbl.** = barrel (carburetor); **PFI** = port (multi-point) fuel injection; **TBI** = throttle-body (single-point) fuel injection; **rpm** = revolutions per minute; **OD** = overdrive transmission; **S** = standard; **O** = optional; **NA** = not available.

Prices are accurate at time of printing; subject to manufacturer's change.

	Retail Price	Dealer Invoice	Low Price
Limited 4-door notchback	16730	14438	15584
Limited 2-door notchback	16630	14352	15491
Destination charge	505	505	505

Standard Equipment:

Base and Custom: 3.8-liter PFI V-6, 4-speed automatic transmission, power steering and brakes, air conditioning, tilt steering column, cloth split bench seat with armrest, trip odometer, tinted glass, AM/FM ST ET, automatic front seatbelts, outboard rear lap/shoulder belts, 205/75R14 all-season SBR tires. **Limited** adds: 55/45 front seat with storage armrest, reclining seatbacks, wide lower bodyside moldings.

Optional Equipment:

Anti-lock braking system	925	786	851
HD battery	26	22	24
HD engine & transmission cooling	40	34	37
Rear defogger	145	123	133
California emissions pkg.	100	85	92
Engine block heater	18	15	17
Gauges & tachometer	110	94	101
Decklid luggage rack	115	98	106
AM/FM ST ET cassette	132	112	121
AM/FM ST ET cassette w/EQ	367	312	338
Power antenna	70	60	64
Bodyside stripes	45	38	41
Automatic level control	175	149	161
Full vinyl top, 4-doors	200	170	184
55/45 seat, base & Custom	183	156	168
Leather/vinyl 55/45 seat, Ltd	425	361	391
Leather/vinyl 45/45 seat, base & T Type . .	325	276	299
Alloy wheels, 14″	255	217	235
Styled steel wheels	99	84	91
Locking wire wheel covers, exc. T Type . . .	199	169	183
Gran Touring Pkg.	548	466	504

Gran Touring Suspension, 215/65R15 tires on alloy wheels, 2.97 axle ratio, HD cooling, leather-wrapped steering wheel.

T Type Pkg., base	1902	1617	1750

Gran Touring Pkg., cruise control, intermittent wipers, gauges, 45/45 seat, operating console, AM/FM ST ET cassette with Concert Sound speakers.

T Type Popular Pkg. SF	2422	2059	2228

T Type Pkg., rear defogger, rear carpet savers, power windows and locks, front seatback recliners.

T Type Premium Pkg. SG	2623	2230	2413

Pkg. SF plus power driver's seat, power antenna, power mirrors.

T Type Luxury Pkg. SH	2771	2355	2549

Pkg. SG plus lighted right visor mirror, power passenger seat.

Popular Pkg. SB, base & Custom	618	525	569

Rear defogger, intermittent wipers, cruise control, 205/75R14 WSW tires, bodyside molding, rear bumper guards, 55/45 front seat with storage armrest.

Premium Pkg. SC, base 2-door	989	841	910
Custom 4-door	1039	883	956

Pkg. SB plus carpet savers, power door locks, AM/FM ST ET cassette, wire wheel covers.

Luxury Pkg. SD, base 2-door	1454	1236	1338
Custom 4-door	1589	1351	1462

Pkg. SC plus power windows, door edge guards, power driver's seat, front seatback recliners.

Prestige Pkg. SE, base 2-door	1708	1452	1571
Custom 4-door	1843	1567	1696

Pkg. SD plus power trunk release, lighted visor mirror, power antenna, power mirrors, Concert Sound speakers.

Special Pkg. SJ, base 2-door	823	700	757
Custom 4-door	948	806	872

Rear defogger, intermittent wipers, cruise control, 205/75R14 WSW tires, bodyside molding, rear bumper guards, 55/45 front seat with storage armrest, power windows and locks.

Special Pkg. SK, base 2-door	955	812	879
Custom 4-door	1080	918	994

Pkg. SJ plus AM/FM ST ET cassette.

	Retail Price	Dealer Invoice	Low Price
Popular Pkg. SB, Ltd 2-door	1073	912	987
Ltd 4-door	1198	1018	1102

Rear defogger, intermittent wipers, cruise control, 205/75R14 WSW tires, carpet savers, power door locks, AM/FM ST cassette, power windows, power driver's seat.

Premium Pkg. SC, Ltd 2-door	1265	1075	1164
Ltd 4-door	1400	1190	1288

Pkg. SB plus wire wheel covers, door edge guards, power trunk release, lighted right visor mirror, power antenna.

Luxury Pkg. SD, Ltd 2-door	1592	1353	1465
Ltd 4-door	1797	1527	1653

Pkg. SC plus power mirrors, power passenger seatback recliner, deluxe trunk trim, automatic climate control, front and rear reading and courtesy lights, door courtesy and warning lights.

Prestige Pkg. SD, Ltd 2-door	1967	1672	1810
Ltd 4-door	2102	1787	1934

Pkg. SD plus power passenger seat, AM/FM ST ET cassette w/EQ.

Buick LeSabre/ Electra Estate

Buick LeSabre Estate

What's New

A new 5000-pound trailer-towing package is optional for both versions of the full-size, rear-drive Estate wagons, and leather seats are a new option for the ritzier Electra model. The new towing package includes heavy-duty engine and transmission cooling, automatic level control, limited slip differential, and a 3.23 final drive ratio. The LeSabre and Electra Estates are built from the same design and use the same 140-horsepower 5.0-liter V-8 engine; they differ only in interior and exterior trim, and the Electra's longer standard equipment list. The design for these rear-drive wagons dates to the 1977 model year, when GM first downsized its full-size cars. The Chevrolet Caprice sedan and wagon, Oldsmobile Custom Cruiser wagon and Pontiac Safari wagon share this design.

For

- Room/comfort • Cargo room
- Trailer towing capacity

Against

- Fuel economy • Size and weight
- Handling/maneuverability

Prices are accurate at time of printing; subject to manufacturer's change.

Summary

The rear-drive Estate wagons appeal to buyers with large families and those who need this much cargo space and trailer-towing ability. With 8-passenger seating and nearly 90 cubic foot of cargo space, these big wagons exceed the capabilities of mid-size, front-drive rivals, and nearly match the minivans. In addition, they may come out a little cheaper than a fully equipped minivan. GM's full-size wagons have impressive reputations for occupant protection. At well over 4000 pounds, they're too heavy for their carbureted V-8 engine, contributing to their poor gas mileage and lackluster acceleration. They also have clumsy, boat-like handling and a mushy, poorly controlled ride, though the optional trailering package will cure some of that. While these aren't our favorites, they serve a market niche that calls for a luxurious beast of burden in a traditional station wagon format.

Specifications

	5-door wagon
Wheelbase, in.	115.9
Overall length, in.	220.5
Overall width, in.	79.3
Overall height, in.	59.3
Front track, in.	62.2
Rear track, in.	64.0
Turn diameter, ft.	41.4
Curb weight, lbs.	4209
Cargo vol., cu. ft.	87.9
Fuel capacity, gal.	22.0
Seating capacity	8
Front headroom, in.	39.6
Front shoulder room, in.	60.9
Front legroom, max., in.	52.2
Rear headroom, in.	39.3
Rear shoulder room, in.	61.0
Rear legroom, min., in.	37.2

Powertrain layout: longitudinal front engine/rear-wheel drive.

Engines

	ohv V-8
Size, liters/cu. in.	5.0/307
Fuel delivery	4 bbl.
Horsepower @ rpm	140 @ 3200
Torque (lbs./ft.) @ rpm	255 @ 2000
Availability	S

EPA city/highway mpg

4-speed OD automatic	17/24

Prices

Buick Le Sabre/ Electra Estate	Retail Price	Dealer Invoice	Low Price
LeSabre Estate 5-door wagon	$16770	$14473	$15622
Electra Estate 5-door wagon	19860	17139	18500
Destination charge	550	550	550

Standard Equipment:

LeSabre: 5.0-liter 4bbl. V-8, 4-speed automatic transmission, power steering and brakes, air conditioning, 55/45 cloth front seat with armrest, reclining passenger-side backrest, headlamps-on tone, trip odometer, tinted glass, outboard rear lap/shoulder belts, low fuel and washer fluid indicators, AM/FM ST ET, 225/75R15 all-season SBR tires. **Electra** adds: power windows and locks, roof rack, power driver's seat, woodgrain vinyl applique.

Optional Equipment:	Retail Price	Dealer Invoice	Low Price
Rear defogger	145	123	133
California emissions pkg	100	85	92
AM/FM ST ET cassette, LeSabre	132	112	121
AM/FM ST ET cassette w/EQ, Electra	202	210	259
Power passenger seat, Electra	240	204	221
Trailer Towing Pkg.	315	268	290
3.23 axle ratio, HD engine and transmission cooling, automatic level control, limited-slip differential.			
Leather/vinyl 55/45 front seat, LeSabre	525	446	483
Electra	425	361	391
Locking wire wheel covers	199	169	183
Vinyl woodgrain applique, LeSabre	345	293	317
Popular Pkg. SB, LeSabre	870	740	800
Power door locks, remote tailgate release, tilt steering column, roof rack, cruise control, intermittent wipers, rear defogger.			
Premium Pkg. SC, LeSabre	1747	1485	1607
Pkg. SB plus power windows, door edge guards, AM/FM ST ET cassette, air deflector, carpet savers, Molding Pkg., power driver's seat.			
Luxury Pkg. SD, LeSabre	2030	1726	1868
Pkg. SC plus power antenna, lighted right visor mirror, halogen headlamps, power seatback recliner.			
Popular Pkg. SB, Electra	790	672	727
Cruise control, AM/FM ST ET cassette, air deflector, carpet savers, power antenna, lighted right visor mirror, rear defogger, moldings (rocker panel, lower fender, bodyside).			
Premium Pkg. SC, Electra	1233	1048	1134
Pkg. SB plus halogen headlamps, power seatback recliner, automatic climate control, door courtesy and warning lights, front light monitors, power mirrors.			
Luxury Pkg. SD, Electra	1578	1341	1452
Pkg. SC plus cornering lights, Twilight Sentinel, illuminated driver's door lock and interior light control, AM/FM ST ET cassette w/EQ.			

Buick Reatta

What's New

A convertible version of the 2-seat Reatta is planned, but probably won't arrive until the 1990 model year. Until then the coupe caries over with a handful of equipment changes. Reatta, introduced in January with a $25,000 base price, was a surprise success story for Buick as the 3000-unit 1988 production was sold out (to dealers, not retail customers) by July. The main changes for 1989 are that the base price is now $26,700, and a new remote keyless entry system is standard. The radio-frequency system uses a hand-held transmitter to lock or unlock the doors and trunk from distances up to 30 feet. Reatta is built on a shortened version of Riviera's front-drive platform, and uses the same drivetrain: 165-horsepower 3.8-liter V-6 and 4-speed overdrive automatic transmission. Reatta also shares the Riviera's instrument panel, including the Electronic Control Center, which uses a touch-sensitive display screen instead of conventional controls for the climate system, stereo and other functions. However, Reatta has its own exterior styling. The only two extra-cost options are an electric sliding steel sunroof and 16-way power driver's seat. Buick has replaced the electrically operated glovebox door that Reatta inherited from the Riviera with a manually opening door, and moved the remote trunk and fuel door

> **KEY: ohv** = overhead valve; **ohc** = overhead cam; **dohc** = double overhead cam; **I** = inline cylinders; **V** = cylinders in V configuration; **flat** = horizontally opposed cylinders; **bbl.** = barrel (carburetor); **PFI** = port (multi-point) fuel injection; **TBI** = throttle-body (single-point) fuel injection; **rpm** = revolutions per minute; **OD** = overdrive transmission; **S** = standard; **O** = optional; **NA** = not available.

Prices are accurate at time of printing; subject to manufacturer's change.

Buick Reatta

releases from the instrument panel to the glovebox. Cloth upholstery is no longer available, so if you want a Reatta, you have to take leather.

For

- Performance - Handling/roadholding
- Anti-lock brakes

Against

- Ride - Instruments/controls

Summary

Reatta is an attractive luxury 2-seater that can sprint to 60 mph in about 10 seconds and deliver strong passing power on the highway. It corners with more athletic skill than you expect from a Buick, thanks partly to high-performance tires that have an all-season tread design, making for good traction even in rain or snow. However, these tires also are stiff and noisy, and Reatta's short chassis has two-thirds of its weight at the front. As a result, there's marked body pitch on wavy roads, and the suspension feels too stiff over bumpy pavement, creating a jiggly ride. The standard anti-lock disc brakes provide short, true stops. Reatta has adequate passenger and cargo room for a 2-seater, and a tilt steering wheel and 6-way power seats are included among the many standard features. Among features we could do without are complicated controls for lights and wiper/washer; an inflexible automatic climate control system; and the distracting Electronic Control Center. Despite Reatta's faults, most of which are in the interior, we still like it. It's competitively priced against the Acura Legend Coupe and Lincoln Mark VII LSC.

Specifications

	2-door notchback
Wheelbase, in.	98.5
Overall length, in.	183.5
Overall width, in.	73.0
Overall height, in.	51.2
Front track, in.	60.3
Rear track, in.	60.3

KEY: ohv = overhead valve; **ohc** = overhead cam; **dohc** = double overhead cam; **I** = inline cylinders; **V** = cylinders in V configuration; **flat** = horizontally opposed cylinders; **bbl.** = barrel (carburetor); **PFI** = port (multi-point) fuel injection; **TBI** = throttle-body (single-point) fuel injection; **rpm** = revolutions per minute; **OD** = overdrive transmission; **S** = standard; **O** = optional; **NA** = not available.

	2-door notchback
Turn diameter, ft.	38.0
Curb weight, lbs.	3394
Cargo vol., cu. ft.	10.3
Fuel capacity, gal.	18.2
Seating capacity	2
Front headroom, in.	36.9
Front shoulder room, in.	57.0
Front legroom, max., in.	43.1
Rear headroom, in.	—
Rear shoulder room, in.	—
Rear legroom, min., in.	—

Powertrain layout: transverse front engine/front-wheel drive.

Engines	ohv V-6
Size, liters/cu. in.	3.8/231
Fuel delivery	PFI
Horsepower @ rpm	165 @ 4800
Torque (lbs./ft.) @ rpm	210 @ 2000
Availability	S

EPA city/highway mpg

4-speed OD automatic	19/28

Prices

Buick Reatta	Retail Price	Dealer Invoice	Low Price
2-door notchback	$26700	$23042	$26000
Destination charge	550	550	550

Standard Equipment:

3.8-liter PFI V-6 engine, 4-speed automatic transmission, power steering, power 4-wheel disc brakes, anti-lock braking system, automatic climate control air conditioning, power windows and locks, power front bucket seats with recliners, leather upholstery, remote keyless entry, touch-sensitive Electronic Control Center, trip odometer, tinted glass, intermittent wipers, AM/FM ST ET cassette with EQ and power antenna, full-length console, lighted visor mirrors, intermittent wipers, tilt steering column, leather-wrapped steering wheel, rear defogger, digital clock, cruise control, headlamps-on and turn-signal reminder tones, remote fuel door and trunk releases, cornering lamps, fog lamps, theft deterrent system, map lights, illuminated driver's door lock and interior light control, 215/65R15 Goodyear Eagle GT+4 tires on alloy wheels.

Optional Equipment:

California emissions pkg.	100	85	92
Power sunroof	895	761	823
16-way power driver's seat	680	578	626

Buick Regal

What's New

All-new for 1988, the mid-size Regal 2-door coupe gets several significant changes for its second season, including a new standard engine at mid-year. Anti-lock brakes, a power sunroof and remote keyless entry system are new options for the front-drive Regal, built under the GM W-body program that also produced the Oldsmobile Cutlass Supreme and Pontiac Grand Prix coupes last year. While none of the three new coupes met GM's 1988 sales projections, Regal sales were higher than in 1987. A 4-door version of Regal is planned for 1990 or 1991. The new engine

Prices are accurate at time of printing; subject to manufacturer's change.

Buick Regal Custom

sion, which was too stiff and suffered too much impact harshness. The '89 GT suspension is much more compliant without any apparent loss in handling or roadholding, so we recommend you try it. The new Dynaride base suspension doesn't seem to be much different than last year's, either in ride (which was good) or handling (which was competent). Regal's remaining obstacles include lack of a 4-door body style (one is coming) and a poorly designed instrument cluster that is a narrow strip with minimal information in standard form. An optional $299 electronic cluster can be substituted, but is squeezed into the same small space and has poorly designed graphics. Regal and the similar GM coupes offer nothing exceptional, though we're happy to see anti-lock brakes available as an option. While other options may be more entertaining, anti-lock brakes could save your life. Also check out the Cutlass Supreme and Grand Prix; if you need a 4-door, the Chevrolet Lumina due next spring will be built from this design

that will arrive at mid-year is a 3.1-liter V-6; it will replace the standard 2.8-liter V-6, from which it was derived. The 3.1 produces 140 horsepower (10 more than the 2.8) and 190 pounds/feet torque (20 more than the 2.8); like the 2.8, it will come only with a 4-speed overdrive automatic. Four-wheel disc brakes are standard on Regal and for '89 an anti-lock feature is a $925 option. Base prices have soared nearly $2000 since last fall, mostly from new standard features: air conditioning, rear shoulder belts (now standard on all GM cars), AM/FM stereo radio, tilt steering column and front seatback recliners for Custom models. Limited models gain a center armrest with storage compartment. There's only a $525 price difference between Custom and Limited models, but the Custom has more option packages to choose from. A compact disc player and auxiliary radio controls mounted in the steering wheel are new options, plus the optional power windows include a new automatic down feature for the driver's side. The new remote keyless entry system uses radio signals to lock or unlock the doors and trunk from up to 30 feet away. Regal also gains Buick's Dynaride suspension this year, which relies on deflected disc shock absorber valving to provide a soft, comfortable ride without hurting handling, according to Buick. The optional Gran Touring suspension included in the Gran Touring Package and Regal Gran Sport package also has been tuned for a softer ride (though Buick doesn't call it Dynaride), while the tires that come with those packages are a size larger in diameter and tread width than last year's: 215/60R16 instead of 195/70R15.

For

• Ride • Passenger room • Trunk space
• Anti-lock brakes

Against

• Performance • Instrumentation • Engine noise

Summary

Our biggest disappointment with the new Regal has been the mediocre performance of the 2.8-liter V-6, whose 170 pounds/feet of torque struggles with 3000 pounds of curb weight. The new 3.1-liter V-6 due at mid-year adds 20 pounds/feet—not nearly enough to make this car a threat to Porsche but enough to provide improved acceleration and quicker passing response, with less noise. The biggest improvement for 1989 may be in the Gran Touring suspension. Despite this year's larger tires, softer shock absorbers allow a far more comfortable ride than the 1988 GT suspen-

Specifications

	2-door notchback
Wheelbase, in.	107.5
Overall length, in.	192.2
Overall width, in.	72.5
Overall height, in.	53.0
Front track, in.	59.3
Rear track, in.	58.0
Turn diameter, ft.	39.0
Curb weight, lbs.	3144
Cargo vol., cu. ft.	15.5
Fuel capacity, gal.	16.5
Seating capacity	6
Front headroom, in.	37.8
Front shoulder room, in.	57.6
Front legroom, max., in.	42.3
Rear headroom, in.	37.1
Rear shoulder room, in.	56.8
Rear legroom, min., in.	34.8

Powertrain layout: transverse front engine/front-wheel drive.

Engines

	ohv V-6	ohv V-6
Size, liters/cu. in.	2.8/173	3.1/190
Fuel delivery	PFI	PFI
Horsepower @ rpm	130 @ 4500	140 @ 4500
Torque (lbs./ft.) @ rpm	170 @ 1600	190 @ 3600
Availability	S	S
EPA city/highway mpg		
4-speed OD automatic	20/29	NA

Prices

Buick Regal	Retail Price	Dealer Invoice	Low Price
Custom 2-door notchback	$14214	$12267	$13241
Limited 2-door notchback	14739	12720	13730
Destination charge	455	455	455

Standard Equipment:

2.8-liter PFI V-6, 4-speed automatic transmission, power steering and brakes, cloth split bench seat with armrest and recliners, headlamps-on tone, tilt steering column, tinted glass, digital speedometer, optical horn, left remote and right manual mirrors, black lower bodyside moldings, AM/FM ST ET, automatic front seatbelts, outboard rear lap/shoulder belts, 205/70R14 all-season SBR WSW tires. **Limited** adds: upgraded carpet, bright wide bodyside moldings, 55/45 seat with storage armrest.

Prices are accurate at time of printing; subject to manufacturer's change.

Optional Equipment:

	Retail Price	Dealer Invoice	Low Price
Anti-lock braking system	925	786	851
High-capacity cooling system	40	34	37
Rear defogger	145	123	133
California emissions pkg	100	85	92
Electronic digital/graphic instruments	299	254	275
Remote keyless entry, Gran Sport	125	106	115
Decklid luggage rack, Custom	115	98	106
Wide black molding w/red stripe, Custom .	70	60	64
Lower accent paint	205	174	189
AM/FM ST ET cassette w/EQ	272	231	250
Power antenna	70	60	64
CD player	396	337	364
Power glass sunroof	650	553	598
Steering-wheel-mounted radio controls			
w/o Four Seater, Gran Touring,			
or Gran Sport	125	106	115
w/Four Seater, Gran Touring or Gran Sport	29	25	27
Bodyside stripes	45	38	41
Cloth 55/45 seat w/storage armrest, Custom	183	156	168
Cloth reclining buckets w/console, Custom .	105	89	97
Leather/vinyl 55/45 seat w/recliners, Ltd . .	395	336	363
Locking wire wheel covers	199	169	183
Styled aluminum wheels	199	169	183
Styled steel wheels	99	84	91
Gran Touring Pkg., w/o Four Seater	600	510	552
w/Four Seater	504	428	464

Gran Touring Suspension, 215/60R16 tires on alloy wheels, leather-wrapped steering wheel.

	Retail Price	Dealer Invoice	Low Price
Four Seater Pkg., Custom	700	595	644
Limited	600	510	552
Gran Sport	409	348	376

Front and rear bucket seats, rear armrest with storage, console with cassette storage, leather-wrapped steering wheel, rear headrests.

Gran Sport Pkg., Custom	1205	1024	1109

Gran Touring Pkg., bucket seats with console, blackout exterior, fog lamps, aero rocker panels, wide bodyside moldings, front spoiler.

Popular Pkg. SB, Custom	171	145	157

Rear defogger, white stripe tires.

Premium Pkg. SC, Custom	761	647	700

Pkg. SB plus AM/FM ST ET cassette w/EQ, intermittent wipers, cruise control, carpet savers, door edge guards, 55/45 front seat with storage armrest, power antenna, power door locks.

Luxury Pkg. SD, Custom	908	772	835

Pkg. SC plus power windows, wire wheel covers, lighted right visor mirror.

Prestige Pkg. SE, Custom	1499	1274	1379

Pkg. SD plus power driver's seat, power mirrors, power trunk release, Concert Sound speakers, steering-wheel-mounted redundant accessory controls.

Special Pkg. SJ, Custom	486	413	447

Rear defogger, white stripe tires, intermittent wipers, cruise control, door edge guards, power windows.

Premium Pkg. SC, Ltd	373	317	343

Rear defogger, white stripe tires, AM/FM ST ET cassette, intermittent wipers, cruise control.

Luxury Pkg. SD, Ltd	1165	990	1072

Pkg. SC plus carpet savers, door edge guards, power antenna, power windows and locks, wire wheel covers, lighted right visor mirror, power driver's seat.

Prestige Pkg. SE, Ltd	1441	1225	1326

Pkg. SD plus power mirrors, power trunk release, Concert Sound speakers, steering-wheel-mounted redundant accessory controls, remote keyless entry.

Special Pkg. SM, Gran Sport	1547	1315	1423

Rear defogger, AM/FM ST ET cassette, intermittent wipers, cruise control, carpet savers.

	Retail Price	Dealer Invoice	Low Price
Popular Pkg. SF, Gran Sport	2405	2044	2213

Pkg. SM plus electronic digital/graphic instruments, power antenna, power windows and locks, lighted right visor mirror, power driver's seat, power mirrors, power trunk release, Concert Sound speakers.

Buick Riviera

Buick Riviera

What's New

Buick tries to rekindle interest in the slow-selling Riviera by adding 11 inches to its overall length and recapturing the look of previous models. Overall length has increased to 198.3 inches, making Riviera the longest front-drive Buick, stretching 1.4 inches longer than the Electra; wheelbase remains the same as 108 inches. The longer body accounts for about 75 pounds of the Riviera's 3350-pound curb weight. Nearly all the additional length has gone into the rear, where a sloping rump with large horizontal taillamps has replaced the 1986-88 models' shorter, chopped posterior. The rear roof pillars are wider and the rear window has more slant than before. At the front, the new grille has larger vertical bars, and chrome trim is more prominent on all sides. In another bow to Riviera tradition, whitewall tires and wire wheel covers are standard, replacing blackwalls and aluminum wheels; wheel diameter has been increased from 14 inches to 15. Buick has dropped the T Type option package with its firm Gran Touring suspension and made a softer Dynaride suspension standard. New hydraulic engine mounts and additional road isolation measures have been added this year. A firmer suspension is available as part of the $49 Gran Touring Package. Inside, the touch-screen Electronic Control Center remains, while imitation wood has replaced much of the vinyl trim on the dashboard and console. A conventional manual glovebox door has replaced an electrically released door, and the standard trunk and fuel door power releases have been moved from the dashboard to the glovebox. A remote keyless entry system is a new option. Little has changed under the hood, except that heavy duty engine and transmission cooling are standard. A 165-horsepower 3.8-liter V-6 and 4-speed overdrive automatic transmission comprise the only available drivetrain.

For

- Performance
- Anti-lock brakes
- Quietness

Against

- Instruments/controls
- Fuel economy

Prices are accurate at time of printing; subject to manufacturer's change.

Summary

Riviera shares its basic design with the Cadillac Eldorado and Oldsmobile Toronado luxury coupes, both of which have sold better than Buick's luxury coupe. Cadillac added three inches to the Eldorado's length last year, and sales increased by more than 30 percent. Buick is hoping this year's new styling reverses three years of dismal sales for Riviera. The styling differences are dramatic. Whereas the 1986-88 models could be mistaken for a GM compact from a distance, the 1989 edition immediately stands out as a Riviera. Whether this is enough to turn things around remains to be seen. Styling aside, Riviera doesn't offer anything exceptional. You get plenty of luxury and convenience features for the $22,540 base price, and adequate space for four people and their luggage. Acceleration is brisk, while ride and handling are at least competent and the optional anti-lock brakes give it commendable stopping ability. Riviera's most unusual feature remains the touch-screen Electronic Control Center, which we still judge more complicated and less useful than conventional controls for the stereo, climate system and other functions.

Specifications

	2-door notchback
Wheelbase, in.	108.0
Overall length, in.	198.3
Overall width, in.	71.7
Overall height, in.	53.6
Front track, in.	59.9
Rear track, in.	59.9
Turn diameter, ft.	37.5
Curb weight, lbs.	3436
Cargo vol., cu. ft.	13.9
Fuel capacity, gal.	18.2
Seating capacity	5
Front headroom, in.	37.8
Front shoulder room, in.	57.9
Front legroom, max., in.	42.7
Rear headroom, in.	37.8
Rear shoulder room, in.	57.4
Rear legroom, min., in.	35.6

Powertrain layout: transverse front engine/front-wheel drive.

Engines

	ohv V-6
Size, liters/cu. in.	3.8/231
Fuel delivery	PFI
Horsepower @ rpm	165 @ 4800
Torque (lbs./ft.) @ rpm	210 @ 2000
Availability	S
EPA city/highway mpg	
4-speed OD automatic	19/28

Prices

Buick Riviera	Retail Price	Dealer Invoice	Low Price
2-door notchback	$22540	$19452	$20996
Destination charge	550	550	550

Standard Equipment:

3.8-liter PFI V-6, 4-speed automatic transmission, power steering, power 4-wheel disc brakes, power windows and locks, automatic level control, cloth reclining bucket seats, power driver's seat, AM/FM ST ET cassette, power antenna, outboard rear lap/shoulder belts, carpet savers, full-length console, cruise control, rear defogger, intermittent wipers, door edge guards, touch-sensitive Electronic Control Center, remote fuel door and trunk releases, tinted glass, tilt steering column, leather-wrapped steering wheel, illuminated driver's door lock and interior light control, digital instruments, trip odometer, low fuel indicator, cornering lamps, coach lamps, door courtesy and warning lights, power mirrors, lighted visor mirrors, 205/70R15 all-season SBR tires, locking wire wheel covers.

Optional Equipment:	Retail Price	Dealer Invoice	Low Price
Anti-lock braking system	925	786	851
California emissions pkg.	100	85	92
Firemist paint	210	179	193
Lower accent paint	190	162	175
Delco/Bose music system	703	598	647
Astroroof	1230	1046	1132
Cellular telephone	1975	1679	1817
Heavily padded vinyl roof	695	591	639
Leather & suede trim	487	414	448
16-way leather & suede driver's seat	1167	992	1074
Styled alloy wheels	NC	NC	NC
Gran Touring Pkg.	49	42	45

Gran Touring Suspension, 2.97 axle ratio, 215/65R15 Goodyear Eagle GT + 4 tires on alloy wheels, leather-wrapped sport steering wheel and shift handle, fast-ratio power steering.

	Retail Price	Dealer Invoice	Low Price
Appearance Pkg.	150	128	138

Platinum beige Firemist lower accent paint, bodyside stripe, painted aluminum wheels.

	Retail Price	Dealer Invoice	Low Price
Popular Pkg. SB, w/o 16-way seat	435	370	400
w/16-way seat	360	306	331

AM/FM ST ET cassette w/EQ, power passenger seat.

	Retail Price	Dealer Invoice	Low Price
Premium Pkg. SC, w/o 16-way seat	840	714	773
w/16-way seat	765	650	704

Pkg. SB plus power decklid pulldown, remote keyless entry and automatic power door locks, Twilight Sentinel.

	Retail Price	Dealer Invoice	Low Price
Luxury Pkg. SD, w/o 16-way seat	1189	1011	1094
w/16-way seat	1114	947	1025

Pkg. SC plus theft deterrent system, power mirrors with heated left, automatic rear-view mirror, electronic compass.

Buick Skyhawk

Buick Skyhawk 4-door

What's New

Rear shoulder belts, additional sound insulation and Buick's Dynaride suspension are standard for 1989 on Skyhawk, a front-drive subcompact built from GM's J-car design. All five GM divisions introduced versions of the J-car for 1982. The Cadillac Cimarron and Oldsmobile Firenza have been dropped, but the Chevrolet Cavalier and Pontiac Sunbird continue for 1989. The new Dynaride suspension uses deflected disc shock absorber valving to provide a

Prices are accurate at time of printing; subject to manufacturer's change.

soft ride and, Buick says, no compromise in handling. Deflected disc technology lets the shock absorbers adjust automatically to harsh roads to prevent excessive bouncing or suspension bottoming. Skyhawk is again available in three body styles: 2-door coupe, 4-door sedan and 5-door wagon. Only one price series is available and all models use a 90-horsepower 2.0-liter overhead valve 4-cylinder engine that's also standard in the Cavalier. Last year Skyhawk came with a 96-horsepower overhead cam engine of the same size. A 5-speed manual transmission is standard and a 3-speed automatic optional. S/E option packages available for the 2-door include hidden headlamps, sport exterior trim, tachometer and gauge cluster, Gran Touring suspension, and high performance tires mounted on 14-inch aluminum wheels.

For

●Fuel economy   ●Handling/roadholding (S/E)

Against

●Performance (automatic transmission)   ●Driveability

Summary

Small cars don't generate big sales for Buick, but Skyhawk stays around as an entry-level model to draw young buyers into the Buick fold. The same body styles are offered at lower prices by Chevrolet and Pontiac, and you're likely to find a bigger selection at Chevy and Pontiac dealers. Skyhawk's 2.0-liter engine is economical, yet lacks the performance, smoothness and quietness of newer, more sophisticated Japanese engines of the same size. Acceleration feels adequate with the standard manual transmission, but most Skyhawks are sold with automatic, and acceleration is tedious with that transmission. By comparison, a Mazda 323 with automatic transmission feels brisk. Handling is uninspiring with the base suspension and tires, improved with the sporty S/E packages. Since this is the eighth model year for the J-cars, they're old next to Japanese subcompacts such as the Honda Civic and Toyota Corolla, and it shows in their less efficient interior packaging. In Skyhawk's favor, however, it is priced competitively, thanks to changes in the U.S. dollar-Japanese yen relationship, so you can get a well-equipped version for around $11,000. You can get the same thing from Chevy or Pontiac.

Specifications

	2-door notchback	4-door notchback	5-door wagon
Wheelbase, in.	101.2	101.2	101.2
Overall length, in.	179.3	181.7	181.7
Overall width, in.	65.0	65.0	65.0
Overall height, in.	54.0	54.0	54.4
Front track, in.	55.6	55.6	55.4
Rear track, in.	55.2	55.2	55.2
Turn diameter, ft.	34.3	34.3	34.3
Curb weight, lbs.	2420	2469	2551
Cargo vol., cu. ft.	12.6	13.5	64.4
Fuel capacity, gal.	13.6	13.6	13.6
Seating capacity	5	5	5
Front headroom, in.	37.4	38.2	38.3
Front shoulder room, in.	53.7	53.7	53.7
Front legroom, max., in.	42.1	42.2	42.2
Rear headroom, in.	36.4	37.6	38.8
Rear shoulder room, in.	52.6	53.8	53.8
Rear legroom, min., in.	31.8	34.3	33.7

Powertrain layout: transverse front engine/front-wheel drive.

Engines

	ohv I-4
Size, liters/cu. in.	2.0/121
Fuel delivery	TBI
Horsepower @ rpm	90 @ 5600
Torque (lbs./ft.) @ rpm	108 @ 3200
Availability	S

EPA city/highway mpg
5-speed OD manual	26/36
3-speed automatic	25/32

Prices

Buick Skyhawk	Retail Price	Dealer Invoice	Low Price
4-door notchback	$9285	$8570	$8928
2-door notchback	9285	8570	8928
5-door wagon	10230	9442	9836
Destination charge	425	425	425

Standard Equipment:

2.0-liter TBI 4-cylinder engine, 5-speed manual transmission, power steering and brakes, reclining cloth front bucket seats, left remote and right manual mirrors, tinted glass, full-length console, coin holder, headlamps-on tone, AM/FM ST ET, outboard rear lap/shoulder belts, 185/80R13 all-season SBR WSW tires. **Wagon** has: air deflector, luggage rack, split folding rear seatback.

Optional Equipment:

3-speed automatic transmission	415	353	382
Air conditioning	675	574	621
Rear defogger	145	123	133
California emissions pkg.	100	85	92
S/E Pkg., 2-door	1095	931	1007

Gran Touring Suspension, 215/60R14 Goodyear Eagle GT + 4 tires on alloy wheels, leather-wrapped steering wheel, carpet savers, gauges and tachometer, blackout exterior moldings, concealed headlamps, fog lamps.

Popular Pkg. SB	617	524	568

3-speed automatic transmission, tilt steering column, intermittent wipers, cruise control, AM/FM ST ET cassette, 4-way manual driver's seat, console armrest, rear defogger.

Premium Pkg. SC, 2-door	617	524	568
4-door	667	567	614
Wagon	717	609	660

Deletes rear defogger from Pkg. SB and adds power door locks.

Luxury Pkg. SD, 2-door	662	563	609
4-door	712	605	655
Wagon	762	648	701

Pkg. SC plus rear defogger.

S/E Popular Pkg. SF, 2-door	1412	1200	1299

S/E Pkg., 3-speed automatic transmission, tilt steering column, intermittent wipers, cruise control, AM/FM ST ET cassette, 4-way manual driver's seat, console armrest, rear defogger.

S/E Premium Pkg. SG, 2-door	1412	1200	1299

Deletes rear defogger from Pkg. SF and adds power door locks.

S/E Luxury Pkg. SH, 2-door	1457	1238	1340

Pkg. SG plus rear defogger.

Buick Skylark

What's New

A larger, more powerful V-6 engine is optional this year for Skylark, Buick's N-body compact that's similar to the Oldsmobile Cutlass Calais and Pontiac Grand Am. The new V-6, called the "3300" by GM, is a 3.3-liter engine derived from last year's optional 3.0 V-6, which has been dropped. The

Buick Skylark Limited 4-door

3.3 makes 160 horsepower and 185 pounds/feet torque, 35 more than last year's 3.0 on both counts. The 3300 has multi-point fuel injection, roller valve lifters and GM's Computer Controlled Coil Ignition, which eliminates the distributor. A 2.5-liter 4-cylinder engine remains standard in Skylark, but horsepower has been increased to 110 from 98. The 150-horsepower Quad 4 double overhead cam 4-cylinder also is optional. All engines are available only with a 3-speed automatic transmission. Last year's lineup of 2-door coupes and 4-door sedans in base Custom and upscale Limited price levels continues. Sedans now have flush composite headlamps, same as the coupes. A split front bench seat is standard on all models in place of bucket seats, which are now optional. To accommodate the 3-place bench seat, the shift lever has been moved from the floor to the steering column and a pedal-operated parking brake has replaced the hand brake. As with all '89 GM cars, rear shoulder belts are standard. The standard suspension this year is Buick's Dynaride system, which uses deflected disc shock absorber valving for a softer ride without, Buick claims, sacrificing handling.

For

●Performance (V-6, Quad 4) ●Visibility

Against

●Noise (Quad 4) ●Rear seat room ●Trunk space

Summary

Buick's front-drive compact certainly doesn't lack for horsepower with a choice of three engines offering 110, 150 or 160. Too bad all of them have to work through a 3-speed automatic that doesn't tap all of the available power. Blame the lack of an overdrive fourth gear, which forces Buick to use extremely tall final drive ratios to achieve better EPA economy figures. The economy gearing (2.39:1) is especially evident with the new V-6, which lacks snap and strong passing response. In Skylark, the transmission is slow to downshift when you need more power for passing. The V-6 feels livelier in the heavier Century sedan, where it's teamed with a more responsive 4-speed overdrive automatic and a shorter final drive ratio (2.73:1). Even so, the V-6 gives Skylark ample pickup, and smooth, quiet cruising. Sadly, the potent Quad 4 engine also suffers with the 3-speed automatic transmission since it requires higher engine speeds to produce maximum horsepower and torque. The Quad 4 also has a loud, nasty growl when worked hard. The standard 110-horsepower 2.5-liter 4-cylinder is adequate for this car. Skylark's interior and trunk space aren't as generous as some Japanese rivals, such as the Honda Accord and Toyota Camry. With a flat front bench seat, column-mounted shift lever and foot-operated parking brake, Skylark has a conservative American flavor to its packaging. There's nothing

special here to make us want to buy one, yet prices are reasonable and dealers should be discounting. Also check out the similar Calais and Grand Am, and the Chevrolet Corsica/Beretta.

Specifications

	2-door notchback	4-door notchback
Wheelbase, in.	103.4	103.4
Overall length, in.	180.0	180.0
Overall width, in.	66.6	66.6
Overall height, in.	52.1	52.1
Front track, in.	55.6	55.6
Rear track, in.	55.2	55.2
Turn diameter, ft.	37.8	37.8
Curb weight, lbs.	2583	2640
Cargo vol., cu. ft.	13.1	13.1
Fuel capacity, gal.	13.6	13.6
Seating capacity	6	6
Front headroom, in.	37.7	37.7
Front shoulder room, in.	54.6	54.3
Front legroom, max., in.	42.9	42.9
Rear headroom, in.	37.1	37.1
Rear shoulder room, in.	55.2	54.1
Rear legroom, min., in.	34.3	34.3

Powertrain layout: transverse front engine/front-wheel drive.

Engines

	ohv I-4	dohc I-4	ohv V-6
Size, liters/cu. in.	2.5/151	2.3/138	3.3/204
Fuel delivery	TBI	PFI	PFI
Horsepower @ rpm	110 @ 5200	150 @ 5200	160 @ 5200
Torque (lbs./ft.) @ rpm	135 @ 3200	160 @ 4000	185 @ 2000
Availability	S	O	O

EPA city/highway mpg

3-speed automatic	23/30	23/32	20/27

Prices

Buick Skylark	Retail Price	Dealer Invoice	Low Price
Custom 4-door notchback	$11115	$9926	$10521
Custom 2-door notchback	11115	9926	10521
Limited 4-door notchback	12345	11024	11685
Limited 2-door notchback	12345	11024	11685
Destination charge	425	361	391

Standard Equipment:

Custom: 2.5-liter TBI 4-cylinder engine, 3-speed automatic transmission, power steering and brakes, reclining cloth split bench seat, headlamps-on tone, remote fuel door release, tinted glass, trip odometer, AM/FM ST ET, automatic front seatbelts, outboard rear lap/shoulder belts, right visor mirror, 185/80R13 all-season SBR tires. **Limited** adds: front and rear center armrests, front and rear courtesy lights, narrow rocker panel moldings, wheel opening moldings.

Optional Equipment:

2.3-liter Quad 4 4-cylinder engine	660	561	607
3.3-liter PFI V-6	710	604	653
Rear defogger	145	123	133
California emissions pkg.	100	85	92

> **KEY: ohv** = overhead valve; **ohc** = overhead cam; **dohc** = double overhead cam; **I** = inline cylinders; **V** = cylinders in V configuration; **flat** = horizontally opposed cylinders; **bbl.** = barrel (carburetor); **PFI** = port (multi-point) fuel injection; **TBI** = throttle-body (single-point) fuel injection; **rpm** = revolutions per minute; **OD** = overdrive transmission; **S** = standard; **O** = optional; **NA** = not available.

Prices are accurate at time of printing; subject to manufacturer's change.

	Retail Price	Dealer Invoice	Low Price
Engine block heater	18	15	17
Decklid luggage rack	115	98	106
Lower accent paint	195	166	179
Bodyside stripes	45	38	41
205/70R13 blackwall tires	124	105	114
205/70R13 WSW tires	190	162	175
Cloth bucket seats & console	180	153	166
13″ styled alloy wheels	229	195	211
Styled hubcaps & trim rings	38	32	35
Locking wire wheel covers	199	169	183
Gran Touring Pkg.	592	503	545

215/60R14 Goodyear Eagle GT + 4 tires on alloy wheels, Gran Touring Suspension, leather-wrapped steering wheel.

Exterior Sport Pkg., Custom 2-door	374	318	344
Ltd 2-door	318	270	293

Blackout moldings, front air dam, aero rocker moldings, wide bodyside moldings, Touring Suspension, leather-wrapped steering wheel.

S/E Pkg., Custom	1123	955	1033

Bucket seats and console, left remote and right manual mirrors, bodyside and wheel opening moldings, AM/FM ST ET cassette, leather-wrapped steering wheel, 215/60R14 Goodyear Eagle GT + 4 tires on alloy wheels, Gran Touring Suspension.

Popular Pkg. SB, Custom	576	490	530

Air conditioning, tilt steering column, 185/80R13 WSW tires, left remote and right manual mirrors, rear defogger, wide bodyside moldings, wheel opening moldings.

Premium Pkg. SC, Custom	1025	871	943

Pkg. SB plus cruise control, intermittent wipers, carpet savers, AM/FM ST ET cassette, wire wheel covers, 4-way manual driver's seat.

Luxury Pkg. SD, Custom 2-door	1180	1003	1086
Custom 4-door	1305	1109	1201

Pkg. SC plus power windows and locks.

Prestige Pkg. SE, Custom 2-door	1357	1153	1248
Custom 4-door	1482	1260	1363

Pkg. SD plus power driver's seat, power antenna, lighted right visor mirror, power trunk release, front and rear courtesy lights.

Special Pkg. SJ, Custom	639	543	588

Air conditioning, tilt steering column, 185/80R13 WSW tires, left remote and right manual mirrors, rear defogger, wide bodyside moldings, wheel opening moldings, cruise control, intermittent wipers, carpet savers.

Premium Pkg. SC, Ltd	606	515	558

Air conditioning, tilt steering column, 185/80R13 WSW tires, variable intermittent wipers, carpet savers, cruise control, wide bodyside moldings, rear defogger.

Luxury Pkg. SD, Ltd 2-door	1197	1017	1101
Ltd 4-door	1322	1124	1216

AM/FM ST ET cassette, wire wheel covers, power windows and locks, 4-way manual driver's seat.

Prestige Pkg. SE, Ltd 2-door	1310	1114	1205
Ltd 4-door	1435	1220	1320

Pkg. SD plus power driver's seat, power antenna, lighted right visor mirror, power trunk release, wide rocker panel moldings.

S/E Popular Pkg. SF	1568	1333	1443

S/E Pkg., air conditioning, tilt steering column, rear defogger.

S/E Premium Pkg. SG	1766	1501	1625

Pkg. SF plus cruise control, intermittent wipers, carpet savers, 4-way manual driver's seat.

S/E Luxury Pkg. SH, 2-door	2091	1777	1924
4-door	2216	1884	2039

Pkg. SG plus power windows and locks, power antenna.

Cadillac Allante

What's New

Cadillac bolsters the road manners of its ultra-luxury 2-seater, plagued by slow sales since its 1987 introduction, with a larger, more potent engine, variable-assist power steering, automatically adjusting shock absorbers and bigger tires for 1989. A 4.5-liter V-8 replaces the 4.1 that

Cadillac Allante

powered the 1987-88 Allante. Though the new 4.5 uses the same aluminum block found in other front-drive Cadillacs, the Allante engine benefits from sequential multi-point fuel injection (one injector for each cylinder) and other performance enhancements to produce 200 horsepower and 265 pounds/feet of torque. By comparison, last year's 4.1 V-8 made 170 horsepower and 230 pounds/feet torque. Premium unleaded gas is recommended for the 4.5. Cadillac now claims a 0-60 mph time of 8.3 seconds, compared to 9.8 seconds for the 1987-88 models. Of equal note is that Allante will carry a gas guzzler tax, the first GM car to do so. With projected fuel economy of 15 mpg city and 22 mpg highway, Allante falls short of the federal government's overall standard of 22.5 mpg combined, so it will carry a guzzler tax of $650 on top of its $56,533 base price, which is unchanged from last year. Chassis revisions include shock absorbers with deflected disc valving and speed-dependent electronic controls that provide three distinct damping modes. Below 25 mph, damping is set for a compliant ride; above 25 mph damping automatically changes to a "normal" ride; above 60 mph and in hard acceleration or braking it automatically changes to firm damping. A new variable-assist power steering system provides more assist at low speeds than higher speeds. Wheel and tire diameter have been increased from 15 inches to 16, and suspension bushings and mounts have been revised for less impact harshness. Allante's standard manual folding soft top has been redesigned for easier operation; it can now be raised or lowered by one person in about 20 seconds. The outside rearview mirrors have been moved forward 2.5 inches. Inside, the Recaro leather seats are reshaped and softer, and charcoal has been added as a new interior color (tan and maroon are the only others). A theft-deterrent system similar to the one used on the Corvette is now standard. A sensor in the ignition lock has to detect the correct resistor, imbedded in the ignition key, or the starter and electric fuel pump are disabled.

For

- Performance • Ride • Anti-lock brakes
- Luxury Appointments

Against

- Price • Fuel economy

Summary

Dismal sales forced Cadillac to halt Allante production for more than two months this summer to give dealers time to clear out leftover 1988 models, and even some dusty '87s, before the substantially revised 1989 edition arrives. We drove 1988 and 1989 models back-to-back and found this

Prices are accurate at time of printing; subject to manufacturer's change.

year's changes dramatically improve Allante's performance and feel. The engine is more substantial and more responsive. Cadillac correctly reasons that customers in this price range want performance more than fuel economy, so they won't mind paying the gas guzzler tax. Allante's ride is now more stable at high speeds, while at the same time more absorbent over bumpy pavement. The steering is more direct and responsive, while the body feels more rigid than before. To top it off, the new seats are more comfortable and supportive than the old ones. Add the standard anti-lock brakes to the mechanical refinements, and Allante has the performance credentials it needs to compete in the ultra-luxury league against the likes of the Mercedes-Benz 560SL. Does that make it worth $56,000? Not for us, but you're buying prestige more than value in this league. Allante should have been this good when it first came out. That it took Cadillac two years to bring it up to this level shows they were just feeling their way along with this car. However, they've obviously learned some important lessons.

Specifications

	2-door conv.
Wheelbase, in.	99.4
Overall length, in.	178.6
Overall width, in.	73.5
Overall height, in.	52.2
Front track, in.	60.4
Rear track, in.	60.4
Turn diameter, ft.	36.1
Curb weight, lbs.	3492
Cargo vol., cu. ft.	16.2
Fuel capacity, gal.	22.0
Seating capacity	2
Front headroom, in.	37.3
Front shoulder room, in.	57.6
Front legroom, max., in.	43.1
Rear headroom, in.	—
Rear shoulder room, in.	—
Rear legroom, min., in.	—

Powertrain layout: transverse front engine/front-wheel drive.

Engines

	ohv V-8
Size, liters/cu. in.	4.5/273
Fuel delivery	PFI
Horsepower @ rpm	200 @ 4300
Torque (lbs./ft.) @ rpm	270 @ 3200
Availability	S
EPA city/highway mpg	
4-speed OD automatic	15/23

Prices

Cadillac Allante	Retail Price	Dealer Invoice	Low Price
2-door convertible	$57183	$48873	$51700
Destination charge	550	550	550

Price includes $650 Gas Guzzler Tax.

Standard Equipment:

4.5-liter PFI V-8 engine, 4-speed automatic transmission, power steering, power 4-wheel disc brakes, anti-lock braking system, removable hardtop, folding convertible top, ten-way power Recaro seats with leather upholstery and driver's side position memory, Delco-GM/Bose Symphony music sys-

tem, tilt/telescopic steering column, dual power mirrors, intermittent wipers, automatic day/night mirror, power decklid pulldown, theft deterrent system, 225/55VR16 Goodyear Eagle VL tires on alloy wheels.

Optional Equipment:	Retail Price	Dealer Invoice	Low Price
California emissions pkg.	99	84	91
Analog instrument cluster	NC	NC	NC
Mobile cellular telephone	NA	NA	NA
Provisions for above	NA	NA	NA
California emissions pkg.	100	84	92
Cellular mobile telephone	1975	1659	1807
Provision for above	395	332	361
Analog instrument cluster	NC	NC	NC

Cadillac Brougham

Cadillac Brougham

What's New

The biggest Cadillac, and the only one with rear-wheel drive, gets a restyled grille and a few more standard features this year, but otherwise is a carryover. New standard features for Brougham are cruise control, intermittent wipers and a power trunk lid, as Cadillac continues to trim its optional equipment lists. Brougham was last redesigned for the 1977 model year in GM's first round of downsizing. It is offered only as a 4-door sedan with a carbureted 5.0-liter V-8 that produces 140 horsepower. With an overall length of 221 inches, it is the longest production car made in America, edging the arch-rival Lincoln Town Car by nearly two inches.

For

- Passenger and cargo room ●Luxury appointments
- Quietness

Against

- Performance ●Fuel economy ●Size and weight

Summary

While Cadillac claims its front-drive De Ville/Fleetwood can hit 60 mph in less than 10 seconds, the rear-drive Brougham requires 14.3 seconds, which is slower than many economy cars that get twice as much gas mileage. You don't have to be a performance fanatic to realize that's a sizeable difference. It's a struggle to get one of these up to

> **KEY: ohv** = overhead valve; **ohc** = overhead cam; **dohc** = double overhead cam; **I** = inline cylinders; **V** = cylinders in V configuration; **flat** = horizontally opposed cylinders; **bbl.** = barrel (carburetor); **PFI** = port (multi-point) fuel injection; **TBI** = throttle-body (single-point) fuel injection; **rpm** = revolutions per minute; **OD** = overdrive transmission; **S** = standard; **O** = optional; **NA** = not available.

Prices are accurate at time of printing; subject to manufacturer's change.

Cadillac

speed to safely merge into expressway traffic. Though Brougham still appeals to fans of traditional American luxury cars because it's big, posh, and can carry six people and their golf clubs, the tepid performance, dismal fuel economy and clumsy road manners leave us cold. This year's De Ville/Fleetwood sedans are now closer in exterior and interior dimensions to the Brougham, and far superior in acceleration, ride and handling. In addition, anti-lock brakes and a driver's-side air bag are available on the De Ville and Fleetwood, two important safety features not available on Brougham. Our advice: look to one of the newer front-drive models if you're drawn to the Cadillac crest.

Specifications

	4-door notchback
Wheelbase, in.	121.5
Overall length, in.	221.0
Overall width, in.	76.5
Overall height, in.	56.7
Front track, in.	61.7
Rear track, in.	60.7
Turn diameter, ft.	40.5
Curb weight, lbs.	4190
Cargo vol., cu. ft.	19.5
Fuel capacity, gal.	25.0
Seating capacity	6
Front headroom, in.	39.0
Front shoulder room, in.	59.4
Front legroom, max., in.	42.0
Rear headroom, in.	38.1
Rear shoulder room, in.	59.4
Rear legroom, min., in.	41.2

Powertrain layout: longitudinal front engine/rear-wheel drive.

Engines

	ohv V-8
Size, liters/cu. in.	5.0/307
Fuel delivery	4 bbl.
Horsepower @ rpm	140 @ 3200
Torque (lbs./ft.) @ rpm	255 @ 2000
Availability	S

EPA city/highway mpg

4-speed OD automatic	17/24

Prices

Cadillac Brougham	Retail Price	Dealer Invoice	Low Price
4-door notchback	$25699	$21921	$23510
Destination charge	550	550	550

Standard Equipment:

5.0-liter 4bbl. V-8 engine, 4-speed automatic transmission, power steering and brakes, Dual Comfort 55/45 front seats, illuminated entry system, front and rear center armrests, headlights-on warning, power windows and locks, automatic climate control, outboard rear lap/shoulder belts, front and rear lamp monitors, cornering lights, sunshade support and door courtesy lights, power decklid release, cruise control, carpeted litter receptacle, low fuel

KEY: **ohv** = overhead valve; **ohc** = overhead cam; **dohc** = double overhead cam; **I** = inline cylinders; **V** = cylinders in V configuration; **flat** = horizontally opposed cylinders; **bbl.** = barrel (carburetor); **PFI** = port (multi-point) fuel injection; **TBI** = throttle-body (single-point) fuel injection; **rpm** = revolutions per minute; **OD** = overdrive transmission; **S** = standard; **O** = optional; **NA** = not available.

and washer fluid indicators, dual remote mirrors, front seatback pockets, power driver's seat, tinted glass, trip odometer, padded sunvisors, AM/FM ST ET with power antenna, tilt/telescopic steering column, full padded vinyl roof, front bumper guards, door sill plates, full carpeting, opera lamps, automatic parking brake release, P225/75R15 Uniroyal Royal Seal all-season SBR tires.

Optional Equipment:

	Retail Price	Dealer Invoice	Low Price
Astroroof	1255	1054	1148
California emissions pkg.	100	84	92
Rear defogger	170	143	156
D'Elegance Pkg., w/cloth upholstery	2146	1803	1964
w/leather	2706	2273	2476
50/50 dual comfort front seats, power passenger seat, lighted visor mirrors, illuminated entry, rear reading lamps.			
Automatic power door locks	185	155	169
Engine block heater	45	38	41
Remote fuel door release	65	55	59
HD Ride Pkg.	120	101	110
Leather seating area	560	470	512
Automatic day/night mirror	80	67	73
Firemist paint	240	202	220
AM/FM ST ET cassette	309	260	283
Power passenger seatback recliner	160	134	146
w/d'Elegance	95	80	87
Premier Formal Vinyl Roof	1095	920	1002
Leather-trimmed steering wheel	115	97	105
Theft deterrent system	200	168	183
Trailer Towing Pkg.	299	251	274
HD springs and shock absorbers, upgraded transmission cooling, 3.23 axle ratio, trailer wiring harness.			
Locking wire wheel discs	400	336	366
Wire wheels	940	790	860
Option Pkg. B	385	323	352
Door edge guards, floormats, power passenger seat, trunk mat.			
Option Pkg. C	743	624	680
Pkg. B plus lighted visor mirrors, rear reading lamps, power decklid pulldown, Twilight Sentinel.			

Cadillac De Ville/ Fleetwood

Cadillac Sedan de Ville

What's New

De Ville and Fleetwood are restyled and longer to reflect more traditional Cadillac appearances, while the 4-door sedans also get a 3-inch-longer wheelbase. The 2-door coupes are 5.8 inches longer at 202.2 overall; wheelbase remains the same as before at 110.8. Sedans have been stretched 8.8 inches to 205.2, and they now ride a 113.8-inch wheelbase. The longer wheelbase for the sedans all goes to the rear seat, and Cadillac boasts that back-seat passengers now enjoy more than 43 inches of leg room. The sedans gain about 100 pounds in curb weight with the longer body. Front fenders are made of a "nylon composite alloy" material that weighs less than steel and won't rust. Trunk capacity has increased from 16.1 cubic feet to 18.1 on the coupe and 18.4 on the sedan. This year's lineup includes the Coupe and Sedan de Ville, plusher Fleetwood Coupe (back after a 2-year absence) and Sedan, and limited-production Fleetwood Sixty Special. The Sixty Special actually loses two inches from its wheelbase (formerly 115.8 inches) to share the one used for the De Ville and Fleetwood sedans, but overall length is increased 3.7 inches to match that of its siblings. All models have standard rear shoulder belts, while the sedans also get integral rear seat headrests and a storage compartment in the rear parcel shelf. The Sixty Special wears rear fender skirts (as do the Fleetwoods) and gains new Giugiaro-styled leather seats with electrical heating elements and new power controls for the split front seat. Anti-lock brakes, previously standard only on the Sixty Special, also are standard on the Fleetwood Coupe and Sedan this year. They remain optional on De Ville. All models are powered by a 155-horsepower 4.5-liter V-8 with single-point fuel injection, introduced last year. A driver's-side air bag, ElectriClear heated windshield, and a theft-deterrent system deactivated by a resistor imbedded in the ignition key are new options for all models.

For

- Performance
- Driveability
- Anti-lock brakes
- Air bag option
- Quietness

Against

- Fuel economy
- Price

Summary

Downsizing and converting these cars to front-wheel drive for 1985 was a move that didn't sell well among some of Cadillac's regular customers, who equate size and weight with value. Cadillac was encouraged by increased sales of the 1988 Eldorado after a less dramatic restyling that added three inches to the Eldo's overall length, so it's expecting even better results with the De Ville and Fleetwood. We've had no chance to drive the longer sedans, but jumping into a prototype showed that the rear seat has substantially more leg room, so all passengers can stretch out. Trunk space also has grown, addressing another shortcoming of the 1985-88 models. We did drive 1989 coupes and found them much like the '88s: Acceleration is fairly lively, and the engine and transmission respond promptly to the throttle. Gas mileage has never been a Cadillac strong point; these cars uphold that tradition. Expect 14-17 mpg around town,

and maybe low-to mid-20s on the open road. These cars uphold other Cadillac traditions as well, offering plenty of refinement and all the luxury that $25,000 or more can buy. More importantly, Cadillac is apparently listening to its owners, who complained loudly in recent years that they weren't getting their money's worth. The result is that Cadillac is building better cars, as shown by its fourth-place finish in the latest J.D. Power customer satisfaction index. Cadillac for the most part still builds softly sprung luxury cruisers rather than European-style road cars, but that's not all bad. Along with traditional Cadillac luxury and loads of convenience features, you can also get modern safety features such as anti-lock brakes and a driver's-side air bag.

Specifications

	2-door notchback	4-door notchback
Wheelbase, in.	110.8	113.8
Overall length, in.	202.3	205.3
Overall width, in.	72.5	72.5
Overall height, in.	55.0	55.0
Front track, in.	60.3	60.3
Rear track, in.	59.8	59.8
Turn diameter, ft.	40.0	42.6
Curb weight, lbs.	3397	3470
Cargo vol., cu. ft.	18.1	18.4
Fuel capacity, gal.	18.0	18.0
Seating capacity	6	6
Front headroom, in.	39.3	39.3
Front shoulder room, in.	59.0	59.0
Front legroom, max., in.	42.4	42.4
Rear headroom, in.	38.0	38.1
Rear shoulder room, in.	57.6	59.3
Rear legroom, min., in.	40.3	43.3

Powertrain layout: transverse front engine/front-wheel drive.

Engines

	ohv V-8
Size, liters/cu. in.	4.5/273
Fuel delivery	TBI
Horsepower @ rpm	155 @ 4000
Torque (lbs./ft.) @ rpm	240 @ 2800
Availability	S

EPA city/highway mpg

4-speed OD automatic	17/25

Prices

Cadillac De Ville/Fleetwood	Retail Price	Dealer Invoice	Low Price
Coupe De Ville 2-door notchback	$24960	$21291	$22626
Sedan De Ville 4-door notchback	25435	21696	23066
Fleetwood 2-door notchback	29825	25441	27133
Fleetwood 4-door notchback	30300	25846	27573
Fleetwood Sixty Special 4-door notchback	34230	29198	31214
Destination charge	550	550	550

Standard Equipment:

4.5-liter TBI V-8 engine, 4-speed automatic transmission, power steering and brakes, automatic level control, air adjustable rear struts, power driver's seat, manual recliners, power windows and locks, tilt/telescopic steering column, cruise control, Dual Comfort seat with power driver's side, intermittent wipers, power trunk release, automatic climate control, AM/FM ST ET, power mirrors, sail panel courtesy/reading lights, rear head restraints, Miche-

Cadillac Coupe de Ville

Prices are accurate at time of printing; subject to manufacturer's change.

Cadillac

lin 205/70R15 all-season SBR tires. **Fleetwood** adds: anti-lock braking system, padded vinyl roof (2-door), formal cabriolet roof (4-door), power seats, storage armrest, power decklid release, power decklid pulldown, illuminated entry, Twilight Sentinel, lighted visor mirrors, trunk mat, power mirrors, rear reading lamps, overhead assist handles, AM/FM ST ET cassette w/EQ, digital instrument cluster, locking wire wheel discs. **Sixty Special** adds: leather interior, heated front seats with power lumbar, thigh and lateral supports, automatic day/night mirror, rear passenger footrests, rear overhead console, vinyl top.

Cadillac Eldorado

Optional Equipment:	Retail Price	Dealer Invoice	Low Price
Anti-lock braking system, De Ville	925	777	846
Supplemental Inflatable Restraint	NA	NA	NA
Front storage armrest, De Ville	70	59	64
Astroroof (NA Sixty Special)	1255	1054	1148
California emissions pkg	100	84	92
Rear defogger	170	143	156
Heated windshield	250	210	229
Automatic power door locks	185	155	169
Engine block heater	45	38	41
Rear floormats, De Ville	25	21	23
Digital instruments, De Ville	238	200	218
Leather seating area (std. Sixty Special) . .	525	441	480
Memory power driver's seat (NA De Ville) .	235	197	215
Automatic day/night mirror	80	67	73
Firemist paint	240	202	220
Delco-Bose sound system w/cassette	576	484	527
w/CD player	872	732	798
Power recliners, Fleetwood (each)	95	80	87
Formal cabriolet roof, De Ville 2-door . . .	713	599	652
Padded vinyl roof, De Ville 4-door	713	599	652
Full cabriolet roof, De Ville 2-door	995	836	910
Leather-trimmed steering wheel, De Ville . .	115	97	105
Theft deterrent system	200	168	183
Trunk mat, De Ville	36	30	33
Wire wheel discs, De Ville	320	269	293
Alloy wheels, De Ville	435	365	398
Fleetwood	115	97	105
Accent striping, De Ville	65	55	59
De Ville Option Pkg. B	324	272	296
Door edge guards, floormats, power passenger seat.			
De Ville Option Pkg. C	739	621	676
Pkg. B plus illuminated entry, lighted visor mirrors, power decklid pulldown, Twilight Sentinel.			
De Ville Option Pkg. D	894	751	818
Pkg. C plus remote fuel door release, manual driver's seatback recliner, trumpet horn.			

Cadillac Eldorado/Seville

What's New

The sporty Seville Touring Sedan (STS), introduced as a limited-production model at the end of 1988, becomes a regular production option for 1989, though only 3000 or so will be built (twice as many as last year). Among the STS's features are monochrome exterior treatment in one of four colors, shorter final-drive ratio for quicker acceleration, Touring Suspension, high-performance tires on alloy wheels, leather seats, burl elm interior trim and 4-passenger seating. Standard equipment on the base Seville is increased with the addition of 6-way power seats for the driver and front passenger, express-down feature for the

> **KEY: ohv** = overhead valve; **ohc** = overhead cam; **dohc** = double overhead cam; **I** = inline cylinders; **V** = cylinders in V configuration; **flat** = horizontally opposed cylinders; **bbl.** = barrel (carburetor); **PFI** = port (multi-point) fuel injection; **TBI** = throttle-body (single-point) fuel injection; **rpm** = revolutions per minute; **OD** = overdrive transmission; **S** = standard; **O** = optional; **NA** = not available.

power driver's window, maple wood trim for the interior, stereo with cassette player and graphic equalizer, an engine oil life indicator, and a new theft-deterrent system operated by a resistor pellet in the ignition key. If the right key isn't used, the engine computer shuts down the starter and fuel system so the car can't be started. A different theft-deterrent system, which flashes the lights and blows the horn if a car is broken into, remains a $200 option. Eldorado, the 2-door cousin to the 4-door Seville, enjoyed a healthy sales increase during 1988 following an exterior remake that added three inches to its overall length. For '89, Eldo gets the same new standard features as Seville, except the maple interior trim is standard only with the extra-cost Biarritz package and optional on the base model. Eldorado and Seville use Cadillac's 155-horsepower 4.5-liter V-8, introduced last year and shared with the De Ville/Fleetwood line.

For

- ● Performance ● Driveability
- ● Anti-lock brakes ● Luxury appointments

Against

- ● Fuel economy ● Trunk space ● Price

Summary

With the arrival of the 4.5-liter V-8 last year, Seville and Eldorado gained horsepower, improved driveability and respect. The additional power gives these luxury cars ample get-up-and-go, plus this performance is delivered promptly and smoothly. Previous models suffered from an engine with lackadaisical response and a transmission with an annoying tendency to hunt between gears, as if it didn't know where it was supposed to go. The '88s were much improved on both counts, making these cars more enjoyable. An '88 Eldorado with the optional Touring Suspension package is part of our long-term test fleet, and it has performed capably in our hands. While the performance has been impressive, the gas mileage hasn't. We only crack 20 mpg in straight highway driving; otherwise, it's usually under 15 mpg around town. The stiff Touring Suspension needs more absorbency on rough roads and better bounce control at speed, but it gives Eldorado strong cornering ability, helped by the sticky performance tires that come with this package. Try an Eldorado or Seville with and without this suspension before you buy to see which you prefer. Both models remain a little short of interior and trunk space, and they get pricey once you add a few options. The one option we highly recommend is the anti-lock brake system, which provide an extra measure of safety in emergencies. When these cars first came out we were lukewarm about them, as were most buyers. We're fonder of them now, and urge you to take a look if you're shopping in the $30,000 range.

Prices are accurate at time of printing; subject to manufacturer's change.

Specifications

	2-door notchback	4-door notchback
Wheelbase, in.	108.0	108.0
Overall length, in.	190.8	191.4
Overall width, in.	70.9	71.7
Overall height, in.	53.7	53.7
Front track, in.	59.9	59.9
Rear track, in.	59.9	59.9
Turn diameter, ft.	40.3	40.3
Curb weight, lbs.	3422	3422
Cargo vol., cu. ft.	14.1	14.1
Fuel capacity, gal.	18.8	18.8
Seating capacity	5	5
Front headroom, in.	37.8	37.8
Front shoulder room, in.	57.2	57.6
Front legroom, max., in.	42.5	42.5
Rear headroom, in.	37.9	37.8
Rear shoulder room, in.	57.2	57.5
Rear legroom, min., in.	36.1	36.1

Powertrain layout: transverse front engine/front-wheel drive.

Engines

	ohv V-8
Size, liters/cu. in.	4.5/273
Fuel delivery	TBI
Horsepower @ rpm	155 @ 4000
Torque (lbs./ft.) @ rpm	240 @ 2800
Availability	S

EPA city/highway mpg

4-speed OD automatic	17/25

Prices

Cadillac Eldorado/Seville	Retail Price	Dealer Invoice	Low Price
Eldorado 2-door notchback	$26738	$22808	$24473
Seville 4-door notchback	29750	25377	27264
Destination charge	550	550	550

Standard Equipment:

Eldorado: 4.5-liter TBI V-8 engine, power steering, power 4-wheel disc brakes, automatic climate control, power windows and locks, retained accessory power system, cruise control, Twilight Sentinel, Pass Key theft deterrent system, power mirrors, reclining cloth and leather bucket seats w/power adjustments, outboard rear lap/shoulder belts, digital instrument cluster including tachometer, coolant temperature gauge and voltmeter, low fuel warning light, trip computer, outside thermometer, fold-down center armrest with storage bins, rotating cup holder and coin retainer, leather-wrapped steering wheel with tilt and telescope feature, lamp monitors, AM/FM ST ET cassette w/EQ, dual spot/map lights with retainer for garage door opener, reversible floormats, 215/65R15 Goodyear Eagle GT + 4 tires on alloy wheels. **Seville** adds: Birdseye maple trim, two-tone paint.

Optional Equipment:

Anti-lock braking system	925	777	846
Birdseye maple appliques, Eldorado	245	206	224
Astroroof	1255	1054	1148
Biarritz Pkg. w/leather upholstery, Eldorado	3180	2671	2910
w/cloth	2770	2327	2535

Power front recliners and lumbar support adjusters, power passenger seat, seatback pockets, walnut on instrument panel, console and door panels, cabriolet roof with opera lamps, two-tone paint, closed-in backlight treatment, wire wheel discs, accent molding, reversible floormats.

STS Pkg., Seville	5754	4833	5265

Anti-lock braking system, 3.31 axle ratio, leather interior with elm burl accents, rear console, automatic power locks, illuminated entry, lighted visor mirrors, theft deterrent system, 215/65R15 Goodyear Eagle GT + 4 tires.

Cadillac Seville

	Retail Price	Dealer Invoice	Low Price
California emissions pkg.	100	84	92
Cellular mobile telephone	1975	1659	1807
Provision for cellular phone	395	332	361
Rear defogger	170	143	156
Automatic power door locks	185	155	169
Engine block heater	45	38	41
Leather seating area	410	344	375
Automatic day/night mirror	80	67	73
Firemist paint	240	202	220
Firemist paint, primary	190	160	174
Secondary	50	42	46
White diamond paint	240	202	220
Two-tone paint, Seville	225	189	206
Delco-Bose music system w/cassette	576	484	527
w/CD player	872	732	798
Power passenger seatback recliner, Eldorado	95	80	87
Padded vinyl roof, Eldorado	995	836	910
Full cabriolet roof, Eldorado	995	836	910
Phaeton roof, Eldorado	1095	920	1002
Theft deterrent system	200	168	183
215/65R15 tires	76	64	70
Touring Suspension	155	130	142
Locking wire wheel discs	190	160	174

Chevrolet Astro/ GMC Safari

What's New

Anti-lock rear brakes are standard on all passenger versions of Astro and Safari, GM's compact rear-drive vans. Braking pressure to the rear wheels is modulated by the anti-lock system to keep the rear wheels from locking before the vehicle stops to prevent skids. A new electric speedometer has been added to work with the anti-lock system. The standard gauge cluster now includes fuel, volt, oil pressure and water temperature gauges, and a trip odometer. Astro, sold by Chevrolet dealers, and Safari, sold by GM dealers with GMC truck franchises, are identical except for names and series designations. Power steering, a front stabilizer bar and 27-gallon fuel tank, all of which were optional last year, are also standard this year on passenger models, as are shoulder belts for all outboard seating positions except the one next to the sliding side door. Five-passenger seating is standard; the optional 4-seat package has been dropped for '89, leaving 7- and 8-passenger arrangements available. A sport suspension package that was promised last year but not delivered is promised again for mid-1989; it includes gas-pressurized shock absorbers, rear stabilizer bar and 245/60HR15 tires. All passenger Astros and Safaris are powered by a 150-horsepower 4.3-liter V-6, with a choice of 5-speed manual or extra-cost

Prices are accurate at time of printing; subject to manufacturer's change.

Chevrolet

Chevrolet Astro

4-speed automatic transmissions. Base payload is 1000 pounds and the Astro/Safari can be equipped to tow up to 6000 pounds.

For

- Anti-lock rear brakes
- Passenger and cargo room
- Towing capacity

Against

- Fuel economy
- Entry/exit
- Ride

Summary

GM's rear-drive minivans are similar to the Ford Aerostar in design and concept: They're modern, smaller interpretations of the traditional van, whereas the front-drive Dodge Caravan and Plymouth Voyager are more like modern station wagons. Astro and Safari offer bountiful passenger and cargo room, hefty payloads and commendable trailer-towing ability, so they're popular for commercial use. They drive much like traditional vans, which means a truck-like ride, poor traction in rain and snow, and clumsy handling. The tall design makes them prone to being blown around on the open road, and harder to get in or out of. The standard V-6 engine has plenty of torque for hauling heavy loads and towing, but that muscle doesn't translate into brisk acceleration. While Astro and Safari aren't slow, they aren't sprinters either. Gas mileage is disappointing: we usually average around 15 mpg in urban driving and 20 or so mpg on the highway. Astro and Safari are good choices for hauling and towing, but the Caravan and Voyager remain our favorite compact passenger vans. GM plans to introduce a front-drive minivan for the 1990 model year, tailoring it for passenger use.

Specifications

	5-door van
Wheelbase, in.	111.0
Overall length, in.	176.8
Overall width, in.	77.0
Overall height, in.	71.7
Front track, in.	NA
Rear track, in.	NA
Turn diameter, ft.	40.2
Curb weight, lbs.	3084
Cargo vol., cu. ft.	151.8
Fuel capacity, gal.	27.0
Seating capacity	8

	5-door van
Front headroom, in.	NA
Front shoulder room, in.	NA
Front legroom, max., in.	NA
Rear headroom, in.	NA
Rear shoulder room, in.	NA
Rear legroom, min., in.	NA

Powertrain layout: longitudinal front engine/rear-wheel drive.

Engines

	ohv V-6
Size, liters/cu. in.	4.3/262
Fuel delivery	TBI
Horsepower @ rpm	155 @ 4000
Torque (lbs./ft.) @ rpm	230 @ 2400
Availability	S

EPA city/highway mpg

5-speed OD manual	17/24
4-speed OD automatic	17/22

Prices

Chevrolet Astro Passenger Van	Retail Price	Dealer Invoice	Low Price
CS 4-door van	$11900	$10627	$11263
CL 4-door van	12633	11281	11957
LT 4-door van	14144	12631	13388
Destination charge	465	465	465

Standard Equipment:

CS: 4.3-liter TBI V-6 engine, 5-speed manual transmission, anti-lock rear brakes, full instrumentation (coolant temperature and oil pressure gauges, voltmeter, trip odometer), swing-out side windows, high-back bucket seats, 5-passenger seating, rubber floor covering, inside fuel door release, lighted visor mirrors, under-floor spare tire carrier, 195/75R15 tires. **CL** adds: trip odometer, voltmeter, oil pressure and coolant temperature gauges, custom steering wheel, wheel trim rings, auxiliary lighting, carpet. **LT** adds: air dam with fog lamps, luxury velour seat and door panel trim, upgraded front bucket seats with recliners, integrated armrests and adjustable headrests, split-back center bench seat with integrated armrests, headrests, recliners and fold-down center console with convenience tray and cup pockets; right-hand seat folds forward for access to rear.

Optional Equipment:

4-speed automatic transmission	610	519	561
Air conditioning, front	736	626	677
Front & rear	1320	1122	1214
Optional axle ratio	38	32	35
Locking differential	252	214	232
Operating Convenience Pkg	411	349	378
Power windows and locks.			
Power door locks	211	179	194
7-passenger seating, CS	1069	909	983
CL	981	834	903
LT	878	746	808
8-passenger seating, CS	344	292	316
CL	396	337	364
LT	878	746	808
Custom vinyl bucket seats, w/8-pass	158	134	145
w/5-pass	106	90	98
Custom cloth bucket seats, w/8-pass	158	134	145
w/5-pass	106	90	98
Special two-tone paint, CS & CL	251	213	231
LT	172	146	158
Sport two-tone paint, CS & CL	172	146	158
California emissions pkg	100	85	92
Air dam with fog lamps, CS & CL	115	98	106
Deluxe chromed bumpers, CS	128	109	118
CL, LT	76	65	70

Prices are accurate at time of printing; subject to manufacturer's change.

	Retail Price	Dealer Invoice	Low Price
Color-keyed bumpers, CS	52	44	48
Luggage carrier	126	107	116
Engine oil cooler	126	107	116
HD radiator	56	48	52
HD radiator & trans oil cooler, w/o A/C	118	100	109
w/A/C	63	54	58
Cold Climate Pkg	46	39	42
Roof console	83	71	76
Convenience Group	326	277	300
Cruise control, tilt steering column.			
Wheelhouse & floor carpet, CS	117	99	108
Floormats, w/5-pass	47	40	43
w/7- or 8-pass	.69	59	63
Complete body glass	128	109	118
Tinted glass, w/o complete body glass . . .	75	64	69
w/complete body glass	104	88	96
Tinted windshield	40	34	37
Deep tinted glass, w/o complete body glass .	236	201	217
w/complete body glass	365	310	336
w/light tinted rear window, w/o body glass	185	157	170
w/body glass	308	262	283
Swing-out rear door glass, CS & CL . . .	59	50	54
Rear heater	267	227	246
Electronic instruments	88	75	81
Dome & reading lamps	33	28	30
Cigaret lighter	32	27	29
Auxiliary lighting	128	109	118
Dual outside mirrors	52	44	48
Power mirrors	150	128	138
Lighted right visor mirror	43	37	40
Black bodyside moldings, CS	59	50	54
AM radio	122	104	112
AM/FM ST ET	293	249	270
w/seek/scan & clock	347	295	319
AM/FM ST ET cassette	474	403	436
w/EQ	624	530	574
Front recliners & armrests	241	205	222
Power driver's seat	240	204	221
HD shock absorbers	36	31	33
Cruise control	205	174	189
Custom steering wheel, CS	28	24	26
Sport steering wheel, CS	35	30	32
CL	7	6	6
Sport suspension, CS	509	433	468
CL	467	397	430
LT	417	354	384
Puncture sealant tires	184		
HD Trailering Special Equipment, w/o A/C	555	472	511
w/A/C	498	423	458
LD Trailering Special Equipment	109	93	100
Bright wheel covers, CS	42	36	39
Rally wheels, CS	92	78	85
CL	50	43	46
Aluminum wheels, CS or w/o sport suspension	316	269	291
CL	274	233	252
CS w/sport suspension or LT	224	190	206
Intermittent wipers	59	50	54

Chevrolet Camaro

What's New

All Camaros gain a standard theft deterrent system similar to the one used in the Corvette, while a new RS model replaces the Sport Coupe as a lower-cost running mate to the high-performance IROC-Z. Both the RS and IROC-Z are available as a 3-door hatchback or 2-door convertible. Standard on all versions is GM's "pass-key" theft deterrent system, which debuted on the 1986 Corvette. A resistor

Chevrolet Camaro IROC-Z

pellet in the ignition key must match special coding in the ignition lock, or the starter is temporarily disabled. Chevrolet says the system has helped lower theft rates and reduce insurance premiums on 1986-88 Corvettes. According to the most recent federal statistics, more Camaros are stolen than any other car. The RS coupe imitates the IROC's exterior styling, but has a 135-horsepower 2.8-liter V-6 standard. RS debuted last spring as a California model equipped to have lower insurance premiums for young drivers compared to the IROC. A 170-horsepower 5.0-liter V-8 with single-point fuel injection is standard on the IROCs and the RS convertible, optional on the RS coupe. Optional V-8s for the IROC are a 220-horsepower 5.0-liter with multi-point injection and a 230-horsepower 5.7-liter, also with multi-point injection. By December, dual catalytic converters feeding into a single muffler are to become available with the IROC's optional engines at extra cost. The system is supposed to reduce back pressure and increase horsepower by 10 on both engines. The Pontiac Firebird shares Camaro's rear-drive design and offers a similar engine lineup, but doesn't come as a convertible.

For

● Performance (V-8s)　● Handling　● Resale value

Against

● Fuel economy　● Ride　● Interior room

Summary

Camaro used to be the most popular "pony car" but that title is now claimed by the Ford Mustang GT, which offers comparable performance at lower cost. Camaro still appeals to performance fans and younger drivers (who help keep resale values high by eagerly buying used Camaros) because of its extroverted styling and V-8 muscle. Before you buy, check with your insurance agent. Despite the new standard theft-deterrent system, we suspect insurance rates will still be quite high for most buyers. Camaro offers traditional muscle-car performance: V-8 engine, lousy gas mileage, blaring exhaust, a stiff ride and poor wet-weather traction. When we say "stiff" we mean jarring, punishing.

KEY: ohv = overhead valve; **ohc** = overhead cam; **dohc** = double overhead cam; **I** = inline cylinders; **V** = cylinders in V configuration; **flat** = horizontally opposed cylinders; **bbl.** = barrel (carburetor); **PFI** = port (multi-point) fuel injection; **TBI** = throttle-body (single-point) fuel injection; **rpm** = revolutions per minute; **OD** = overdrive transmission; **S** = standard; **O** = optional; **NA** = not available.

Prices are accurate at time of printing; subject to manufacturer's change.

And when we say "poor" traction, we mean sideways in snow. In addition, passenger and cargo room are skimpy for Camaro's overall length and width. We find newer, tamer coupes, such as the Chevy Beretta and Acura Integra, far easier to live with as everyday vehicles. If this macho-style performance is your meat, then we suggest you compare it to the Mustang GT.

Specifications

	3-door hatchback	2-door conv.
Wheelbase, in.	101.1	101.1
Overall length, in.	192.0	192.0
Overall width, in.	72.8	72.8
Overall height, in.	50.0	50.3
Front track, in.	60.7	60.7
Rear track, in.	61.6	61.6
Turn diameter, ft.	36.9	36.9
Curb weight, lbs.	3082	3116
Cargo vol., cu. ft.	31.0	5.2
Fuel capacity, gal.	15.5	15.5
Seating capacity	4	4
Front headroom, in.	37.0	37.1
Front shoulder room, in.	57.5	58.6
Front legroom, max., in.	43.0	42.9
Rear headroom, in.	35.6	36.1
Rear shoulder room, in.	56.3	48.1
Rear legroom, min., in.	29.8	28.3

Powertrain layout: longitudinal front engine/rear-wheel drive.

Engines	ohv V-6	ohv V-8	ohv V-8	ohv V-8
Size, liters/cu. in.	2.8/173	5.0/305	5.0/305	5.7/350
Fuel delivery	PFI	TBI	PFI	PFI
Horsepower @ rpm	135 @ 4900	170 @ 4000	220 @ 4400	230 @ 4400
Torque (lbs./ft.) @ rpm	160 @ 3900	255 @ 2400	290 @ 3200	330 @ 3200
Availability	S[1]	S[2]	O[3]	O[3]

EPA city/highway mpg				
5-speed OD manual	18/27	17/25	17/26	
4-speed OD automatic	19/27	17/26	16/25	17/25

1. RS coupe   2. RS conv., IROC-Z; optional RS coupe   3. IROC-Z

Prices

Chevrolet Camaro	Retail Price	Dealer Invoice	Low Price
RS 3-door hatchback	$11495	$10265	$10780
RS 2-door convertible	16995	15177	15986
IROC-Z 3-door hatchback	14145	12631	13288
IROC-Z 2-door convertible	18945	16918	17832
Destination charge	439	439	439

Standard Equipment:

RS: 2.8-liter PFI V-6 engine (hatchback; convertible has 5.0-liter TBI V-8), 5-speed manual transmission, power steering and brakes, AM/FM ST ET, left remote and right manual mirrors, reclining front bucket seats, cloth upholstery, outboard rear lap/shoulder belts, headlamps-on tone, folding rear seat, automatic hatch pulldown, tachometer, coolant temperature and oil pressure gauges, voltmeter, 215/65R15 all-season SBR tires on alloy wheels. **IROC-Z** adds: 5.0-liter TPI V-8 engine, right visor mirror, fog lamps, performance tires.

Optional Equipment:

	Retail	Dealer	Low
5.0-liter PFI (TPI) V-8	745	633	685
5.0-liter TBI V-8, RS 3-door	400	340	368
5.7-liter PFI V-8	1045	888	961
4-speed automatic transmission	490	417	451
Custom cloth bucket seats	277	235	255

	Retail Price	Dealer Invoice	Low Price
Custom leather bucket seats	750	638	690
Air conditioning	775	659	713
Limited-slip differential	100	85	92
4-wheel disc brakes	179	152	165
Locking rear storage cover	80	68	74
Rear defogger	145	123	133
Power door locks	145	123	133
Power hatch release	50	43	46
Engine block heater	20	17	18
Rear window louvers	210	179	193
Deluxe luggage compartment trim, RS 3-door	164	139	151
IROC-Z 3-door	84	71	77
Power mirrors	91	77	84
Door edge guards	15	13	14
Removable glass roof panels	866	736	797
Split folding rear seatbacks	50	43	46
245/50VR16 tires on alloy wheels	468	398	431
AM/FM ST ET cassette, RS 3-door	122	104	112
w/EQ	272	231	250
w/Delco/Bose	885	752	814

Sound system prices are for base models; prices vary with option package content.

	Retail	Dealer	Low
RS Coupe Preferred Equipment Group 1	255	217	235

HD battery, tinted glass, auxiliary lighting, bodyside moldings.

| RS Coupe Preferred Equipment Group 2 | 1687 | 1434 | 1552 |

Group 1 plus air conditioning, power door locks, floormats, AM/FM ST ET cassette, cruise control, tilt steering column, intermittent wipers.

| RS Coupe Preferred Equipment Group 3 | 2014 | 1712 | 1853 |

Group 2 plus cargo cover, halogen headlamps, reading lamps, power windows.

| IROC-Z Preferred Equipment Group 1, w/o 5.7 | 255 | 217 | 235 |
| w/5.7 | 229 | 195 | 211 |

Tinted glass, auxiliary lighting, bodyside moldings.

| IROC-Z Preferred Equipment Group 2, w/o 5.7 | 1737 | 1476 | 1598 |
| w/5.7 | 1711 | 1454 | 1574 |

Group 1 plus air conditioning, power door locks, floormats, power hatch release, AM/FM ST ET cassette, cruise control, tilt steering column, intermittent wipers.

| IROC-Z Preferred Equipment Group 3, w/o 5.7 | 2545 | 2163 | 2341 |
| w/5.7 | 2519 | 2141 | 2317 |

Group 2 plus cargo cover, halogen headlamps, reading lamps, power mirrors, power driver's seat, power windows.

| RS Conv. Preferred Equipment Group 1 | 206 | 175 | 190 |

HD battery, tinted glass, bodyside moldings.

| RS Conv. Preferred Equipment Group 2 | 1638 | 1392 | 1507 |

Group 1 plus air conditioning, power door locks, floormats, AM/FM ST ET cassette, cruise control, tilt steering column, intermittent wipers.

| RS Conv. Preferred Equipment Group 3 | 1873 | 1592 | 1723 |

Group 2 plus power windows, halogen headlamps.

| IROC-Z Conv. Preferred Equipment Group 1 | 206 | 175 | 190 |

HD battery, tinted glass, bodyside moldings.

| IROC-Z Conv. Preferred Equipment Group 2 | 1638 | 1392 | 1507 |

Group 1 plus air conditioning, power door locks, floormats, AM/FM ST ET cassette, cruise control, tilt steering column, intermittent wipers.

| IROC-Z Conv. Preferred Equipment Group 3 | 2354 | 2001 | 2166 |

Halogen headlamps, power mirrors, AM/FM ST ET cassette w/EQ, power driver's seat, power windows.

Chevrolet Caprice

What's New

All Caprices come with air conditioning and a V-8 engine this year, and the V-8 in the 4-door sedan has fuel injection instead of a carburetor. A 4.3-liter V-6 that had been standard in the sedan has been dropped. Instead, a 5.0-liter (305-cubic-inch) V-8 is now standard and it comes with single-point (or throttle-body) fuel injection instead of a 4-barrel carburetor. Two injectors are mounted in a single throttle-body housing; horsepower is still listed as 170,

Prices are accurate at time of printing; subject to manufacturer's change.

Chevrolet Caprice Classic 4-door

while torque is 255 pounds/feet, up by five. The Caprice station wagon still comes with a carbureted 5.0-liter (307-cubic-inch) V-8 rated at 140 horsepower. Air conditioning, optional on 4-doors last year, is standard, accounting for part of the nearly $2000 increase in this year's base prices. Other changes for 1989 are that 3-point rear shoulder belts are standard and an elasticized cargo restraint net is a new option for sedans. Last year's lineup for the full-size, rear-drive Caprice returns: The sedan comes in base, Classic, Brougham and LS trim levels, while the 3-seat/8-passenger wagon comes only in Classic trim. Buick, Oldsmobile and Pontiac offer similar wagons, but Caprice is the only rear-drive family sedan in the GM stable. The design for these cars debuted for 1977 in GM's first wave of downsizing. Caprice is supposed to be redesigned for 1990, but retain rear-wheel drive.

For

- ●Performance (4-doors)   ●Trailer towing
- ●Passenger and cargo room   ●Quietness

Against

- ●Fuel economy   ●Size and weight
- ●Handling/maneuverability

Summary

As GM's last full-size, rear-drive family sedan, Caprice serves buyers who want a big car with V-8 power at reasonable cost, but an increasing portion of its sales are to police departments and taxi companies. The only direct rivals are the Ford LTD Crown Victoria and nearly identical Mercury Grand Marquis, which we find a little more pleasant to drive, thanks to their refined V-8 engine. Caprice runs a close second. The sedan now uses a fuel-injected V-8 that's nearly as smooth and responsive as Ford's, and slightly more powerful. The Caprice wagon's carbureted V-8 isn't as refined or potent, but is still sufficient for a 4100-pound vehicle. With either engine, gas mileage won't be a strong point: 15-17 mpg around town and 21-23 mpg on the highway. Interior room and luggage space remain major reasons for buying this car, plus insurance statistics show these cars historically have the best ratings for collision damage and occupant protection in the industry, even without features such as anti-lock brakes or air bags. There's more good news: a base or Classic sedan can be comfortably furnished for $15,500-17,000, using suggested retail prices. That's cheaper than many mid-size cars and even some Japanese compacts. It shouldn't be hard to get most dealers to discount their prices, giving you even better value in a full-size car.

Specifications

	4-door notchback	5-door wagon
Wheelbase, in.	116.0	116.0
Overall length, in.	212.2	215.7
Overall width, in.	75.4	79.3
Overall height, in.	56.4	58.2
Front track, in.	61.7	60.0
Rear track, in.	60.7	64.1
Turn diameter, ft.	38.7	39.7
Curb weight, lbs.	3693	3770
Cargo vol., cu. ft.	20.9	87.9
Fuel capacity, gal.	24.5	22.0
Seating capacity	6	8
Front headroom, in.	39.5	39.6
Front shoulder room, in.	60.5	60.5
Front legroom, max., in.	42.2	42.2
Rear headroom, in.	38.2	39.3
Rear shoulder room, in.	60.9	60.9
Rear legroom, min., in.	39.1	37.2

Powertrain layout: longitudinal front engine/rear-wheel drive.

Engines

	ohv V-8	ohv V-8
Size, liters/cu. in.	5.0/305	5.0/307
Fuel delivery	TBI	4 bbl.
Horsepower @ rpm	170 @ 4000	140 @ 3200
Torque (lbs./ft.) @ rpm	255 @ 2400	255 @ 2000
Availability	S[1]	S[2]

EPA city/highway mpg

4-speed OD automatic	17/24	17/24

1. Sedans   2. Wagon

Prices

Chevrolet Caprice	Retail Price	Dealer Invoice	Low Price
4-door notchback	$13865	$11966	$12916
Classic 4-door notchback	14445	12466	13456
Classic Brougham 4-door notchback	15615	13476	14546
Classic Brougham LS 4-door notchback	16835	14529	15682
Classic 5-door wagon	15025	12967	13996
Destination charge	505	505	505

Standard Equipment:

5.0-liter (305-cid) TBI V-8 engine (4-doors; 307-cid 4bbl. on wagon), 4-speed automatic transmission, power steering and brakes, air conditioning, left remote and right manual mirrors, tinted glass, outboard rear lap/shoulder belts, AM/FM ST ET, knit cloth bench seat, 205/75R15 all-season SBR tires (4-doors; wagon has 225/75R15). **Classic** adds: wheel opening moldings, Quiet Sound Group, vinyl door pull straps, bright wide lower bodyside moldings, headlamps-on tone, hood ornament, carpeted lower door panels. **Brougham** adds: upgraded carpet, front door courtesy lights, vinyl roof, 55/45 cloth front seat with center armrest. **LS** adds: Landau-style vinyl roof, sport mirrors, tinted glass.

Optional Equipment:

Limited-slip differential	100	85	92
Performance axle ratio	21	18	19

> **KEY: ohv** = overhead valve; **ohc** = overhead cam; **dohc** = double overhead cam; **I** = inline cylinders; **V** = cylinders in V configuration; **flat** = horizontally opposed cylinders; **bbl.** = barrel (carburetor); **PFI** = port (multi-point) fuel injection; **TBI** = throttle-body (single-point) fuel injection; **rpm** = revolutions per minute; **OD** = overdrive transmission; **S** = standard; **O** = optional; **NA** = not available.

Prices are accurate at time of printing; subject to manufacturer's change.

	Retail Price	Dealer Invoice	Low Price
Vinyl bench seat, base	28	24	26
Wagon (credit)	(172)	(146)	(146)
50/50 vinyl seat, base	305	259	281
Wagon	103	88	95
50/50 cloth seat	275	234	253
45/55 leather seat	550	468	506
Custom two-tone paint	141	120	130
Vinyl roof	200	170	184
Air deflector, wagon	65	55	60
Trunk cargo net	30	26	28
HD cooling	40	34	37
Rear defogger	145	123	133
Power door locks, 4-doors	195	166	179
Wagon	255	217	235
Estate equipment	307	261	282
Engine block heater	20	17	18
Pinstriping	61	52	56
HD suspension, F40 (std. wagon)	26	22	24
Sport suspension, F41	49	42	45
Inflatable rear shock absorbers	64	54	59
WSW tires	76	65	70
225/70R15 WSW tires	188	160	173
Custom wheel covers, base	65	55	60
Locking wire wheel covers, Brougham . . .	134	114	123
Others	199	169	183
AM/FM ST ET cassette	122	104	112
w/EQ	232	197	213

Sound system prices are for base model; prices vary with option package content.

Preferred Group 1, base	196	167	180

HD battery, floormats, bodyside and wheel opening moldings, extended-range speakers.

Preferred Group 2, base	796	677	732

Group 1 plus power door locks, auxiliary lighting, cruise control, tilt steering column, intermittent wipers.

Classic Sedan Preferred Group 1	709	603	652

HD battery, floormats, auxiliary lighting, remote mirrors, bodyside moldings, extended-range speakers, cruise control, tilt steering column, WSW tires, power decklid release, intermittent wipers.

Classic Sedan Preferred Group 2,

w/50/50 seat	1970	1675	1812
w/bench seat	1490	1267	1371

Group 1 plus power door locks, deluxe luggage compartment trim, lighted right visor mirror, AM/FM ST ET cassette, power windows, power seats (w/50/50 seat).

Wagon Preferred Group 1	237	201	218

HD battery, load floor carpet, auxiliary lighting, remote mirrors, bodyside moldings.

Wagon Preferred Group 2	1484	1261	1365

Group 1 plus roof carrier, floormats, AM/FM ST ET cassette, cruise control, tilt steering column, power windows, intermittent wipers.

Wagon Preferred Group 3, w/50/50 seat . .	2343	1992	2156
w/bench seat	1863	1584	1714

Group 2 plus gauge package with trip odometer, Twilight Sentinel, cornering lamps, lighted right visor mirror, deluxe rear compartment decor.

Brougham Preferred Group 1	496	422	456

HD battery, floormats, gauge package with trip odometer, Twilight Sentinel, lighted right visor mirror, remote mirrors, bodyside moldings, extended-range speakers, WSW tires, power decklid release.

Brougham Preferred Group 2	2251	1913	2071

Group 1 plus power door locks, cornering lamps, deluxe luggage compartment trim, AM/FM ST ET cassette, power seats, cruise control, tilt steering column, wire wheel covers, power windows, intermittent wipers.

LS Preferred Group 1	2137	1816	1966

HD battery, power door locks, floormats, gauge package with trip odometer, Twilight Sentinel, lighted right visor mirror, remote mirrors, bodyside moldings, AM/FM ST ET cassette, power seats, cruise control, tilt steering column, WSW tires, power decklid release, wire wheel covers, power windows, intermittent wipers.

LS Preferred Group 2	2361	2007	2172

Group 1 plus cornering lamps, deluxe luggage compartment trim.

Chevrolet Cavalier

Chevrolet Cavalier Z24

What's New

A new self-aligning steering wheel designed to reduce chest injuries in crashes debuts as standard equipment in Cavalier, Chevrolet's version of the J-car front-drive subcompact. The steering wheel has an energy-absorbing hub with six deformable metal legs that bring the wheel parallel to the driver's chest in a crash, spreading the force of the impact to reduce injury potential. Similar technology has been used by Volvo for several years. Cavalier was restyled last year, but this year's changes are less dramatic. The RS 2-door coupe and 4-door sedan have been dropped, and instead RS option packages are offered on base versions of the 2-door coupe, 4-door sedan and 5-door wagon. RS packages include sport suspension, 14-inch tires on rally wheels, power steering and a gauge package with tachometer. Other models include a VL ("value leader") 2-door, and high-performance Z24 2-door coupe and convertible. A 90-horsepower 2.0-liter 4-cylinder engine is standard on all models except the Z24, which comes with a 130-horsepower 2.8-liter V-6. The V-6 is optional on the station wagon. Rear shoulder belts become standard on all models for 1989. Z24s gain gas-pressurized shock absorbers and, on the coupe, a split folding rear seatback. Cloth interiors are standard on base Cavaliers and the Z24's cloth trim and folding rear seat are optional. Cavalier was introduced for the 1982 model year, as were similar J-body subcompacts at the other four GM divisions. Cadillac and Oldsmobile have dropped their versions for 1989, but the Buick Skyhawk and Pontiac Sunbird are back for another season.

For

● Performance (V-6)   ●Fuel economy (4-cylinder)
●Handling/roadholding (Z24)

Against

●Performance (4-cylinder/automatic)   ●Rear seat room

Summary

Now seven years old, Cavalier isn't as roomy or technically sophisticated as newer Japanese subcompacts, yet it offers good value for the money. While Japanese cars have jumped in price the past few years, Cavalier has inched higher by comparison. For example, a Cavalier 4-door with automatic transmission, air conditioning, power steering and several other comfort and convenience features is around $10,700 at suggested retail price. That's less than most comparably equipped Japanese sedans. Cavalier isn't as competitive in performance. The standard 4-cylinder engine is economical and delivers decent performance with manual transmission, but loses a lot of zip and fuel economy with the more popular automatic transmission. Note also that the automatic is a 3-speed transmission, while most Japanese rivals offer automatics with an overdrive fourth gear, allowing quieter, more economical cruising. The V-6 engine is a different story. The Z24 is a mini-muscle car with brisk performance and a reasonable price, while the V-6 gives the wagon enough muscle to handle a full load of passengers or cargo. Cavalier is by no means the best choice among subcompacts, yet it has enough plus points to deserve consideration.

Specifications

	2-door notchback	4-door notchback	5-door wagon	2-door conv.
Wheelbase, in.	101.2	101.2	101.2	101.2
Overall length, in.	178.4	178.4	174.5	178.7
Overall width, in.	66.0	66.3	66.3	66.0
Overall height, in.	52.0	52.1	52.8	52.7
Front track, in.	55.4	55.4	55.4	55.4
Rear track, in.	55.2	55.2	55.2	55.2
Turn diameter, ft.	34.7	34.7	34.7	34.7
Curb weight, lbs.	2418	2423	2478	2729
Cargo vol., cu. ft.	13.2	13.7	64.4	10.4
Fuel capacity, gal.	13.6	13.6	13.6	13.6
Seating capacity	5	5	5	4
Front headroom, in.	37.9	39.7	38.3	39.1
Front shoulder room, in.	53.7	53.7	53.7	53.7
Front legroom, max., in.	42.9	42.2	42.2	42.9
Rear headroom, in.	36.1	37.9	38.8	37.4
Rear shoulder room, in.	52.0	53.7	53.7	38.0
Rear legroom, min., in.	30.5	34.3	33.7	31.1

Powertrain layout: transverse front engine/front-wheel drive.

Engines

	ohv I-4	ohv V-6
Size, liters/cu. in.	2.0/121	2.8/173
Fuel delivery	TBI	PFI
Horsepower @ rpm	90 @ 5600	130 @ 4500
Torque (lbs./ft.) @ rpm	108 @ 3200	160 @ 3600
Availability	S	S[1]

EPA city/highway mpg

5-speed OD manual	26/36	18/29
3-speed automatic	25/32	20/27

1. Z24; optional, wagon

Prices

Chevrolet Cavalier	Retail Price	Dealer Invoice	Low Price
VL 2-door notchback	$7375	$6955	$7165
Base 2-door notchback	8395	7749	8072
Base 4-door notchback	8595	7933	8264
Base 5-door wagon	8975	8284	8630
Z24 2-door notchback	11325	10113	10719
Z24 2-door convertible	16615	14837	15726
Destination charge	425	425	425

Standard Equipment:

VL: 2.0-liter TBI 4-cylinder engine, 5-speed manual transmission, power brakes, vinyl reclining front bucket seats, outboard rear lap/shoulder belts, 185/80R13 all-season SBR tires on styled steel wheels. **Base adds:** easy-entry passenger seat (2-door), ashtray and glovebox lights, headlights-on tone, AM/FM ST ET. **Z24 adds:** 2.8-liter PFI V-6 engine, custom cloth interior with contoured front seats, split folding rear seatback (3-door), center console, FE3 sport suspension, 215/60R14 Goodyear Eagle GT+4 tires on aluminum wheels.

Optional Equipment:	Retail Price	Dealer Invoice	Low Price
2.8-liter PFI V-6, wagon	660	561	607
3-speed automatic transmission	415	353	382
Air conditioning	675	574	621
Cloth bucket seats	20	24	20
Sport cloth bucket seats, 2-door	459	390	422
4-door	483	411	444
Wagon	383	326	352
Roof luggage carrier, wagon	115	98	106
Decklid luggage carrier, 4-door	115	98	106
Rear defogger	145	123	133
Power door locks, 2-door	145	123	133
4-door	195	166	179
Tinted glass	105	89	97
Engine block heater	20	17	18
Electronic instrument cluster	156	133	144
Bodyside moldings	50	43	46
Removable sunroof	350	298	322
FE2 sport suspension	27	23	25
WSW tires	68	58	63
195/70R14 tires	129	110	119
215/60R14 OWL tires, Z24	102	87	94
Power decklid release	50	43	46
Wheel trim rings	39	33	36
Alloy wheels	212	180	195
Power windows (std. conv.)	285	242	262
AM/FM ST ET	332	282	305
AM/FM ST ET cassette	454	386	418
Sound system prices are for VL; prices vary with model and option package content.			
VL Preferred Group 1, w/5-speed	303	258	279
w/automatic	277	235	255
Auxiliary lighting, power steering.			
VL Preferred Group 2, w/5-speed	544	462	500
w/automatic	518	440	477
Group 1 plus floormats, tinted glass, left remote and right manual mirrors, bodyside moldings.			
VL Preferred Group 3, w/5-speed	1574	1338	1448
w/automatic	1548	1316	1424
Group 2 plus air conditioning, cruise control, tilt steering column, intermittent wipers.			
Base Preferred Group 1, 2- & 4-door			
w/5-speed	303	258	279
Wagon w/2.8	309	263	284
2- & 4-door w/automatic	277	235	255
Wagon w/2.0 & automatic	283	241	260
Auxiliary lighting, power steering.			
Base Preferred Group 2, 2- & 4-door			
w/5-speed	544	462	500
Wagon w/2.8	550	468	506
2- & 4-door w/automatic	518	440	477
Wagon w/2.0 & automatic	524	445	482
Group 1 plus floormats, tinted glass, left remote and right manual mirrors, bodyside moldings.			
Base Preferred Group 3, 2- & 4-door			
w/5-speed	1574	1338	1448
Wagon w/2.8	1695	1441	1559
2- & 4-door w/automatic	1548	1316	1424
Wagon w/2.0 & automatic	1669	1419	1535
Group 2 plus air conditioning, cruise control, tilt steering column, intermittent wipers, roof rack (wagon).			
Base Preferred Group 4, 2-door w/5-speed	2051	1743	1887
4-door w/5-speed	2176	1850	2002
2-door w/automatic	2025	1721	1863
4-door w/automatic	2150	1828	1978
Group 3 plus power windows and locks, AM/FM ST ET cassette.			

KEY: ohv = overhead valve; **ohc** = overhead cam; **dohc** = double overhead cam; **I** = inline cylinders; **V** = cylinders in V configuration; **flat** = horizontally opposed cylinders; **bbl.** = barrel (carburetor); **PFI** = port (multi-point) fuel injection; **TBI** = throttle-body (single-point) fuel injection; **rpm** = revolutions per minute; **OD** = overdrive transmission; **S** = standard; **O** = optional; **NA** = not available.

Prices are accurate at time of printing; subject to manufacturer's change.

	Retail Price	Dealer Invoice	Low Price
RS Pkg., 2-door	695	591	639
4-door	705	599	649

Black exterior accents, dual mirrors, FE2 sport suspension, power steering, gauges including tachometer and trip odometer, 195/70R14 SBR tires on rally wheels.

RS Preferred Group 1, 2-door w/5-speed .	911	774	838
4-door w/5-speed	921	783	847
Wagon w/5-speed or 2.8	927	788	853
2-door w/automatic	885	752	814
4-door w/automatic	895	761	823
Wagon w/2.0 & automatic	901	766	829

RS Pkg. plus floormats, tinted glass, auxiliary lighting.

RS Preferred Group 2, 2-door w/5-speed .	1941	1650	1786
4-door w/5-speed	1951	1658	1795
Wagon w/5-speed or 2.8	1957	1663	1800
2-door w/automatic	1915	1628	1762
4-door w/automatic	1925	1636	1771
Wagon w/2.0 & automatic	1931	1641	1777

Group 1 plus air conditioning, cruise control, tilt steering column, intermittent wipers.

RS Preferred Group 3, 2-door w/5-speed .	2583	2196	2376
4-door w/5-speed	2603	2213	2395
Wagon w/5-speed or 2.8	2674	2273	2460
2-door w/automatic	2557	2173	2352
4-door w/automatic	2577	2190	2371
Wagon w/2.0 & automatic	2648	2251	2436

Group 2 plus decklid luggage carrier (2-door), roof luggage carrier (wagon), power windows and locks, AM/FM ST ET cassette, power decklid release.

Z24 Coupe Preferred Group 1	850	723	782

Air conditioning, HD battery, tinted glass, auxiliary lighting.

Z24 Coupe Preferred Group 2	1360	1156	1251

Group 1 plus AM/FM ST ET cassette, cruise control, tilt steering column, intermittent wipers.

Z24 Coupe Preferred Group 3	1915	1628	1762

Group 2 plus power windows and locks, power decklid release, AM/FM ST ET cassette w/EQ.

Z24 Convertible Preferred Group 1	838	712	771

Air conditioning, HD battery, auxiliary lighting, AM/FM ST ET cassette.

Z24 Convertible Preferred Group 2	1426	1212	1312

Group 1 plus AM/FM ST ET cassette w/EQ, power decklid release, cruise control, tilt steering column, intermittent wipers.

Chevrolet Celebrity

What's New

Celebrity loses its 2-door coupe this year, but the standard 2.5-liter 4-cylinder engine gets another round of internal revisions later in the year and 12 more horsepower. The cylinder head has larger intake and exhaust ports and longer valve stems, and the single-point fuel-injection system has a larger-bore throttle-body. Horsepower increases from 98 to 110 when these changes arrive around mid-year. A 125-horsepower 2.8-liter V-6 remains optional. A 3-speed automatic transmission is standard with either engine and a 4-speed overdrive automatic is optional with the V-6. A 5-speed manual transmission that had been a credit option with the V-6 has been dropped because only a few hundred were sold during the 1987 and 1988 model years. Celebrity is a front-drive intermediate built from the GM A-body design that includes the Buick Century, Oldsmobile Cutlass Ciera and Pontiac 6000. With the demise of the slow-selling 2-door (only 5 percent of 1988 sales), remaining body styles are a 4-door sedan and 5-door wagon. The rear-drive Monte Carlo coupe was dropped last year, leaving Chevy without a 2-door entry in the mid-size market. With the scheduled arrival of the new Lumina 4-door sedan

Chevrolet Celebrity Eurosport 4-door

next spring, this should be the last year for the Celebrity sedan, introduced as a 1982 model. The Celebrity wagon, however, is expected to remain in production.

For

- Passenger and cargo room
- Quietness
- Handling (Eurosport)

Against

- Performance (4-cylinder)
- Handling (base suspension)

Summary

Celebrity's age and sedate styling are no help against the arch-rival Ford Taurus, which is newer and fresher looking. However, Celebrity is just as roomy inside and offers comparable performance with the optional V-6 engine. There's room for six adults (and two children in the 3-seat wagon). Land speed records won't be within your reach with the V-6, but there's enough power for a mid-size sedan or wagon to accomplish most chores without strain. The standard 4-cylinder is hard pressed to handle a full load of people and/or cargo. Taurus' standard suspension and tires offer better handling and a more controlled ride than Celebrity's, but the $230 Eurosport package upgrades handling and cornering ability through firmer suspension pieces and larger tires, without a major penalty in ride comfort. A 4-door with the V-6 and Eurosport Group 3 comes to a reasonable $15,000 or so, and that's including air conditioning, power locks and windows, premium stereo, tilt steering wheel and other amenities. Priced any Japanese compacts with similar equipment lately? By comparison, the mid-size Celebrity emerges a bargain.

Specifications	4-door notchback	5-door wagon
Wheelbase, in.	104.9	104.9
Overall length, in.	188.3	190.8
Overall width, in.	69.3	69.3
Overall height, in.	54.1	54.3
Front track, in.	58.7	58.7
Rear track, in.	57.0	57.0
Turn diameter, ft.	38.7	38.7
Curb weight, lbs.	2751	2888
Cargo vol., cu. ft.	16.2	75.1
Fuel capacity, gal.	15.7	15.7
Seating capacity	6	8
Front headroom, in.	38.6	38.6
Front shoulder room, in.	56.2	56.2
Front legroom, max., in.	42.1	42.1

Prices are accurate at time of printing; subject to manufacturer's change.

	4-door notchback	5-door wagon
Rear headroom, in.	38.0	38.9
Rear shoulder room, in.	56.2	56.2
Rear legroom, min., in.	36.4	35.6

Powertrain layout: transverse front engine/front-wheel drive

Engines	ohv I-4	ohv V-6
Size, liters/cu. in.	2.5/151	2.8/173
Fuel delivery	TBI	PFI
Horsepower @ rpm	98 @ 4800	125 @ 4500
Torque (lbs./ft.) @ rpm	135 @ 3200	160 @ 3600
Availability	S	O
EPA city/highway mpg		
3-speed automatic	23/30	20/27
4-speed OD automatic		20/29

Prices

Chevrolet Celebrity	Retail Price	Dealer Invoice	Low Price
4-door notchback	$11495	$9920	$10708
5-door 2-seat wagon	11925	10291	11108
5-door 3-seat wagon	12175	10507	11341
Destination charge	450	450	450

Standard Equipment:

2.5-liter TBI 4-cylinder engine, 3-speed automatic transmission, power steering and brakes, front bench seat with center armrest, cloth upholstery, outboard rear lap/shoulder belts, AM/FM ST ET, bodyside moldings, day/night mirror, left remote and right manual mirrors, headlamps-on tone, 185/75R14 all-season SBR tires.

Optional Equipment:

	Retail	Dealer	Low
2.8-liter PFI V-6	610	519	561
4-speed automatic transmission (2.8 req.) .	175	149	161
Air conditioning	775	659	713
Cloth bucket seats w/console	257	218	236
Cloth 55/45 seat	133	113	122
Custom cloth 55/45 seat, 4-door	385	327	354
w/split folding middle seat, wagon . .	435	370	400
Custom cloth 45/45 seat w/console, 4-door .	335	285	308
w/split folding middle seat, wagon . . .	385	327	354
Custom two-tone paint, w/option pkg	93	79	86
w/o option pkg	148	126	136
Air deflector, wagon	40	34	37
HD cooling, w/o A/C	70	60	64
w/A/C	40	34	37
Rear defogger	145	123	133
Power door locks	195	166	179
California emissions pkg	100	85	92
Gauge Pkg	64	54	59
Coolant temperature gauge, voltmeter, trip odometer.			
Engine block heater	20	17	18
Exterior Molding Pkg	55	47	51
Deluxe rear compartment decor	40	34	37
Power driver's seat	240	204	221
Reclining front seatbacks	90	77	83
Cargo Area Security Pkg	44	37	40
Inflatable rear shock absorbers	64	54	59
HD suspension, F40	26	22	24
WSW tires	68	58	63
195/70R14 tires	90	77	83
Sport wheel covers	65	55	60
Locking wire wheel covers	199	169	183
Alloy wheels	143	122	132
Rally wheels	56	48	52
Intermittent wipers	125	106	115
AM/FM ST ET cassette	122	104	112

	Retail Price	Dealer Invoice	Low Price
Preferred Group 1, 4-door w/2.5	911	774	838
4-door w/2.8	937	796	862
Wagon w/2.5	903	768	831
Wagon w/2.8	929	790	855
Air conditioning, floormats, auxiliary lighting, Exterior Molding Pkg.			
Preferred Group 2, 4-door w/2.5	1525	1296	1403
4-door w/2.8	1551	1318	1427
Wagon w/2.5	1682	1430	1547
Wagon w/2.8	1708	1452	1571
4-door: Group 1 plus power door locks, Gauge Pkg., cruise control, tilt steering column, intermittent wipers. Wagon adds luggage rack, power liftgate release.			
Preferred Group 3, 4-door w/2.5	2012	1710	1851
4-door w/2.8	2038	1732	1875
Wagon w/2.5	2203	1873	2027
Wagon w/2.8	2229	1895	2051
4-door: Group 2 plus remote mirrors, AM/FM ST ET cassette, power decklid release, power windows. Wagon adds deluxe rear compartment decor, Cargo Area Security Pkg.			
Eurosport Equipment Group	230	196	212
Sport suspension, quick-ratio steering, blackout exterior trim, padded sport steering wheel, 195/75R14 all-season SBR tires on rally wheels.			
Eurosport Preferred Group 1, 4-door w/2.5 .	1141	970	1050
4-door w/2.8	1167	992	1074
Wagon w/2.5	1133	963	1042
Wagon w/2.8	1159	985	1066
Eurosport Equipment plus air conditioning, floormats, auxiliary lighting, Exterior Molding Pkg.			
Eurosport Preferred Group 2, 4-door w/2.5 .	1755	1492	1615
4-door w/2.8	1781	1514	1639
Wagon w/2.5	1912	1625	1759
Wagon w/2.8	1938	1647	1783
4-door: Group 1 plus power door locks, Gauge Pkg., cruise control, tilt steering column, intermittent wipers. Wagon adds luggage rack, power liftgate release.			
Eurosport Preferred Group 3, 4-door w/2.5 .	2242	1906	2063
4-door w/2.8	2268	1928	2087
Wagon w/2.5	2433	2068	2238
Wagon w/2.8	2459	2090	2262
4-door: Group 2 plus remote mirrors, AM/FM ST ET cassette, power decklid release, power windows. Wagon adds deluxe rear compartment decor, Cargo Area Security Pkg.			

Chevrolet Corsica/Beretta

What's New

A new 5-door hatchback joins the Corsica 4-door sedan and Beretta 2-door coupe as Chevy's front-drive compacts, which halfway through calendar 1988 ranked as the best-selling car line in the U.S. when counted together. Corsica and Beretta are built on the same platform, but differ in exterior styling, interior furnishings and chassis components. Also new for Corsica is a top-line LTZ sporty 4-door. The base Beretta sports last year's GT exterior trim, while the 1989 Beretta GT is now a regular model instead of an option package, and it gets new 15x7-inch aluminum wheels standard. The Corsica hatchback, which goes on sale in late fall, has a rear seatback that folds down for a 39-cubic foot cargo area. A cargo tray and window-shade type cover are standard. A new 60/40 split rear seatback is

> **KEY: ohv** = overhead valve; **ohc** = overhead cam; **dohc** = double overhead cam; **I** = inline cylinders; **V** = cylinders in V configuration; **flat** = horizontally opposed cylinders; **bbl.** = barrel (carburetor); **PFI** = port (multi-point) fuel injection; **TBI** = throttle-body (single-point) fuel injection; **rpm** = revolutions per minute; **OD** = overdrive transmission; **S** = standard; **O** = optional; **NA** = not available.

Prices are accurate at time of printing; subject to manufacturer's change.

Chevrolet

Chevrolet Corsica 5-door

standard on the LTZ 4-door and is supposed to become optional on the hatchback and base 4-door. The new LTZ model also includes the basic ingredients of the Beretta GT: V-6 engine, sport suspension, high-performance tires and 15-inch aluminum wheels, plus a trunk lid luggage rack. Base engine remains a 90-horsepower 2.0-liter 4-cylinder; the 130-horsepower 2.8-liter V-6 is standard on the Beretta GT and Corsica LTZ, optional in the others. New 195/70R14 all-season tires are standard on the base Beretta, while the GT gets new cloth interior trim and a 60/40 split folding rear seatback with an armrest. All models come with 3-point rear shoulder belts and front shoulder belts that can be left buckled for automatic deployment.

For

- Performance (V-6) • Handling/roadholding

Against

- Performance (4-cylinder) • Rear seat room

Summary

Neither Corsica nor Beretta has any outstanding features, but both are generally competent and offer enough value for the money for us to recommend them. First, though, you have to forget the base 4-cylinder engine because it's too weak for even adequate acceleration. Go with the V-6 and optional automatic transmission; the standard 5-speed manual lacks precision and the V-6 seems to suffer from excessive hesitation and driveability problems with that transmission. We also prefer the sport suspension and 195/70R14 tires, standard on the base Beretta and optional on Corsica. You'll get the best balance between handling ability and ride quality with that arrangement. The larger tires and firmer suspensions on the Beretta GT and Corsica LTZ elevate handling a notch or two, but with a noticeable loss in ride comfort. A Corsica with the V-6, automatic and the most expensive LT option package is $14,000; a base Beretta with the same powertrain and option group 3 is about $13,400. Good cars, reasonable prices. Check them out.

Specifications

	2-door notchback	4-door notchback	5-door hatchback
Wheelbase, in.	103.4	103.4	103.4
Overall length, in.	187.2	183.4	183.4
Overall width, in.	68.2	68.2	68.2
Overall height, in.	55.3	56.2	56.2
Front track, in.	55.6	55.6	55.6
Rear track, in.	56.6	55.1	55.1
Turn diameter, ft.	34.8	34.8	34.8
Curb weight, lbs.	2631	2595	2648
Cargo vol., cu. ft.	13.5	13.5	39.1
Fuel capacity, gal.	13.6	13.6	13.6

	2-door notchback	4-door notchback	5-door hatchback
Seating capacity	5	5	5
Front headroom, in.	38.0	38.1	38.1
Front shoulder room, in.	55.3	55.4	55.4
Front legroom, max., in.	43.4	43.4	43.4
Rear headroom, in.	36.6	37.4	37.4
Rear shoulder room, in.	55.1	55.6	55.6
Rear legroom, min., in.	34.6	35.0	35.0

Powertrain layout: transverse front engine/front-wheel drive.

Engines

	ohv I-4	ohv V-6
Size, liters/cu. in.	2.0/121	2.8/173
Fuel delivery	TBI	PFI
Horsepower @ rpm	90 @ 5600	130 @ 4700
Torque (lbs./ft.) @ rpm	108 @ 3200	160 @ 3600
Availability	S	O[1]

EPA city/highway mpg

5-speed OD manual	24/34	18/29
3-speed automatic	24/31	20/27

[1] Standard Corsica LTZ and Beretta GT

Prices

Chevrolet Corsica	Retail Price	Dealer Invoice	Low Price
4-door notchback	$9985	$8917	$9601
5-door hatchback	10375	9265	9970
LTZ 4-door notchback	12825	11453	12289
Destination charge	425	425	425

Standard Equipment:

2.0-liter TBI 4-cylinder engine, 5-speed manual transmission, power steering, power brakes, cloth reclining front bucket seats, door pockets, automatic front seatbelts, outboard rear lap/shoulder belts, headlamps-on warning, AM/FM ST ET, P185/75R14 all-season SBR tires on steel wheels; hatchback has sliding package tray, cargo cover. **LTZ** adds: 2.8-liter PFI V-6, FE3 sport suspension, luggage rack, 205/60R15 Goodyear Eagle GT tires, on alloy wheels, sport steering wheel, custom cloth interior with center console, 60/40 split rear seatback with center armrest, gauge package with tachometer, overhead console.

Optional Equipment:

2.8-liter PFI V-6, base & LT	660	561	607
3-speed automatic transmission	490	417	451
Air conditioning, base & LT	750	638	690
Custom cloth CL bucket seats, 4-door	425	361	391
5-door	275	234	253
Custom two-tone paint	123	105	113
HD battery	26	22	24
Decklid luggage carrier	115	98	106
Floor console	60	51	55
Rear defogger	145	123	133
Power door locks	195	166	179
California emissions pkg	100	85	92
Gauge Pkg. (std. LTZ)	139	118	128
Tachometer, voltmeter, coolant temperature and oil pressure gauges, trip odometer.			
Engine block heater	20	17	18
Auxiliary lighting, 4-door	64	54	59
5-door	56	48	52
Bodyside striping	57	48	52
F41 sport suspension, base	49	42	45
195/70R14 tires, base	93	79	86
WSW tires, base	68	58	63
Styled wheels, base	56	48	52
Alloy wheels, base	215	183	198
LT	159	135	146
Intermittent wipers	55	47	51

Prices are accurate at time of printing; subject to manufacturer's change.

	Retail Price	Dealer Invoice	Low Price
AM/FM ST ET cassette	122	104	112
AM/FM ST ET cassette w/EQ	272	231	250
Sound system prices vary with option package content.			
Preferred Group 1, base 4 door	243	207	224
Base 5-door	235	200	216
HD battery, floormats, tinted glass, auxiliary lighting			
Preferred Group 2, base 4-door	1348	1146	1240
Base hatchback	1340	1139	1233
Group 1 plus air conditioning, cruise control, tilt steering column, intermittent wipers.			
Preferred Group 3, base 4-door	2000	1700	1840
Base 5-door	1992	1693	1833
Group 2 plus power windows and locks, AM/FM ST ET cassette, power decklid release.			
LT Equipment Group, base	244	207	224
F41 sport suspension, 195/70R14 Goodyear Eagle GA tires, blackout exterior trim, sport steering wheel.			
LT Preferred Group 1, 4-door	487	414	448
5-door	479	407	441
LT Group plus HD battery, floormats, tinted glass, auxiliary lighting.			
LT Preferred Group 2, 4-door	1731	1471	1593
5-door	1723	1465	1585
Group 1 plus air conditioning, Gauge Pkg., cruise control, tilt steering column, intermittent wipers.			
LT Preferred Group 3, 4-door	2383	2026	2192
5-door	2375	2019	2185
Group 2 plus power windows and locks, AM/FM ST ET cassette, power decklid release.			
Preferred Group 1, LTZ	534	454	491
HD battery, floormats, tinted glass, cruise control, tilt steering column, intermittent wipers.			
Preferred Group 2, LTZ	1186	1008	1091
Group 1 plus power windows and locks, AM/FM ST ET cassette, power decklid release.			

Chevrolet Beretta

	Retail Price	Dealer Invoice	Low Price
2-door notchback	$10575	$9443	$10159
GT 2-door notchback	12685	11328	12157
Destination charge	425	425	425

Standard Equipment:

2.0-liter TBI 4-cylinder engine, 5-speed manual transmission, power steering and brakes, cloth reclining front bucket seats, automatic front seatbelts, console with storage armrest, outboard rear lap/shoulder belts, tachometer, coolant temperature and oil pressure gauges, voltmeter, trip odometer, headlamps-on warning, left remote and right manual mirrors, AM/FM ST ET, console, tinted glass, F41 sport suspension, P195/70R14 Goodyear Eagle GA all- season SBR tires on steel wheels. **GT** adds: 2.8-liter PFI V-6, FE3 sport suspension, sport steering wheel, 60/40 split folding rear seatback, custom cloth trim, overhead consolette, power decklid release, 205/60R15 Goodyear Eagle GT tires on styled steel wheels.

Optional Equipment:

	Retail	Dealer	Low
2.8-liter PFI V-6, base	660	561	607
3-speed automatic transmission	490	417	451
Air conditioning, base	750	638	690
Custom two-tone paint	123	105	113
HD battery	26	22	24
Decklid luggage carrier	115	98	106
Rear defogger	145	123	133
Power door locks	145	123	133
California emissions pkg	100	85	92
Engine block heater	20	17	18
Electronic instrumentation	156	133	144
Auxiliary lighting	32	27	29
Removable sunroof	350	298	322
Alloy wheels	159	135	146
Intermittent wipers	55	47	51
AM/FM ST ET cassette	122	104	112
AM/FM ST ET cassette w/EQ	272	231	250
Sound system prices vary with option package content.			

	Retail Price	Dealer Invoice	Low Price
Preferred Group 1, base	91	77	84
HD battery, floormats, auxiliary lighting.			
Preferred Group 2, base	1196	1017	1100
Group 1 plus air conditioning, cruise control, tilt steering column, intermittent wipers.			
Preferred Group 3, base	1723	1466	1585
Group 2 plus power windows and locks, AM/FM ST ET cassette, power decklid release.			
Preferred Group 1, GT	446	379	410
HD battery, floormats, auxiliary lighting, cruise control, tilt steering column, intermittent wipers.			
Preferred Group 2, GT	973	827	895
Group 1 plus AM/FM ST ET cassette, power windows and locks, power decklid release.			

Chevrolet Corvette

Chevrolet Corvette

What's New

Chevrolet goes after the title of world's fastest production car with the Corvette ZR1, a limited production version of the 2-seat sports car that uses an aluminum double-overhead cam, 32-valve 5.7-liter V-8 jointly developed by Chevrolet and Group Lotus, the British car company now owned by GM. The engine, designated the LT5, will be assembled in Oklahoma City by Mercury Marine under a contract with GM. The ZR1, unofficially dubbed the "King of the Hill" Corvette, is scheduled to go on sale in February; only 2000-3000 will be built per year, all of them coupes. Chevrolet wouldn't say what the ZR1's top speed will be, but engineers hinted it was 180 mph, or about 25 mph more than the base Corvette. Chevrolet is estimating horsepower at 385 and torque at 370 pounds/feet. By comparison, the base Corvette's L98 engine makes 245 horsepower and 340 pounds/feet torque. Despite the ZR1's performance, it will not carry a gas guzzler tax. Chevrolet says it can meet the federal EPA standard of 22.5 mpg city and highway combined with the new ZF 6-speed manual transmission that will be on all ZR1s; an automatic transmission isn't planned. In light throttle applications, the computer-assisted ZF 6-speed activates a pin in the shift linkage to block shifts from first to second gear; instead, the pin bumps the shift lever

---

KEY: ohv = overhead valve; **ohc** = overhead cam; **dohc** = double overhead cam; **I** = inline cylinders; **V** = cylinders in V configuration; **flat** = horizontally opposed cylinders; **bbl.** = barrel (carburetor); **PFI** = port (multi-point) fuel injection; **TBI** = throttle-body (single-point) fuel injection; **rpm** = revolutions per minute; **OD** = overdrive transmission; **S** = standard; **O** = optional; **NA** = not available.

Prices are accurate at time of printing; subject to manufacturer's change.

directly to fourth gear. With a heavier throttle foot, the transmission allows normal shifting through all the gears. Fifth and sixth gears are overdrive. ZR1s will have a convex instead of concave rear body panel with rectangular instead of round taillamps, wider rear fenders and huge 315/35VR17 tires. Chevy hasn't issued a price, but most estimates are it will be at least $50,000. The base Corvette coupe and convertible also get the new ZF 6-speed manual, though about 80 percent of these cars are sold with automatic transmission, a no-cost option. The 6-speed replaces a Doug Nash 4-speed manual with electronic overdrive for the top three gears. Coupes with the manual transmission and Z51 Performance Handling Package also will be available with the new FX3 Delco-Bilstein Selective Ride Control, which lets the driver select damping control by a console-mounted switch. Choices are Touring, Sport or Competition, plus the system automatically selects an appropriate damping level at higher speeds. Later in the year, a fiberglass removable hardtop with heated glass rear window will become available for convertibles. A removable hardtop was a factory option on 1955-1975 Corvettes.

For

● Performance ● Handling ● Anti-lock brakes

Against

● Ride ● Noise ● Fuel economy ● Insurance costs

Summary

While Corvette is one of the most impractical cars that more than $30,000 can buy, the even more outrageous ZR1 is actually a practical move by Chevrolet. It will generate tremendous attention for Chevy and GM as a shining example of what the No. 1 U.S. automaker is capable of doing. Aside from the ZR1, which we haven't driven yet, Chevrolet still has to sell the "garden-variety" Corvette and convertible, which hasn't been easy the past couple of years. Most people either love them or hate them; we do both. We love the image it creates and the first-class performance, while we hate the stiff, uncompromising ride and treacherous wet-road handling. In its favor, a fully tricked-out Corvette costs little more than a 4-cylinder Porsche; or, you can buy a Corvette coupe and a convertible for the price of one Ferrari.

Specifications

	3-door hatchback	2-door conv.
Wheelbase, in.	96.2	96.2
Overall length, in.	176.5	176.5
Overall width, in.	71.0	71.0
Overall height, in.	46.7	46.4
Front track, in.	59.6	59.6
Rear track, in.	60.4	60.4
Turn diameter, ft.	40.4	40.4
Curb weight, lbs.	3229	3269
Cargo vol., cu. ft.	17.9	6.6
Fuel capacity, gal.	20.0	20.0
Seating capacity	2	2
Front headroom, in.	36.4	36.5
Front shoulder room, in.	54.1	54.1
Front legroom, max., in.	42.6	42.6
Rear headroom, in.	—	—
Rear shoulder room, in.	—	—
Rear legroom, min., in.	—	—

Powertrain layout: longitudinal front engine/rear-wheel drive.

Engines

	ohv V-8
Size, liters/cu. in.	5.7/350
Fuel delivery	PFI
Horsepower @ rpm	245 @ 4300
Torque (lbs./ft.) @ rpm	340 @ 3200
Availability	S

EPA city/highway mpg

6-speed OD manual	16/25
4-speed OD automatic	17/25

Prices

Chevrolet Corvette	Retail Price	Dealer Invoice	Low Price
3-door hatchback	$31545	$26592	$29069
2-door convertible	36785	31010	33898
Destination charge	500	500	500

Standard Equipment:

5.7-liter PFI V-8 engine, 6-speed manual or 4-speed automatic transmission, power steering, power 4-wheel disc brakes, anti-lock braking system, Pass Key theft deterrent system, air conditioning, AM/FM ST ET cassette, power antenna, tinted glass, power mirrors, tilt/telescopic steering column, power windows, intermittent wipers, 275/40ZR17 Goodyear Eagle GT tires on cast alloy wheels.

Optional Equipment:

Z51 Performance Handling Pkg	.NA	NA	NA
FX3 Selective Ride Control System	NA	NA	NA
Leather seats	400	332	368
Electronic control air conditioning	150	125	138
Performance axle ratio	22	18	20
Engine oil cooler	110	91	101
California emissions pkg	100	83	92
Radiator cooling boost fan	75	62	69
Engine block heater	20	17	18
Low tire pressure warning indicator	325	270	299
Lighted right visor mirror	58	48	53
HD radiator	40	33	37
Removable roof panel	615	510	566
Dual removable roof panels	915	759	842
Power seats (each)	240	199	221
Delco/Bose music system	773	642	711
Preferred Group 1	1163	965	1070

Electronic air conditioning, Delco/Bose music system, power driver' seat.

Chevrolet Lumina

What's New

Lumina is a new front-drive intermediate that is supposed to arrive late next spring as the first 4-door sedan from the GM10 or W-body design that produced the Buick Regal, Oldsmobile Cutlass Supreme and Pontiac Grand Prix 2-door coupes, introduced as 1988 models. Lumina, which will be introduced as an early 1990 model, will ride the same 107.5-inch wheelbase as the W-body coupes and come with the 140-horsepower 3.1-liter V-6 and 4-speed overdrive automatic transmission also used in those cars. Lumina's wheelbase is three inches longer than Celebrity's, the front-drive, mid-size Chevrolet since 1982. The Celebrity 4-door will disappear after this year to make room for Lumina, but the Celebrity wagon is expected to

Prices are accurate at time of printing; subject to manufacturer's change.

CONSUMER GUIDE®

1990 Chevrolet Lumina

continue. By 1990, Chevy will get a 2-door version of the W-body to replace the rear-drive Monte Carlo, and the other three divisions will get 4-door sedan versions, though styling is supposed to be different on all models. Chevrolet hasn't released additional information and we haven't driven Lumina, so we can't comment on its performance.

Prices not available at time of publication.

Chevrolet S10 Blazer/ GMC S15 Jimmy

Chevrolet S10 Blazer 2WD

What's New

Anti-lock rear brakes and rear shoulder belts are standard on the S10 Blazer and identical S15 Jimmy, the most popular compact sport utility vehicles. The anti-lock system is designed to prevent the rear wheels from locking before the vehicle stops, thereby preventing skids. A new electric speedometer generates the speed signal for the anti-lock system. The S10 Blazer/S15 Jimmy twins come only in a 3-door body style (a 5-door is planned for the future), but with a choice of 2- or 4-wheel drive. A 125-horsepower 2.8-liter V-6 and 5-speed manual transmission are standard, and a 160-horsepower 4.3-liter V-6, available only with a 4-speed overdrive automatic transmission, is optional. The 4.3 V-6, introduced last spring, was installed in about 60 percent of the models produced late in the 1988 model year. GM's Insta-Trac 4WD system has a new transfer case for use with the 2.8 V-6 that has a longer shift lever and redesigned gear sets that are supposed to be quieter and smoother. Automatic locking front hubs are standard on 4x4s and Insta-Trac, a part-time system not for use on dry pavement, has full shift-on-the-fly capability. Other

changes are that power steering, formerly optional on 2WD models, is now standard, and a rear wiper/washer and an electronic instrument cluster have been added as options.

For

- Performance (4.3) • 4WD traction
- Anti-lock rear brakes • Cargo room

Against

- Fuel economy • Ride

Summary

GM's compact 4x4s lack some of the features of the Jeep Cherokee, such as a 5-door body style and a full-time 4WD system, plus Jeep this year offers a 4-wheel anti-lock brake system, while GM's anti-lock feature operates only on the rear wheels (which is better than no anti-lock system). However, the S10 Blazer and S15 Jimmy have features we like, such as an easy-to-use part-time 4WD system and the muscular 4.3-liter V-6 engine option. The standard 2.8-liter V-6, which comes only with the 5-speed manual, has adequate acceleration. If you want automatic transmission, you have to order the 4.3, and those two options add up to $1050. In either case, gas mileage won't be anything to rave about. The 4.3 develops considerably more torque at lower engine speeds for stronger acceleration and greater trailer towing capacity (5000 pounds versus 4000 on 4WD models), though these vehicles are still no threat to Corvette in a 0-60 mph sprint. We recommend the optional heavy-duty shock absorbers for a more comfortable ride, and suggest you buy the optional off-road suspension only if you actually do a lot of off-road driving; the on-road ride can be punishing with that option. These vehicles get quite expensive as you add options, though not as expensive as a loaded Cherokee, which is another reason to look at these as alternatives to a Jeep.

Specifications

	3-door wagon
Wheelbase, in.	100.5
Overall length, in.	170.3
Overall width, in.	65.7
Overall height, in.	64.0
Front track, in.	55.7
Rear track, in.	54.2
Turn diameter, ft.	NA
Curb weight, lbs.	3150
Cargo vol., cu. ft.	62.7
Fuel capacity, gal.	20.0
Seating capacity	4
Front headroom, in.	39.0
Front shoulder room, in.	53.9
Front legroom, max., in.	42.5
Rear headroom, in.	NA
Rear shoulder room, in.	NA
Rear legroom, min., in.	NA

Powertrain layout: longitudinal front engine/rear-wheel drive or on-demand 4WD.

KEY: ohv = overhead valve; **ohc** = overhead cam; **dohc** = double overhead cam; **I** = inline cylinders; **V** = cylinders in V configuration; **flat** = horizontally opposed cylinders; **bbl.** = barrel (carburetor); **PFI** = port (multi-point) fuel injection; **TBI** = throttle-body (single-point) fuel injection; **rpm** = revolutions per minute; **OD** = overdrive transmission; **S** = standard; **O** = optional; **NA** = not available.

Prices are accurate at time of printing; subject to manufacturer's change.

Engines

	ohv V-6	ohv V-6
Size, liters/cu. in.	2.8/173	4.3/262
Fuel delivery	TBI	TBI
Horsepower @ rpm	125 @	160 @
	4800	400
Torque (lbs./ft.) @ rpm	150 @	230 @
	2400	2800
Availability	S	O
EPA city/highway mpg		
5-speed OD manual	18/25	
4-speed OD automatic		17/21

Prices

Chevrolet S10 Blazer	Retail Price	Dealer Invoice	Low Price
2WD 3-door wagon	$11738	$10482	$11110
4WD 3-door wagon	13313	11889	12601
Destination charge	435	435	435

Standard Equipment:

2.8-liter TBI V-6, 5-speed manual transmission, anti-lock rear brakes, power steering and brakes, front and rear chromed bumpers with black rub strips, tinted glass, bright metal hubcaps, dual outside mirrors, front armrests, front and rear dome lamps, door sill plates, rubber floormat, headlamp warning buzzer, highback vinyl front bucket seats with easy-entry feature on passenger side, 195/75R15 all-season SBR tires on painted argent wheels.

Optional Equipment:

4.3-liter TBI V-6	255	217	235
4-speed automatic transmission	795	676	731
Optional rear axle ratio	38	32	35
Locking differential	252	214	232
Air conditioning, w/2.8	680	578	626
w/4.3	736	626	677
HD 630-amp battery	56	48	52
Luggage carrier	126	107	116
Outside spare tire carrier	192	163	177
w/Sport or Tahoe	159	135	146
Cold-Climate Pkg., w/o Sport or Tahoe . .	156	133	144
w/Sport or Tahoe	113	96	104
Tahoe Pkg., 2WD	715	608	658
4WD w/o High Country	683	581	628
4WD w/High Country	473	402	435
Upgraded interior and exterior trim, Gauge Pkg.			
Sport Pkg., 2WD	1068	908	983
4WD w/o High Country	1038	882	955
4WD w/High Country	671	570	617
Reclining seats, console, upgraded interior and exterior trim, visor mirror, Gauge Pkg.			
High Country Pkg., w/Tahoe	1025	871	943
w/Sport	925	786	851
Deep tinted glass, gold-painted aluminum wheels, blackout exterior trim, dual mirrors.			
Custom two-tone paint, w/o Tahoe	324	275	298
w/Tahoe	172	146	158
Special two-tone paint, w/o Sunshine Striping	212	180	195
w/Sunshine Striping	163	139	150
Deluxe two-tone paint, w/o Tahoe	329	280	303
w/Tahoe	177	150	163
Front console	114	97	105
Engine oil cooler	126	107	116
HD radiator	56	48	52
HD radiator & trans oil cooler, w/o A/C . .	118	100	109
w/A/C	63	54	58
ZM7 Driver Convenience Pkg	180	153	166
Intermittent wipers, tilt steering column.			
ZM8 Driver Convenience Pkg	197	167	181
Rear defogger and tailgate window release.			

	Retail Price	Dealer Invoice	Low Price
Spare tire cover	33	28	30
Air deflector	43	37	40
Gauge Pkg	62	53	57
Coolant temperature and oil pressure, voltmeter.			
Deep tinted glass, w/o High Country	200	170	184
w/High Country	56	48	52
w/light tinted rear window	144	122	132
Halogen headlamps	24	20	22
Engine block heater	33	28	30
Electronic instruments, w/o Tahoe or Sport .	358	304	329
w/Tahoe or Sport	296	252	272
Dual mirrors, painted	52	44	48
Bright	87	74	80
Inside rear-view mirror w/reading lights . .	26	22	24
Bodyside & wheel opening moldings	152	129	140
Bright wheel opening moldings	31	26	29
Black wheel opening moldings, w/o Tahoe .	43	37	40
w/Tahoe	13	11	12
Operating Convenience Pkg	.344	292	316
Power windows and locks.			
AM/FM ST ET	275	234	253
AM/FM ST ET cassette	454	386	418
AM/FM ST ET cassette w/EQ	604	513	556
Folding rear bench seat	409	348	376
Reclining front seatbacks	74	63	68
Shield Pkg., 4WD	75	64	69
HD shock absorbers	36	31	33
Cruise control	205	174	189
Bodyside striping	49	42	45
Sunshine Striping	116	99	107
Glass sunroof	250	213	230
HD suspension (snowplow use)	63	54	58
Off-Road Suspension Pkg.,			
w/o Sport or Tahoe	220	187	202
w/Sport or Tahoe	160	136	147
Front tow hooks	38	32	35
Deadweight trailer hitch	68	58	63
HD Trailering Special Pkg	211	179	194
LD Trailering Special Pkg	109	93	100
Wheel trim rings	60	51	55
Rally wheels, 2WD	92	78	85
Fifth wheel for above	15	13	14
Alloy wheels, 4WD	308	262	283
w/Sport, Tahoe or Off-Road Pkg . .	252	214	232
Fifth wheel for above	80	68	74
Sliding rear quarter windows	257	218	236
Rear wiper/washer	125	106	115
Deluxe cloth highback bucket seats	26	22	24
Leather reclining bucket seats	386	328	355
w/folding rear seat	486	413	447
w/Sport	312	265	287
w/Sport, w/folding rear seat	412	350	379
California emissions pkg	100	85	92

Chrysler Conquest TSi

What's New

Chrysler calls this captive import a "sports specialty coupe," and makes few changes in it for '89. Power windows now have an automatic-down feature, rear passengers get shoulder belts, the standard seats have new patterned velour inserts and there's a higher capacity rear-window defroster. This rear-drive, 4-passenger, 3-door hatchback is made by Mitsubishi, which is partly owned by Chrysler. The Japanese automaker sells a similar car in the U.S. under the Starion name. Conquest has a turbocharged and intercooled 2.6-liter 4-cylinder engine mated to a 5-speed manual transmission. A 4-speed automatic is optional. Standard equipment includes 4-wheel disc brakes with anti-lock rear

Prices are accurate at time of printing; subject to manufacturer's change.

Chrysler Conquest TSi

brakes, cruise control, power windows and door locks and a cassette stereo with graphic equalizer. Motorized automatic front shoulder belts are standard; manual lap belts must be buckled separately. The suspension is all independent and carries 205/55VR16 tires on 7-inch-wide wheels in front and 225/50VR16 tires on 8-inch-wide wheels in the rear. An optional handling package adds adjustable shock absorbers and 225/45VR16 tires on 8-inch-wide front wheels and 245/50VR16 tires on 9-inch-wide rears.

For

- Performance
- Handling/roadholding
- Anti-lock rear brakes

Against

- Interior room
- Cargo space
- Price

Summary

With its rear-drive layout and muscular looks, Conquest might be described as a Japanese version of the American pony car. The broad-shouldered stance, low seating, taut ride and throaty exhaust lend a masculine air. Acceleration is satisfying and turbocharger lag is minimal; boost comes up as low as 2000 rpm. The engine revs willingly to its 6000 rpm redline, and though it's not especially silky, the big four (which uses two balance shafts) doesn't sound overtaxed cruising in the upper rev ranges. Clutch feel and gearshift action are top notch. The steering is light but precise. Wide, low-profile tires and a firm suspension mean little compliance over rough roads, but good control at speed and in turns. Plus, ample power and rear-wheel drive mean genuine high-performance-car cornering characteristics. The supportive front bucket seats have adjustable side bolsters and the driver's positioning before the pedals and tilt steering wheel is businesslike. Conquest is really a 2+2, with a rear seat suitable only for children and subpar cargo room, despite the hatchback design. The price is high, but so is Conquest's performance.

Specifications

	3-door hatchback
Wheelbase, in.	95.9
Overall length, in.	173.2
Overall width, in.	66.3
Overall height, in.	51.8
Front track, in.	54.9
Rear track, in.	55.1
Turn diameter, ft.	31.5
Curb weight, lbs.	2822
Cargo vol., cu. ft.	NA
Fuel capacity, gal.	19.8
Seating capacity	4
Front headroom, in.	36.6

	3-door hatchback
Front shoulder room, in.	52.4
Front legroom, max., in.	40.7
Rear headroom, in.	35.4
Rear shoulder room, in.	51.2
Rear legroom, min., in.	29.1

Powertrain layout: longitudinal front engine/rear-wheel drive.

Engines	Turbo ohc I-4
Size, liters/cu. in.	2.6/156
Fuel delivery	TBI
Horsepower @ rpm	188 @ 5000
Torque (lbs./ft.) @ rpm	223 @ 2500
Availability	S
EPA city/highway mpg	
5-speed OD manual	18/22
4-speed OD automatic	18/24

Prices

Chrysler Conquest TSi	Retail Price	Dealer Invoice	Low Price
3-door hatchback	$18974	$15938	$17456
Destination charge	255	255	255

Standard Equipment:

2.6-liter turbocharged, intercooled TBI 4-cylinder engine, 5-speed manual transmission, power steering, power brakes, anti-lock rear braking system, limited-slip differential, power windows and locks, cruise control, tilt steering column, digital clock, tachometer, trip odometer, coolant temperature gauge and oil pressure gauges, voltmeter, cloth reclining front bucket seats with adjustable side bolsters, split folding rear seatback, cargo area security panel, rear defogger, anti-theft system, AM/FM ST ET cassette w/EQ and power antenna, fog lamps, remote fuel filler and hatch releases, bronze tinted glass, optical horn, automatic headlamp shutoff, warning lights (low fuel and washer fluid, door or liftgate ajar), lighted visor mirrors, heated power mirrors, variable intermittent wipers, 205/55VR16 front and 225/50VR16 rear tires on aluminum wheels.

Optional Equipment:

4-speed automatic transmission	718	596	657
Automatic air conditioning	955	793	874
Leather seats	372	309	341
Sunroof	292	242	267
Performance Handling Pkg.	205	170	188
225/50VR16 front and 245/45VR16 rear tires on alloy wheels, adjustable shock absorbers.			
Carpet protectors	28	23	26

Chrysler Fifth Avenue

What's New

A standard driver's-side air bag was added in mid-1988 and continues this fall as Chrysler's only rear-drive sedan heads into its swan-song season. Chrysler is preparing to introduce a front-drive replacement for the Fifth Avenue as

> **KEY: ohv** = overhead valve; **ohc** = overhead cam; **dohc** = double overhead cam; **I** = inline cylinders; **V** = cylinders in V configuration; **flat** = horizontally opposed cylinders; **bbl.** = barrel (carburetor); **PFI** = port (multi-point) fuel injection; **TBI** = throttle-body (single-point) fuel injection; **rpm** = revolutions per minute; **OD** = overdrive transmission; **S** = standard; **O** = optional; **NA** = not available.

Prices are accurate at time of printing; subject to manufacturer's change.

Chrysler

Chrysler Fifth Avenue

a 1990 model. It will be based on a stretched platform taken from the front-drive New Yorker. The 1990 Fifth Avenue's body will be about six inches longer than the current model's and it will use the New Yorker's Mitsubishi-made V-6 engine. Its sister ship will be a new Imperial model. On the current Fifth Avenue, the air bag supplements a conventional lap/shoulder belt and is contained in the steering wheel hub. It is one of the few such passive restraints used in conjunction with a tilt steering column. Changes for 1989 are that a new leather-wrapped steering wheel is included in the optional Luxury Equipment package and a power front passenger seat, formerly included only in the LE package, is now a free-standing option. The only available powertrain is a carbureted 5.2-liter V-8 and a 3-speed automatic transmission. The Fifth Avenue seats five and can tow up to 2000 pounds. It's plusher than its Dodge Diplomat and Plymouth Gran Fury siblings, which also are scheduled to be discontinued after 1989.

For

● Air bag ● Driveability ● Luxury appointments

Against

● Fuel economy ● Handling/roadholding ● Ride

Summary

Though there are some who will mourn the Fifth Avenue's passing, it won't be missed by us. It does heap on the comfort and convenience features, but that does nothing to hide its ancient rear-drive chassis, soft suspension and sloppy ride and handling. The Fifth Avenue bounces and bucks over bumps, and has slow, uncertain responses to the steering wheel. Its body leans heavily in turns and its nose dives in normal braking. Interior room is adequate for five, but there's not much luggage space. The proven drivetrain is among the few redeeming features. What the V-8 lacks in muscle, it compensates with a generally smooth power delivery. Alas, the venerable Chrysler 318-cubic inch engine is a heavy drinker and there isn't an overdrive top gear on the transmission to improve highway mileage. Drive the new New Yorker to see how far Chrysler has come in its interpretation of the American luxury car. If your goal is still a V-8 engine in a rear-drive chassis, start with the Ford LTD Crown Victoria/Mercury Grand Marquis.

KEY: **ohv** = overhead valve; **ohc** = overhead cam; **dohc** = double overhead cam; **I** = inline cylinders; **V** = cylinders in V configuration; **flat** = horizontally opposed cylinders; **bbl.** = barrel (carburetor); **PFI** = port (multi-point) fuel injection; **TBI** = throttle-body (single-point) fuel injection; **rpm** = revolutions per minute; **OD** = overdrive transmission; **S** = standard; **O** = optional; **NA** = not available.

Specifications

	4-door notchback
Wheelbase, in.	112.6
Overall length, in.	206.7
Overall width, in.	72.4
Overall height, in.	55.0
Front track, in.	60.5
Rear track, in.	60.0
Turn diameter, ft.	40.7
Curb weight, lbs.	3741
Cargo vol., cu. ft.	15.6
Fuel capacity, gal.	18.0
Seating capacity	6
Front headroom, in.	39.3
Front shoulder room, in.	56.0
Front legroom, max., in.	42.5
Rear headroom, in.	37.7
Rear shoulder room, in.	55.5
Rear legroom, min., in.	37.0

Powertrain layout: longitudinal front engine/rear-wheel drive.

Engines

	ohv V-8
Size, liters/cu. in.	5.2/318
Fuel delivery	2 bbl.
Horsepower @ rpm	140 @ 3600
Torque (lbs./ft.) @ rpm	265 @ 2000
Availability	S

EPA city/highway mpg
3-speed automatic	16/22

Prices

Chrysler Fifth Avenue	Retail Price	Dealer Invoice	Low Price
4-door notchback	$18045	$15538	$16792
Destination charge	495	495	495

Standard Equipment:

5.2-liter 2bbl. V-8 engine, 3-speed automatic transmission, power steering and brakes, driver's side air bag, automatic air conditioning, power windows, AM & FM ST ET, rear defogger, tinted glass, 60/40 bench seat with folding armrest and driver's seatback recliner, tilt steering column, power mirrors, cloth upholstery, coolant temperature gauge, ammeter, trip odometer, headlamps-on tone, intermittent wipers, padded landau roof, P205/75R15 SBR WSW tires.

Optional Equipment:

Power Convenience Pkg	680	578	626
Cruise control, power decklid release, power locks, power driver's seat.			
Luxury Equipment Discount Pkg.	1921	1633	1767
Power Convenience Pkg. plus vinyl bodyside molding, floormats, illuminated entry, leather/vinyl seats, leather-wrapped steering wheel, overhead console with compass and outside temperature readouts, power antenna, lighted visor mirrors, wire wheel covers.			
Pearl coat paint	41	35	38
California emissions pkg.	102	87	94
Power passenger seat	262	223	241
Requires Luxury Pkg.			
AM & FM ST ET cassette	262	223	241
Conventional spare tire	96	82	88
Power sunroof	928	789	854
Requires Luxury Pkg.; deletes overhead console.			
HD suspension	27	23	25
Undercoating	44	37	40
Wire wheel covers	212	180	195

Prices are accurate at time of printing; subject to manufacturer's change.

CONSUMER GUIDE®

Chrysler LeBaron

Chrysler LeBaron GTC Coupe

What's New

The 4-door notchback sedan and the station wagon have been dropped from the restructured LeBaron lineup. The carried-over front-drive coupe, convertible and 5-door hatchback get a newly available 2.5-liter turbocharged engine. The coupe and convertible also get standard 4-wheel disc brakes and a driver's-side air bag. Chrysler has pared its use of the LeBaron badge. Last year, in addition to the sedan and wagon, there was a LeBaron Coupe, a LeBaron Convertible and a Lebaron GTS 5-door hatchback. Those three body styles are now simply known as the coupe, convertible or hatchback sedan, all under a single LeBaron nameplate. Performance versions are identified by a GTC suffix for the coupe and convertible; GTS for the hatchback. They come with special exterior graphics, low-profile tires on alloy wheels and a handling suspension. The standard GTC/GTS engine is a turbocharged 2.2-liter 4-cylinder rated at 174-horsepower. That's a 28-horsepower increase over last year's 2.2 turbo, thanks in part to the addition of an intercooler. Chrysler designates this the Turbo II engine and makes it available only with a 5-speed manual transmission. A credit option on GTC/GTS cars is the Turbo I engine, a 150-horsepower 2.5 mated to a 3-speed automatic transmission. This engine also is optional on all other LeBarons. A naturally aspirated 2.5 is rated at 100 horsepower, up four from last year. This engine is standard on premium coupes and convertibles with automatic transmission, and on base coupes and convertibles and premium sedans with manual transmission. The base sedan has a naturally aspirated 2.2, with the 2.5 optional.

For

- Air bag (coupe and convertible) • Handling/roadholding
- Passenger and cargo room (hatchback)
- Performance (turbos)

Against

- Engine noise (turbos) • Ride • Manual shift linkage
- Rear seat and cargo room (coupe and convertible)

Summary

Too bad the road manners and the quality of assembly and interior materials of these front-drive cars don't measure up to their attractive looks. Coupes and convertibles are short on substance and long on style. And while the hatchback is unrefined compared to other mid-size sedans, foreign and domestic, its roomy, practical design and reasonable price are enough to keep it on our shopping list. On all LeBarons, the turbocharged engines provide good acceleration—very good in the case of the Turbo II. The naturally aspirated 2.5 is smoother and much easier to live with than the raucous turbos. GTC/GTS models improve upon already respectable handling, but their stiff suspensions don't offer much comfort and can cause the cars to skitter in fast driving on rough pavement. Chrysler has improved the 5-speed's shift action, though it's still not up to that of most imports, so we recommend automatic transmission for all versions.

Specifications

	2-door notchback	2-door conv.	5-door hatchback
Wheelbase, in.	100.3	100.3	103.1
Overall length, in.	184.9	184.9	180.4
Overall width, in.	68.4	68.4	68.3
Overall height, in.	50.9	52.2	53.0
Front track, in.	57.5	57.5	57.6
Rear track, in.	57.6	57.6	57.2
Turn diameter, ft.	38.1	38.1	36.2
Curb weight, lbs.	2769	2860	2659
Cargo vol., cu. ft.	33.4	14.0	42.6
Fuel capacity, gal.	14.0	14.0	14.0
Seating capacity	5	4	5
Front headroom, in.	37.6	38.3	38.3
Front shoulder room, in.	55.9	55.9	55.8
Front legroom, max., in.	43.1	43.1	41.1
Rear headroom, in.	36.3	37.0	37.9
Rear shoulder room, in.	56.3	45.7	55.9
Rear legroom, min., in.	33.0	33.0	36.5

Powertrain layout: transverse front engine/front-wheel drive.

Engines

	ohc I-4	ohc I-4	Turbo ohc I-4	Turbo ohc I-4
Size, liters/cu. in.	2.2/135	2.5/153	2.5/153	2.2/135
Fuel delivery	TBI	TBI	PFI	PFI
Horsepower @ rpm	93 @ 4800	100 @ 4800	150 @ 4800	174 @ 5200
Torque (lbs./ft.) @ rpm	122 @ 3200	135 @ 2800	180 @ 2000	200 @ 2400
Availability	S	S	O	S[1]
EPA city/highway mpg				
5-speed OD manual	24/34	24/34	20/25	20/28
3-speed automatic	24/28	23/29	18/23	

1. LeBaron GTC/GTS with 5-speed manual trans.

Prices

Chrysler LeBaron Coupe & Convertible	Retail Price	Dealer Invoice	Low Price
Highline 2-door notchback	$11695	$10325	$11110
Highline 2-door convertible	14195	12500	13448
Premium 2-door notchback	14695	12935	13915
Premium 2-door convertible	18195	15980	17188
GTC 2-door notchback	17519	15392	16556
GTC 2-door convertible	19680	17272	18576
Destination charge	440	440	440

Standard Equipment:

Highline: 2.5-liter TBI 4-cylinder engine, 5-speed manual transmission, power steering, power 4-wheel disc brakes, cloth reclining front bucket seats, tachometer, coolant temperature and oil pressure gauges, voltmeter, trip odometer, trip computer, center console, remote fuel door and decklid releases, tinted glass, variable intermittent wipers, message center, remote mirrors, visor mirrors, wide bodyside moldings, AM & FM ST ET, 195/70R14 SBR tires. **Convertible** adds: power top, power windows. **Premium** adds to Highline coupe: 3-speed automatic transmission, automatic air conditioning, overhead console with compass and outside temperature readouts,

Prices are accurate at time of printing; subject to manufacturer's change.

Chrysler

power locks, electronic instrument cluster, floormats, cornering lights, heated power mirrors, leather-wrapped steering wheel, two-tone paint. **Premium convertible** deletes overhead console and adds: illuminated entry, trunk dress-up, lighted visor mirrors, AM & FM ST ET cassette with Infinity speakers and power antenna, power driver's seat, cruise control, tilt steering column, undercoating. **GTC** deletes electronic instruments and cornering lights and adds: 2.2-liter intercooled Turbo II 4-cylinder engine and heavy-duty 5-speed manual transmission (2.5-liter Turbo I and automatic may be substituted at no charge), performance suspension, quick-ratio power steering, decklid luggage rack, 205/55R16 unidirectional SBR tires on alloy wheels.

Optional Equipment:

2.5-liter turbo engine, Highline & Premium .	698	593	642
GTC	NC	NC	NC
3-speed automatic transmission, Highline .	552	469	508
GTC	NC	NC	NC
Manual air conditioning, Highline	798	678	734
Deluxe Convenience Pkg.	309	263	284

Cruise control, tilt steering column; standard on Premium convertible and GTC.

Light Pkg., Highline, Premium coupe . . .	203	173	187
Power Convenience Pkg. I, Highline . . .	243	207	224

Power locks, heated power mirrors; requires Popular Pkg.

Power Convenience Pkg. II, Highline . . .	459	390	422

Pkg. I plus power windows.

Popular Equipment Pkg., Highline coupe .	1028	874	946
Highline conv.	1368	1163	1259

Coupe: Manual air conditioning, floormats, cruise control, tilt steering column, undercoating. Convertible adds: headlights with time delay, illuminated entry, lighted visor mirrors, power locks, heated power mirrors.

Luxury Equipment Pkg., Highline coupe . .	1898	1613	1746
Premium coupe	697	592	641

Manual air conditioning, floormats, headlights with time delay, illuminated entry, lighted visor mirrors, power windows and locks, heated power mirrors, power driver's seat, cruise control, tilt steering column, undercoating, leather-wrapped steering wheel.

Electronic Discount Pkg., Premium	494	420	454

Electronic Monitor, Electronic Navigator; coupe requires Luxury Pkg.

Two-tone paint, Highline	233	198	214
Leather seats, Premium coupe	646	549	594

Requires Luxury Pkg.

Premium leather seats w/power left, Premium conv.	1112	945	1023
Enthusiast leather power seats, GTC cpe .	646	549	594
Electronic Navigator, GTC	280	238	258
Power driver's seat, Highline, Premium coupe	247	210	227
Power windows, Premium coupe	216	184	199
AM & FM ST ET cassette, Highline	157	133	144
w/seek/scan, Premium coupe	569	484	523
CD player, Premium & GTC coupes	412	350	379
AM & FM ST ET cassette w/EQ, Highline .	754	641	694
w/seek/scan & EQ, Premium coupe . . .	785	667	722
Premium conv., GTC	216	184	199
Removable glass sunroof, Highline coupe .	409	348	376
w/o overhead console, Premium coupe .	229	195	211
Overhead console, Premium coupe	268	228	247
195/75R14 WSW touring tires, Highline & Premium	74	63	68
14" alloy wheels, Highline & Premium . . .	322	274	296
Sport suspension, Highline, Premium conv.	59	50	54
w/premium wheel covers, Highline	329	280	303
w/alloy wheels, Premium	651	553	599

Chrysler LeBaron

5-door hatchback	$11495	$10266	$10731
Premium 5-door hatchback	13495	12026	12611
GTS 5-door hatchback	17095	15194	15995
Destination charge	440	440	440

Standard Equipment:

2.2-liter TBI 4-cylinder engine, 5-speed manual transmission, power steering and brakes, cloth reclining bucket seats, tachometer, coolant temperature and oil pressure gauges, voltmeter, trip odometer, front console with storage, tinted glass, variable intermittent wipers, rear defogger, message center, remote mirrors, visor mirrors, remote fuel door and liftgate releases, wide bodyside moldings, AM & FM ST ET, removable shelf panel, 195/70R14 all-season SBR touring tires. **Premium** adds: 2.5-liter engine, air conditioning, power locks, power mirrors, lighted visor mirrors, floormats, Light Group. **GTS** adds: 2.2-liter intercooled Turbo II engine, power windows, illuminated entry, overhead console with compass and outside temperature readout, AM & FM ST ET cassette, cruise control, rear spoiler, tilt steering column, leather-wrapped steering wheel, sport suspension, 205/60R15 performance tires on alloy wheels.

Optional Equipment:

	Retail Price	Dealer Invoice	Low Price
2.5-liter engine, base	287	244	264
2.5-liter turbo engine, base & Premium . .	698	593	642
GTS w/automatic	NC	NC	NC

Base requires Popular or Luxury Pkg.

3-speed automatic transmission, base & Premium	552	469	508
GTS (credit w/2.5 Turbo I)	(180)	(153)	(153)
Air conditioning, base	798	678	734
Basic Equipment Pkg., base	899	764	827

Air conditioning, console extension with armrest, Light Group.

Popular Equipment Pkg., base w/5-speed .	1495	1271	1375
Base incl. automatic or turbo	1947	1655	1791
Base incl. automatic & turbo	2399	2039	2207

Basic Pkg. plus 2.5-liter engine, cruise control, floormats, illuminated entry, AM & FM ST ET cassette, tilt steering column, undercoating.

Luxury Equipment Pkg., base	2139	1818	1968
Premium	1423	1210	1309
Premium incl. 2.5 turbo	1996	1697	1836

Base: Popular Pkg. plus power windows and locks, heated power mirrors, leather-wrapped steering wheel, lighted visor mirrors. Premium adds: overhead console with compass and outside temperature readout, AM & FM ST ET cassette with Infinity speakers, power driver's seat.

Power Convenience Pkg., base	530	451	488

Power windows and locks, heated power mirrors.

Deluxe Convenience Pkg., base & Premium .	309	263	284

Cruise control, tilt steering column; base requires Basic Pkg.

Sport Handling Pkg., base	158	134	145
Base w/Luxury Pkg., Premium	97	82	89

Sport suspension, leather-wrapped steering wheel, all-season performance tires; base requires Popular or Premium Pkg.

Power locks, base	201	171	185

Requires Basic, Luxury or Sport Handling Pkg.

Electronic instruments, Premium	308	262	283

Requires Electronic Navigator.

California emissions pkg.	102	87	94
Electronic Navigator, Premium & GTS . . .	280	238	258

Premium requires Luxury Pkg.

Power driver's seat	247	210	227

Base requires Basic, Popular or Luxury Pkg.

Power windows, Premium	294	250	270
AM & FM ST ET cassette, base	157	133	144

Requires Basic, Popular or Luxury Pkg.

AM & FM ST ET cassette, Premium	472	401	434
Premium w/Popular or Luxury Pkg.	315	268	290
Cassette stereo w/EQ, Premium & GTS . .	216	184	199

Premium requires Luxury Pkg.

Power sunroof, base w/Luxury Pkg., Premium	799	679	735
Premium w/Luxury Pkg., GTS	619	526	569
205/60R15 tires on alloy wheels, Premium .	592	503	545
Premium w/Sport Handling Pkg.	522	444	480

Requires Luxury Pkg.

14" alloy wheels, base & Premium	322	274	296

Base requires Popular or Luxury Pkg.

Liftgate wiper/washer	130	111	120

Base requires Basic, Popular or Premium Pkg.

Prices are accurate at time of printing; subject to manufacturer's change.

CONSUMER GUIDE®

Chrysler New Yorker

Chrysler New Yorker Landau

What's New

Chrysler's flagship is refined for '89 with more power, a new 4-speed overdrive automatic transmission and some new convenience options. New Yorker was reintroduced as an all-new, front-drive 6-passenger sedan for 1988. It continues in a single 4-door notchback body style in base and Landau trim. A redesigned intake manifold and exhaust-flow improvements help boost the horsepower of its Mitsubishi-made 3.0-liter V-6 from 136 to 141 this year. Replacing a 3-speed automatic as the only available transmission is a new electronically controlled 4-speed overdrive automatic. Four-wheel disc brakes with an anti-lock feature continue as an option in place of the standard front disc/rear drum setup. Among new options are an 8-way power driver's seat with memory feature and a new anti-theft system. The optional electronic information center housed in an overhead console includes among its new warnings a "turn signal on" alert that's activated when the car has traveled more than half a mile at a speed of at least 15 mph with the turn signal on. The Landau has a vinyl landau roof, automatic power door locks that activate when the car reaches about 15 mph, 6-way power driver's seat, cruise control and a leather-wrapped steering wheel with a tilt column.

For

●Ride ●Anti-lock brakes ●Interior room

Against

●Performance ●Handling

Summary

A conservative sedan that evokes Chrysler's big-car luxury heritage, the New Yorker is smooth, quiet and unexciting. The V-6 has adequate power for most driving, but is severely taxed in the 3400-pound Landau on an upgrade with a full load and the air conditioner on. Likewise, the car's front-drive manners feature good around-town handling, but succumb to tire-squealing body lean at the first hint of spirited cornering. The chassis filters out most bumps, yet the car doesn't float excessively at speed. The available anti-lock brakes are a big plus. The interior will seat five adults in true comfort, through three in front is a bit of a squeeze. Soft, flat seats coddle your body rather than support it. Dashboard controls are generally easy to reach and well labeled. Trunk space is good. Overall, Chrysler took no chances with the New Yorker's styling or performance. Much could have been done with this basic package to appeal to younger, driving-oriented buyers. But the traditional-luxury audience will be well served by a car that's vastly more competent than the old E-body New

Yorker Turbo or the aged rear-drive Chrysler Fifth Avenue. In stretched form, the New Yorker will form the basis for the 1990 Fifth Avenue and Chrysler Imperial.

Specifications

	4-door notchback
Wheelbase, in.	104.3
Overall length, in.	193.6
Overall width, in.	68.5
Overall height, in.	53.5
Front track, in.	57.0
Rear track, in.	57.6
Turn diameter, ft.	40.5
Curb weight, lbs.	3214
Cargo vol., cu. ft.	16.5
Fuel capacity, gal.	16.0
Seating capacity	6
Front headroom, in.	38.3
Front shoulder room, in.	56.4
Front legroom, max., in.	41.9
Rear headroom, in.	37.8
Rear shoulder room, in.	55.9
Rear legroom, min., in.	38.7

Powertrain layout: transverse front engine/front-wheel drive.

Engines

	ohc V-6
Size, liters/cu. in.	3.0/181
Fuel delivery	PFI
Horsepower @ rpm	141 @ 5000
Torque (lbs./ft.) @ rpm	171 @ 2800
Availability	S

EPA city/highway mpg

4-speed OD automatic	18/26

Prices

Chrysler New Yorker	Retail Price	Dealer Invoice	Low Price
4-door notchback	$17416	$15004	$16310
Landau 4-door notchback	19509	16783	18246
Destination charge	495	495	495

Standard Equipment:

3.0-liter PFI V-6 engine, 4-speed automatic transmission, power steering and brakes, automatic air conditioning, 50/50 cloth bench seat with armrest, front seatback storage pockets, power windows, tinted glass, heated power mirrors, intermittent wipers, optical horn, coolant temperature and oil pressure gauges, voltmeter, low-fuel light, message center, visor mirrors, AM & FM ST ET, cornering lights, 195/75R14 SBR WSW tires. **Landau** adds: landau vinyl roof, automatic rear load leveling, power driver's seat, speed-activated power door locks, cruise control, tilt steering column, leather-wrapped steering wheel, electronic instruments, trip computer, upgraded wheel covers.

Optional Equipment:

Anti-lock 4-wheel disc brakes	954	811	878
Base requires Luxury Pkg.			

KEY: ohv = overhead valve; ohc = overhead cam; dohc = double overhead cam; I = inline cylinders; V = cylinders in V configuration; flat = horizontally opposed cylinders; bbl. = barrel (carburetor); PFI = port (multi-point) fuel injection; TBI = throttle-body (single-point) fuel injection; rpm = revolutions per minute; OD = overdrive transmission; S = standard; O = optional; NA = not available.

	Retail Price	Dealer Invoice	Low Price
Luxury Equipment Pkg., base	1557	1323	1432
Landau	1962	1668	1805

Base: automatic power locks, bodyside molding, electronic instruments, cruise control, floormats, illuminated entry, lighted visor mirrors, leather-wrapped steering wheel, automatic rear load leveling, tilt steering column, trip computer, undercoating, wire wheel covers. Landau adds: overhead console with electronic information center, power antenna, memory power seats.

	Retail Price	Dealer Invoice	Low Price
Mark Cross Edition, base	2140	1819	1969
Landau	2545	2163	2341

Luxury Pkg. plus leather/vinyl 50/50 power front seats, upgraded door panels.

	Retail Price	Dealer Invoice	Low Price
Interior Illumination Pkg.	197	167	181

Illuminated entry, lighted visor mirrors.

Deluxe Convenience Pkg., base	309	263	284

Cruise control, tilt steering column.

Overhead console w/electronic info center, base	690	587	635

Requires Luxury Pkg.

Automatic power locks, base	294	250	270
AM/FM ST ET cassette	262	223	241
AM/FM Cassette Pkg., Landau w/Luxury Pkg.	497	422	457
Base or Landau w/o Luxury Pkg.	569	484	523

Includes 6 Infinity speakers and power antenna.

AM/FM Cassette w/EQ Pkg., Landau	713	606	656
Base	785	667	722

Requires Luxury or Mark Cross Pkg.

Security alarm	150	128	138

Requires power locks.

Power driver's seat, base	247	210	227
Memory power seats, base	351	298	323

Requires Luxury Pkg.

Power sunroof	799	679	735

Base requires Luxury Pkg.

Road handling suspension, base	59	50	54
Automatic rear load leveling, base	185	157	170
Conventional spare tire	85	72	78
Wire wheel covers, Landau	231	196	213
Alloy wheels, base w/Luxury or Mark Cross Pkg.	42	36	39
Landau	273	232	251

Chrysler's TC by Maserati

Chrysler's TC by Maserati

What's New

After several years of development, Chrysler is finally set this fall to introduce its first-ever 2-seat luxury coupe. Intended as an image leader, the $30,000 front-drive TC will be assembled in Milan, Italy, by Chrysler's Italian partner, Maserati. A folding convertible top and removable hardtop, leather interior and a full complement of luxury equipment are standard. There are no extra-cost options, but a pair of engine/transmission teams is available. The most sporting is a 200-horsepower turbocharged and intercooled 2.2-liter four. Its dual-overhead cam, 16-valve cylinder head was designed by Maserati and it's available only with a Getrag 5-speed manual transmission. When ordered with a Chrysler 3-speed automatic, the TC gets a 160-horsepower turbocharged and intercooled 8-valve 2.2 four. This is essentially the Turbo II engine that's used with manual transmissions in several domestic Chrysler cars, where it is rated at 174 horsepower. All TCs have a Maserati-tuned suspension and anti-lock, 4-wheel disc brakes. The convertible top is manually operated, but an electric motor pulls the fabric taut when the top is up. When down, the top is stowed beneath a metal tonneau cover. TC's removable fiberglass hardtop is notable for the circular glass opera windows in its rear pillars. It also has a glass rear window with an electric defroster.

For

● Exclusivity ● Styling

Against

● Price ● Performance ● Refinement

Summary

Chrysler describes the TC not as its bid for the ultimate European road car, but as a "beautiful little high-performance convertible with a designer label on it." After a test drive at Chrysler's Michigan proving grounds, we doubt whether the TC's performance or its label will score a hit in the demanding sports-luxury market. As for its beauty, the car is attractive, but it looks very much like a Chrysler LeBaron convertible, which lists for about $8000 less, fully equipped. The pre-production 200-horsepower TC we drove was indeed quick, with an engine that pressed us back in the seat and sounded great above 3000 rpm. Precise steering, strong brakes and good handling were evident, though the limp seats offered little support in hard cornering. The clutch was extremely heavy, however, and the notchy shifter had to be forced through most gear changes. Mechanical shortcomings were common in pre-production examples we drove, but other problems soured the car's luxury-touring air. The cowl shook noticeably over bumps and the dashboard controls appeared borrowed from the Chrysler parts bin. Most TCs will be sold with automatic transmission, and Chrysler reportedly is considering offering a version of its Mitsubishi-made 3.0-liter V-6 for that application. The TC needs more than a Japanese V-6 to earn its stripes in this class.

Specifications

	2-door conv.
Wheelbase, in.	93.3
Overall length, in.	175.8
Overall width, in.	68.5
Overall height, in.	51.9
Front track, in.	57.5
Rear track, in.	57.6
Turn diameter, ft.	39.2
Curb weight, lbs.	3250
Cargo vol., cu. ft.	NA
Fuel capacity, gal.	14.0
Seating capacity	2
Front headroom, in.	37.4
Front shoulder room, in.	55.9
Front legroom, max., in.	42.8
Rear headroom, in.	—

Prices are accurate at time of printing; subject to manufacturer's change

	2-door conv.
Rear shoulder room, in.	—
Rear legroom, min., in.	—

Powertrain layout: transverse front engine/front-wheel drive.

Engines	Turbo ohc I-4	Turbo dohc I-4
Size, liters/cu. in.	2.2/135	2.2/135
Fuel delivery	PFI	PFI
Horsepower @ rpm	160 @ 5200	200 @ 5500
Torque (lbs./ft.) @ rpm	171 @ 3600	220 @ 3400
Availability	O	S
EPA city/highway mpg		
5-speed OD manual		21/28
3-speed automatic		18/22

Prices not available at time of publication.

Daihatsu Charade

1988 Daihatsu Charade CSX

What's New

Daihatsu had a limited presence in 1988, its first year in the U.S., when imports were limited to 11,500 Charade mini-compacts available in Hawaii and 10 continental states in the south and west. For 1989, Daihatsu's allotment under Japan's voluntary quota system grows to 17,000 cars, and the company plans to expand into several southeastern states. The 1989 Charade gets a revised lineup with more drivetrain choices. Next summer, Daihatsu plans to introduce a 4-door sedan that's larger than Charade, followed by a new 4-wheel-drive sport utility vehicle called Rocky II. Eventually, Daihatsu plans to offer a full line of vehicles in the U.S. A 1.3-liter 4-cylinder engine and a 3-speed automatic transmission are scheduled to arrive in January for the front-drive Charade. Until then, all Charades will come only with a 53-horsepower 1.0-liter 3-cylinder engine and a 5-speed manual transmission. In January, the lineup changes as follows: A base CES model will use the 53-horsepower 3-cylinder engine and the 5-speed manual. A mid-level CLS will be available with either the 3-cylinder or the new 4-cylinder engine, rated at 90 horsepower. A top-line CLX will come only with the 4-cylinder engine. The new automatic transmission will be optional with the 4-cylinder engine on the CLS and CLX.

For

- Fuel economy   • Instruments/controls

Against

- Performance   • Ride   • Passenger and cargo room
- Noise

Summary

Daihatsu has positioned Charade as an upscale small car, not an entry-level, low-budget model. Base prices on 1988 models ranged from nearly $6500 to nearly $9500, with the top-line CSX model having air conditioning and a stereo with cassette player as standard equipment. That will change for 1989 because some buyers don't want that much equipment on a small car, Daihatsu discovered. The same money buys a Honda Civic hatchback with a smoother and livelier 4-cylinder engine, sportier road manners and equal or better economy. Or, you can have a loaded Dodge Omni/Plymouth Horizon America and enjoy more room and power. Charade's 3-cylinder engine is willing, but too underpowered to be acceptable all-around transportation. You have to rev the engine hard even when going less than all-out. The tiny engine zings to its recommended 6000-rpm limit with ease, though with a lot of noise and throbbing. Charade hops and skates through bumpy bends on its firm suspension, which combines with the short wheelbase for a choppy ride on all but glass-smooth roads.

1988 Specifications	3-door hatchback
Wheelbase, in.	92.1
Overall length, in.	144.9
Overall width, in.	63.6
Overall height, in.	54.5
Front track, in.	54.5
Rear track, in.	53.7
Turn diameter, ft.	31.5
Curb weight, lbs.	1775
Cargo vol., cu. ft.	33.0
Fuel capacity, gal.	10.6
Seating capacity	4
Front headroom, in.	37.5
Front shoulder room, in.	51.1
Front legroom, max., in.	41.1
Rear headroom, in.	35.7
Rear shoulder room, in.	51.9
Rear legroom, min., in.	31.8

Powertrain layout: transverse front engine/front-wheel drive.

Engines	ohc I-3
Size, liters/cu. in.	1.0/61
Fuel delivery	PFI
Horsepower @ rpm	53 @ 5200
Torque (lbs./ft.) @ rpm	58 @ 3600
Availability	S

> **KEY: ohv** = overhead valve; **ohc** = overhead cam; **dohc** = double overhead cam; **I** = inline cylinders; **V** = cylinders in V configuration; **flat** = horizontally opposed cylinders; **bbl.** = barrel (carburetor); **PFI** = port (multi-point) fuel injection; **TBI** = throttle-body (single-point) fuel injection; **rpm** = revolutions per minute; **OD** = overdrive transmission; **S** = standard; **O** = optional; **NA** = not available.

EPA city/highway mpg	ohc I-3
5-speed OD manual	38/42

Prices

Daihatsu Charade (1988 prices)	Retail Price	Dealer Invoice	Low Price
CLS 3-door hatchback	$6397	$5822	$6110
CLX 3-door hatchback	7650	6733	7192
w/Automatic Restraint System	7725	6798	7262
CSX 3-door hatchback	9232	8124	8678
Destination charge	257	257	257

Standard Equipment:

1.0-liter PFI 3-cylinder engine, 5-speed manual transmission, power brakes, cloth reclining front bucket seats, folding rear seat, tinted windshield, remote fuel door and liftgate releases, visor mirrors, rear defogger, intermittent wipers, trip odometer, 145/80R13 tires. **CLX** adds: digital clock, door pockets, power mirrors, bodyside moldings, front mud guards, sport front seats, tachometer, rear wiper/washer, 155/80R13 tires, wheel covers. **CSX** adds: air conditioning, AM/FM ST ET cassette, 165/70R13 tires on alloy wheels.

Optional Equipment:

Power Option Pkg.	305	253	279
Power windows and locks.			
Air conditioning, CLS & CLX	695	591	643
Air conditioning delete (credit), CSX	(695)	(591)	(591)
AM/FM ST ET cassette, CLS, CLX	295	245	270
Premium AM/FM cassette, CLS, CLX	395	328	362

Dodge Aries America

Dodge Aries 4-door

What's New

Dodge drops the Aries station wagon and makes only minor changes in its K-car sedans for '89. The front-drive 2-and 4-door notchbacks now wear the "America" label, denoting Chrysler's program to reduce manufacturing and retail costs by offering a single trim level with most options grouped into packages. Aries' base engine remains a 93-horsepower 2.2-liter 4-cylinder. It comes with a 5-speed manual transmission standard; a 3-speed automatic is optional. Available at extra cost is a 2.5-liter four, which goes from 96 to 100 horsepower this year. It's available only with the automatic transmission. Both engines have been treated to slight internal modifications that Dodge says will make them run quieter. And Dodge says new front-suspension components will reduce ride harshness. Among the few new features for '89 are identifying paint markings for underhood service points such as the dipstick and coolant overflow-bottle cap. There's also a new 4-speaker stereo

system. Front bucket seats are standard; a front bench, air conditioning and deluxe stereos are among the few free-standing options.

For

- ●Value   ●Passenger room   ●Fuel economy

Against

- ●Handling   ●Engine noise (2.2)

Summary

K-cars offer few frills and even fewer thrills, providing instead functional transportation at a reasonable price. The standard 2.2-liter is adequate, but the optional 2.5 provides stronger acceleration without using much more gas. It's also smoother and quieter, thanks to its internal twin balance shafts. The 2.5 is available only with automatic transmission as part of the optional Popular or Premium Discount packages. That raises the price, but the automatic is preferable to the poor-shifting 5-speed manual. Aries has modest handling ability and is most composed when driven gently. Even then, a freeway cloverleaf will expose its poor cornering ability and rough pavement will reveal its undisciplined suspension control. You'll be most comfortable with the standard front bucket seats and full center console. The 3-place front bench seat (a no-cost option) increases passenger capacity to six, but it has a fixed seatback that's so reclined it forces an uncomfortable driving position and drastically reduces rear-seat leg room. The bucket seats have adjustable seatbacks that eliminate those problems. While K-cars are not the most solidly built sedans, they're still a fine value.

Specifications

	2-door notchback	4-door notchback
Wheelbase, in.	100.3	100.3
Overall length, in.	178.9	178.9
Overall width, in.	67.9	67.9
Overall height, in.	52.5	52.5
Front track, in.	57.6	57.6
Rear track, in.	57.2	57.2
Turn diameter, ft.	34.8	34.8
Curb weight, lbs.	2317	2323
Cargo vol., cu. ft.	15.0	15.0
Fuel capacity, gal.	14.0	14.0
Seating capacity	6	6
Front headroom, in.	38.2	38.6
Front shoulder room, in.	55.0	55.4
Front legroom, max., in.	42.4	42.2
Rear headroom, in.	37.0	37.8
Rear shoulder room, in.	58.8	55.9
Rear legroom, min., in.	35.1	35.4

Powertrain layout: transverse front engine/front-wheel drive.

Engines	ohc I-4	ohc I-4
Size liters/cu. in.	2.2/135	2.5/153
Fuel delivery	TBI	TBI
Horsepower @ rpm	93 @ 4800	100 @ 4800
Torque (lbs./ft.) @ rpm	122 @ 3200	135 @ 2800
Availability	S	O

EPA city/highway mpg		
5-speed OD manual	25/34	
3-speed automatic	24/30	23/28

Prices are accurate at time of printing; subject to manufacturer's change

Prices

Dodge Aries	Retail Price	Dealer Invoice	Low Price
2-door notchback	7595	6935	7365
4-door notchback	7595	6935	7365
Destination charge	454	454	454

Standard Equipment:

2.2-liter TBI 4-cylinder engine, 5-speed manual transmission, power brakes, reclining front bucket seats (bench seat may be substituted at no charge, but requires Popular Equipment Discount Pkg.), cloth and vinyl upholstery, intermittent wipers, optical horn, left remote mirror, right visor mirror with map/reading light, 185/70R13 SBR tires.

Optional Equipment:

2.5-liter engine	279	237	257
Requires Basic or Popular Pkg.			
Power steering	240	204	221
Air conditioning	775	659	713
Requires tinted glass.			
Tinted glass	120	102	110
Basic Equipment Pkg	776	660	714
3-speed automatic transmission, power steering.			
Basic Radio Discount Pkg	929	790	855
Basic Pkg. plus AM & FM ST ET.			
Popular Equipment Pkg., w/bucket seats .	1292	1098	1189
w/bench seat	1192	1013	1097
Basic Pkg. (including radio) plus tinted glass, remote mirrors, bodyside tape stripes, added sound insulation, floor console (with bucket seats), trunk dress up, 185/70R14 tires, wheel covers.			
Premium Pkg., 2-door w/bucket seats . .	1730	1471	1592
2-door w/bench seat	1630	1386	1500
4-door w/bucket seats	1780	1513	1638
4-door w/bench seat	1680	1428	1546
Popular Pkg. plus cruise control, tilt steering column, power door locks, luxury steering wheel, Light Group.			
500-amp battery, w/2.2	44	37	40
Rear defogger	145	123	133
California emissions pkg	100	85	92
AM & FM ST ET cassette	152	129	140
Requires Popular Pkg.			
Conventional spare tire, 13"	73	62	67
14"	83	71	76
185/70R14 WSW tires	68	58	63
Requires Popular Pkg.			

Dodge Caravan

What's New

A turbocharged engine and a new 4-speed automatic transmission are Caravan newsmakers. Caravan, nearly identical to the Plymouth Voyager, is available in base, SE and LE trim in a 175.9-inch-long body on a 112-inch wheelbase, and as the Grand Caravan in SE and LE trim in a 190.5-inch-body on a 119-inch wheelbase. The new turbo 2.5-liter 4-cylinder packs 150 horsepower, 50 more than its naturally aspirated cousin. It's optional on short-wheelbase SE and LE models with either a 5-speed manual transmission or a 3-speed automatic. Chrysler's new 4-speed overdrive automatic is available only with the 141-horsepower, Mitsubishi-made 3.0-liter V-6. The V-6 is standard in the Grand Caravan LE and optional in the Grand Caravan SE and the short-wheelbase LE. Among other Caravan additions for '89, Dodge says ride harshness is reduced through use of some front suspension components from the luxury Dodge Dynasty and SE models have improved interior space thanks to thinner high-back front bucket

Dodge Grand Caravan

seats. Leather seating is a new option on LE models and a power motor to open and close rear-quarter vent windows is a new option for SEs and LEs. A tachometer is now included in the optional gauge package, air conditioning is standard on LE models, and a power liftgate release is now standard on SE and LE models. An optional power sunroof is due late in the 1989 model year.

For

● Passenger room ● Cargo room ● Ride/handling

Against

● Performance (4-cylinder) ● Fuel economy

Summary

We drove a turbocharged Caravan at Chrysler's Michigan proving grounds and found the engine somewhat mismatched to minivan duty. It provides good off-the-line acceleration and fine highway passing power, but due to turbo lag the engine tends to deliver power in bursts more suited to sporty cars. And though it could return better mileage than the V-6, the turbo four is not nearly as quiet or as smooth-running. The V-6 is the best choice, justifying its extra cost over the base 2.5-liter four with significantly more horsepower and torque. The new automatic transmission's principal benefit is its overdrive fourth gear, which should make for more relaxed and economical highway driving. Chrysler has wisely done little else to alter the character of its front-drive minivans, which remain our favorites. They have better traction than their rear-drive rivals, superior highway stability and more car-like ride and handling. The short-wheelbase models accommodate five, six or seven passengers, depending on seating configuration. The stretched versions seat up to eight and exceed nearly all competitors in roominess. The main drawback to the Caravan is a steep price fueled by strong demand.

Specifications	4-door van	4-door van
Wheelbase, in.	112.0	119.1
Overall length, in.	179.5	190.5
Overall width, in.	69.6	69.6
Overall height, in.	64.2	65.0

KEY: ohv = overhead valve; **ohc** = overhead cam; **dohc** = double overhead cam; **I** = inline cylinders; **V** = cylinders in V configuration; **flat** = horizontally opposed cylinders; **bbl.** = barrel (carburetor); **PFI** = port (multi-point) fuel injection; **TBI** = throttle-body (single-point) fuel injection; **rpm** = revolutions per minute; **OD** = overdrive transmission; **S** = standard; **O** = optional; **NA** = not available.

Prices are accurate at time of printing; subject to manufacturer's change

Dodge

	4-door van	4-door van
Front track, in.	59.9	59.9
Rear track, in.	62.1	62.1
Turn diameter, ft.	41.0	43.2
Curb weight, lbs.	3003	3304
Cargo vol., cu. ft.	125.0	150.0
Fuel capacity, gal.	15.0[1]	15.0[1]
Seating capacity	7	8
Front headroom, in.	39.0	39.0
Front shoulder room, in.	58.4	58.4
Front legroom, max., in.	38.2	38.2
Rear headroom, in.	37.7	37.6
Rear shoulder room, in.	61.3	61.3
Rear legroom, min., in.	37.7	37.8

1. 20.0 gal opt.

Powertrain layout: transverse front engine/front-wheel drive.

Engines

	ohc I-4	Turbo ohc I-4	ohc V-6
Size, liters/cu. in.	2.5/153	2.5/153	3.0/181
Fuel delivery	TBI	PFI	PFI
Horsepower @ rpm	100 @ 4800	150 @ 4800	141 @ 5000
Torque (lbs./ft.) @ rpm	135 @ 2800	180 @ 2000	171 @ 2800
Availability	S	O	O[1]

EPA city/highway mpg

5-speed OD manual	21/28	18/25	
3-speed automatic	21/23	18/21	18/22
4-speed OD automatic			18/23

1. Standard, Grand Caravan LE

Prices

Dodge Caravan	Retail Price	Dealer Invoice	Low Price
Base SWB 4-door van	$11312	$10105	$10909
SE SWB 4-door van	12039	10744	11592
LE SWB 4-door van	13987	12459	13423
Grand SE 4-door van	13061	11644	12553
Grand LE 4-door van	16362	14549	15656
Destination charge	500	500	500

SWB denotes short-wheelbase models.

Standard Equipment:

2.5-liter TBI engine, 5-speed manual transmission, power steering, power brakes, liftgate wiper/washer, headlamps-on tone, 5-passenger seating, variable intermittent wipers, tinted glass, left remote mirror, AM & FM ST ET, P185/75R14 SBR tires. **SE** adds: highback reclining front seats, rear seat (Grand), front folding center armrests, upgraded door panels, power liftgate release. **LE** adds: front air conditioning, added sound insulation, remote mirrors, bodyside moldings, woodgrain exterior applique, upgraded steering wheel. **Grand LE** adds: 3.0-liter PFI V-6, 4-speed automatic transmission.

Optional Equipment:

2.5-liter turbo engine, SWB	680	578	626
3.0-liter V-6, SE & LE SWB	680	578	626
3-speed automatic transmission (NA 7-pass.)	565	480	520
4-speed automatic, LE SWB & Grand SE	735	625	676
Requires 3.0 V-6.			
Front air conditioning, base & SE	840	714	773
Rear air conditioning w/heater, Grand	560	476	515
Cloth lowback seats, base	45	38	41
Leather highback seats, LE	671	570	617
Requires Luxury Equipment or LE Decor Pkg.			
7-pass. seating, SWB	389	331	358
8-pass. seating, SE Grand	120	102	110
SE Grand w/Popular Pkg	5	4	5

	Retail Price	Dealer Invoice	Low Price
Converta-Bed, SWB (NA base)	542	461	499
Value Wagon Pkg., base w/5-speed	1045	888	961
Base incl. automatic	1283	1091	1180
SE SWB w/5-speed	994	845	914
SE SWB incl. automatic	1232	1047	1133

Front air conditioning, rear defogger, dual horns, deluxe sound insulation, Light Group, high-back cloth reclining front seats.

Popular Equipment Discount Pkg., SE SWB	1661	1412	1528
SE SWB w/Value Wgn Pkg	743	632	684
SE SWB w/7/8-pass.	1683	1431	1548
SE SWB w/7/8-pass. & Value Wgn Pkg	765	650	704
SE Grand	1683	1431	1548
LE SWB	981	834	903
LE SWB w/7-pass, LE Grand	1003	853	923

SE: front air conditioning, forward storage console, overhead console, floormats, gauges, deluxe sound insulation, Light Group, power rear quarter vent windows, conventional spare tire, cruise control, tilt steering column. LE adds: lighted visor mirror, power door locks.

Luxury Equipment Discount Pkg., SE SWB	2348	1996	2160
SE SWB w/Value Wgn Pkg	1485	1262	1366
SE SWB w/7/8-pass., SE Grand	2371	2015	2181
SE SWB w/7-pass. & Value Wgn Pkg	1507	1281	1386
LE SWB	1603	1363	1475
LE SWB w/7-pass, LE Grand	1626	1382	1496

SE: Popular Pkg. plus lighted visor mirror, power door locks, power mirrors, power front windows, AM & FM ST ET cassette, Eurosport steering wheel. LE adds: power driver's seat.

Turbo Sport Pkg., SE SWB	1073	912	987

2.5-liter turbo engine, gauges, 205/70R15 all-season tires on alloy wheels.

HD Trailer Tow Pkg., SE Grand	392	333	361

HD suspension, 120-amp alternator, conventional spare tire, trailer wiring harness, 205/70R15 all-season tires on styled steel wheels.

LE Decor Pkg., LE SWB	1449	1232	1333

2.5-liter turbo engine, warm silver fascia, moldings, 7-passenger seating, tape stripes, 205/70R15 all-season tires on alloy wheels.

Rear defogger	165	140	152
California emissions pkg	99	84	91
Sunscreen glass	406	345	374
Rear heater, SE & LE SWB	329	280	303
120-amp alternator, Grand	62	53	57
Luggage rack, base & SE	144	122	132
Power door locks	203	173	187
Requires Value Wgn or Popular Pkg.			
AM & FM ST ET cassette, base & SE	152	129	140
Requires Value Wagon or Popular Pkg.			
Ultimate Sound stereo	214	182	197
Requires Luxury or LE Decor Pkg.			
AM & FM ST ET delete (credit)	(136)	(116)	(116)
Cruise control, base	207	176	190
Requires tilt steering column and Value Wgn Pkg.			
Tilt steering column, base	122	104	112
Requires cruise control and Value Wgn Pkg.			
HD suspension	68	58	63
Conventional spare tire	104	88	96
Wire wheel covers, LE (Luxury Pkg. req.)	239	203	220
Sport road wheels (NA base)	415	353	382
Pearl coat paint	46	39	42
Two-tone paint, SE	236	201	217

Dodge/Plymouth Colt

What's New

An all-new 3-door hatchback and a full-time 4-wheel drive version of the carried-over 5-door station wagon are major additions to the Colt line. The lineup no longer includes a 4-door body style. Colts are built by Chrysler's Japanese partner, Mitsubishi, and sold in identical form under the Dodge and Plymouth nameplates. Styling of the new front-drive hatchback is much more rounded than the previous version. Its subcompact 94-inch wheelbase is virtually unchanged, though the body is longer, wider and taller. Rear-

Prices are accurate at time of printing; subject to manufacturer's change

Dodge Colt 3-door

seat leg room is up by nearly two inches and there's more cargo room. The hatchback line ranges from an entry-level model with a 4-speed manual transmission, all-vinyl interior and mandatory manual steering, through the Colt E and sporty GT models. The latter two offer a choice of a 5-speed manual or a 3-speed automatic. All three use a fuel-injected 1.5-liter 4-cylinder engine rated at 81 horsepower, 13 more than the carbureted version in the previous-generation Colt. Top of the line is the GT DOHC Turbo. Its turbocharged 16-valve 1.6-liter four is rated at 135 horsepower—30 more than its 1988 counterpart. A 5-speed, power steering, sport suspension and aero body add-ons are standard. Inside, there's a tachometer and a tilt and telescope steering column. Mitsubishi dealers market the new hatchback, and a 4-door, under the Mirage name. The front-drive, 5-passenger, 5-door wagon gets the hatchback's 1.5-liter four and new interior trim. The 4-wheel-drive wagon features a permanently engaged 4WD system that uses a center differential with a viscous coupling to split power among the four wheels as needed to maintain traction. Its only powertrain is an 87-horsepower 1.8-liter four mated to a 5-speed manual. The 4WD wagon's body is raised about two inches to provide additional ground clearance, its rear track is an inch wider and it has 14-inch all-season tires versus the front-driver's 13s.

For

- Performance (Turbo)   •Fuel economy
- Handling   •4WD traction

Against

- Performance (automatic transmission)
- Cargo room (hatchbacks)

Summary

We drove a selection of Colts at Chrysler's Michigan proving grounds and found the hatchbacks vastly improved over the previous generation, which was introduced for 1985. Their trim, aeroshapes, build quality and over-the-road feel all recall the pace-setting Honda Civic hatchbacks. The base model with the 4-speed and manual steering has a pleasing, no-nonsense manner. The GT DOHC Turbo is slick and quick, with fluid clutch and gearshift action, direct steering, a rev-happy engine and tenacious handling. Both provide decent space for four adults, though rear headroom is at a premium. We accelerated hard in loose gravel and on wet pavement, trying in vain to spin a tire on the 4WD wagon. But its viscous coupling maintained traction by quickly distributing power to the wheels that needed it. Power is adequate with liberal use of the 5-speed manual, and the wagon seems free of the drivetrain friction that plagues some other 4WD passenger cars.

Dodge

Specifications

	3-door hatchback	5-door wagon
Wheelbase, in.	93.9	93.7
Overall length, in.	158.7	169.3
Overall width, in.	65.5	64.4
Overall height, in.	54.1	55.9
Front track, in.	56.3	55.5
Rear track, in.	56.3	52.8
Turn diameter, ft.	32.4	32.4
Curb weight, lbs.	2195	2271[1]
Cargo vol., cu. ft.	34.7	60.4
Fuel capacity, gal.	34.7	60.4
Seating capacity	5	5
Front headroom, in.	38.3	37.7
Front shoulder room, in.	53.5	52.8
Front legroom, max., in.	41.9	40.6
Rear headroom, in.	36.9	38.2
Rear shoulder room, in.	52.1	52.8
Rear legroom, min., in.	32.5	34.1

1. 2568 lbs., 4WD wagon.

Powertrain layout: transverse front engine/front-wheel drive or permanent 4WD.

Engines

	ohc I-4	Turbo dohc I-4	ohc I-4
Size, liters/cu. in.	1.5/90	1.6/97	1.8/110
Fuel delivery	PFI	PFI	PFI
Horsepower @ rpm	81 @ 5500	135 @ 6000	87 @ 5000
Torque (lbs./ft.) @ rpm	91 @ 3000	141 @ 3000	102 @ 3000
Availability	S	O	S[1]

EPA city/highway mpg

5-speed OD manual	30/35	23/29	23/28
3-speed automatic	27/29		

1. Colt DL 4WD wagon.

Prices

Dodge/Plymouth Colt	Retail Price	Dealer Invoice	Low Price
3-door hatchback	$6678	$6110	$6494
E 3-door hatchback	7505	6822	7164
GT 3-door hatchback	8863	7711	8287
Destination charge	255	217	235

Standard Equipment:

1.5-liter PFI 4-cylinder engine, 4-speed manual transmission, power brakes, vinyl bucket seats, center console with storage, split folding rear seatback, coolant temperature gauge, trip odometer, motorized front shoulder belts and manual lap belts, outboard rear lap/shoulder belts, locking fuel-filler door, 145/80R13 SBR tires. **E** adds: 5-speed manual transmission, rear defogger, cloth seat inserts, bodyside moldings, 155/80R13 tires. **GT** adds: wide bodyside moldings, remote left mirror, assist grips, cloth upholstery, rear security panel.

Optional Equipment:

3-speed automatic transmission, GT	499	414	459
Power steering, E & GT	259	215	238

KEY: ohv = overhead valve; **ohc** = overhead cam; **dohc** = double overhead cam; **I** = inline cylinders; **V** = cylinders in V configuration; **flat** = horizontally opposed cylinders; **bbl.** = barrel (carburetor); **PFI** = port (multi-point) fuel injection; **TBI** = throttle-body (single-point) fuel injection; **rpm** = revolutions per minute; **OD** = overdrive transmission; **S** = standard; **O** = optional; **NA** = not available.

Prices are accurate at time of printing; subject to manufacturer's change

Dodge

	Retail Price	Dealer Invoice	Low Price
Air conditioning	739	613	680
Turbo Pkg., GT	2819	2340	259?

1.6-liter DOHC turbo engine, power steering, 4-wheel disc brakes, 1?? 60R14 tires, tachometer, low washer fluid light, sport steering wheel, sp?? suspension, upgraded seat trim, tilt/telescopic steering column, power mirrors, intermittent wipers, remote fuel door and hatch releases, foot rest, cargo lamp, digital clock, rear wiper/washer, front air dam, sill extensions, rear spoiler.

	Retail Price	Dealer Invoice	Low Price
AM/FM ST ET, base & E	273	227	251
GT	298	247	274
AM/FM ST ET cassette, E & GT	465	386	428
AM/FM ST ET cassette w/EQ, GT w/Turbo Pkg	734	609	675
Carpet protectors, E & GT	28	23	26
Digital clock, GT	54	45	50
Cruise control & intermittent wipers, GT . .	208	173	191
Tinted glass	63	52	58
Power mirrors	87	72	80
Power windows, GT	211	175	194
Rear shelf, base & E	51	42	47
Rear defogger, base	66	55	61
Rear wiper/washer, E & GT	133	110	122
Wheel trim rings, E	55	46	51
Intermittent wipers, GT	28	23	26
13"alloy wheels, GT	293	243	270
14"alloy wheels, GT w/Turbo Pkg	275	228	253
Two-tone paint	157	130	144

Dodge/Plymouth Colt DL Wagon

	Retail Price	Dealer Invoice	Low Price
5-door wagon, 2WD	$9316	$8105	$8711
5-door wagon, 4WD	11145	9696	10421
Destination charge	255	255	255

Standard Equipment:

1.5-liter PFI 4-cylinder engine, 5-speed manual transmission, power brakes, cloth reclining front bucket seats, outboard rear lap/shoulder belts, trip odometer, coolant temperature gauge, rear defogger, locking fuel-filler door, rear wiper/washer, 175/70R13 SBR tires. **4WD** adds: 1.8-liter engine, permanent, full-time 4WD, power steering, 185/70R14 tires.

Optional Equipment:

	Retail Price	Dealer Invoice	Low Price
3-speed automatic transmission, 2WD . . .	499	414	459
Power steering	259	215	238
Air conditioning	739	613	680
Tinted glass	63	52	58
AM/FM ST ET	273	227	251
AM/FM ST ET cassette	424	352	390
Digital clock	54	45	50
Power mirrors	81	67	75
Intermittent wipers	52	43	48
Alloy wheels	299	248	275
Custom Pkg	516	428	475

Velour sport seats, power mirrors, intermittent wipers, rear heat ducts, remote fuel and liftgate releases, tape stripes.

	Retail Price	Dealer Invoice	Low Price
Limited-slip differential, 4WD	218	181	201
Driver's seat height control	19	16	17
Carpet protectors	28	23	26
Luggage rack	128	106	118

Dodge/Plymouth Colt Vista

What's New

These 7-passenger front-drive and 4-wheel-drive wagons continue virtually unaltered. The only changes are the addition of shoulder belts for rear outboard passengers and an auto-down control for the optional power driver's window.

Dodge Colt Vista 2WD

Vista is made by Chrysler's Japanese partner, Mitsubishi. Power is provided by a 96-horsepower 2.0-liter 4-cylinder engine. The front-drive version comes standard with a 5-speed manual transmission; a 3-speed automatic is optional. It also has 13-inch tires and independent rear suspension. The optional on-demand 4WD system can be engaged while underway and can be used on smooth, dry pavement. The 4WD model comes with a 5-speed manual transmission only, 14-inch tires and a torsion-bar rear suspension. Both have a standard adjustable roof rack, front bucket seats and two sets of 50/50 split folding rear seats. With the second and third rear seats folded forward, a carpeted cargo area of 78 cubic feet is formed. Fold them back and their cushions form a single or double bed. The 4WD model substitutes a lap belt for a shoulder belt in the third seat's right-side position; the front-drive model has shoulder belts at all outboard seating positions.

For

●Passenger room   ●Cargo space   ●4WD traction

Against

●Performance   ●Noise

Summary

Colt Vista is one of the most unusual and versatile wagons on the market. It can seat up to seven people, the seats can be folded for use as beds, or you can haul a good deal of luggage by folding the rear seat and expanding the rear cargo area. The 2-place rear seat is cramped, so it's better suited to children than adults, and there isn't much room for climbing in or out of the middle or rear seats. The change to fuel injection for 1988 boosted horsepower by eight, which made Colt Vista feel peppier and sound less harried. But with 96 horsepower for nearly 2600 pounds of curb weight on the front-drive model (and nearly 2900 on the 4WD model), you still don't have that much power for climbing hills or accelerating briskly out of turns. Add four or five people and the engine is overmatched, especially with automatic transmission. The convenient 4WD system is engaged with the push of a button to give you extra traction when you need it without having to stop the vehicle or shift a transfer case lever. Trailer towing is limited to 1500 pounds on the front-drive model and 2000 pounds on the 4x4. Base prices have gone up considerably the past two years, so Colt Vista is less attractive from a value standpoint.

Specifications

	5-door wagon	5-door 4WD wagon
Wheelbase, in.	103.3	103.3
Overall length, in.	174.6	174.6

Prices are accurate at time of printing; subject to manufacturer's change

Dodge Daytona

	5-door wagon	5-door 4WD wagon
Overall width, in.	64.6	64.6
Overall height, in.	57.3	59.4
Front track, in.	55.5	55.5
Rear track, in.	54.1	54.1
Turn diameter, ft.	34.8	34.8
Curb weight, lbs.	2557	2888
Cargo vol., cu. ft.	63.9	63.9
Fuel capacity, gal.	13.2	14.5
Seating capacity	7	7
Front headroom, in.	38.3	38.3
Front shoulder room, in.	53.1	53.1
Front legroom, max., in.	38.8	38.8
Rear headroom, in.	38.3	38.3
Rear shoulder room, in.	53.2	53.2
Rear legroom, min., in.	36.5	36.5

Powertrain layout: transverse front engine/front-wheel drive or on-demand 4WD.

Engines

	ohc I-4
Size, liters/cu. in.	2.0/122
Fuel delivery	PFI
Horsepower @ rpm	96 @ 5000
Torque (lbs./ft.) @ rpm	113 @ 3500
Availability	S

EPA city/highway mpg

5-speed OD manual	22/28
3-speed automatic	22/23

Prices

Dodge/Plymouth Colt Vista	Retail Price	Dealer Invoice	Low Price
5-door wagon, 2WD	$11518	$9790	$10654
5-door wagon, 4WD	12828	10904	11866
Destination charge	255	255	255

Standard Equipment:

2.0-liter PFI 4-cylinder engine, 5-speed manual transmission, power steering (4WD), power brakes, reclining front bucket seats, cloth and vinyl upholstery, trip odometer, coolant temperature gauge, left remote mirror, variable intermittent wipers, outboard rear lap/shoulder belts, rear defogger, remote fuel filler release, tinted glass, optical horn, rear seat heat ducts, wide bodyside moldings, 165/80R13 SBR tires (2WD), 185/70R14 SBR tires (4WD).

Optional Equipment:

3-speed automatic transmission, 2WD	499	414	459
Air conditioning	739	613	680
Limited-slip differential, 4WD	218	181	201
Power steering, 2WD	259	215	238
AM/FM ST ET	273	227	251
AM/FM ST ET cassette	431	358	397
Cruise control	183	152	168
Rear wiper/washer	152	126	140
Power windows	260	216	239
Power door locks	173	144	159
Alloy wheels, 2WD w/o Custom Pkg	311	258	286
2WD w/Custom Pkg	269	223	247
4WD	299	248	275
Custom Pkg., 2WD	445	369	409
4WD	365	303	336

Power mirrors, velour upholstery, visor mirrors, digital clock, map lights, remote liftgate release, tachometer, courtesy lights, tape stripes, 185/70R13 SBR tires (2WD), wheel covers (2WD).

Luggage rack	128	106	118
Two-tone paint, w/o Custom Pkg	305	253	281
w/Custom Pkg	252	209	232

Dodge Daytona

What's New

Dodge turns up the performance wick on its front-drive sport coupes and freshens their styling with a new front fascia and wraparound taillights. Last year's Pacifica model is gone, replaced by an ES model. Daytona gained a driver's-side air bag as standard equipment late in 1988, and this year, all versions—base, ES, ES Turbo and Daytona Shelby—get 4-wheel disc brakes as standard. ES Turbos have as standard a new 150-horsepower turbocharged 2.5-liter four known as the Turbo I. Last year's Turbo I was a 146-horsepower 2.2-liter four. Daytona Shelby loses its previous "Z" suffix, but keeps its Turbo II engine, a turbocharged and intercooled 174-horsepower 2.2-liter four. Shelbys this year get a unique paint treatment that graduates from charcoal in the lower panels to body color in the upper panels. They also get new-style "pumper" alloy wheels and 205/55VR16 unidirectional performance tires (last year's were 225/50VR15). Shelbys ordered with automatic transmission must use the Turbo I engine. A removable interior sunshade is furnished this year with the optional T-bar roof. A new C/S Competition Package, available on base models only, includes many of the Shelby's mechanicals, but cuts about 200 pounds from the weight of a Shelby. The C/S package includes the Turbo II engine, 225/50VR15 tires and the Shelby's upgraded brakes, suspension, special bucket seats and rear spoiler. A tilt steering column is standard, but the only option available with the C/S package is a premium cassette radio.

For

● Air bag ● Performance (turbos)
● Handling/roadholding

Against

● Ride ● Engine noise (turbos) ● Rear seat room

Summary

Dodge's interpretation of the front-drive sport coupe has improved each year since its 1984 introduction. The styling has matured and standard 4-wheel disc brakes and the new C/S package show that performance is taken seriously.

KEY: ohv = overhead valve; **ohc** = overhead cam; **dohc** = double overhead cam; **I** = inline cylinders; **V** = cylinders in V configuration; **flat** = horizontally opposed cylinders; **bbl.** = barrel (carburetor); **PFI** = port (multi-point) fuel injection; **TBI** = throttle-body (single-point) fuel injection; **rpm** = revolutions per minute; **OD** = overdrive transmission; **S** = standard; **O** = optional; **NA** = not available.

Prices are accurate at time of printing; subject to manufacturer's change

Dodge

Daytona's trim-level and power choices match up well with those of such Japanese competitors as the Toyota Celica. Dodge gives the imports a big edge in refinement and assembly quality, but makes up for some of that with lower prices. Handling is quite impressive, even on wet pavement, though the stiff sport suspensions have a ride that's jittery, jolting and lacking in absorbency. Either turbocharged engine gives satisfying acceleration; the Turbo II is downright aggressive. Their nature is coarse, however, and the noise they generate drains pleasure from long drives. We prefer the automatic transmission no matter which engine you choose. The 5-speed manual requires too much shift effort and the heavy clutch pedal on turbocharged models can wear your leg out in traffic.

Specifications

	3-door hatchback
Wheelbase, in.	97.0
Overall length, in.	175.0
Overall width, in.	69.3
Overall height, in.	50.4
Front track, in.	57.6
Rear track, in.	57.6
Turn diameter, ft.	34.3
Curb weight, lbs.	2676
Cargo vol., cu. ft.	33.0
Fuel capacity, gal.	14.0
Seating capacity	4
Front headroom, in.	37.1
Front shoulder room, in.	55.9
Front legroom, max., in.	42.4
Rear headroom, in.	34.3
Rear shoulder room, in.	53.6
Rear legroom, min., in.	30.0

Powertrain layout: transverse front engine/front-wheel drive.

Engines	ohc I-4	Turbo ohc I-4	Turbo ohc I-4
Size, liters/cu. in.	2.5/153	2.5/153	2.2/135
Fuel delivery	TBI	PFI	PFI
Horsepower @ rpm	100 @ 4800	150 @ 4800	174 @ 5200
Torque (lbs./ft.) @ rpm	135 @ 2800	180 @ 2000	200 @ 2400
Availability	S	O	S[1]

EPA city/highway mpg

5-speed OD manual	24/34	20/29	20/28
3-speed automatic	23/29	18/23	

1. Daytona Shelby with 5-speed manual trans.

Prices

Dodge Daytona	Retail Price	Dealer Invoice	Low Price
3-door hatchback	$9395	$8512	$8954
ES 3-door hatchback	10495	9491	9993
ES Turbo 3-door hatchback	11995	10826	11411
Shelby 3-door hatchback	13295	11983	12639
Destination charge	429	429	429

Standard Equipment:

2.5-liter TBI 4-cylinder engine, 5-speed manual transmission, power steering, power 4-wheel disc brakes, driver's side air bag, passenger side motorized shoulder belt and manual lap belt, center console with armrest and storage compartment, rear defogger, remote fuel filler and liftgate releases, tinted glass, trip odometer, tachometer, gauges (coolant temperature, oil pressure, voltmeter), message center with warning lights, head-

lamps-on tone, variable intermittent wipers, visor mirrors, remote mirrors, bodyside moldings, AM & FM ST ET, reclining front seats, cloth and vinyl upholstery, folding rear seats, P185/70R14 SBR tires. **ES** adds: front air dam with fog lights, sill extensions, rear spoiler, 205/60HR15 tires on alloy wheels. **ES Turbo** adds: 2.5-liter turbocharged engine, Light Group, tilt steering column, cloth performance seats, AM & FM ST ET cassette, tonneau cover. **Shelby** adds: 2.2-liter intercooled Turbo II engine, P205/55VR16 SBR tires on alloy wheels.

Optional Equipment:	Retail Price	Dealer Invoice	Low Price
3-speed automatic transmission	552	469	508
Popular Equipment Discount Pkg., base & ES	940	799	865
ES Turbo & Shelby	937	796	862
w/o A/C, base	332	282	305
Base & ES: air conditioning, front floormats, Light Group, heated power mirrors, tilt steering column. ES Turbo and Shelby add: power windows and locks.			
Lights & Locks Discount Pkg., base & ES	296	252	272
Headlight extinguish delay, illuminated entry, power door locks, lighted visor mirrors; requires Popular Pkg.			
C/S Performance Discount Pkg., base	1354	1151	1246
2.5-liter turbo engine, rear spoiler, performance handling suspension, 205/60R15 SBR performance tires on alloy wheels.			
C/S Competition Pkg., base	2825	2401	2599
2.2-liter Turbo II engine, performance front seats, rear spoiler, tilt steering column, maximum performance suspension, 225/50VR15 SBR unidirectional tires on alloy wheels, uprated transmission and brakes.			
Electronic Equipment Pkg., ES Turbo & Shelby	600	510	552
Overhead console with compass and outside temperature readouts, Electronic Navigator, headlight extinguish delay, illuminated entry, lighted visor mirrors; requires Popular Pkg.			
T-bar Roof Pkg., ES Turbo & Shelby	1324	1125	1218
T-bar roof and sunshades, lighted visor mirrors; requires Popular Pkg.			
Automatic Transmission Pkg., Shelby	NC	NC	NC
2.5-liter Turbo I engine, 3-speed automatic transmission, cruise control.			
Pearl/clear coat paint	41	35	38
Enthusiast seats, ES Turbo & Shelby	403	343	371
Requires Popular Pkg.			
Leather enthusiast seats, Shelby	905	769	833
ES Turbo	966	821	889
Requires Popular Pkg. on ES Turbo.			
Electronic instruments, ES Turbo & Shelby	320	272	294
Requires Electronic Equipment Pkg.			
California emissions pkg	102	87	94
Power windows, base & ES	216	184	199
Requires Popular Pkg.			
AM & FM ST ET cassette, base & ES	157	133	144
AM & FM ST ET cassette w/EQ, ES Turbo & Shelby	216	184	199
CD player, ES Turbo & Shelby	412	350	379
Requires factory cassette stereo.			
Cruise control (Popular Pkg. req)	180	153	166
Rear-window sun louver	216	184	199
Removable sunroof	383	326	352
Requires Popular Pkg. on base & ES.			
Tonneau cover, base & ES	71	60	65
Requires Popular Pkg.			
Rear wiper/washer	130	111	120
Alloy wheels, base	322	274	296

Dodge Diplomat

What's New

Dodge will drop this perennial darling of police departments after 1989 and no replacement is planned. Diplomat continues unchanged for '89 as a 4-door notchback sedan with a 5.2-liter V-8 engine and 3-speed automatic transmission. Salon and SE trim levels are offered. Both have 6-pas-

Prices are accurate at time of printing; subject to manufacturer's change.

Dodge Diplomat

	4-door notchback
Front headroom, in.	39.3
Front shoulder room, in.	56.0
Front legroom, max., in.	42.5
Rear headroom, in.	37.7
Rear shoulder room, in.	55.9
Rear legroom, min., in.	36.6

Powertrain layout: longitudinal front engine/rear-wheel drive.

Engines

	ohv V-8
Size, liters/cu. in.	5.2/318
Fuel delivery	2 bbl.
Horsepower @ rpm	140 @ 3600
Torque (lbs./ft.) @ rpm	265 @ 1600
Availability	S

EPA city/highway mpg

3-speed automatic	16/22

senger seating, tilt steering columns and a driver's-side air bag standard. Automatic air conditioning and cruise control are standard on SE versions. Diplomat's rear-drive chassis is derived from the one introduced for the 1976 model year in the Dodge Aspen/Plymouth Volare. Diplomat debuted in its present form for 1980, along with the similar Chrysler Fifth Avenue. Coupe and wagon body styles were dropped in 1982, the same year a Plymouth version was launched under the Gran Fury name. This also will be the last year for the rear-drive Fifth Avenue and Gran Fury, leaving Chrysler with no domestically produced rear-drive cars.

For

●Air bag ●Driveability

Against

●Fuel economy ●Ride ●Handling/roadholding

Summary

Most Diplomats are sold as taxicabs or police cars, but Dodge kept it in the retail fleet to satisfy buyers who want a rear-drive sedan with a V-8 engine. If that's what you want, there are better alternatives—the Ford LTD Crown Victoria/Mercury Grand Marquis and Chevrolet Caprice. We think Diplomat's time-tested drivetrain is its best selling point, but even that has a catch. The V-8 and automatic transmission combine to consume a lot of gas by today's standards, and Dodge recommends using expensive premium unleaded. The engine is smooth and quiet, and produces plenty of low-end torque for good response in most situations, plus the 3-speed automatic shifts cleanly. Unfortunately, Diplomat's soft suspension fosters sloppy cornering, imprecise steering response, and too much bouncing on rough pavement. The full-size Fords and Chevys have more interior and trunk space, comparable performance, and much better road manners.

Specifications

	4-door notchback
Wheelbase, in.	112.7
Overall length, in.	204.6
Overall width, in.	74.2
Overall height, in.	55.3
Front track, in.	60.0
Rear track, in.	59.5
Turn diameter, ft.	40.7
Curb weight, lbs.	3556
Cargo vol., cu. ft.	15.6
Fuel capacity, gal.	18.0
Seating capacity	6

Prices

Dodge Diplomat	Retail Price	Dealer Invoice	Low Price
Salon 4-door notchback	$11995	$10706	$11351
SE 4-door notchback	14795	12726	13761
Destination charge	495	495	495

Standard Equipment:

Salon: 5.2-liter 2bbl. V-8 engine, 3-speed automatic transmission, power steering, power brakes, driver's-side air bag, tinted glass, AM/FM ST ET, tilt steering column, bench seat with center armrest, cloth and vinyl upholstery, ammeter, coolant temperature gauge, headlamps-on tone, intermittent wipers, P205/75R15 tires. **SE** adds: automatic air conditioning, vinyl roof, 60/40 front seat with driver's seatback recliner, cloth upholstery, rear center armrest, rear defogger, cruise control, remote mirrors, lighted right visor mirror, trunk dress-up, upgraded sound insulation.

Optional Equipment:

Automatic air conditioning, Salon	855	727	787
Popular Equipment Discount Pkg., Salon	1147	975	1055
Automatic air conditioning, rear defogger, power mirrors, cruise control, premium wheel covers.			
Protection Pkg., Salon	185	157	170
Bodyside moldings, rear bumper guards, floormats, undercoating.			
Light Group	133	113	122
Luxury Equipment Discount Pkg., SE	1202	1022	1106
Light group, power decklid release, power windows and locks, power driver's seat, leather-wrapped steering wheel, lighted visor mirrors, wire wheel covers.			
Pearl/clear coat paint	41	35	38
Rear defogger, Salon	149	127	137
California emissions pkg	102	87	94
Illuminated entry, SE	78	66	72
Requires Luxury Pkg.			
Power mirrors, Salon	164	139	151
Power door locks	201	171	185
Requires power mirrors or Popular Pkg.			
Power windows	294	250	270
Requires power locks; Salon requires power mirrors or Popular Pkg.			
AM & FM ST ET cassette	262	223	241

> **KEY: ohv** = overhead valve; **ohc** = overhead cam; **dohc** = double overhead cam; **I** = inline cylinders; **V** = cylinders in V configuration; **flat** = horizontally opposed cylinders; **bbl.** = barrel (carburetor); **PFI** = port (multi-point) fuel injection; **TBI** = throttle-body (single-point) fuel injection; **rpm** = revolutions per minute; **OD** = overdrive transmission; **S** = standard; **O** = optional; **NA** = not available.

Prices are accurate at time of printing; subject to manufacturer's change.

	Retail Price	Dealer Invoice	Low Price
Power antenna, SE	72	61	66
Requires Luxury Pkg.			
HD suspension	27	23	25
Power sunroof, SE	1108	942	1019
Trunk dress-up, Salon	58	49	53
Requires Popular Pkg.			
Conventional spare tire	96	82	88
Vinyl roof	206	175	190
Salon requires Popular Pkg.			
Wire wheel covers, SE	231	196	213

Dodge Dynasty

Dodge Dynasty

What's New

This is Dodge's version of the C-body car that Chrysler uses for its New Yorker. Introduced last year, both are front-drive, 6-passenger 4-door notchback sedans. Both are back for 1989 with slightly more power and a newly available 4-speed automatic transmission. Dynasty comes in base and LE trim levels. A 2.5-liter 4-cylinder engine is standard on the base model. It's rated at 100 horsepower, up from 96 last year, and is available only with a 3-speed automatic transmission. The top-line LE comes standard with a Mitsubishi-made 3.0-liter V-6. A redesigned intake manifold and freer-flowing exhaust raise its horsepower from 136 to 141. This engine is available only with Chrysler's new 4-speed overdrive automatic transmission. The V-6 powertrain is optional on the base Dynasty. Four-wheel disc brakes with an anti-lock feature are optional in place of the standard front disc/rear drum setup. Among 1989's new options are a 6-way power driver's seat with memory and an anti-theft system.

For

- Ride
- Anti-lock brakes option
- Passenger and cargo room

Against

- Performance
- Handling

Summary

Since Dynasty is so similar to the New Yorker, we encourage you to also read our comments for the Chrysler version. We find the V-6 engine to be a bit sluggish in these cars, especially from a standing start. Passing response is better, but it's still no ball of fire. The ride is soft and generally quite comfortable, without excessive bouncing at high

speeds. Handling is a different story. The body leans heavily in turns and the soft all-season tires roll over onto their sidewalls, scrubbing off speed with noticeable squealing. The best feature available on Dynasty is the anti-lock brake option, which can stop this car safely from high speeds on water, snow or ice. We strongly recommend you order ABS, even if it means sacrificing another expensive option like a power sunroof. Dynasty has ample interior room, so fitting six people isn't a major squeeze play. The functional design provides plenty of leg room in the rear seat and adequate head room. Cargo space also is more than adequate. Other than the anti-lock brakes, there's nothing technically interesting, and the styling isn't visually exciting. Dynasty still seems to be a well-designed family sedan with plenty of room and tolerable engine performance.

Specifications

	4-door notchback
Wheelbase, in.	104.3
Overall length, in.	192.0
Overall width, in.	68.5
Overall height, in.	53.5
Front track, in.	57.6
Rear track, in.	57.6
Turn diameter, ft.	40.5
Curb weight, lbs.	3000
Cargo vol., cu. ft.	16.0
Fuel capacity, gal.	16.0
Seating capacity	6
Front headroom, in.	37.6
Front shoulder room, in.	56.4
Front legroom, max., in.	41.9
Rear headroom, in.	37.5
Rear shoulder room, in.	55.9
Rear legroom, min., in.	37.9

Powertrain layout: transverse front engine/front-wheel drive.

Engines

	ohc I-4	ohc V-6
Size, liters/cu. in.	2.5/153	3.0/181
Fuel delivery	TBI	PFI
Horsepower @ rpm	100 @ 4800	141 @ 5000
Torque (lbs./ft.) @ rpm	135 @ 2800	171 @ 2800
Availability	S	O

EPA city/highway mpg

3-speed automatic	22/28	
4-speed OD automatic		18/26

Prices

Dodge Dynasty	Retail Price	Dealer Invoice	Low Price
4-door notchback	$12295	$10601	$11548
LE 4-door notchback	13595	11706	12751
Destination charge	495	495	495

Standard Equipment:

2.5-liter TBI 4-cylinder engine, 3-speed automatic transmission, power steering and brakes, bench seat, cloth upholstery, headlamps-on tone, front cup holder, rear defogger, optical horn, trip odometer, voltmeter, coolant temperature and oil pressure gauges, intermittent wipers, tinted glass, dual remote mirrors, visor mirrors, bodyside moldings, AM/FM ST ET, remote decklid and fuel filler releases, P195/75R14 SBR tires. **LE** adds: 3.0-liter PFI V-6, 4-speed automatic transmission, 50/50 bench seat, leather-wrapped steering wheel, upgraded wheel covers.

Prices are accurate at time of printing; subject to manufacturer's change.

Optional Equipment:

	Retail Price	Dealer Invoice	Low Price
3.0-liter V-6 & 4-speed auto trans, base . .	774	658	712
Anti-lock 4-wheel disc brakes, LE	954	811	878
Requires Popular Pkg.			
Air conditioning	798	678	734
Power Convenience Pkg., base	349	297	321
Power windows and mirrors; requires Popular Pkg.			
Popular Equipment Discount Pkg., base . .	1266	1076	1165
LE	1604	1363	1476
Base: air conditioning, cruise control, floormats, power locks, tilt steering column, undercoating. LE adds: heated power mirrors, power windows.			
Interior Illumination Pkg.	197	167	181
Illuminated entry, lighted visor mirrors; requires Popular Pkg.			
Luxury Equipment Discount Pkg., LE	2669	2269	2455
Popular Pkg. plus illuminated entry, lighted visor mirrors, power front seats, security alarm, wire wheel covers, leather-wrapped steering wheel.			
Deluxe Convenience Pkg., LE	309	263	284
Cruise control, tilt steering column.			
Power locks	294	250	270
AM & FM ST ET cassette, base	157	133	144
w/seek & scan, LE	262	223	241
Sound Pkg	569	484	523
w/EQ, LE	785	667	722
AM & FM ST ET cassette, six Infinity speakers, power antenna, digital clock.			
Power driver's seat, LE	247	210	227
Memory power seats, LE w/Luxury Pkg . .	351	298	323
Security alarm	150	128	138
Requires power locks.			
Conventional spare tire	85	72	78
Power sunroof, LE	799	679	735
Requires Popular Pkg.			
Road Handling Suspension	59	50	54
Requires Popular Pkg.; not available with auto load leveling suspension.			
Auto load leveling suspension	185	157	170
Wire wheel covers	231	196	213
Requires Popular Pkg.			
Alloy wheels, LE w/Luxury Pkg	42	36	39
LE w/o Luxury Pkg	273	232	251

Dodge Lancer

Dodge Lancer Shelby

What's New

The high-performance Shelby package becomes a distinct Lancer model for 1989 and a new turbocharged engine joins the line. Dodge's front-drive 5-passenger compact is the same 5-door that Chrysler sells as the LeBaron hatchback sedan. Standard on base Lancers is a 93-horsepower 2.2-liter 4-cylinder engine. A naturally aspirated 100-horsepower 2.5-liter four is optional, as is the Turbo I engine, a 150-horsepower turbocharged 2.5. It replaces the previous Turbo I, a 146-horsepower 2.2 four. Lancer ES gets blackout exterior trim and wider standard and optional tires. The Turbo I engine is standard on ES; the naturally aspirated 2.5 is optional. Lancer Shelby is named for former race

driver and Dodge performance consultant Carroll Shelby. Standard is the 176-horsepower 2.2-liter turbocharged and intercooled Turbo II engine. Shelbys have a special sport-handling suspension, aero body add-ons and they come in monochrome paint schemes of white, red or black. Leather-trimmed upholstery and a 6-way power driver's seat are standard. Air conditioning is standard on ES and Shelby models. Turbo II Shelbys are available only with the 5-speed; ordering the automatic as a no-cost option mandates the Turbo I engine. All Lancers have front-disc and rear-drum brakes.

For

- Handling/roadholding ● Passenger room
- Cargo room ● Performance (turbos)

Against

- Engine noise (turbo) ● Ride (Shelby)
- Manual shift linkage

Summary

While its near-twin LeBaron hatchback highlights luxury, Lancer emphasizes performance, a theme taken to the extreme in the Shelby model. While the Shelby handles quite well and has great acceleration, the suspension is extremely stiff and the exhaust is quite loud. A turbocharged ES is a little easier to live with, but the noisy Turbo I engine also grates on your nerves after a while. If you can live with less performance, opt for the quieter and smoother naturally aspirated 2.5. We also prefer the 3-speed automatic to the 5-speed manual, which requires a high shift effort and patience to find the right gear. Performance aside, Lancer is a spacious, versatile hatchback that carries five in reasonable comfort with ample room for their luggage. Like the LeBaron hatchback, the stylish Lancer is short on refinement, but long on practicality, plus it's available at fairly low prices.

Specifications

	5-door hatchback
Wheelbase, in.	103.1
Overall length, in.	180.4
Overall width, in.	68.3
Overall height, in.	53.0
Front track, in.	57.6
Rear track, in.	57.2
Turn diameter, ft.	36.2
Curb weight, lbs.	2643
Cargo vol., cu. ft.	42.6
Fuel capacity, gal.	14.0
Seating capacity	5
Front headroom, in.	38.3
Front shoulder room, in.	55.8
Front legroom, max., in.	41.1
Rear headroom, in.	37.9
Rear shoulder room, in.	55.9
Rear legroom, min., in.	36.5

Powertrain layout: transverse front engine/front-wheel drive.

KEY: ohv = overhead valve; **ohc** = overhead cam; **dohc** = double overhead cam; **I** = inline cylinders; **V** = cylinders in V configuration; **flat** = horizontally opposed cylinders; **bbl.** = barrel (carburetor); **PFI** = port (multi-point) fuel injection; **TBI** = throttle-body (single-point) fuel injection; **rpm** = revolutions per minute; **OD** = overdrive transmission; **S** = standard; **O** = optional; **NA** = not available.

Prices are accurate at time of printing; subject to manufacturer's change.

Dodge

Engines

	ohc I-4	ohc I-4	Turbo ohc I-4	Turbo ohc I-4
Size, liters/cu. in.	2.2/135	2.5/153	2.5/153	2.2/135
Fuel delivery	TBI	TBI	PFI	PFI
Horsepower @ rpm	93 @ 4800	100 @ 4800	150 @ 4800	174 @ 5200
Torque (lbs./ft.) @ rpm . . .	122 @ 3200	135 @ 2800	180 @ 2000	200 @ 2400
Availability	S	O	S[1]	S[2]

EPA city/highway mpg

5-speed OD manual	24/34	24/34	20/29	20/28
3-speed automatic	24/28	23/29	19/24	

1. ES.   2. Shelby with 5-speed manual trans.

Prices

Dodge Lancer	Retail Price	Dealer Invoice	Low Price
5-door hatchback	$11195	$10002	$10499
ES 5-door hatchback	13695	12202	12849
Shelby 5-door hatchback	17395	15458	16327
Destination charge	440	440	440

Standard Equipment:

2.2-liter TBI 4-cylinder engine, 5-speed manual transmission, power steering and brakes, reclining front bucket seats, cloth and vinyl upholstery, mini console, cup holder, rear defogger, remote mirrors, remote fuel filler and liftgate releases, tinted glass, intermittent wipers, headlamps-on reminder, optical horn, trip odometer, tachometer, coolant temperature and oil pressure gauges, voltmeter, message center with warning lights, wide bodyside moldings, AM/FM ST ET, removable shelf panel, P195/70R14 SBR tires. **ES** adds: 2.5-liter turbo engine, air conditioning, power door locks, heated power mirrors, leather-wrapped steering wheel, alloy wheels. **Shelby** adds: 2.2-liter intercooled Turbo II engine, front and rear air dams, monochrome exterior treatment, fog lights, overhead console with compass and outside temperature readouts, map lights, premium leather bucket seats, power driver's seat, power windows, AM & FM ST ET cassette w/EQ, 205/60R15 SBR tires on alloy wheels.

Optional Equipment:

2.5-liter engine, base	287	244	264
Requires Basic Pkg.			
2.5-liter turbo engine, base	839	713	772
Requires Popular or Luxury Pkg.			
3-speed automatic transmission, base & ES .	552	469	508
Shelby (credit; incl. 2.5-liter Turbo I) . .	(180)	(153)	(153)
Air conditioning, base	798	678	734
Basic Equipment Discount Pkg., base	899	764	827
Air conditioning, console extension with armrest, Light Group.			
Popular Equipment Discount Pkg., base . .	1495	1271	1375
Base incl. automatic or turbo	1947	1655	1791
Base incl. automatic & turbo	2399	2039	2207
ES	1446	1229	1330
2.5-liter engine, air conditioning, console extension with armrest, cruise control, floormats, Light Group, illuminated entry, AM & FM ST ET cassette, tilt steering column, undercoating.			
Luxury Equipment Discount Pkg., base . .	2139	1818	1968
ES	1446	1229	1330
Base: Popular Pkg. plus power windows and locks, heated power mirrors, leather-wrapped steering wheel, lighted visor mirrors. ES adds: console with compass and outside temperature readout, fog lamps, AM & FM ST ET cassette with Infinity speakers, power driver's seat.			
Sport Appearance Discount Pkg., base . . .	1490	1267	1371
ES	1749	1487	1609
Base: 2.5-liter turbo engine, ground effects addenda, precision-feel steering, 205/60R15 tires on alloy wheels; requires Popular or Luxury Pkg. ES adds fog lamps, illuminated entry, AM & FM ST ET cassette with Infinity speakers, power sunroof.			
Power Convenience Discount Pkg., base . .	530	451	488
Power windows and locks, heated power mirrors; requires Popular Pkg.			

	Retail Price	Dealer Invoice	Low Price
Sport Handling Pkg., base	158	134	145
Base or ES w/Luxury Pkg	97	82	89
Leather-wrapped steering wheel, sport handling suspension, 195/70R14 all-season performance tires; base requires Popular or Luxury Pkg.; ES requires Luxury Pkg.			
Deluxe Convenience Pkg., base & ES . . .	309	263	284
Cruise control, tilt steering column; base requires Basic Pkg.			
Power locks, base	201	171	185
Requires Basic Pkg.			
Electronic instruments, ES	308	262	283
Requires Electronic Navigator.			
California emissions pkg	102	87	94
Electronic Navigator	280	238	258
Power driver's seat, base & ES	247	210	227
Requires Basic, Popular or Luxury Pkg. on base.			
Power windows, ES	294	250	270
AM & FM ST ET cassette, base	157	133	144
Requires Basic, Popular or Luxury Pkg.			
AM & FM ST ET cassette w/seek & scan, ES .	472	401	434
Base w/Popular or Luxury Pkg	.315	268	290
AM & FM ST ET cassette w/EQ, ES	216	184	199
Requires Sport Appearance or Luxury Pkg.			
Power sunroof, base & ES	799	679	735
ES w/Luxury Pkg	619	526	569
Base requires Luxury Pkg.			
205/60R15 tires on alloy wheels, ES	270	230	248
ES w/Sport Handling Pkg	200	170	184
14"alloy wheels, base	322	274	296
Requires Popular or Luxury Pkg.			
Rear wiper/washer, base & ES	130	111	120
Base requires Basic, Popular or Luxury Pkg.			

Dodge Omni America

Dodge Omni

What's New

Dodge's twin to the front-drive Plymouth Horizon enters its 12th model year largely unchanged. The 5-door hatchback earned the "America" suffix in 1987 when a cost-cutting program consolidated the line into a single trim level with limited options. For '89, its 2.2-liter 4-cylinder engine gets slight internal modifications that Dodge says will make it run quieter. Engine-bay service items such as the dipstick, power-steering reservoir cap and coolant overflow-bottle cap have been highlighted with paint for easier identification. Standard equipment includes a 5-speed manual transmission, a tachometer and oil pressure and voltage gauges. Tinted glass, a luggage compartment light and a one-piece fold-down rear seat are also standard. Among items included in options packages are power-assisted steering, AM/FM stereo radio, high-back bucket seats with recliners and a 3-speed automatic transmission. Air condi-

Prices are accurate at time of printing; subject to manufacturer's change.

tioning, a new stereo radio with cassette player and a full-size spare tire are the only free-standing options.

For

● Value   ● Fuel economy   ● Performance

Against

● Rear seat room   ● Driving position

Summary

Shop for a new car and you'll quickly learn that you don't get much equipment for less than $10,000. The Omni and identical Plymouth Horizon are exceptions. For about $9000, they deliver automatic transmission, power steering, rear window defogger, reclining cloth bucket seats, a center console, rear window wiper/washer and an AM/FM stereo radio. Among comparable cars, only the Hyundai Excel/Mitsubishi Precis can compete with that kind of value. Those Korean imports fall short in some key areas, however. Chrysler's 93-horsepower engine gives Omni/Horizon peppy acceleration with automatic transmission, even with the air conditioner on. The Excel/Precis has a 68-horsepower 1.5-liter that's sluggish with automatic and anemic with the air conditioner on. Omni/Horizon's cabin is cramped and its interior is decidedly dated next to contemporary rivals such as the Mazda 323 and Toyota Tercel. The driving position is awkward, with the steering wheel too low, the pedals too close and the seatback too reclined. The handling is unsporting, the ride is sometimes harsh. But with its generous standard equipment and budget prices even on "loaded" versions, we'd be hard pressed to find a better small-car value than the Omni/Horizon.

Specifications

	5-door hatchback
Wheelbase, in.	99.1
Overall length, in.	163.2
Overall width, in.	66.8
Overall height, in.	53.0
Front track, in.	56.1
Rear track, in.	55.7
Turn diameter, ft.	38.1
Curb weight, lbs.	2237
Cargo vol., cu. ft.	33.0
Fuel capacity, gal.	13.0
Seating capacity	5
Front headroom, in.	38.1
Front shoulder room, in.	51.7
Front legroom, max., in.	42.1
Rear headroom, in.	36.9
Rear shoulder room, in.	51.5
Rear legroom, min., in.	33.3

Powertrain layout: transverse front engine/front-wheel drive.

Engines	ohc I-4
Size, liters/cu. in.	2.2/135
Fuel delivery	TBI
Horsepower @ rpm	93 @ 4800
Torque (lbs./ft.) @ rpm	122 @ 3200
Availability	S
EPA city/highway mpg	
5-speed OD manual	26/35
3-speed automatic	24/30

Prices

Dodge Omni	Retail Price	Dealer Invoice	Low Price
5-door hatchback	$6595	$6001	$6350
Destination charge	348	348	348

Standard Equipment:

2.2-liter 4-cylinder TBI engine, 5-speed manual transmission, power brakes, rear defogger, rear wiper/washer, trip odometer, tachometer, coolant temperature, oil pressure and voltage gauges, tinted glass, luggage compartment light, black bodyside moldings, left remote mirror, right visor mirror, folding shelf panel, intermittent wipers, cloth and vinyl upholstery, P165/80R13 tires on styled steel wheels.

Optional Equipment:

Basic Pkg	776	660	714
3-speed automatic transmission, power steering.			
Manual Transmission Discount Pkg	705	599	648
Console, power steering, AM & FM ST ET, cloth reclining sport seats, trunk dress-up.			
Automatic Transmission Discount Pkg	1186	1008	1091
Adds 3-speed automatic transmission to Manual Transmission Pkg.			
Air conditioning	694	590	638
Requires Transmission Discount Pkg. and conventional spare tire.			
California emissions pkg	100	85	92
AM & FM ST ET cassette	152	129	140
Requires Transmission Discount Pkg.			
Conventional spare tire	73	62	67
Tinted glass	105	89	97

Dodge Raider

Dodge Raider

What's New

This compact 4-wheel-drive sport-utility vehicle gets an optional V-6 engine and a host of new options. Raider is built in Japan by Chrysler's partner, Mitsubishi, and the new V-6 is a Mitsubishi 141-horsepower 3.0-liter. It's available in place of a 109-horsepower 2.6-liter four. A 5-speed manual transmission is standard with either engine; a 4-

> **KEY: ohv** = overhead valve; **ohc** = overhead cam; **dohc** = double overhead cam; **I** = inline cylinders; **V** = cylinders in V configuration; **flat** = horizontally opposed cylinders; **bbl.** = barrel (carburetor); **PFI** = port (multi-point) fuel injection; **TBI** = throttle-body (single-point) fuel injection; **rpm** = revolutions per minute; **OD** = overdrive transmission; **S** = standard; **O** = optional; **NA** = not available.

Prices are accurate at time of printing; subject to manufacturer's change.

Dodge

speed overdrive automatic is optional only with the V-6. Ordering the V-6 brings a 19.8-gallon fuel tank in place of a 15.9-gallon tank, and slightly wider tires. A black grille and black headlamp bezels are new for all Raiders. The optional rear seat, which now folds down and tips up for added cargo space, gets shoulder belts for outboard passengers in '89. New reclining front bucket seats have see-through head restraints. Interior door panels are also new, as is the inclusion of a visor vanity mirror on the passenger side. Optional power windows, power door locks and cruise control are available for the first time, but only with the V-6 option. Raider features an on-demand, part-time 4WD system (not for use on dry pavement) with automatic locking front hubs. The vehicle has to be stopped to engage or disengage 4WD, but once engaged, it allows intermittent 4WD operation without having to stop.

For

● 4WD traction   ● Maneuverability   ● Performance (V-6)

Against

● Ride/handling   ● Entry/exit

Summary

With its short wheelbase, narrow track and trim overall length, Raider is more maneuverable than larger 4x4s on the road and off. The penalty is a lot of bouncing and pitching over uneven surfaces, and hopping over tar strips. Body lean and understeer (resistance to turning) warn you to slow around corners. Plus, generous ground clearance makes it a high step up into the cabin. Raiders with the 2.6-liter four have ample pickup around town and surprising passing ability with the 5-speed manual, though the engine sounds thrashy above 3000 rpm. There's even better power with the V-6, naturally, and an additional bonus in the competent 4-speed automatic. It upshifts smoothly and changes down quickly for rapid overtaking. The V-6 can't do much to make up for the inconvenience of a 4WD system that lacks the complete shift-on-the-fly capability of some domestic rivals. Raider's rear seat holds only two and cargo capacity is modest, even with the rear seat folded. Raider lacks the on-road comfort and passenger convenience of some other compact 4x4s, such as the Jeep Cherokee/Wagoneer, but it feels quite sturdy.

Specifications

	3-door wagon
Wheelbase, in.	92.5
Overall length, in.	157.3
Overall width, in.	66.1
Overall height, in.	72.4
Front track, in.	55.1
Rear track, in.	54.1
Turn diameter, ft.	34.1
Curb weight, lbs.	3175
Cargo vol., cu. ft.	NA
Fuel capacity, gal.	15.9[1]
Seating capacity	4

KEY: ohv = overhead valve; ohc = overhead cam; dohc = double overhead cam; I = inline cylinders; V = cylinders in V configuration; flat = horizontally opposed cylinders; bbl. = barrel (carburetor); PFI = port (multi-point) fuel injection; TBI = throttle-body (single-point) fuel injection; rpm = revolutions per minute; OD = overdrive transmission; S = standard; O = optional; NA = not available.

	3-door wagon
Front headroom, in.	36.4
Front shoulder room, in.	55.1
Front legroom, max., in.	39.6
Rear headroom, in.	33.9
Rear shoulder room, in.	57.1
Rear legroom, min., in.	34.4

1. 19.8 with V-6

Powertrain layout: longitudinal front engine/rear-wheel drive or on-demand 4WD.

Engines

	ohc I-4	ohc V-6
Size, liters/cu. in.	2.6/156	3.0/181
Fuel delivery	2 bbl.	PFI
Horsepower @ rpm	109 @ 5000	141 @ 5000
Torque (lbs./ft.) @ rpm	142 @ 3000	171 @ 2800
Availability	S	O

EPA city/highway mpg

	ohc I-4	ohc V-6
5-speed OD manual	16/19	15/18
4-speed OD automatic		17/17

Prices

Dodge Raider	Retail Price	Dealer Invoice	Low Price
3-door 4WD wagon	$12550	$10831	$11691
Destination charge	255	255	255

Standard Equipment:

2.6-liter 2bbl. 4-cylinder engine, 5-speed manual transmission, part-time 4WD, 2-speed transfer case, automatic locking front hubs, power steering and brakes, skid plates (engine, fuel tank, transfer case), cloth/vinyl reclining front bucket seats, adjustable steering column, tachometer, coolant temperature gauge, trip odometer, tinted glass, intermittent wipers, digital clock, console, remote fuel filler and liftgate releases, low-mount manual mirrors, sliding rear quarter windows, right visor mirror, 225/75R15 SBR tires, outside spare tire carrier with lock.

Optional Equipment:

3.0-liter V-6	1727	1433	1580
Includes rear coil spring suspension, larger fuel tank, 235/75R15 M + S tires.			
4-speed automatic transmission, V-6	733	608	671
Air conditioning	781	648	715
Off-Road Pkg.	316	262	289
Limited-slip differential, headlamp washers, halogen headlamps, driver's suspension seat.			
Cruise control	187	155	171
Rear defogger	66	55	60
Power locks, V-6	149	124	136
Dual low-mount mirrors	59	49	54
AM/FM ST ET	273	227	250
w/cassette	424	352	388
Folding rear seat	488	405	447
Driver's suspension seat	63	52	58
Power windows, V-6	199	165	182
Rear wiper/washer	160	133	146

Dodge Shadow

What's New

A new optional competition package, a new performance engine and a new grille have been added to Dodge's front-drive subcompact. Shadow shares the Plymouth Sundance's 3- and 5-door hatchback body styles and most of

Prices are accurate at time of printing; subject to manufacturer's change.

Dodge Shadow 3-door

sion, air conditioning, stereo and other options goes for around $11,000. It's not the best subcompact around, but at today's prices, that's a pretty good value.

its mechanicals, but the Dodge is marketed as the sportier car. All Shadows get a new body-color grille this year. Near-flush aero-style headlamps supplant recessed ones and a cross-hair pattern replaces the grille's previous egg-crate center. New front bucket seats have thinner backs that Dodge says increase rear-seat knee room one half inch. Outboard rear-seat passengers now get shoulder belts, and there are five new exterior colors. Among new options for '89 is a 6-way power driver's seat. To Shadow's lineup of naturally aspirated 2.2- and 2.5-liter 4 cylinder engines comes Chrysler's new Turbo I, a 150-horsepower turbocharged 2.5 four. It replaces the previous Turbo I, a 146-horsepower 2.2-liter four, and is available on Shadows with the optional ES package. That package adds fog lamps and wide tires on new "pumper"-style alloy wheels. Dodge says the sporty ES's suspension is improved for '89 with stiffer front and rear springs and performance-tuned rear shock absorbers and front struts. Special front jounce bumpers designed to provide a soft stop for the suspension on severe bumps have also been added. The new competition package is available only on 3-door models and basic-ally adds the ES tires, suspension components and spoilers and the Turbo I engine to a base Shadow.

For

- Performance (turbo) • Ride/handling • Cargo room

Against

- Engine noise • Rear seat room • Manual shift linkage

Summary

Shadow uses the same powertrains as most other front-drive Chrysler cars and shares its basic platform with the Dodge Daytona, so only its exterior styling and cabin lay-out are unique. Overall, the interior is nicely packaged, with reclining front bucket seats, folding rear seatback, a tachometer and other gauges among the amenities. The standard suspension is firm for a domestic car and results in competent handling and a stable highway ride. (We haven't tested the new ES suspension.) With base curb weights over 2500 pounds, Shadow is portly for its external dimensions, and performance is listless with the base 2.2-liter four. Acceleration is much stronger with the turbo, but so is the commotion from under the hood and from the exhaust system. A good compromise is the naturally aspi-rated 2.5. Performance is reasonable and it runs smoother and quieter. The standard 5-speed manual lacks precision and can be tiring to wrestle from gear to gear in traffic. We prefer the optional automatic. Rear seat room is slightly below par, though dropping the rear seatbacks creates a generous cargo hold. A Shadow with automatic transmis-

Specifications

	3-door hatchback	5-door hatchback
Wheelbase, in.	97.0	97.0
Overall length, in.	171.7	171.7
Overall width, in.	67.3	67.3
Overall height, in.	52.7	52.7
Front track, in.	57.6	57.6
Rear track, in.	57.2	57.2
Turn diameter, ft.	36.2	36.2
Curb weight, lbs.	2520	2558
Cargo vol., cu. ft.	33.3	33.0
Fuel capacity, gal.	14.0	14.0
Seating capacity	5	5
Front headroom, in.	38.3	38.3
Front shoulder room, in.	54.4	54.7
Front legroom, max., in.	41.5	41.5
Rear headroom, in.	37.4	37.4
Rear shoulder room, in.	52.5	54.5
Rear legroom, min., in.	20.1	20.1

Powertrain layout: transverse front engine/front-wheel drive.

Engines

	ohc I-4	ohc I-4	Turbo ohc I-4
Size, liters/cu. in.	2.2/135	2.5/153	2.5/153
Fuel delivery	TBI	TBI	PFI
Horsepower @ rpm	93 @ 4800	100 @ 4800	150 @ 4800
Torque (lbs./ft.) @ rpm	122 @ 3200	135 @ 2800	180 @ 2000
Availability	S	O	O
EPA city/highway mpg			
5-speed OD manual	24/34	24/34	20/29
3-speed automatic	24/28	23/29	19/24

Prices

Dodge Shadow	Retail Price	Dealer Invoice	Low Price
3-door hatchback	$8495	$7746	$8196
5-door hatchback	8695	7926	8386
Destination charge	415	415	415

Standard Equipment:

2.2-liter TBI 4-cylinder engine, 5-speed manual transmission, power steering and brakes, reclining front bucket seats, cloth upholstery, motorized front shoulder belts and manual lap belts, outboard rear lap/shoulder belts, one-piece folding rear seatback, removable shelf panel, dual remote mirrors, intermittent wipers, trip odometer, tachometer, coolant temperature gauge, voltmeter, headlamps-on tone, optical horn, mini console with storage bin, remote liftgate release, bodyside moldings, AM & FM ST ET, 185/70R14 SBR tires.

Optional Equipment:

2.5-liter engine	287	244	264
w/ES or Turbo Engine Pkg. (credit)	(553)	(470)	(470)
Requires Popular Pkg.			
3-speed automatic transmission	552	469	508
Air conditioning	715	608	658
Requires tinted glass.			
Tinted glass	108	92	99
Popular Equipment Discount Pkg., 3-door	262	223	241
5-door	274	233	252

Full console, rear defogger, Light Group, four speakers.

Prices are accurate at time of printing; subject to manufacturer's change.

	Retail Price	Dealer Invoice	Low Price
Deluxe Convenience Pkg	396	337	364
w/ES Pkg	474	403	436
Conventional spare tire, floormats, cruise control, tilt steering column; requires Popular or ES Pkg.			
Power Assist Pkg. I, 3-door	151	128	139
5-door	225	191	207
w/ES Pkg	49	42	45
Power locks and mirrors; requires Popular or ES Pkg.			
Power Assist Pkg. II, 3-door	417	354	384
5-door	487	414	448
Power windows, power driver's seat; requires Pkg. I, Popular or ES Pkg.			
ES Pkg., 3-door	2040	1734	1877
5-door	2104	1788	1936
2.5-liter turbo engine (non-turbo 2.5 may be substituted for credit), AM & FM ST ET cassette, performance front seats, full console, rear defogger, fog lights, Light Group, power locks, premium interior, 195/60HR15 all-season performance tires on alloy wheels, message center with warning lights, sill moldings, front fender flares, tape stripes, four speakers, rear spoiler, sport suspension, leather-wrapped steering wheel, 125-mph speedometer, turbo boost gauge.			
Turbo Engine Pkg	950	808	874
2.5-liter turbo engine, 185/70R14 all-season performance tires, message center with warning lights, turbo boost gauge, 125-mph speedometer; requires Popular Pkg.			
Competition Pkg., 3-door	1550	1318	1426
2.5-liter turbo engine, engine dress-up, heel/toe pedals (w/5-speed), 195/60R15 all-season performance tires on alloy wheels, four speakers, rear spoiler, sport suspension, turbo boost gauge, message center with warning lights			
Pearl/clear coat paint	41	35	38
500-amp battery	45	38	41
Rear defogger	149	127	137
California emissions pkg	102	87	94
Power locks, 3-door	149	127	137
5-door	201	171	185
AM & FM ST ET cassette	207	176	190
AM & FM ST ET cassette w/seek/scan . . .	262	223	241
Requires Popular Pkg.			
Removable sunroof	383	326	352
Requires Popular or ES Pkg.			
14"alloy wheels, w/o Deluxe Convenience Pkg	306	260	282
w/Deluxe Convenience Pkg	383	326	352
Not available with 2.2-liter engine or ES; requires Popular Pkg.			

Dodge Spirit

Dodge Spirit

What's New

This is Dodge's version of Chrysler's new line of front-drive, 4-door notchback sedans, scheduled to go on sale by January. Spirit is a near twin to the new Plymouth Acclaim. Both are known as A-cars under Chrysler's product code. Chrysler calls these mid-size sedans; they're bigger than the compact Dodge Aries and Plymouth Reliant, and about the same size as the Plymouth Caravelle and Dodge 600, which they replace. Standard on base and mid-line LE Spirits is a 100-horsepower 2.5-liter 4-cylinder engine. A 150-horsepower turbocharged version is optional. A 5-speed manual is standard; a 3-speed automatic is optional with either 2.5. The sporty Spirit ES has the turbo engine as standard. Optional on the ES only is a Mitsubishi-made 141-horsepower 3.0-liter V-6 mated to Chrysler's new 4-speed automatic transmission. Base models come with front bucket seats, a tachometer, outboard rear-seat shoulder belts, remote trunk and fuel-door releases and 14-inch tires. A front bench seat and a folding split-back rear seat are optional. The LE gets the folding split rear seat as standard, plus an adjustable lumbar bolster in the driver's bucket seat, cruise control, tilt steering wheel, rear defroster and a mini trip computer. The LE is available with optional 15-tires on alloy wheels. Spirit ES combines the LE's standard features with sport bucket seats and adds aero body panels, fog lamps and a monochromatic paint scheme in red, black, silver or white; the standard 15-inch alloy wheels are painted to match white cars. A pop-up, removable sunroof is optional on all Spirits.

For

● Interior room   ● Cargo space   ● Ride
● Performance (V-6 engine)

Against

● Rear seat comfort   ● Shift action (5-speed)
● Noise (turbo engine)

Summary

While Plymouth tailors its A-car for family duty, Dodge tilts the Spirit toward the sporty. The difference shows up in engine and trim choices. Spirits also have what Chrysler calls a "handling suspension" with shocks that are firmer than those on the base and mid-level Acclaim. Indeed, the Spirit we drove at Chrysler's Michigan proving grounds rode much more firmly than the Acclaims we tried. Its handling was sharper, too, with better response to the nicely weighted steering. Control was quite good and Spirit felt composed in tight corners and on high-speed straights. The 2.5 turbo in our test vehicle suffered from turbo lag that hampered acceleration below 3000 rpm, but power was delivered promptly above that. The 5-speed manual's shift action was somewhat notchy, and the lever demanded a sure hand during most gear changes. All Spirits have front disc and rear drum brakes, and though there was some nosedive in harder stops, control and pedal feel were top-notch. Spirit has flush headlamps and a slightly rounded nose, but it avoids the obvious aerodynamics of a Ford Taurus or Chevrolet Corsica. Its tall, square greenhouse increases interior room, affords excellent outward visibility and makes it easy to get into and out of. The readable gauges have white numerals on black backgrounds and dashboards are generally well laid out, though the climate-system buttons are too small. The base cloth front buckets are comfortable, but the tradeoff for adequate rear head and knee room is a back-seat cushion that's too low and short. Cabin storage benefits from front-door map pockets, a center console and cup holders for rear-seat passengers. The large trunk has a flat floor and a low liftover. Spirit breaks no new ground, but emerges as a domestic sedan that's solid, roomy and competent.

Prices are accurate at time of printing; subject to manufacturer's change.

Specifications

	4-door notchback
Wheelbase, in.	103.3
Overall length, in.	181.2
Overall width, in.	68.1
Overall height, in.	55.5
Front track, in.	57.5
Rear track, in.	57.2
Turn diameter, ft.	37.6
Curb weight, lbs.	2770
Cargo vol., cu. ft.	14.4
Fuel capacity, gal.	16.0
Seating capacity	6
Front headroom, in.	38.4
Front shoulder room, in.	54.3
Front legroom, max., in.	41.8
Rear headroom, in.	37.9
Rear shoulder room, in.	54.9
Rear legroom, min., in.	38.5

Powertrain layout: transverse front engine/front-wheel drive.

Engines	ohc I-4	Turbo ohc I-4	ohc V-6
Size, liters/cu. in.	2.5/153	2.5/153	3.0/181
Fuel delivery	TBI	PFI	PFI
Horsepower @ rpm	100 @ 4800	150 @ 4800	141 @ 5000
Torque (lbs./ft.) @ rpm	135 @ 2800	180 @ 2000	171 @ 2800
Availability	S	O[1]	O[2]
EPA city/highway mpg			
5-speed OD manual	24/34	20/29	
3-speed automatic	23/29	19/24	
4-speed OD automatic			18/26

1. Standard, ES. 2. ES only.

Equipment Summary

Dodge Spirit
(prices not available at time of publication)

Standard Equipment:

2.5-liter TBI 4-cylinder engine, 5-speed manual transmission, power steering and brakes, cloth reclining front bucket seats, visor mirrors, AM & FM ST ET, remote fuel filler and decklid releases, 195/70R14 all-season SBR tires. **LE** adds: tinted glass, driver's seat lumbar support adjustment, split folding rear seatback, dual remote mirrors, rear defogger, cruise control, tilt steering column, message center. **ES** adds: 2.5-liter turbo engine, AM & FM ST ET cassette, lighted visor mirrors, sill extensions, front air dam with fog lights, trip computer, 205/60R14 Goodyear Eagle GT all-season tires on alloy wheels.

Eagle Medallion

What's New

A redesigned instrument panel and the prospect of a V-6 engine option are news for these French-built 4-door sedans and 5-door wagons. Chrysler Motors acquired the Medallion in August 1987 when it purchased AMC, which had been importing the Renault-made car since the spring of '87. Chrysler reintroduced the front-drive compact in May 1988, replacing the Renault badge with the Eagle nameplate. They are now sold through Chrysler's Jeep-Eagle division. The redesigned instrument panel places controls closer to the driver, but it does not include a glove compartment; one will be built into a new dash due before

Eagle Medallion LX 4-door

1990. Also anticipated is an optionally available 150-horse-power V-6 mated to a 4-speed automatic transmission. That's the same powertrain used in the larger Eagle Premier. For now, Medallions have a Renault-built all-aluminum 2.2-liter four mounted longitudinally. A 5-speed manual transmission is standard on the sedan; a 3-speed automatic is optional on the sedan and standard on the wagon. Wagons have 5-passenger seating standard and room for seven with an optional third seat. Motorized front shoulder belts are standard.

For

- Passenger room
- Cargo space
- Handling/roadholding
- Ride

Against

- Heat/vent controls
- Interior materials

Summary

Medallion is comfortable, roomy and fun to drive. The smooth-running 2.2 is spirited with manual transmission and returns at least adequate performance with automatic. Handling is above average for this class, with great maneuverability and fine grip. Body lean in turns is noticeable, but is a fair tradeoff for the ample wheel travel and capable suspension that contribute to a supple, controlled ride. Supportive seats and a comfortable driving position add to the enjoyment. The sedan has ample room for four adults, and a fifth can fit with a little squeezing. Trunk volume is quite good for a compact. The longer wagon has more passenger room and a long, wide cargo area with a flat floor. Our main complaints have been with the poorly designed heat/vent controls, which are oddly positioned and laid out, and the flimsy-feeling plastic interior trim, which doesn't appear to be designed for the long haul. While we've found no major flaws in the Medallions we've tested, we won't say their quality equals that of such Japanese compacts as the Honda Accord or Toyota Camry. Mechanical woes and assembly problems did much to doom previous Renault products in America, and that remains a concern. Still, we find much to like about the Medallion, including its price.

> **KEY: ohv** = overhead valve; **ohc** = overhead cam; **dohc** = double overhead cam; **I** = inline cylinders; **V** = cylinders in V configuration; **flat** = horizontally opposed cylinders; **bbl.** = barrel (carburetor); **PFI** = port (multi-point) fuel injection; **TBI** = throttle-body (single-point) fuel injection; **rpm** = revolutions per minute; **OD** = overdrive transmission; **S** = standard; **O** = optional; **NA** = not available.

Prices are accurate at time of printing; subject to manufacturer's change.

Specifications

	4-door notchback	5-door wagon
Wheelbase, in.	102.3	108.2
Overall length, in.	183.2	189.7
Overall width, in.	67.5	67.5
Overall height, in.	55.7	56.3
Front track, in.	57.2	57.2
Rear track, in.	55.3	55.3
Turn diameter, ft.	NA	NA
Curb weight, lbs.	2588	2736
Cargo vol., cu. ft.	15.2	89.5
Fuel capacity, gal.	17.4	17.4
Seating capacity	5	5
Front headroom, in.	37.8	37.8
Front shoulder room, in.	55.1	55.1
Front legroom, max., in.	42.2	42.2
Rear headroom, in.	37.5	37.4
Rear shoulder room, in.	54.5	54.5
Rear legroom, min., in.	40.8	40.8

Powertrain layout: longitudinal front engine/front-wheel drive.

Engines

	ohc I-4
Size, liters/cu. in.	2.2/132
Fuel delivery	PFI
Horsepower @ rpm	103 @ 5000
Torque (lbs./ft.) @ rpm	124 @ 2500
Availability	S

EPA city/highway mpg

5-speed OD manual	23/31
3-speed automatic	21/27

Prices

Eagle Medallion

	Retail Price	Dealer Invoice	Low Price
DL 4-door notchback	$10405	$8944	$9675
DL 5-door wagon	11649	10002	10826
LX 4-door notchback	10938	9397	10168
LX 4-door notchback	12275	10534	11405
Destination charge	425	425	425

Standard Equipment:

DL: 2.2-liter PFI 4-cylinder engine, 5-speed manual transmission (4-doors; wagons have 3-speed automatic), power steering and brakes, reclining front bucket seats, cloth upholstery, driver's seat height adjustment, passive front seatbelts, outboard rear lap/shoulder belts, intermittent wipers, tilt steering column, rear wiper/washer (wagon), tachometer, coolant temperature gauge, trip odometer, digital clock, rear defogger, tinted glass, door pockets, floormats, power mirrors, AM/FM ST ET, roof rack (wagon), 185/65R14 all-season SBR tires. **LX** adds: rear head restraints and courtesy lamps, rocker passenger seat, upgraded upholstery, split folding rear seat with center armrest.

Optional Equipment:

3-speed automatic transmission, 4-doors	490	402	446
Air conditioning	799	655	727

> **KEY: ohv** = overhead valve; **ohc** = overhead cam; **dohc** = double overhead cam; **I** = inline cylinders; **V** = cylinders in V configuration; **flat** = horizontally opposed cylinders; **bbl.** = barrel (carburetor); **PFI** = port (multi-point) fuel injection; **TBI** = throttle-body (single-point) fuel injection; **rpm** = revolutions per minute; **OD** = overdrive transmission; **S** = standard; **O** = optional; **NA** = not available.

Eagle Medallion LX wagon

	Retail Price	Dealer Invoice	Low Price
Driving Group, LX	404	331	368
Leather-wrapped steering wheel, 195/60HR14 tires on alloy wheels.			
Power Windows and Door Locks Pkg., LX	758	622	690
Power windows and locks, heated power mirrors, remote trunk release or tailgate lock, lighted right visor mirror, locking fuel door.			
Premium Audio Group, LX	515	422	469
AM/FM ST ET cassette with graphic EQ and six speakers, power antenna.			
Cargo cover, DL wagon	72	59	66
Cruise control	175	144	160
Power locks	366	300	333
Includes power trunk release or tailgate lock, locking fuel door.			
AM/FM ST ET cassette	360	295	328
Power sunroof	599	491	545
Third seat, DL wagon	280	230	255
LX wagon	208	171	190

Eagle Premier

Eagle Premier LX

What's New

A sporty ES Limited model is due in April, but this front-drive 4-door sedan is otherwise carried over with only minor changes. Designed jointly by AMC and French automaker Renault, Premier is built at a former AMC plant in Toronto. It's sold through Chrysler Motors' Jeep-Eagle division, which Chrysler formed after buying AMC. LX and upgraded ES versions are offered. An AMC 2.5-liter 4-cylinder is standard in the LX. Optional on LX and standard on ES is an all-aluminum 3.0-liter V-6. A 4-speed overdrive automatic is the only available transmission. Both cars have 14-inch wheels and all-independent suspensions, though ES's suspension is set for a firmer ride. Six-passenger seating with a 55/45 reclining front-bench seat is standard. Models with the optional 5-passenger seating have

Prices are accurate at time of printing; subject to manufacturer's change.

front bucket seats and a center console, and they are set to get a floor-mounted shift lever during the '89 model year. Premier's electronic turn-signal beep has been changed to a chime tone for 1989, and the standard power steering gets increased assist. Previously optional, a rear-window defogger is now standard on all Premiers while the LX gains the rallye instrument package as standard. The ES Limited model will feature a white monochromatic exterior, leather seat trim, V-6 engine and the floor-shift automatic transmission. It will be the first Premier with 4-wheel disc brakes. Among its standard features will be body colored 15-inch alloy wheels, air conditioning, power locks, windows, mirrors and seats.

For

- ●Performance (V-6) ●Handling/roadholding
- ●Ride ●Passenger room ●Cargo space

Against

- ●Performance (4 cylinder) ●Instruments/controls

Summary

This may be the best driving, most efficiently designed sedan in the Chrysler stable. The smooth V-6 is teamed with a precise-shifting transmission to deliver ample power, even in the demanding midrange speeds of everyday driving. Premier's ride is firm, but pliant and well controlled, with good road feel through the accurate and nicely weighted power steering. Inside, there are supportive seats and room for six in an airy cabin of pleasing and contemporary design. The trunk is roomy. And prices are reasonable. There is a down side. The quirkiness of some of the controls compromises their function. Turn signals are activated not by a traditional stalk that clicks into place, but by a springy lever that sets off a soft electronic tone. The shift lever that snakes awkwardly from behind the steering wheel is not easy to work. Consumers complained about both, Chrysler says, and the new turn-signal chime and the coming floor-mounted shift lever are their responses. But these shortcomings are minor compared to Premier's main hurdle: can a basically sound car and a good value for the money overcome ancestral ties to Renault and AMC, companies with tarnished reputations for quality and resale value? We encourage you to take a look.

Specifications

	4-door notchback
Wheelbase, in.	105.5
Overall length, in.	192.7
Overall width, in.	69.8
Overall height, in.	55.9
Front track, in.	58.1
Rear track, in.	57.0
Turn diameter, ft.	36.9
Curb weight, lbs.	2862
Cargo vol., cu. ft.	16.3
Fuel capacity, gal.	17.0
Seating capacity	5
Front headroom, in.	38.5
Front shoulder room, in.	57.8
Front legroom, max., in.	43.7
Rear headroom, in.	37.5
Rear shoulder room, in.	56.9
Rear legroom, min., in.	39.4

Powertrain layout: longitudinal front engine/front-wheel drive.

Engines

	ohv I-4	ohc V-6
Size, liters/cu. in.	2.5/150	3.0/182
Fuel delivery	TBI	PFI
Horsepower @ rpm	111 @ 4750	150 @ 5000
Torque (lbs./ft.) @ rpm	142 @ 2500	171 @ 3750
Availability	S	S[1]

EPA city/highway mpg

4-speed OD automatic	22/31	18/27

1. ES; optional, LX

Prices

Eagle Premier	Retail Price	Dealer Invoice	Low Price
LX 4-door notchback	$13276	$11435	$12356
ES 4-door notchback	15259	13120	14190
Destination charge	440	440	440

Standard Equipment:

LX: 2.5-liter TBI 4-cylinder engine, 4-speed automatic transmission, power steering and brakes, 55/45 reclining front seats, rear armrest, AM/FM ST ET, tachometer, coolant temperature and oil pressure gauges, voltmeter, trip odometer, intermittent wipers, rear defogger, tinted glass, digital clock, door and seatback pockets, map lights, right visor mirror, leather-wrapped steering wheel, remote mirrors, 195/70R14 tires. **ES** adds: 3.0-liter PFI V-6, touring suspension, 45/45 velour front seats, full-length console with armrest, trip computer/vehicle maintenance monitor, 8 Jensen speakers, 205/70HR14 tires on polycast wheels.

Optional Equipment:

3.0-liter PFI V-6, LX	680	578	626
Air conditioning	855	727	787
Popular Option Group	1339	1138	1232
Air conditioning, cruise control, tilt steering column.			
Luxury Option Group, LX	2261	1922	2080
ES	2014	1712	1853
Popular Group plus power mirrors, lighted visor mirrors, power windows, premium audio system.			
Electronic Information Pkg., LX	309	263	284
Trip computer, vehicle maintenance monitor; requires Popular or Luxury Group.			
Enthusiast Group, LX	648	551	596
Velour bucket seats, touring suspension, 205/70HR14 tires on alloy wheels; requires Popular or Luxury Group.			
Convenience Group	309	263	284
Cruise control, tilt steering column.			
Decklid luggage rack	118	100	109
Power Lock Group	415	353	382
Power locks with remote control and illuminated entry, power decklid release, remote fuel door release.			
Power windows	294	250	270
Requires Power Lock Group.			
Power front seats	494	420	454
Requires Popular or Luxury Group.			
Velour bucket seats, LX	103	88	95
Leather/vinyl bucket seats, ES	583	496	536
Bucket seats include console and passive restraints (motorized shoulder belts/manual lap belts).			
AM/FM ST ET cassette	241	205	222
Premium audio system, LX	620	527	570
ES	447	380	411
AM/FM ST ET cassette w/EQ, eight Accusound by Jensen speakers, power antenna.			
Conventional spare tire	85	72	78
Alloy wheels, base	370	315	340
ES	295	251	271
Air conditioning, rear defogger.			

Prices are accurate at time of printing; subject to manufacturer's change.

Eagle Summit

Eagle Summit DL

What's New

This front-drive, 4-door subcompact sedan debuts for 1989 as the entry-level Eagle. Designed by Mitsubishi Motors Corporation, Chrysler's Japanese partner, Summit is a 4-door version of the Mitsubishi hatchback that Dodge and Plymouth dealers sell as the Colt. Mitsubishi dealers market a similar 4-door sedan as the Mirage. Summit is offered in two trim levels, DL and LX, with a performance package available on the LX version. An 81-horsepower 1.5-liter 4-cylinder is the base engine. A 5-speed manual transmission is standard; a 3-speed automatic is optional. LXs upgrade DL equipment with velour seats, a height-adjustable driver's seat, split fold-down rear seat with center armrest, power steering and a tilt and telescope steering column. The performance package is called the LX DOHC and is built around a 113-horsepower, dual overhead cam, 16-valve 1.6-liter four. A sport suspension, 4-wheel disc brakes and sport front-bucket seats are included. LX DOHC versions get an optional 4-speed overdrive automatic and 14-inch wheels, versus the others' 13 inchers. Alloy wheels are optional for all three models. All Summits have motorized front shoulder belts, as well as shoulder belts for outboard rear passengers.

For

●Fuel economy ●Maneuverability
●Performance (DOHC engine)

Against

●Interior room ●Noise
●Performance (automatic transmission)

Summary

Summit initially will be built by Mitsubishi in Japan, but eventually will be assembled at the Diamond-Star Motors plant in Normal, Ill. The plant is a joint venture between Chrysler and Mitsubishi. Chrysler's decision to include Summit in this lineup is a precursor of the Diamond-Star-built, 4-wheel-drive sport coupe to be sold through Eagle late in 1989. A front-drive version of the sleek coupe will be marketed as the Plymouth Laser and Mitsubishi Eclipse. The Summit, meanwhile, is a pleasant example of the contemporary Japanese subcompact. Its main competition is the Toyota Corolla and Honda Civic, and the Summits we drove at Chrysler's Michigan proving grounds stacked up pretty well against both. Construction is solid, road man-

ners competent. The cabins are relatively roomy and airy, with storage bins and pockets placed thoughtfully. Usable trunk space is above average for the class. Blemishes included a noisy highway ride: Wind roar around the roof's front pillars is much louder than in a Civic, and the 1.5-liter engine with the 3-speed-automatic runs at a buzzy 3400 rpm at 65 mph. You'll get quieter cruising and better fuel economy with the 5-speed. The LX DOHC seems like a budget sports sedan worth considering, but not with automatic transmission.

Specifications

	4-door notchback
Wheelbase, in.	96.7
Overall length, in.	170.1
Overall width, in.	65.7
Overall height, in.	52.8
Front track, in.	56.3
Rear track, in.	56.3
Turn diameter, ft.	30.8
Curb weight, lbs.	2271
Cargo vol., cu. ft.	10.3
Fuel capacity, gal.	13.2
Seating capacity	5
Front headroom, in.	39.1
Front shoulder room, in.	53.5
Front legroom, max., in.	41.9
Rear headroom, in.	37.5
Rear shoulder room, in.	53.1
Rear legroom, min., in.	34.3

Powertrain layout: transverse front engine/front-wheel drive.

Engines	ohc I-4	dohc I-4
Size, liters/cu. in.	1.5/90	1.6/97
Fuel delivery	PFI	PFI
Horsepower @ rpm	81 @ 5500	113 @ 6500
Torque (lbs./ft.) @ rpm	91 @ 3000	99 @ 5000
Availability	S	O[1]

EPA city/highway mpg

5-speed OD manual	29/35	23/28
3-speed automatic	27/30	
4-speed OD automatic		23/28

1. LX

Prices

Eagle Summit	Retail Price	Dealer Invoice	Low Price
DL 4-door notchback	$9347	$8419	$8883
w/Pkg. IFB	10028	8998	9513
w/Pkg. IFC	10247	9184	9716
w/Pkg. IFD	10549	9441	9995
LX 4-door notchback	10364	9324	9844
w/Pkg. IFB	10714	9621	10168
w/Pkg. IFC	11324	10140	10732
w/Pkg. IFF	11596	10372	10984
w/DOHC Pkg. IFD	11695	10455	11075
w/DOHC Pkg. IFE	12920	11497	12209
Destination charge	255	255	255

Standard Equipment:

DL: 1.5-liter PFI 4-cylinder engine, 5-speed manual transmission, power brakes, cloth bucket seats, trip odometer, coolant temperature gauge, motorized front shoulder belts and manual lap belts, outboard rear lap/shoulder belts, locking fuel filler door, remote mirrors, rear defogger, intermittent wipers, 155/80R13 all-season SBR tires. **DL Pkg. IFB** adds: power steering,

Prices are accurate at time of printing; subject to manufacturer's change.

digital clock, tinted glass, AM/FM ST ET. **DL Pkg. IFC** adds: AM/FM ST ET cassette, power mirrors. **DL Pkg. IFD** adds: alloy wheels. **LX** adds to base DL: power steering, velour seats with driver's side height adjustment, split folding rear seat with center armrest, digital clock, remote fuel door and decklid releases, tachometer, tinted glass, full wheel covers, 175/70R13 tires. **LX Pkg. IFB** adds: power mirrors, AM/FM ST ET. **LX Pkg. ISC** adds: AM/FM ST ET cassette, power windows and locks. **LX Pkg. IFF** adds: alloy wheels. **LX DOHC Pkg. IFD** adds to base LX: DOHC Pkg., power mirrors, AM/FM ST ET cassette. **LX DOHC Pkg. IFE** adds: premium audio system w/EQ, power windows and locks, cruise control, alloy wheels.

Optional Equipment:	Retail Price	Dealer Invoice	Low Price
3-speed automatic transmission, DL & LX .	505	429	465
4-speed automatic transmission, LX w/DOHC	682	580	627
Power steering, DL	262	223	241
DOHC Pkg., LX	805	684	741
DOHC 16-valve engine, 4-wheel disc brakes, sport suspension, 195/60R14 tires, driver's footrest, sport seats, sport steering wheel, blackout exterior treatment.			
Air conditioning	748	636	688
Decklid luggage rack	94	80	86
Front floormats	28	24	26
Cruise control & variable int. wipers, LX .	211	179	194
Digital clock, DL	55	47	51
Power mirrors, DL & LX	49	42	45
LX w/DOHC	55	47	51
Power windows and locks, LX	440	374	405
AM/FM ST ET	301	256	277
AM/FM ST ET cassette	471	400	433
Premium audio system w/EQ, LX	743	632	684
Tinted glass, DL	63	54	58
Euro Cast alloy wheels	272	231	250
Sport alloy wheels, LX w/DOHC	302	257	278
Two-tone paint, LX	159	135	146

Ford Aerostar

Ford Aerostar Eddie Bauer

What's New

A stretched model joins Ford's rear-drive compact van line for '89. Extended-length Aerostars, scheduled to go on sale by the end of December, retain the 118.9-inch wheelbase of the regular models, but have 15.4 inches added to their rear cargo areas for a 28-cubic-foot boost in cargo volume. They also gain one inch of leg room for the second-row seats and two inches for the third-row seats. Also new for '89 is a sporty body-trim package under the XL Sport Wagon badge. It features front and rear bumper cladding, lower rocker-panel moldings and a monochromatic paint scheme. Aerostar is available as a 2-seat cargo hauler, called the Aerostar Van, or as the Wagon, a passenger model with standard seating for five and optional seating

for seven. Besides a base model, wagon trim levels include XL, XL Plus, XLT, XLT Plus and Eddie Bauer, all available as "preferred equipment packages." All Aerostars are powered by a 3.0-liter V-6. A 5-speed manual transmission is standard, a 4-speed overdrive automatic is optional. Among other changes for '89: a new grille in a revised horizontal-bar design joins a matching new front bumper with an air intake slot, and the fuel tank grows from 17 to 21 gallons. A series of changes will come on line later in the year: The Wagon's second- and third-row bench seats will get seatbacks that fold flat for improved cargo loading. Ford says the seat-removal/replacement mechanisms will also be improved. Wagons will get revised headliner and rear-quarter trim panel armrests, and extended-length models will get additional cabin storage bins.

For

- Passenger room
- Cargo space
- Trailer towing ability

Against

- Fuel economy
- Entry/exit
- Ride

Summary

Aerostar is better suited to heavy-duty work like hauling hefty payloads or towing trailers than the front-drive Chrysler minivans. It shares this advantage with its rear-drive rivals, the Chevrolet Astro/GMC Safari. Now, with the extended-length model, Aerostar joins the Dodge Grand Caravan and Plymouth Grand Voyager in the stretched-minivan arena. Aerostar's V-6 engine has more than adequate muscle, though it pays for it with generally poor fuel economy. Even with the long wheelbase, ride quality deteriorates markedly when the going gets rough. Cabins are roomy and the XLT and Eddie Bauer models' interior furnishings are plush and comfortable. Getting into the front seats requires a high step up compared to Caravan/Voyager, but the doorways are wider than in the Toyota Van or VW Vanagon. On the standard-size model, cargo space is unimpressive with all seats in place. The rear seat is quite heavy, so removing it is a chore, but once accomplished there's room aplenty. On balance, we like Aerostar more than the rear-drive Japanese minivans, and a little better than the Chevy Astro/GMC Safari. Caravan and Voyager are still tops for passenger use, but Aerostar has the edge in brawn.

Specifications	4-door van	4-door van
Wheelbase, in.	118.9	118.9
Overall length, in.	174.9	190.3
Overall width, in.	71.7	72.0
Overall height, in.	72.9	73.2
Front track, in.	61.5	61.5
Rear track, in.	60.0	60.0
Turn diameter, ft.	39.8	39.8
Curb weight, lbs.	3500	NA
Cargo vol., cu. ft.	140.4	168.4

> **KEY: ohv** = overhead valve; **ohc** = overhead cam; **dohc** = double overhead cam; **I** = inline cylinders; **V** = cylinders in V configuration; **flat** = horizontally opposed cylinders; **bbl.** = barrel (carburetor); **PFI** = port (multi-point) fuel injection; **TBI** = throttle-body (single-point) fuel injection; **rpm** = revolutions per minute; **OD** = overdrive transmission; **S** = standard; **O** = optional; **NA** = not available.

Prices are accurate at time of printing; subject to manufacturer's change.

	4-door van	4-door van
Fuel capacity, gal.	21.0	21.0
Seating capacity	7	7
Front headroom, in.	39.5	39.5
Front shoulder room, in.	60.0	60.0
Front legroom, max., in.	41.4	41.4
Rear headroom, in.	38.1	38.3
Rear shoulder room, in.	NA	NA
Rear legroom, min., in.	37.9	38.8

Powertrain layout: longitudinal front engine/rear-wheel drive.

Engines

	ohv V-6
Size, liters/cu. in.	3.0/182
Fuel delivery	PFI
Horsepower @ rpm	145 @ 4800
Torque (lbs./ft.) @ rpm	165 @ 3600
Availability	S

EPA city/highway mpg

5-speed OD manual	16/23
4-speed OD automatic	17/22

Prices

Ford Aerostar Wagon	Retail Price	Dealer Invoice	Low Price
4-door van	$11567	$10324	$11046
Destination charge	450	450	450

Standard Equipment:

XL: 3.0-liter PFI V-6, 5-speed manual transmission, power steering and brakes, lowback front bucket seats, 3-passenger middle seat, remote fuel door release, tinted glass, dual outside mirrors, visor mirrors, AM ET, 215/70R14SL tires. **XL Plus** trim adds: front air conditioning, intermittent wipers, electronic instrument cluster, front captain's chairs, cruise control, tilt steering column. **XLT Plus** trim adds: front captain's chairs with power lumbar support and seatback pockets, 2-passenger middle seat, 3-passenger rear seat, power windows and locks, leather-wrapped steering wheel, luggage rack, power mirrors, nylon liftgate net, Light Group, AM/FM ST ET cassette, overhead console with electronic day/night mirror, lighted visor mirrors, full wheel covers. **Eddie Bauer** trim adds: front and rear air conditioning with high-capacity heater, upgraded upholstery, middle and rear seat/bed, two-tone paint, alloy wheels.

Optional Equipment:

4-speed automatic transmission	607	516	558
Limited-slip axle	248	210	228
Air conditioning	846	719	778
High-capacity A/C & aux. heater, base	1422	1209	1308
w/any Preferred Pkg	576	489	530
XL Plus Preferred Pkg. 401A	2084	1771	1917
XLT Preferred Pkg. 402A	3162	2688	2909
XLT Plus Preferred Pkg. 403A	4543	3861	4180
Eddie Bauer Preferred Pkg. 404A	6503	5527	5983
California emissions pkg	99	84	91
WSW tires	72	61	66
7-pass. seating w/bucket seats	296	252	272
5-pass. seating w/2 captain's chairs	480	408	442
7-pass. seating w/2 captain's chairs &			
2- & 3-pass. seat/bed, base	1366	1161	1257
XL Plus	539	458	496
7-pass. seating w/2 captain's chairs	827	723	761
w/4 captain's chairs, XLT	585	498	538
w/4 captain's chairs, Eddie Bauer	NC	NC	NC
Rear defogger, base	224	191	206
XL Plus	166	141	153

	Retail Price	Dealer Invoice	Low Price
Electronics Group, base	1027	873	945
w/XLT Pkg. 402A	745	633	685
Overhead console with trip computer, electronic day/night mirror and map lights, electronic instruments, Super Sound System.			
Exterior Appearance Group, base	632	537	581
w/XL Plus Pkg. 401A	263	224	242
w/XLT Pkg. 402A or 403A	184	157	169
Privacy glass, two-tone paint.			
Engine block heater	33	28	30
Light Group	159	135	146
Luggage rack	141	120	130
Black swing-away mirrors	52	45	48
Bodyside moldings, w/XL trim	63	54	58
w/XLT or Eddie Bauer	35	30	32
Power Convenience Group	501	426	461
Power windows, locks and mirrors.			
Cruise control & tilt steering column	296	252	272
Trailer Towing Pkg	282	239	259
Class I wiring harness, HD turn signal flasher, limited-slip axle.			
Rear wiper/washer, w/rear defogger	139	118	128
w/o rear defogger	198	168	182
Forged aluminum wheels, w/XL trim	309	262	284
w/XLT or Exterior Appearance Group	270	229	248
Intermittent wipers	59	50	54
AM/FM ST ET	211	179	194
AM/FM ST ET cassette	313	266	288
AM radio delete (credit)	(61)	(52)	(52)

Ford Bronco II

Ford Bronco II XLT

What's New

Ford's compact sport utility vehicle looks more like its big-brother Bronco and F-Series pickup trucks for '89 thanks to a restyled front end. There's a new hood and front fenders, aerodynamic headlamps with integral turn signals and parking lights, and a new wraparound front bumper. New inside is an F-Series-style instrument panel and a standard AM/FM stereo radio with digital clock. The automatic transmission shifter has been relocated from the floor to the steering column and the door trim panels are new. Mechanically, a modified power-steering valve is designed to enhance steering feel, a new front axle that allows caster adjustment is intended to reduce tire wear, and a revised tire-tread pattern on 4-wheel-drive models is adopted in an effort to reduce road noise and tire wear. Standard tires for '89 are a size wider, at 205/75R15SL. Bronco II is available with 2- or 4-wheel drive. Manual locking front hubs are standard on the 4WD model. The optional Touch Drive system includes automatic locking front hubs and allows changing in or out of 4WD on the fly. A 2-door body style in four trim

Prices are accurate at time of printing; subject to manufacturer's change.

levels is currently the only one offered, though Ford is reportedly readying a 4-door version for sale during the 1990 model year. The only engine is a 2.9-liter V-6. A 5-speed manual transmission is standard, a 4-speed overdrive automatic is optional. Towing capacity is 5200 pounds.

For

●4WD traction ●Performance ●Anti-lock rear brakes

Against

●Ride ●Fuel economy ●Rear seat room

Summary

Bronco II's standard V-6 has enough power to merge safely with expressway traffic, pass quickly on the highway or tow a trailer weighing up to 5200 pounds. It also works well with the optional automatic transmission, which shifts crisply and doesn't hurry into overdrive or balk at changing down a gear or two for passing. The 4WD model comes with manual locking front hubs standard, so you have to get out and lock them by hand, then engage 4WD by shifting a transfer case lever. The optional Touch Drive system, which includes automatic locking front hubs, is far more convenient since it allows changing in or out of 4WD on the fly by simply pushing a button from the driver's seat. Bronco II's short 94-inch wheelbase aids maneuverability, but makes for a bouncy ride and pronounced body lean in turns. Its short body limits rear-seat leg room and cargo space. And its elevated ride height makes for a tall step up into the interior. We like the new interior, the engine, the Touch Drive 4WD system and the standard anti-lock rear brakes, but Bronco II compromises more in convenience and cabin space than do such rivals as the Chevy S10 Blazer and Jeep Cherokee. It is competitively priced, however, and backed by Ford's generous warranties.

Specifications

	3-door wagon
Wheelbase, in.	94.0
Overall length, in.	158.2
Overall width, in.	68.0
Overall height, in.	68.2
Front track, in.	56.9
Rear track, in.	56.9
Turn diameter, ft.	32.2
Curb weight, lbs.	3166
Cargo vol., cu. ft.	64.9
Fuel capacity, gal.	23.0
Seating capacity	4
Front headroom, in.	39.5
Front shoulder room, in.	55.6
Front legroom, max., in.	42.4
Rear headroom, in.	38.5
Rear shoulder room, in.	56.7
Rear legroom, min., in.	35.2

Powertrain layout: longitudinal front engine/rear-wheel drive or on-demand 4WD.

Engines

	ohv V-6
Size, liters/cu. in.	2.9/177
Fuel delivery	PFI
Horsepower @ rpm	140 @ 4600
Torque (lbs./ft.) @ rpm	170 @ 2600
Availability	S

EPA city/highway mpg	ohv V-6
5-speed OD manual	17/22
4-speed OD automatic	16/20

Prices

Ford Bronco II (1988 prices)	Retail Price	Dealer Invoice	Low Price
3-door 2WD wagon	$11707	$10435	$11071
3-door 4WD wagon	13178	11729	12454
Destination charge	405	405	405

Standard Equipment:

XL: 2.9-liter PFI V-6 engine, 5-speed manual transmission, power steering and brakes, manual locking freewheeling front hubs (4WD), fuel tank skid plate, transfer case skid plate (4WD), intermittent wipers, tinted glass, vinyl door panels, Gauge Package, AM ET radio, vinyl seat trim, P195/75R15SL tires. **XL Sport** trim adds: Sport Appearance Package (brush/grille guards, fog lamps, tubular rear bumper, spats), tachometer, P205/75R15SL RWL tires on alloy wheels. **XLT** trim deletes Sport Appearance Package and adds: air conditioning, cloth door panels, Light Group (cargo area, ashtray, glovebox and engine compartment), right visor mirror, AM/FM ST ET, cloth and vinyl seat trim, leather-wrapped steering wheel, pivoting front vent windows, P195/75R15SL tires on steel wheels. **Eddie Bauer** Trim adds: front captain's chairs, cruise control, tilt steering column, rear quarter privacy glass, bright low-mount swingaway mirrors, two-tone paint, P205/75R15SL RWL tires on alloy wheels.

Optional Equipment:

	Retail	Dealer	Low
4-speed automatic transmission	896	762	824
Limited-slip rear axle	267	227	246
Air conditioning	750	638	690
Touch Drive electric shift, 4WD			
Base or w/Pkg. 921	162	138	149
w/any other package	104	88	96
Preferred Equipment Pkg. 921A	241	205	222
XL Sport Preferred Equipment Pkg. 922A	1213	1031	1116
XLT Preferred Equipment Pkg. 923A	888	755	817
XLT Preferred Equipment Pkg. 931A	1442	1227	1327
Eddie Bauer Preferred Equipment Pkg. 934A	3452	2935	3176
Manual Transmission Pkg. 920A	NC	NC	NC
Manual Transmission Pkg. 921A	242	205	222
Manual Transmission Pkg. 922A	1214	1031	1117
Manual Transmission Pkg. 923A	888	755	817
Manual Transmission Pkg. 931A	1443	1226	1328
Cloth & vinyl seat trim, XL	100	85	92
Cloth 60/40 split bench seat, XL	232	197	213
Cloth & vinyl captain's chairs, XLT	437	371	402
Cloth 60/40 split bench seat, XLT	140	119	129
Floor console	174	148	160
Super engine cooling	57	49	52
Privacy glass	144	122	132
Bright low-mount swingaway mirrors	87	74	80
Power Window/Lock Group, XL	518	441	477
XLT or Eddie Bauer	310	263	285
Luggage rack	126	107	116
Snow Plow Special Pkg.	118	101	109
Cruise control & tilt steering col.	294	250	270
Sport Appearance Pkg.	1213	1031	1116
Tachometer	59	50	54
Two-tone paint, XL	240	204	221
XLT or w/Special Value Pkg.	191	163	176

> **KEY: ohv** = overhead valve; **ohc** = overhead cam; **dohc** = double overhead cam; **I** = inline cylinders; **V** = cylinders in V configuration; **flat** = horizontally opposed cylinders; **bbl.** = barrel (carburetor) **PFI** = port (multi-point) fuel injection; **TBI** = throttle-body (single-point) fuel injection; **rpm** = revolutions per minute; **OD** = overdrive transmission; **S** = standard; **O** = optional; **NA** = not available.

Prices are accurate at time of printing; subject to manufacturer's change.

	Retail Price	Dealer Invoice	Low Price
Aluminum wheels, base	309	262	284
w/XLT, Special Value Pkg. or Pkg. 921	207	176	190
Deluxe wheel trim	110	93	101
Flip-open rear window	90	77	83
Rear defogger & wiper/washer, base	226	192	208
w/Pkg. 931	135	115	124
AM/FM ST ET	93	79	86
AM/FM ST ET cassette	100	85	92
Premium Sound System	121	103	111
AM delete (credit)	(61)	(52)	(52)

Ford Escort

Ford Escort GT

What's New

America's top-selling car received a subtle facelift in mid-1988 and continues virtually unchanged for '89, save for discontinuation of the slow-selling EXP. A 2-seat hatchback, EXP was spawned by the front-drive Escort platform in 1981, was dropped in 1985, reintroduced in '86 and survived for '88 only in Luxury Coupe form. EXP's demise leaves three versions of the subcompact Escort for '89: the stripper Escort Pony and sporty Escort GT, both 3-door hatchbacks, and the Escort LX, which is available as a 3- or 5-door hatchback and a 5-door wagon. Retained is the moderate upgrading of mid-'88: plastic bumpers, new fenders, taillamps, bodyside moldings and quarter panels for the hatchback and wagon; a new grille and rear spoiler for the GT; and 14-inch wheels and tires in place of 13s. Added are an engine malfunctioning warning light and gas-pressurized struts to all models. Escort Ponys and LXs have a 90-horsepower 1.9-liter 4-cylinder engine with single-point fuel injection. GTs have a 115-horsepower multi-point injection version. GTs also gets wider standard tires on 15-inch alloy wheels and a firmer suspension. All Escorts have motorized automatic front shoulder belts with manual lap belts.

For

- Fuel economy • Maneuverability

Against

- Driveability (automatic transmission)
- Passenger room

Summary

Escort is average in most ways, but still ranks as a pretty good value by virtue of its fairly low price and good warranties. Most optional equipment is available in money-saving

option packages, which also simplifies ordering. However, Escort is still priced quite a bit higher than the Dodge Omni/Plymouth Horizon subcompacts. You can probably pick up an Escort for less than suggested retail, though, either through direct dealer discounts or factory incentives. Like the Omni/Horizon, Escort shows its age in some key areas. The interior is cramped compared to newer Japanese designs and the control layout isn't as convenient. The 1.9-liter engine has adequate power and good mileage, but it's not nearly as strong as Chrysler's 2.2 or the smaller, multi-valve engines offered in the Toyota and Honda subcompacts. The optional 3-speed automatic transmission doesn't work well with this engine. It rushes into top gear too soon for good around-town performance, and then balks at downshifting for passing power when you need it. You'll pay less for an Escort with manual shift, and get better performance and mileage in the bargain. There are several newer Japanese small cars with more to offer than Escort, but they're also more expensive, so don't cross this Ford off your list.

Specifications

	3-door hatchback	5-door hatchback	5-door wagon
Wheelbase, in.	94.2	94.2	94.2
Overall length, in.	166.9	166.9	168.0
Overall width, in.	65.9	65.9	65.9
Overall height, in.	53.5	53.5	53.3
Front track, in.	54.7	54.7	54.7
Rear track, in.	56.0	56.0	56.0
Turn diameter, ft.	35.7	35.7	35.7
Curb weight, lbs.	2180	2222	2274
Cargo vol., cu. ft.	38.5	38.5	58.8
Fuel capacity, gal.	13.0	13.0	13.0
Seating capacity	5	5	5
Front headroom, in.	37.9	37.9	37.9
Front shoulder room, in.	51.3	51.3	51.3
Front legroom, max., in.	41.5	41.5	41.5
Rear headroom, in.	37.3	37.3	38.2
Rear shoulder room, in.	51.6	51.4	51.6
Rear legroom, min., in.	35.1	35.1	35.1

Powertrain layout: transverse front engine/front-wheel drive.

Engines

	ohc I-4	ohc I-4
Size, liters/cu. in.	1.9/114	1.9/114
Fuel delivery	TBI	PFI
Horsepower @ rpm	90 @ 4600	115 @ 5200
Torque (lbs./ft.) @ rpm	106 @ 3400	120 @ 4400
Availability	S	S[1]

EPA city/highway mpg

4-speed OD manual	32/42	
5-speed OD manual	27/36	25/32
3-speed automatic	26/31	

1. GT

Prices

Ford Escort	Retail Price	Dealer Invoice	Low Price
Pony 3-door hatchback	$6964	$6418	$6691
LX 3-door hatchback	7349	6621	6985
LX 5-door hatchback	7679	6915	7297
LX 5-door wagon	8280	7450	7865
GT 3-door hatchback	9315	8371	8843
Destination charge	335	335	335

Prices are accurate at time of printing; subject to manufacturer's change.

Standard Equipment:

Pony: 1.9-liter TBI 4-cylinder engine, 4-speed manual transmission, power brakes, cloth and vinyl reclining front bucket seats, folding rear seat, motorized front shoulder belts with manual lap belts, removable cargo cover, 175/70R14 tires. **LX** adds: AM ET radio, full cloth upholstery, door pockets; wagon has 5-speed manual transmission and retractable cargo cover. **GT** adds: PFI engine, 5-speed manual transmission, handling suspension, sport seats, remote fuel door and liftgate releases, power mirrors, tachometer, coolant temperature gauge, trip odometer, visor mirrors, overhead console with digital clock and stopwatch, map lights, center console with graphic monitor and folding armrest, AM/FM ST ET, split folding rear seatback, 195/60HR15 tires on alloy wheels.

Optional Equipment:

	Retail Price	Dealer Invoice	Low Price
5-speed manual trans, LX exc. wagon . . .	76	64	70
3-speed auto trans, Pony, LX exc. wagon .	490	417	451
LX wagon	415	352	382
Air conditioning	688	585	633
Power steering	235	200	216
WSW tires, LX	73	62	67
HD battery	27	23	25
Digital clock/overhead console	82	69	75
Rear defogger	145	123	133
Tinted glass	105	89	97
Instrumentation Group	87	74	80
Tachometer, coolant temperature gauge, trip odometer.			
Light/Security Group, LX	91	78	84
GT	67	57	62
Misc. lights, remote fuel door and liftgate releases.			
Luggage rack	115	97	106
Power mirrors (std. GT)	88	75	81
Wide bodyside moldings	50	43	46
Clearcoat paint, LX	91	78	84
GT (incl. two-tone paint)	183	155	168
Two-tone paint	91	78	84
AM ET radio, Pony	54	46	50
AM/FM ST ET, Pony	206	175	190
LX	152	130	140
AM/FM ST ET cassette, Pony	343	291	316
LX	289	246	266
GT	137	116	126
Premium sound system	138	117	127
AM delete, LX (credit)	(54)	(46)	(46)
AM/FM ST ET delete, GT (credit)	(206)	(175)	(175)
Cruise control	182	154	167
Split folding rear seat	50	42	46
Tilt steering column	124	106	114
Luxury wheel covers	71	60	65
Vinyl trim	37	31	34
Polycast wheels	162	138	149
Intermittent wipers	55	47	51
Rear wiper/washer	126	107	116
Conventional spare tire	73	62	67
Special Value Pkg. 330A, GT	815	692	750
Manual Transaxle Pkg., LX exc. wagon .	560	476	515
LX wagon	484	412	445
5-speed manual transmission, digital clock/overhead console, rear defogger, tinted glass, Instrumentation Group, Light/Security Group, power mirrors, wide bodyside moldings, AM/FM ST ET, power steering, luxury wheel covers, intermittent wipers.			
Automatic Transaxle Pkg., LX exc. wagon .	938	798	863
LX wagon	863	733	794
Manual Transmission Pkg. plus 3-speed automatic transmission.			

Ford Festiva

What's New

Built for Ford in South Korea by Kia Motors, this diminutive front-driver is available with automatic transmission for the first time in 1989. Festiva is a 3-door hatchback based on a design by Mazda, which is owned partly by Ford Motor Co.

Ford Festiva L Plus

It debuted on the West Coast in May 1987 as a 1988 model and was available nationwide by August. Festiva comes in three trim levels: L, L Plus and LX. All are powered by a 1.3-liter 4-cylinder engine. It has a 2-barrel carburetor and is rated at 58 horsepower with manual transmission. Order the optional 3-speed automatic transmission and the engine gains multi-point fuel injection and five additional horsepower. The automatic is available only on L Plus and LX models. An overdrive 4-speed manual transmission is standard on L and L Plus models; LXs get a 5-speed with overdrive fourth and fifth gears. Air conditioning and aluminum wheels are among Festiva's few options. Among cars sold in the U.S., only the Suzuki-built Geo Metro and the Subaru Justy have base curb weights lower than Festiva's 1713 pounds.

For

- Fuel economy   • Maneuverability

Against

- Performance   • Handling/roadholding   • Noise
- Passenger room

Summary

Festiva has adequate acceleration, but no more. Light, precise manual shift linkage helps get the most from the little 1.3-liter, which doesn't produce much of its power at low engine speeds, yet thrashes above 4000 rpm or so. High gas mileage compensates some for the mediocre performance and lack of refinement. We have not yet tested the automatic transmission model, but we fear performance and economy will suffer with it. The manual steering is surprisingly heavy for parking and negotiating tight turns, and you have to fight the wheel in bumpy corners to keep the little car on course. Body roll and understeer (resistance to turning) are prominent, a result of the tall body, forward weight bias and skinny 12-inch tires. Head room is adequate all around, but most adults will consent to use the back seat only for short trips. Ride comfort is good, all things considered. The suspension feels a little stiff and hoppy over freeway expansion joints but it's reasonably

---

KEY: ohv = overhead valve; **ohc** = overhead cam; **dohc** = double overhead cam; **I** = inline cylinders; **V** = cylinders in V configuration; **flat** = horizontally opposed cylinders; **bbl.** = barrel (carburetor); **PFI** = port (multi-point) fuel injection; **TBI** = throttle-body (single-point) fuel injection; **rpm** = revolutions per minute; **OD** = overdrive transmission; **S** = standard; **O** = optional; **NA** = not available.

Prices are accurate at time of printing; subject to manufacturer's change.

absorbent on rough patches. Even with the added sound insulation on the LX there's plenty of noise in equal doses from wind, road and engine. Festiva has a short interior with little cargo room behind the back seat; folding the rear seatback increases cargo capacity only to the mediocre range. In most respects, Festiva is competent and efficient, though a little harsh mechanically and no trend setter. Low base prices make it a decent entry-level buy.

Specifications

	3-door hatchback
Wheelbase, in.	90.2
Overall length, in.	140.5
Overall width, in.	63.2
Overall height, in.	55.3
Front track, in.	55.1
Rear track, in.	54.5
Turn diameter, ft.	28.2
Curb weight, lbs.	1713
Cargo vol., cu. ft.	26.5
Fuel capacity, gal.	10.0
Seating capacity	4
Front headroom, in.	38.8
Front shoulder room, in.	51.9
Front legroom, max., in.	40.6
Rear headroom, in.	37.7
Rear shoulder room, in.	50.9
Rear legroom, min., in.	35.7

Powertrain layout: transverse front engine/front-wheel drive.

Engines	ohc I-4	ohc I-4
Size, liters/cu. in.	1.3/81	1.3/81
Fuel delivery	2 bbl.	PFI
Horsepower @ rpm	58 @ 5000	63 @ 5000
Torque (lbs./ft.) @ rpm	73 @ 3500	73 @ 3000
Availability	S	O[1]

EPA city/highway mpg

4-speed OD manual	38/40	
5-speed OD manual	39/43	
3-speed automatic		30/31

1. With automatic transmission

Prices

Ford Festiva	Retail Price	Dealer Invoice	Low Price
L 3-door hatchback	$5699	$5300	$5550
L Plus 3-door hatchback	6372	5926	6149
LX 3-door hatchback	7101	6604	6853
Destination charge	255	255	255

Standard Equipment:

L: 1.3-liter 2bbl. 4-cylinder engine, 4-speed manual transmission, power brakes, locking fuel filler door, tethered fuel cap, composite halogen headlamps, wide bodyside moldings, reclining high-back front bucket seats, folding rear seat, cloth upholstery, coolant temperature gauge, rear passenger walk-in device, instrument panel coin bin, shift indicator light, 145/SR12 SBR tires on styled steel wheels. **L Plus** adds: rear defogger, AM/FM ST ET. **LX** adds: 5-speed manual transmission, tachometer, trip odometer, intermittent wipers, rear defogger, tilt steering column, sound insulation package, dual horns, tinted glass, dual power mirrors, low-back reclining front bucket seats, upgraded cloth upholstery, split fold-down rear seat, consolette, full door trim with cloth inserts, door map pockets, underseat stowage, urethane soft-feel steering wheel, package tray, headlamps-on alert, cargo area lamp, driver's side coat hook, 165/70SR12 SBR tires.

Optional Equipment:	Retail Price	Dealer Invoice	Low Price
3-speed automatic transmission (NA L) *Includes PFI engine.*	515	437	476
Air conditioning, L Plus	793	674	734
LX	688	585	637
Rear defogger, L	267	227	247
AM ET radio, L	162	138	150
AM/FM ST ET cassette, L & L Plus	137	116	127
Alloy wheels, L Plus	421	358	390
LX	396	337	367

Ford LTD Crown Victoria

Ford LTD Crown Victoria LX 4-door

What's New

Entering the 11th model year on its current rear-drive platform, Ford's conservative full-size car returns nearly unaltered from the major facelift it received for 1988. An engine-systems warning light replaces the dashboard's low-oil warning light as the only change. The lineup consists of the Crown Victoria 4-door sedan and the Crown Vic and Country Squire 5-door wagons, each in base or LX trim. A 2-door coupe was discontinued last year. A 5.0-liter V-8 with multi-point fuel injection and a 4-speed overdrive automatic transmission is the sole powertrain. Air conditioning, tinted glass and an automatic headlamp on/off system are among the standard features. Ford's Insta-Clear heated windshield is optional. The LTD is among the few cars still using a body-on-frame construction.

For

● Performance ● Driveability ● Trailer towing ability
● Passenger room ● Trunk space ● Quietness

Against

● Fuel economy ● Size and weight ● Maneuverability

Summary

A single direct rival remains for the Crown Victoria line of rear-drive, V-8 powered big cars: the Chevrolet Caprice. Caprice is comparable in most ways, but we prefer the Crown Victoria. Ford's 5.0-liter V-8 is slightly smoother and has more torque for better acceleration than either of the V-8s available in Caprice. Ford's 4-speed automatic transmission doesn't rush into overdrive as fast as GM's, and it downshifts more readily for prompt passing response. Mileage is about the same: around 15 mpg (or less) in city driving and into the low 20s from gentle highway driving. The Crown Vic's ample proportions and considerable

---

Prices are accurate at time of printing; subject to manufacturer's change.

weight give it true 6-passenger capacity and good cargo room, plus insurance-industry statistics show it has an impressive record for occupant protection in crashes. And its body-on-frame construction often means lower repair costs from collision damage. Size and weight also work against the Crown Vic, however. It's cumbersome in tight spots. Body lean is excessive in turns, there's lots of bouncing and swaying at highway speeds, and excessive nose-dive in hard stops. Caprice has similar road manners; that's the way American sedans of this size and vintage behave. But if you want a full-size car with a V-8, there's probably no better choice than the Crown Vic. Mercury's Grand Marquis is a little plusher inside and a little more expensive than the Crown Vic, but is the same otherwise.

Specifications

	4-door notchback	5-door wagon
Wheelbase, in.	114.3	114.3
Overall length, in.	211.0	216.0
Overall width, in.	77.5	79.3
Overall height, in.	55.5	57.0
Front track, in.	62.2	62.2
Rear track, in.	62.0	62.0
Turn diameter, ft.	39.1	39.1
Curb weight, lbs.	3779	3991
Cargo vol., cu. ft.	22.4	88.2
Fuel capacity, gal.	18.0	18.0
Seating capacity	6	6
Front headroom, in.	37.9	38.8
Front shoulder room, in.	61.6	61.6
Front legroom, max., in.	43.5	43.5
Rear headroom, in.	37.2	39.1
Rear shoulder room, in.	61.6	61.6
Rear legroom, min., in.	39.3	37.9

Powertrain layout: longitudinal front engine/rear-wheel drive.

Engines

	ohv V-8
Size, liters/cu. in.	5.0/302
Fuel delivery	PFI
Horsepower @ rpm	150 @ 3200
Torque (lbs./ft.) @ rpm	270 @ 2000
Availability	S

EPA city/highway mpg

4-speed OD automatic	17/24

Prices

Ford LTD Crown Victoria	Retail Price	Dealer Invoice	Low Price
4-door notchback	$15851	$13633	$14842
LX 4-door notchback	16767	14441	15704
5-door wagon	16209	13936	15173
Country Squire 5-door wagon	16527	14207	15467
LX 5-door wagon	17238	14811	16125
LX Country Squire 5-door wagon	17556	15082	16419
Destination charge	480	480	480

Standard Equipment:

5.0-liter PFI V-8 engine, 4-speed automatic transmission, power steering and brakes, air conditioning, Autolamp on/off/delay system, split bench seat with reclining backrests and two center armrests, cloth and vinyl upholstery, automatic parking brake release, trip odometer, intermittent wipers, remote mirrors, padded rear half vinyl roof, digital clock, AM/FM ST ET, front door map pockets, right visor mirror, P215/70R15 all-season SBR WSW tires. Wagons have vinyl upholstery, HD rear brakes, three-way doorgate with

power window, locking side and underfloor storage compartments. **LX** adds: power windows, velour upholstery, dual-facing rear seats (wagon), upgraded door panels with curb lights, Light Group, pivoting front head restraints, lighted visor mirrors, rear center armrest, seatback map pockets. **Country Squire** adds to base wagon: luggage rack, woodtone exterior applique.

Optional Equipment:

	Retail Price	Dealer Invoice	Low Price
Traction-Lok axle	100	85	92
Automatic air conditioning, w/Pkg.			
111A, 112A, 131A or 132A	66	56	61
Others	211	179	194
Pkg. 111A, LX 4-door	383	325	352
Pkg. 112A, LX 4-door	938	796	863
Pkg. 113A, LX 4-door	1514	1285	1393
Pkg. 131A, base 4-door & Country Squire	1280	1088	1178
Pkg. 132A, LX 4-door & wagon	688	584	633
Pkg. 133A, LX 4-door & wagon	1191	1011	1096
High Level Audio System, w/Pkg.			
112A or 132A	335	285	308
w/Pkg. 133A or 133A	167	142	154
Others	472	401	434
HD battery	27	23	25
Bumper guards	62	53	57
Front cornering lamps	68	58	63
Rear defogger	145	123	133
Floormats	43	36	40
Engine block heater	18	16	17
Illuminated entry	82	69	75
Light Group	59	50	54
Power Lock Group	245	208	225
Power locks, remote decklid or tailgate release.			
Bodyside moldings	66	56	61
Clearcoat paint	226	192	208
Two-tone paint/tape treatment	159	135	146
AM/FM ST ET cassette	137	116	126
Power antenna	76	64	70
AM/FM ST ET delete (credit)	(206)	(175)	(175)
Premium Sound System	168	143	155
Power driver's seat	251	214	231
Power front seats, w/Pkg. 112A, 131A			
or 132A	251	214	231
Others	502	427	462
Dual-facing rear seats, wagon	173	147	159
Cruise control	182	154	167
Leather-wrapped steering wheel	59	50	54
Tilt steering column	124	106	114
Automatic load-leveling suspension	195	166	179
HD/handling suspension	26	22	24
HD Trailer Towing Pkg., 4-doors	387	329	356
Wagons	399	339	367
3.55 Traction-Lok axle, HD battery, trailer wiring harness, power steering and transmission fluid oil coolers, HD flasher, conventional spare tire, HD rear brakes (4-doors), HD U-joint, HD radiator, dual exhaust system; for 5000-lb. capacity/750-lb. tongue load.			
Tripminder computer	215	182	198
Pivoting front vent windows	79	67	73
Vinyl roof delete (credit)	(200)	(170)	(170)
Alloy wheels (4)	390	332	359
Power windows & mirrors	379	322	349
Insta-Clear windshield	250	213	230
Brougham half roof treatment	665	565	612
California emissions pkg	99	84	91
All-vinyl seat trim	37	31	34
Duraweave vinyl seat trim	96	82	88
Leather seat trim	415	352	382

> **KEY: ohv** = overhead valve; **ohc** = overhead cam; **dohc** = double overhead cam; **I** = inline cylinders; **V** = cylinders in V configuration; **flat** = horizontally opposed cylinders; **bbl.** = barrel (carburetor); **PFI** = port (multi-point) fuel injection; **TBI** = throttle-body (single-point) fuel injection; **rpm** = revolutions per minute; **OD** = overdrive transmission; **S** = standard; **O** = optional; **NA** = not available.

Prices are accurate at time of printing; subject to manufacturer's change.

Ford Mustang

Ford Mustang GT

What's New

The original pony car celebrates its 25th anniversary in 1989, but Ford's plans to commemorate the occasion had not been announced at publication time. Rumors abound about a possible 25th anniversary Mustang with special performance or styling features. One certainty is that Mustang returns for '89 with the same basic front-engine, rear-drive platform introduced in the fall of 1978. It continues in two basic trim levels, LX and GT, though LXs ordered with the GT's 5.0-liter V-8 engine now get their own model designation: LX 5.0L Sport. The LX comes standard with a 2.3-liter 4-cylinder engine and is available as a 2-door coupe, 3-door hatchback or 2-door convertible. The GT has a 225-horsepower 5.0-liter V-8 in either the hatchback or convertible body style. When the optional V-8 was previously ordered on the LX, it brought the GT's beefed-up suspension, wider tires and alloy wheels, but not its body spoilers and air dams. Those styling distinctions remain, though LX 5.0L Sport hatchbacks and convertibles get the GT's articulated sports seats. Also for '89, power windows and locks are now standard on both convertibles. All models are available with a 5-speed manual transmission or a 4-speed overdrive automatic.

For

● Performance (V-8) ● Handling/roadholding ● Price

Against

● Performance (4-cylinder)
● Fuel economy (V-8) ● Passenger room

Summary

The 2.3-liter four is too weak for this 2800-pound car, so if you want a Mustang, get the V-8. There are plenty of reasons to recommend a Mustang GT or an LX 5.0L Sport. You get blistering acceleration, capable handling, even a fine-shifting manual transmission—all at a reasonable price. We realize a high-powered, rear-drive car with a small interior and a rough ride isn't everyone's cup of tea, but for performance-minded buyers on a budget, nothing else delivers so much bang for so few bucks. An LX 5.0L Sport is less gaudy than the GT, but both have an interior and ride that are more livable than the arch-rival Camaro IROC-Z's. There's not much rear passenger room, but plenty of space in front for two. The hatchback has adequate cargo capacity with the rear seatback folded. A V-8 convertible is loads of fun and,

like the Mustang GT, will no doubt be a collectible car years from now. With this level of performance, we'd prefer genuine high-performance brakes to the front-disc/rear-drum arrangement, and mileage is nothing to crow about, but a V-8 Mustang is still the best of a vanishing breed. Check with your insurance agent before you buy, however. Depending on your age, driving record and where you live, a V-8 Mustang could be costly to insure.

Specifications

	2-door notchback	3-door hatchback	2-door conv.
Wheelbase, in.	100.5	100.5	100.5
Overall length, in.	179.6	179.6	179.6
Overall width, in.	69.1	69.1	69.1
Overall height, in.	52.1	52.1	51.9
Front track, in.	56.6	56.6	56.6
Rear track, in.	57.0	57.0	57.0
Turn diameter, ft.	37.4	37.4	37.4
Curb weight, lbs.	2724	2782	3214
Cargo vol., cu. ft.	12.2	30.0	6.4
Fuel capacity, gal.	15.4	15.4	15.4
Seating capacity	4	4	4
Front headroom, in.	37.0	37.0	37.6
Front shoulder room, in.	55.4	55.4	55.4
Front legroom, max., in.	41.7	41.7	41.7
Rear headroom, in.	35.9	35.6	37.0
Rear shoulder room, in.	54.3	54.3	48.9
Rear legroom, min., in.	30.7	30.7	30.7

Powertrain layout: longitudinal front engine/rear-wheel drive.

Engines

	ohc I-4	ohv V-8
Size, liters/cu. in.	2.3/140	5.0/302
Fuel delivery	PFI	PFI
Horsepower @ rpm	90 @ 3800	225 @ 4200
Torque (lbs./ft.) @ rpm	130 @ 2800	300 @ 3200
Availability	S[1]	S[2]

EPA city/highway mpg

5-speed OD manual	23/29	17/24
4-speed OD automatic	NA	17/25

1. LX   2. GT, LX 5.0L Sport

Prices

Ford Mustang	Retail Price	Dealer Invoice	Low Price
LX 2-door notchback	$9050	$8178	$8714
LX 3-door hatchback	9556	8628	9192
LX 2-door convertible	14140	12707	13524
LX 5.0L Sport 2-door notchback	11410	10278	10944
LX 5.0L Sport 3-door hatchback	12265	11039	11752
LX 5.0L Sport 2-door convertible	17001	15254	16228
GT 3-door hatchback	13272	11935	12704
GT 2-door convertible	17512	15708	16710
Destination charge	374	374	374

Standard Equipment:

LX: 2.3-liter PFI 4-cylinder engine, 5-speed manual transmission, power steering and brakes, cloth reclining bucket seats, tinted glass, tachometer, trip odometer, coolant temperature and oil pressure gauges, voltmeter, intermittent wipers, remote mirrors, console with armrest, cargo area cover (3-door), AM/FM ST ET, 195/70R14 tires. **LX 5.0L Sport** adds: 5.0-liter PFI V-8, Traction-Lok axle, articulated sport seats (hatchback and convertible), 225/60VR15 tires on alloy wheels. **GT** adds: power windows and locks, remote hatch release, tilt steering column, fog lights, driver's foot rest, pivoting map light. Convertibles have power top, power windows and locks, remote decklid release, luggage rack, footwell lights.

Prices are accurate at time of printing; subject to manufacturer's change.

Optional Equipment:	Retail Price	Dealer Invoice	Low Price
4-speed automatic transmission	515	437	474
Custom Equipment Group, LX exc. conv .	1180	1004	1086
LX conv	1080	919	994
Air conditioning, premium sound system, tilt steering column, lighted visor mirrors			
Pkg. 245B (LX 5.0), 249B (GT), exc. conv. .	1006	854	926
LX 5.0L & GT conv.	487	413	448
WSW tires, LX 2.3	82	69	75
Bodyside molding insert stripe	61	52	56
Rear defogger	145	123	133
AM/FM ST ET delete (credit)	(206)	(175)	(175)
Flip-up sunroof	355	302	327
Premium Sound System	168	143	155
Wire wheel covers, LX 2.3	178	151	164
California emissions pkg	99	84	91
Leather articulated sport seats, LX conv . .	780	663	718
GT conv	415	352	382
Vinyl seat trim	37	31	34

Ford Probe

Ford Probe LX

What's New

Once intended to replace Mustang, Ford's new front-drive sporty coupe is instead being sold alongside the venerable rear-drive pony car. Probe was introduced in May 1988 as an '89 model and is unchanged this fall. It shares its chassis and powertrains with Mazda's new MX-6 coupe; both are built at Mazda's plant in Flat Rock, Mich. Ford designed the 3-door hatchback's exterior and interior. Base GL and luxury LX models use Mazda's 110-horsepower 2.2-liter 12-valve 4-cylinder engine. Probe GT comes with a turbocharged and intercooled version rated at 145 horsepower. The GT has rear disc brakes instead of drums, a firmer suspension, alloy wheels, unique front and rear fascias and lower-body cladding. Anti-lock brakes are optional only on the GT. A 5-speed manual transmission is standard on all three models; a 4-speed automatic is optional on the GL and LX.

For

- Performance (GT) • Handling/roadholding
- Cargo space • Anti-lock brakes option

Against

- Performance (GL, LX) • Rear seat room
- Torque steer (GT) • Noise

Summary

Ford stylists did a masterful job of disguising the Probe's origin as a Mazda design. Its strong aero lines are in stark contrast to the more conventional, upright look Mazda gave the MX-6. The Ford-designed interior borrows liberal-

ly from the Mazda, including most controls and the main instrument binnacle that tilts along with the steering wheel. The naturally aspirated Probes have a vastly different personality than the turbocharged GT. With 110 horsepower moving 2720 pounds, the GL and LX are pretty unsporting, especially with automatic transmission. They ride surprisingly well for small coupes, but their handling, braking and roadholding are nothing more than average for the class. By contrast, the Probe GT is quite fast, but it can be an unpredictable handful under power because of excessive torque steer (the front end pulls to one side). At low speeds, turbo boost comes on abruptly with a rush of power that wrenches the front wheels off course, even in second gear. The available anti-lock brakes are a welcome safety advantage and the wide tires and sports-oriented suspension foster good cornering grip for the GT, but quick driving on twisty roads is marred by steering that's artificially heavy off center. Automatic transmission is unavailable for the GT, and the 5-speed favors a firm hand to guide it through the gears. All Probes have ample cargo space, even with the rear seatback up, though the liftover is high. The tiny back seat is suitable only for children. A unique blend of Japanese-American styling and competitive prices have made Probe a hot seller in its first few months on the market.

Specifications

	3-door hatchback
Wheelbase, in.	99.0
Overall length, in.	177.0
Overall width, in.	67.9
Overall height, in.	51.8
Front track, in.	57.3
Rear track, in.	57.7
Turn diameter, ft.	34.8
Curb weight, lbs.	2720
Cargo vol., cu. ft.	40.7
Fuel capacity, gal.	15.1
Seating capacity	5
Front headroom, in.	37.3
Front shoulder room, in.	54.7
Front legroom, max., in.	42.5
Rear headroom, in.	35.0
Rear shoulder room, in.	53.7
Rear legroom, min., in.	29.9

Powertrain layout: transverse front engine/front-wheel drive.

Engines	ohc I-4	Turbo ohc I-4
Size, liters/cu. in.	2.2/133	2.2/133
Fuel delivery	PFI	PFI
Horsepower @ rpm	110 @ 4700	145 @ 4300
Torque (lbs./ft.) @ rpm	130 @ 3000	190 @ 3500
Availability	S[1]	S[2]

EPA city/highway mpg

5-speed OD manual	24/31	21/27
4-speed OD automatic	22/29	

1. GL/LX 2. GT

KEY: **ohv** = overhead valve; **ohc** = overhead cam; **dohc** = double overhead cam; **I** = inline cylinders; **V** = cylinders in V configuration; **flat** = horizontally opposed cylinders; **bbl.** = barrel (carburetor); **PFI** = port (multi-point) fuel injection; **TBI** = throttle-body (single-point) fuel injection; **rpm** = revolutions per minute; **OD** = overdrive transmission; **S** = standard; **O** = optional; **NA** = not available.

Prices are accurate at time of printing; subject to manufacturer's change.

Prices

Ford Probe	Retail Price	Dealer Invoice	Low Price
GL 3-door hatchback	$10943	$9864	$10604
LX 3-door hatchback	11769	10599	11384
GT 3-door hatchback	14077	12653	13565
Destination charge	290	290	290

Standard Equipment:

GL: 2.2-liter PFI 4-cylinder engine, 5-speed manual transmission, power steering and brakes, cloth reclining front bucket seats with driver's seat height adjustment, tachometer, coolant temperature and oil pressure gauges, ammeter, AM/FM ST ET, tinted backlight and quarter windows, cargo cover, full console with storage, split folding rear seatbacks, right visor mirror, digital clock, 185/70SR14 tires. **LX adds:** remote fuel door and liftgate releases, intermittent wipers, rear defogger, full tinted glass, tilt steering column and instrument cluster, power mirrors, overhead console with map light, upgraded carpet and upholstery, door pockets, left visor mirror, folding armrest, driver's seat lumbar and side bolster adjustments. **GT adds:** turbocharged engine, 4-wheel disc brakes, performance suspension with automatic adjustment, passenger lumbar support adjustment, fog lights, 195/60VR15 tires on alloy wheels.

Optional Equipment:

4-speed automatic transmission	515	437	476
Air conditioning	907	771	839
w/option pkg	788	670	729
Pkg. 251A, GL	334	285	310

Tinted glass, intermittent wipers, Light Group, power mirrors, tilt steering column and instrument cluster, rear defogger.

Pkg. 253A, LX	2214	1880	2047

Electronic instrument cluster, electronic air conditioning, illuminated entry, leather-wrapped steering wheel and shift knob, power driver's seat, trip computer, rear wiper/washer, walk-in passenger seat, power windows and locks, cruise control, AM/FM ST ET cassette with power antenna.

Pkg. 261A, GT	2621	2226	2424

Pkg. 253A plus anti-lock braking system, analog instruments.

Rear defogger	145	123	134
Power locks	145	123	134
Cruise control	182	154	168
Flip-up sunroof	355	302	329
Alloy wheels, GL	306	260	283
LX	245	208	227
AM/FM ST ET	168	143	156
w/cassette	344	292	318
w/cassette & CD, w/Pkg. 251A, 252A or 260A	1052	895	974
w/cassette & CD, w/Pkg. 253A or 261A .	708	602	655
Cassette stereos include power antenna.			
AM/FM ST ET delete (credit)	(245)	(208)	(208)
Engine block heater	20	17	19

Ford Taurus

What's New

New to Ford's hot-selling line of front-drive family cars is the Taurus SHO. It's powered by a "Super High Output" 3.0-liter V-6 developed and built in Japan by Yamaha. The 220-horsepower V-6 has dual overhead cams and four valves per cylinder. Ford puts its 0-60 mph acceleration time at about eight seconds and its top speed at around 140 mph. Exclusive to

---

KEY: ohv = overhead valve; **ohc** = overhead cam; **dohc** = double overhead cam; **I** = inline cylinders; **V** = cylinders in V configuration; **flat** = horizontally opposed cylinders; **bbl.** = barrel (carburetor); **PFI** = port (multi-point) fuel injection; **TBI** = throttle-body (single-point) fuel injection; **rpm** = revolutions per minute; **OD** = overdrive transmission; **S** = standard; **O** = optional; **NA** = not available.

Ford Taurus SHO

SHO is a special handling suspension and the first 4-wheel disc brake system available on any Taurus. Optional on other models but standard on SHO are 15-inch performance tires on aluminum wheels. The only available SHO transmission is a 5-speed manual designed by Ford but built by Mazda, which is partly owned by Ford. The SHO is available only as a 4-door sedan and is set apart visually from other Tauruses by its modest ground-effects body panels, including a front air dam with fog lamps. Inside, SHO has unique power front sports seats, power windows, a leather-wrapped steering wheel and analog gauges. Other than SHO, the line's major news is the discontinuation of the MT-5 model, which was previously the only Taurus available with a 5-speed manual transmission. Other Tauruses carry on in 4-door sedan and 5-door station wagon body styles. Minor alterations have been made to the grille, headlamps and taillamp lenses. Inside, all have revised door trim panels.

For

- Performance (SHO and 3.8 V-6) • Passenger room
- Handling/roadholding • Ride • Cargo space

Against

- Performance (4-cylinder) • Fuel economy (V-6s)

Summary

America has taken a liking to Taurus, making it one of the best-selling cars in the land. Its streamlined styling sacrifices little interior room and there's space enough for five or six adults and their luggage. Taurus brought to the domestic mid-size field a new standard of handling that emphasized a taut ride, responsive steering and composed cornering. And Taurus is regularly included in Ford's sales incentives and option group discounts that keep its cost competitive. Under the hood, we much prefer the V-6s to the weak and noisy 2.5-liter four. With its higher torque output, the 3.8-liter six delivers noticeably stronger acceleration from a standstill and better midrange response than the 140-horsepower 3.0 six. As for the new SHO, test drives at Ford's Dearborn, Mich., proving grounds, revealed it performs like a genuine sports sedan, with tenacious handling and a silken, potent engine that revs smoothly to its 7300-rpm red line. Ford says it has expended much engineering effort to minimize the powerful engine's propensity to pull the front wheels off course in hard acceleration, an adverse effect known as torque steer. Indeed, torque steer felt minimal, though the 5-speed occasionally balked going into gear. With good steering, great brakes and fine seats, the SHO shapes up as a solid threat to performance-oriented import sedans costing far more.

Prices are accurate at time of printing; subject to manufacturer's change.

Specifications

	4-door notchback	5-door wagon
Wheelbase, in.	106.0	106.0
Overall length, in.	188.4	191.9
Overall width, in.	70.0	70.6
Overall height, in.	54.3	55.1
Front track, in.	61.5	61.6
Rear track, in.	60.5	59.9
Turn diameter, ft.	39.8	39.8
Curb weight, lbs.	2982	3186
Cargo vol., cu. ft.	17.0	80.7
Fuel capacity, gal.	16.0	16.0
Seating capacity	6	8
Front headroom, in.	38.3	38.6
Front shoulder room, in.	57.5	57.5
Front legroom, max., in.	41.7	41.7
Rear headroom, in.	37.6	38.3
Rear shoulder room, in.	57.5	57.5
Rear legroom, min., in.	37.5	36.6

Powertrain layout: transverse front engine/front-wheel drive.

Engines

	ohv I-4	ohv V-6	ohv V-6	dohc V-6
Size, liters/cu. in.	2.5/153	3.0/182	3.8/232	3.0/182
Fuel delivery	TBI	PFI	PFI	PFI
Horsepower @ rpm	90 @ 4400	140 @ 4800	140 @ 3800	220 @ 6000
Torque (lbs./ft.) @ rpm	130 @ 2600	160 @ 3000	215 @ 2200	200 @ 4800
Availability	S	O[1]	O	S[2]

EPA city/highway mpg

5-speed OD manual				NA
3-speed automatic	21/27			
4-speed OD automatic		21/29	19/28	

1. Std. LX, wagons.   2. SHO

Prices

Ford Taurus	Retail Price	Dealer Invoice	Low Price
L 4-door notchback	$11778	$10152	$10965
L 5-door wagon	13143	11312	12228
GL 4-door notchback	12202	10513	11358
GL 5-door wagon	13544	11653	12599
SHO 4-door notchback	19739	16919	NA
LX 4-door notchback	15282	13130	14206
LX 5-door wagon	16524	14187	15356
Destination charge	450	450	450

Standard Equipment:

L: 2.5-liter TBI 4-cylinder engine, 3-speed automatic transmission (wagon has 3.0-liter PFI V-6 and 4-speed automatic), power steering and brakes, power mirrors, reclining split bench seat, cloth upholstery, outboard rear lap/shoulder belts, 60/40 folding rear seatbacks (wagon), cargo tiedowns (wagon), intermittent wipers, trip odometer, coolant temperature gauge, AM/FM ST ET, tinted glass, luggage rack (wagon), P195/70R14 SBR tires. **GL** adds: split bench or bucket seats with console and recliners, front seatback map pockets, digital clock, rear armrest, rear head restraints (4-door), cargo net (4-door). **LX** adds: 3.0-liter V-6 (4-door; wagon has 3.8-liter V-6), air conditioning, power windows and door locks, power lumbar support, diagnostic alert lights, tilt steering column, remote fuel filler and decklid/liftgate releases, Light Group, automatic parking brake release, lower bodyside cladding, cornering lights, upgraded door panels, lighted visor mirrors, P205/70R14 SBR tires. **SHO** adds: 3.0-liter DOHC 24-valve PFI V-6, 5-speed manual transmission, 4-wheel disc brakes, handling suspension, dual exhausts, sport seats with power lumbar, 8000-rpm tachometer, 140-mph speedometer, fog lamps, special bodyside cladding, wheel spats, rear defogger, cruise control, console with armrest and cup holders, leather-wrapped steering wheel, 215/65R15 performance tires on alloy wheels.

Optional Equipment:	Retail Price	Dealer Invoice	Low Price
3.0-liter V-6, L & GL 4-doors	672	571	618
3.8-liter V-6, GL wagon	400	340	368
Others exc. LX wagon	1072	911	986
Automatic air conditioning, w/Pkg. 202A	971	825	893
SHO, LX or w/Pkg. 204A	183	155	168
Manual air conditioning	788	670	725
Autolamp system	73	62	67
Cargo area cover, wagons	66	56	61
Cornering lamps	68	58	63
Rear defogger	145	123	133
Engine block heater	18	16	17
Remote fuel door & decklid/liftgate release	91	78	84
Extended-range fuel tank	46	39	42
Illuminated entry	82	69	75
Diagnostic instrument cluster	89	76	82
Electronic instruments, LX	239	203	220
GL	351	299	323
Keyless entry, w/Pkg. 207A or 211A	121	103	111
Others	202	172	186
Light Group	59	50	54
Picnic table load floor extension, wagons	66	56	61
Power locks	195	166	179
Lighted visor mirrors	100	85	92
Power moonroof	741	630	682
Automatic parking brake release	12	10	11
High Level Audio System, w/Pkg. 204A	335	285	308
w/Pkg. 207A	167	142	154
Others	472	401	434
AM/FM ST ET cassette	137	116	126
Premium sound system	168	143	155
Power antenna	76	64	70
Rear-facing third seat, wagons	155	132	143
Power driver's seat	251	214	231
Dual power seats, w/Pkg. 204A or 211A	251	214	231
Others	502	427	462
Cruise control	182	154	167
Tilt steering column	124	106	114
Leather-wrapped steering wheel	59	50	54
HD suspension	26	22	24
Rear wiper/washer, wagons	126	107	116
Finned wheel covers	65	55	60
Alloy wheels, GL	227	193	209
w/Pkg. 204A	162	138	149
Styled road wheels, GL	178	151	164
LX or w/Pkg. 204A	113	96	104
Power windows	296	252	272
Insta-Clear heated windshield	250	213	230
California emissions pkg.	100	85	92
Leather seat trim, LX & SHO	415	352	382
GL	518	441	477
Vinyl seat trim, L	51	41	47
GL	37	31	34
Preferred Pkg. 204A, GL	1749	1488	1609
Preferred Pkg. 207A, LX	777	658	715
Preferred Pkg. 208A, LX 4-door	1913	1624	1760
LX wagon	1513	1284	1392
Preferred Pkg. 211A, SHO	533	453	490

Ford Tempo

What's New

Last year, this compact front-drive sedan received the first major facelift since its May 1983 introduction. The revamp made it more closely resemble the aerodynamic Ford Taurus, and it also received a stylish new dashboard. This year's changes are far less significant. Front-wheel drive Tempos continue in GL and GLS trim as 2- and 4-door sedans; the top-of-the line LX is a 4-door sedan only. The All-Wheel Drive model features a part-time 4-wheel drive system (for use on slippery pavement only) and is available

Prices are accurate at time of printing; subject to manufacturer's change.

Ford

Ford Tempo LX 4-door

only as a 4-door. For '89, nitrogen gas-pressurized struts have been added to the GL, all Tempos gain an emissions-system warning light and a slightly larger battery. GLS, LX and All-Wheel Drive models get an elasticized luggage compartment cargo tie-down net. The front center armrest previously optional on the GLS is now standard. GLS 4-doors also get a Sport Appearance Group option that includes body cladding in three monotone color combinations. Powertrains are unchanged. A 98-horsepower 2.3-liter 4-cylinder is standard on the GL and LX; a 100-horsepower version is standard on the GLS and All-Wheel Drive. A 5-speed manual transmission is standard and a 3-speed automatic is optional in all models except the All-Wheel Drive, which has the automatic standard.

For

●4WD traction (All-Wheel Drive sedan) ●Air bag option

Against

●Engine noise ●Driveability

Summary

We're not going to recommend Tempo for its power or performance, which are barely adequate with either engine. Worse, the automatic transmission hurries into top gear and then balks at changing back down for passing. Nor do its passenger room, trunk space, ride or handling rise above average in the compact field. Still, this car is a strong seller and there are solid reasons for that. Prices are attractive and last year's restyle brought a smartly updated exterior and one of the most ergonomically sound dashboards in an American-built car. The All-Wheel Drive sedan is better than run-of-the-mill by virtue of its easy-to-use 4WD system. Engaged or disengaged on the fly by the flip of a switch, the convenient system provides impressive traction in slippery conditions. Another nice feature is the available driver's-side air bag. It became a regular production option on Tempo and similar Topaz for 1987, though fleet buyers for government agencies and insurance companies have made up the bulk of the sales. Ford hasn't done much to promote the safety device, but there's little evidence the average retail customer's buying decision hinges on the availability of an air bag.

Specifications

	2-door notchback	4-door notchback
Wheelbase, in.	99.9	99.9
Overall length, in.	176.5	176.5
Overall width, in.	68.3	68.3
Overall height, in.	52.7	52.7
Front track, in.	54.9	54.9
Rear track, in.	57.6	57.6
Turn diameter, ft.	38.7	38.7
Curb weight, lbs.	2462[1]	2515[2]

	2-door notchback	4-door notchback
Cargo vol., cu. ft.	13.2	13.2
Fuel capacity, gal.	15.4[3]	15.4[3]
Seating capacity	5	5
Front headroom, in.	37.5	37.5
Front shoulder room, in.	53.4	53.4
Front legroom, max., in.	41.5	41.5
Rear headroom, in.	36.9	36.9
Rear shoulder room, in.	54.0	53.4
Rear legroom, min., in.	36.0	36.0

1. 2667 with 4WD   2. 2720 with 4WD   3. 13.7 with 4WD

Powertrain layout: transverse front engine/front-wheel drive or on-demand 4WD.

Engines

	ohv I-4	ohv I-4
Size, liters/cu. in.	2.3/141	2.3/141
Fuel delivery	PFI	PFI
Horsepower @ rpm	98 @ 4400	100 @ 4400
Torque (lbs./ft.) @ rpm	124 @ 2200	130 @ 2600
Availability	S	S[1]

EPA city/highway mpg

5-speed OD manual	23/32	21/28
3-speed automatic	22/26	22/27

1. GLS and 4WD

Prices

Ford Tempo	Retail Price	Dealer Invoice	Low Price
GL 2-door notchback	$9057	$8160	$8609
GL 4-door notchback	9207	8293	8750
GLS 2-door notchback	9697	8729	9213
GLS 4-door notchback	9848	8863	9356
LX 4-door notchback	10156	9138	9647
All Wheel Drive 4-door notchback	10860	9765	10313
Destination charge	425	425	425

Standard Equipment:

GL: 2.3-liter PFI 4-cylinder engine, 5-speed manual transmission, power steering and brakes, cloth reclining front bucket seats, motorized front shoulder belts and manual lap belts, AM/FM ST ET, coolant temperature gauge, tinted glass, intermittent wipers, door pockets, 185/70R14 tires. **GLS** adds: high-output engine, light group, tachometer and trip odometer, leather-wrapped steering wheel, AM/FM cassette, power mirrors, sport seats, luggage tiedown, front center armrest, performance tires on alloy wheels. **LX** adds to GL: illuminated entry, power locks, remote fuel door and decklid releases, tilt steering column, front armrest, upgraded upholstery, seatback pockets, polycast wheels. **All Wheel Drive** adds to GL: high-output engine, 3-speed automatic transmission, part-time 4-wheel drive.

Optional Equipment:

3-speed automatic transmission (std. AWD)	515	437	474
Driver's side air bag, GL	815	692	750
LX	751	639	691
185/70R14 WSW tires	82	69	75
185/70R14 WSW performance tires	82	69	75
Air conditioning	788	670	725
Rear defogger	145	123	133
Sport instrument cluster GL	87	74	80
Tachometer, trip odometer.			
Decklid luggage rack	115	97	106
Power Lock Group, 2-doors	237	201	218
4-doors	287	244	264
Power locks, remote fuel door and decklid releases.			
Power mirrors, GL	111	94	102
AM/FM ST ET cassette (std. GLS)	137	116	126
Power driver's seat	251	214	231

Prices are accurate at time of printing; subject to manufacturer's change.

CONSUMER GUIDE®

	Retail Price	Dealer Invoice	Low Price
Premium sound system	138	117	127
Cruise control	182	154	167
Sports Appearance Group, GLS	1178	1001	1084
Includes bodyside cladding.			
Tilt steering column	124	106	114
Polycast wheels	178	161	164
Power windows, 4-doors	296	252	272
California emissions pkg.	100	85	92
Clearcoat metallic paint	91	78	84
Preferred Pkg. 226A, GL 2-door	449	381	413
GL 4-door	499	424	459
Preferred Pkg. 227A, GL 4-door	1250	1060	1150
Preferred Pkg. 229A, GLS 2-door	1220	1037	1122
GLS 4-door	1270	1079	1168
Preferred Pkg. 233A, LX	863	733	794
Preferred Pkg. 234A, LX	1099	934	1011
Preferred Pkg. 232A, AWD	352	300	324

Ford Thunderbird

Ford Thunderbird LX

What's New

The redesigned 1989 Thunderbird—scheduled to go on sale Dec. 26—retains its rear-drive layout but gets an all-new body that's 3.4 inches shorter, 1.6 inches wider and almost an inch lower than its predecessor. The mid-size coupe's wheelbase, however, has been stretched nine inches and every cabin measurement has been increased versus the '88 model. Underneath is the T-Bird's first all-independent suspension. There are three models: Base and LX Thunderbirds are available only with the 3.8-liter V-6 that was the line's base engine last year. It comes only with a 4-speed overdrive automatic transmission. No longer offered is the 5.0-liter V-8. Gone also is the 2.3-liter turbocharged 4-cylinder that powered the Thunderbird Turbo Coupe. The new performance engine is a supercharged and intercooled version of the 3.8 V-6. It's available only in the new Super Coupe. The SC comes standard with aero body flares and 4-wheel disc brakes with anti-lock feature. The anti-lock, 4-wheel disc brakes are optional on base and LX models in place of their standard front disc/rear drum arrangement. Wheels on base and LX T-Birds increase from 14- to 15-inch diameter for '89; SCs have 16-inch wheels and performance tires. Standard on the supercharged model is the Automatic Adjustable Suspension that allows the driver to adjust shock absorber damping for a soft or a firm ride. Speed-sensitive variable-assist power steering is standard on all but the base T-Bird. LXs get digital instruments, including a tachometer. Base models and SCs get analog gauges; the SC adds a tachometer and a supercharger boost gauge. The SC has power front seats with adjustable backrest wings and inflatable lumbar bolsters. All models have air conditioning and power windows standard.

For

- Performance (Super Coupe)
- Handling/roadholding
- Anti-lock brakes

Against

- Performance (Base and LX)

Summary

Thunderbird's previous redesign, in 1983, introduced aero styling to domestic-car buyers and strongly influenced U.S. car design in the '80s. The '89 version builds on that without rejecting the long-hood, short-deck look T-Bird loyalists love. And significantly, it retains rear-wheel drive, something General Motors shelved in its 1988 revamp of the Grand Prix, Oldsmobile Cutlass Supreme and Buick Regal. With its longer wheelbase, shorter body and reduced overhang, Thunderbird's new design, in effect, pushes the wheels out to the edges of the car, in the European fashion. Unfortunately, curb weight is up by about 300 pounds over comparable 1988 models. It's roomier inside, though, and the rear seat now comfortably accommodates three adults. Controls on the new cockpit-style dashboard are clearer and easier to reach, but buttons on the climate and sound systems have a shiny finish that feels and looks cheap. Ford says the flagship Super Coupe goes from 0-60 mph in about 7.5 seconds with manual transmission; about eight seconds with automatic. A pre-production 5-speed SC suffered little supercharger lag on Ford's Michigan test track. Power blossoms quickly just below 3000 rpm and delivers a solid kick with the slightest throttle tap above 4000 rpm. A supercharged Cougar XR7 with automatic transmission was far less responsive. Both, however, are much stronger than a Thunderbird with the base engine, which runs smoothly but is uninspired. The Mazda-made manual transmission has good shift and clutch action. Body lean in turns is well controlled, grip is great and the SC feels stable. The anti-lock brakes, however, don't respond progressively to pedal pressure in hard stops and don't grab until late. A base Thunderbird soldiers through the same tight test course with tires howling, engine wheezing and body leaning, obviously out of its element. Ford's proving-ground pavement is extremely smooth, so a ride-quality critique awaits a full road test. Ford predicts it will sell 300,000 Thunderbirds and Cougars in the first model year, and we think that is within reach. Our initial impressions are that the styling and packaging make these cars far more attractive than GM's front-drive rivals.

Specifications

	2-door notchback
Wheelbase, in.	113.0
Overall length, in.	198.7
Overall width, in.	72.7
Overall height, in.	52.7
Front track, in.	61.4
Rear track, in.	60.2
Turn diameter, ft.	35.6
Curb weight, lbs.	3542
Cargo vol., cu. ft.	14.7

KEY: ohv = overhead valve; **ohc** = overhead cam; **dohc** = double overhead cam; **I** = inline cylinders; **V** = cylinders in V configuration; **flat** = horizontally opposed cylinders; **bbl.** = barrel (carburetor); **PFI** = port (multi-point) fuel injection; **TBI** = throttle-body (single-point) fuel injection; **rpm** = revolutions per minute; **OD** = overdrive transmission; **S** = standard; **O** = optional; **NA** = not available.

Prices are accurate at time of printing; subject to manufacturer's change.

	2-door notchback
Fuel capacity, gal.	19.0
Seating capacity	5
Front headroom, in.	38.1
Front shoulder room, in.	59.1
Front legroom, max., in.	42.5
Rear headroom, in.	37.5
Rear shoulder room, in.	59.1
Rear legroom, min., in.	35.9

Powertrain layout: longitudinal front engine/rear-wheel drive.

Engines

	ohv V-6	Supercharged ohv V-6
Size, liters/cu. in.	3.8/232	3.8/232
Fuel delivery	PFI	PFI
Horsepower @ rpm	140 @ 3800	210 @ 4000
Torque (lbs./ft.) @ rpm	215 @ 2400	315 @ 2600
Availability	S	S[1]

EPA city/highway mpg

5-speed OD manual		NA
4-speed OD automatic	NA	NA

1. *Super coupe*

Equipment Summary

Ford Thunderbird
(prices not available at time of publication):

Standard Equipment:

3.8-liter PFI V-6, 4-speed automatic transmission, power steering and brakes, air conditioning, cloth reclining front bucket seats, tinted glass, power windows, intermittent wipers, dual remote mirrors, full-length console with armrest and storage bin, coolant temperature gauge, trip odometer, visor mirrors, AM/FM ST ET, 205/70R15 all-season tires. **LX** adds: power driver's seat, illuminated entry, remote fuel door and decklid releases, power locks, cruise control, maintenance monitor, power mirrors, folding rear armrest, electronic instruments, Light Group, lighted visor mirrors, AM/FM ST ET cassette, leather-wrapped steering wheel, tilt steering column, instrument panel storage compartment. **Super Coupe** adds to base: supercharged engine with dual exhaust, 5-speed manual transmission, 4-wheel disc brakes, anti-lock braking system, handling suspension, articulated sport seats, lower bodyside cladding, fog lights, analog instruments with tachometer, soft-feel steering wheel, remote fuel door release, maintenance monitor, power mirrors, folding rear armrest, Light Group, instrument panel storage compartment, 225/60VR16 tires on alloy wheels.

Geo Metro

What's New

The front-drive minicompact previously known as the Chevrolet Sprint has been restyled, renamed, and repositioned as part of Chevrolet's new Geo program. The Geo models are either imported from Japan or built in North America from a Japanese design, and are sold through Chevy dealers who have joined the Geo program. Metro's styling now bears a strong resemblance to that of the Honda Civic 3-door hatchback. Body styles include a 3-door and longer 5-door hatchback, same as with the Sprint. Wheelbases are slightly longer at 89.2 inches for the 3-door

---

KEY: ohv = overhead valve; **ohc** = overhead cam; **dohc** = double overhead cam; **I** = inline cylinders; **V** = cylinders in V configuration; **flat** = horizontally opposed cylinders; **bbl.** = barrel (carburetor); **PFI** = port (multi-point) fuel injection; **TBI** = throttle-body (single-point) fuel injection; **rpm** = revolutions per minute; **OD** = overdrive transmission; **S** = standard; **O** = optional; **NA** = not available.

Geo Metro LSi 3-door

and 93.2 for the 5-door, and overall length is up by five inches on both, to 146.3 and 150.4 inches respectively. The base 3-door is the price- and fuel-economy leader, available only with a 5-speed manual transmission. Upscale LSi trim is available on both body styles with a choice of 5-speed manual or 3-speed automatic transmissions. Power is from a revised 1.0-liter 3-cylinder engine with single-point fuel injection instead of a 2-barrel carburetor; it produces 55 horsepower, seven more than last year's. The turbocharged model of previous years has been dropped, a victim of high insurance rates, according to Chevy.

For

• Fuel economy   • Maneuverability

Against

• Interior room   • Noise   • Size/weight

Summary

Other than the new name and a few more horses under the hood, Metro's main difference from Sprint is its attractive new styling, which should greatly enhance showroom appeal. Metro's main virtue is still high fuel economy. We averaged a commendable 42.3 mpg in a 1988 Sprint with the 5-speed manual and economy gear ratio, which is the best we've seen in a long time. Oddly, there was little difference in economy among our drivers, even though some do mostly city driving while others do mostly suburban expressway driving. You'll lose some mileage and acceleration with the automatic transmission, but will still have sufficient amounts of both. With the 5-speed, Metro feels lively in the lower gears and can easily keep pace with traffic. Trouble is, the engine makes so much noise that it sounds like you're beating the little 3-cylinder to death just driving to work. With an excessive amount of road noise thrown in you'll start wearing ear plugs on your morning commute. The short wheelbases and light curb weights put Metro at the mercy of bumpy pavement, which makes the ride bouncy and jarring. We're also leery of the crash protection cars this small provide, and skeptical about how much collision damage a Metro can sustain. However, if low prices and high fuel economy are important to you, this is a good place to look.

Specifications

	3-door hatchback	5-door hatchback
Wheelbase, in.	89.2	93.2
Overall length, in.	146.3	150.4
Overall width, in.	62.0	62.7
Overall height, in.	53.4	53.5
Front track, in.	53.7	53.7
Rear track, in.	52.8	52.8

	3-door hatchback	5-door hatchback
Turn diameter, ft.	30.2	31.5
Curb weight, lbs.	1591	1640
Cargo vol., cu. ft.	29.1	31.4
Fuel capacity, gal.	10.6	10.6
Seating capacity	4	4
Front headroom, in.	37.8	38.8
Front shoulder room, in.	51.6	51.6
Front legroom, max., in.	42.5	42.5
Rear headroom, in.	36.6	38.0
Rear shoulder room, in.	50.5	50.6
Rear legroom, min., in.	29.8	32.6

Powertrain layout: transverse front engine/front-wheel drive.

Engines

	ohc I-3
Size, liters/cu. in.	1.0/61
Fuel delivery	TBI
Horsepower @ rpm	55 @ 5700
Torque (lbs./ft.) @ rpm	58 @ 3300
Availability	S

EPA city/highway mpg

5-speed OD manual	53/58
3-speed automatic	38/40

Prices

Geo Metro	Retail Price	Dealer Invoice	Low Price
3-door hatchback	$5995	$5755	$5875
LSi 3-door hatchback	6895	6481	6688
LSi 5-door hatchback	7195	6763	6979
Destination charge	255	255	255

Standard Equipment:

1.0-liter TBI 3-cylinder engine, 5-speed manual transmission, power brakes, cloth/vinyl reclining bucket seats, rear lap/shoulder belts, coolant temperature gauge, 145/80R12 SBR tires. **LSi** adds: flush headlights, full cloth upholstery, bodyside molding, dual outside mirrors, full wheel covers.

Optional Equipment:

Air conditioning	655	576	616
AM/FM ST ET	301	265	283
AM/FM ST ET cassette	423	372	398
w/LSi w/Group 2 or 3	122	107	115
Preferred Group 2, base	140	123	132
Rear defogger.			
Preferred Group 2, LSi	476	419	447
Rear defogger, AM/FM ST ET cassette, intermittent wipers, rear wiper/ washer.			
Preferred Group 3, LSi	896	788	842
Group 2 plus 3-speed automatic transmission.			
Console	25	22	24
Left remote and right manual mirrors . . .	20	18	19
Bodyside molding, base	50	44	47
Removable sunroof	250	220	235

Geo Prizm

What's New

Prizm is a new front-drive subcompact that will replace the Chevrolet Nova, but it isn't scheduled to go on sale until next spring as an early 1990 model, Like Nova, Prizm is based on the design for the Toyota Corolla and will be built in California by New United Motor Manufacturing Inc., the

Geo Prizm

joint venture between GM and Toyota. It will be marketed through Chevrolet dealers who have joined GM's new Geo program for Japanese-built imports and North American transplants. Prizm's styling is closer to that of the Japanese Toyota Sprinter than the U.S. Corolla, but Prizm's front-chassis will be the same as Corolla's with a 95.7-inch wheelbase. Body styles will be a 4-door notchback sedan and a 5-door hatchback; Toyota currently offers a Corolla 2-door coupe, 4-door sedan and a 5-door wagon. Two engines will be available: Base and LSi models will have a 95-horsepower double-overhead cam 1.6-liter 4-cylinder with multipoint fuel injection; this engine comes with a 5-speed manual or 3-speed automatic. A sporty GSi model has a 115-horsepower version of the 1.6 with either a 5-speed manual or 4-speed overdrive automatic. Similar engines are available in the Corolla. Chevrolet hasn't released additional information on Prizm. We drove a prototype 4-door briefly at Chevrolet's preview in June, but not enough to sufficiently evaluate its performance.

Equipment Summary

Geo Prizm
(prices not available at time of publication)

Standard Equipment:

1.6-liter DOHC PFI 4-cylinder engine, 5-speed manual transmission, power brakes, velour reclining front bucket seats, tinted glass, door pockets, cup holders, left remote mirror, folding rear seat (hatchbacks), remote fuel door release, bodyside molding, 175/70SR13 SBR tires. **LSi** adds: upgraded upholstery and door panels, visor mirrors, assist grips, split folding rear seat, soft-feel steering wheel, full wheel covers.

Geo Spectrum

What's New

Last year's Chevrolet Spectrum line has been trimmed to a single price series available in front-drive 3-door hatchback and 4-door sedan guise. Other than the new Geo badging, the subcompact Spectrum is unchanged; it is sold through Chevrolet dealers who have joined the new Geo program for marketing imports. Standard equipment is similar to that of last year's entry-level Express model. The turbocharged engine has been dropped, leaving a 70-horsepower 1.5-liter 4-cylinder engine with a 2-barrel carburetor. A 5-speed manual is standard and a 3-speed automatic optional. Spectrum is built in Japan by Isuzu, which is partly owned by GM. Isuzu dealers sell a similar model as

Prices are accurate at time of printing; subject to manufacturer's change.

Geo Spectrum 3-door

the I-Mark, which is offered with three engines this year: the 70-horsepower 1.5, a 110-horsepower turbocharged 1.5 and a 125-horsepower double-overhead cam 1.6. Spectrum and I-Mark will be redesigned for 1990.

For

● Fuel economy   ● Ride

Against

● Performance   ● Handling   ● Noise

Summary

Spectrum was supposed to give Chevrolet dealers a low-cost subcompact to combat the small cars offered by Nissan, Toyota and other Japanese dealers. However, once you add options such as air conditioning, a stereo and automatic transmission, the cost is no longer low: nearly $10,000 for a Spectrum 4-door. At that price, Spectrum's weak, noisy engine and sluggish performance stand out as glaring drawbacks. For the same money, you can get a Mazda 323 with spunkier acceleration and comparable fuel economy, or for less money you can get a Dodge Omni/Plymouth Horizon or a Hyundai Excel. The new Geo program doesn't improve Spectrum's pricing, so we can't get enthused about this car. The performance and refinement are mediocre, while interior room and accommodations are nothing special. You can do better elsewhere.

Specifications

	3-door hatchback	4-door notchback
Wheelbase, in.	94.5	94.5
Overall length, in.	157.4	160.2
Overall width, in.	63.6	63.6
Overall height, in.	52.0	52.0
Front track, in.	54.7	54.7
Rear track, in.	54.3	54.3
Turn diameter, ft.	34.8	34.8
Curb weight, lbs.	1947	1989
Cargo vol., cu. ft.	29.7	11.4
Fuel capacity, gal.	11.1	11.1
Seating capacity	5	5
Front headroom, in.	38.0	37.7
Front shoulder room, in.	52.8	52.8
Front legroom, max., in.	41.7	41.7
Rear headroom, in.	37.1	37.6
Rear shoulder room, in.	52.8	53.8
Rear legroom, min., in.	33.0	33.5

Powertrain layout: transverse front engine/front-wheel drive.

Engines

	ohc I-4
Size, liters/cu. in.	1.5/90
Fuel delivery	2 bbl.
Horsepower @ rpm	70 @ 5400
Torque (lbs./ft.) @ rpm	87 @ 3400
Availability	S

EPA city/highway mpg

5-speed OD manual	37/41
3-speed automatic	31/33

Prices

Geo Spectrum	Retail Price	Dealer Invoice	Low Price
4-door notchback	$7795	$7327	$7561
3-door hatchback	7295	6857	7076
Destination charge	315	315	315

Standard Equipment:

1.5-liter 2bbl. 4-cylinder engine, 5-speed manual transmission, power brakes, cloth/vinyl reclining front bucket seats, split folding rear seatback (hatchback), left remote mirror, center console with storage, headlamps-on tone, coolant temperature gauge, 155/80R13 SBR tires.

Optional Equipment:

Air conditioning	660	614	637
AM/FM ST ET	301	280	290
AM/FM ST ET cassette	423	393	408
Sound system prices vary with option package content.			
Preferred Group 2	421	392	406
Floormats, remote mirrors, AM/FM ST ET, full wheel covers.			
Preferred Group 3	634	590	612
Power steering, 3-speed automatic transmission.			
Preferred Group 4	1055	981	1018
Group 3 plus floormats, remote mirrors, AM/FM ST ET, full wheel covers.			
Cargo area cover	69	64	67
Floormats	25	23	24
Remote mirrors	43	40	41
Cruise control	175	163	169
Bodyside striping	53	49	51

Geo Tracker

What's New

Tracker, built by Suzuki, is a new 4-wheel-drive vehicle available through Chevrolet dealers in only 12 coastal states for 1989, when 10,000 units are scheduled to be imported from Japan. Suzuki dealers will sell a similar model as the Sidekick. By 1990, a joint-venture plant in Canada between GM and Suzuki should be able to produce enough Trackers for national distribution through Chevy dealers in the Geo program. Tracker is larger than the much-maligned Suzuki Samurai, and Chevrolet is trying to separate the designs of the two vehicles. Tracker has an 86.6-inch wheelbase and overall length of 142.5 inches. It's 65.4 inches tall, 64.2 inches wide and has a 55-inch track (the width between the wheels on the same axle), or about the same as the Chevy S10 Blazer compact 4x4. Chevrolet engineers claim Tracker's handling and roadholding ability compares favorably to the S10 Blazer's. Samurai by comparison rides a 79.9-inch wheelbase and is 135 inches long. It's taller than Tracker at 65.6 inches and narrower at 60.2 inches overall, plus it has a narrower track: 51.2 front and 51.6 rear. Convertible and hardtop models are offered in

---

Prices are accurate at time of printing; subject to manufacturer's change.

CONSUMER GUIDE®

Geo Tracker

base trim, both with a folding 2-place rear seat standard. The top-line LSi model, which comes only as a hardtop, has a split folding rear seat. Convertibles have a removable canvas top that can be folded part-way back for a sunroof-like feature. Tracker uses an 80-horsepower 1.6-liter 4-cylinder engine with single-point fuel injection. A 5-speed manual transmission is standard on base models and a 3-speed automatic is standard on the LSi. An on-demand, 4WD system (not for use on dry pavement) with a floor-mounted transfer case is standard. Tracker comes only with manual locking front hubs, which have to be engaged by hand before 4WD can be engaged.

For

●4WD traction ●Driveability

Against

●Inconvenient 4WD system ●Ride ●Rear seat room

Summary

We drove the new Tracker briefly on paved roads at Chevrolet's press preview and found the 1.6-liter engine much quieter and smoother than the Samurai's 1.3-liter 4-cylinder. In addition, Tracker's engine has sufficient power over a broad enough range for decent acceleration and adequate passing response, and the smooth manual shift linkage helps you get the most out of the engine. By contrast, Samurai's engine has to work overtime just to keep pace with traffic. With its wider track and longer wheelbase, Tracker also feels more stable through turns than Samurai and has less body lean. However, there's still plenty of understeer (resistance to turning), which warns you to slow down. Steering response is more direct than in the Samurai. It's much roomier inside and somewhat more comfortable than Samurai, plus it has a handy control layout and an easy-to-read gauge cluster. The rear seat is still better for children than adults. For those serious about off-road driving, Tracker's 4WD system lacks the convenience of many rivals: It comes only with manually locking hubs, and you have to stop the vehicle to engage or disengage 4WD. Overall, this is a much more pleasant on-road vehicle than Samurai, though that's not saying much. We want more experience with Tracker before giving a stronger endorsement.

Specifications

	2-door wagon
Wheelbase, in.	86.6
Overall length, in.	142.5
Overall width, in.	64.2
Overall height, in.	65.4
Front track, in.	54.9
Rear track, in.	55.1

	2-door wagon
Turn diameter, ft.	32.1
Curb weight, lbs.	2238
Cargo vol. cu. ft	31.9
Fuel capacity, gal.	11.1
Seating capacity	4
Front headroom, in.	39.5
Front shoulder room, in.	52.1
Front legroom, max., in.	42.1
Rear headroom, in.	38.3
Rear shoulder room, in.	50.2
Rear legroom, min., in.	31.7

Powertrain layout: longitudinal front engine/rear-wheel drive or on-demand 4WD.

Engines

	ohc I-4
Size, liters/cu. in.	1.6/97
Fuel delivery	TBI
Horsepower @ rpm	80 @ 5500
Torque (lbs./ft.) @ rpm	94 @ 3000
Availability	S

EPA city/highway mpg

5-speed OD manual	28/29
3-speed automatic	25/26

Prices

Geo Tracker	Retail Price	Dealer Invoice	Low Price
2-door wagon	$10495	$9760	$10128
2-door convertible	10195	9481	9838
LSi 2-door wagon	12495	11620	12058
Destination charge	270	270	270

Standard Equipment:

1.6-liter TBI 4-cylinder engine, 5-speed manual transmission, part-time 4-wheel drive, power brakes, cloth reclining front bucket seats, folding rear bench seat, rear lap/shoulder belts, tachometer, coolant temperature gauge, trip odometer, door pockets, assist straps, dual outside mirrors, tow hooks, rear defogger (except convertible), 205/75R15 SBR tires. LSi adds: 3-speed automatic transmission, air conditioning, tinted glass, upgraded upholstery, spare tire cover, fold-and-stow rear bucket seats, intermittent wipers, rear wiper/washer, RWL tires on chrome wheels.

Optional Equipment:

Spare tire cover, base	33	29	31
Floormats	28	25	26
AM/FM ST ET	302	266	284
AM/FM ST ET cassette	423	372	398
Transfer case shield	75	66	71
Trailering Special	109	96	102
Wiring harness and trailer hitch.			

Honda Accord

What's New

The SEi, a luxury version of the Accord sedan, is the only addition to Honda's line of front-drive compacts, which are due for a complete redesign for 1990. The '89 roster retains 2-door coupe, 3-door hatchback and 4-door sedan body styles. All of the coupes and some of the sedans are built at Honda's Marysville, Ohio, assembly plant. In 1988, Honda began shipping coupes to Japan, becoming the first Japanese automaker to export U.S.-built cars back to Japan. DX

Honda Accord DX Coupe

and LX model Accords use a carbureted 2.0-liter 4-cylinder engine. Plusher LXi and the new SEi variations get a fuel-injected version of the 2.0. A 5-speed manual transmission and a 4-speed overdrive automatic are available with either engine. The SEi has LXi equipment, plus standard leather upholstery and a high-power Bose stereo system. A similar SEi sedan was offered in 1985, the last year of the previous-generation Accord. The all-new 1990 Accord reportedly will be about six inches longer and somewhat wider than the present model, and may be offered with an optional 3.0-liter V-6 engine.

For

● Performance (LXi, SEi) ● Handling/roadholding (LXi, SEi) ● Ride ● Fuel economy ● Passenger room ● Resale value

Against

● Road noise ● Price (LXi, SEi)

Summary

A sales winner since its 1986 debut, the third-generation Accord is the class of the compact class, though some rivals, notably the Mazda 626 and Toyota Camry, are closing the gap. Still, Accord is an outstanding all-around buy among family cars. Attractions include decent room for four, adequate cargo space, a comfortable driving position, superb outward vision, exemplary workmanship and good fuel economy. LXi and SEi models have a firmer suspension, thicker stabilizer bars and larger tires than DX and LX versions, giving them sportier handling. Corners are taken with less body lean and the steering feels meatier in the higher-priced Accords. Tire roar on coarse pavement remains high, though, on all models. Most shoppers choose automatic transmission, but sample one with the fine 5-speed manual for its better performance and economy. Despite the LXi's sportier character, we rate the LX sedan as the best Accord buy on a value-per-dollar basis. Though strong resale value and high marks for reliability soothe some of the sticker shock, the Accord is still pricey for a compact. Part of the problem stems from rapid price escalations during 1988 as the dollar weakened against the yen. But mostly, high demand for a very good car makes it a seller's market for Accords.

Specifications

	2-door notchback	3-door hatchback	4-door notchback
Wheelbase, in.	102.4	102.4	102.4
Overall length, in.	179.1	174.8	179.7
Overall width, in.	66.7	66.7	67.4
Overall height, in.	52.6	52.6	53.4
Front track, in.	58.3	58.3	58.3
Rear track, in.	58.1	58.1	58.1
Turn diameter, ft.	34.1	34.8	34.8
Curb weight, lbs.	2493	2513	2482

	2-door notchback	3-door hatchback	4-door notchback
Cargo vol., cu. ft.	14.3	18.9	13.7
Fuel capacity, gal.	15.9	15.9	15.9
Seating capacity	5	5	5
Front headroom, in.	37.9	38.0	38.7
Front shoulder room, in.	54.5	54.5	54.9
Front legroom, max., in.	42.9	42.8	42.8
Rear headroom, in.	36.6	36.6	37.1
Rear shoulder room, in.	53.3	53.1	54.5
Rear legroom, min., in.	30.9	30.2	32.4

Powertrain layout: transverse front engine/front-wheel drive.

Engines

	ohc I-4	ohc I-4
Size, liters/cu. in.	2.0/119	2.0/119
Fuel delivery	2 bbl.	PFI
Horsepower @ rpm	98 @ 5500	120 @ 5800
Torque (lbs./ft.) @ rpm	109 @ 3500	122 @ 4000
Availability	S[1]	S[2]

EPA city/highway mpg

5-speed OD manual	27/34	25/30
4-speed OD automatic	23/30	22/28

1. DX, LX   2. LXi

Prices

Honda Accord	Retail Price	Dealer Invoice	Low Price
DX 3-door hatchback, 5-speed	$11230	$9433	$10730
DX 3-door hatchback, automatic	11840	9945	11340
LXi 3-door hatchback, 5-speed	14530	12205	14030
LXi 3-door hatchback, automatic	15140	12717	14640
DX 2-door notchback, 5-speed	11650	9786	11450
DX 2-door notchback, automatic	12260	10298	12060
LXi 2-door notchback, 5-speed	14690	12339	14490
LXi 2-door notchback, automatic	15300	12852	15100
DX 4-door notchback, 5-speed	11770	9886	11570
DX 4-door notchback, automatic	12380	10399	12180
LX 4-door notchback, 5-speed	14180	11911	13980
LX 4-door notchback, automatic	14790	12423	14590
LXi 4-door notchback, 5-speed	15920	13372	15720
LXi 4-door notchback, automatic	16530	13885	16330
Destination charge	245	245	245

Standard Equipment:

DX: 2.0-liter 2bbl. 4-cylinder engine, 5-speed manual or 4-speed automatic transmission, power steering, cloth reclining front bucket seats, tilt steering column, digital clock, remote fuel door and trunk/hatch releases, cruise control (4-door), tinted glass, rear wiper/washer (3-door), bodyside molding, 185/70R14 tires. **LX** adds: air conditioning, power windows and locks, power mirrors, AM/FM ST ET cassette, one-piece folding rear seatback. **LXi** adds: PFI engine, driver's seat lumbar support adjustment, power moonroof, 50/50 folding rear seatbacks, Michelin MXV 195/60R14 tires on alloy wheels; 2- and 3-door delete power locks.

OPTIONS are available as dealer-installed accessories.

Honda Civic

What's New

Completely redesigned for 1988, these slick subcompacts return unchanged for '89, save for the addition of automatic transmission for the 4-wheel-drive wagon and reintroduction of the sporty Si hatchback. Front-drive Civics come as

Prices are accurate at time of printing; subject to manufacturer's change.

Honda Civic Si 3-door

Specifications

	3-door hatchback	4-door notchback	5-door wagon
Wheelbase, in.	98.4	98.4	98.4
Overall length, in.	156.1	166.5	161.6
Overall width, in.	65.6	65.9	66.1
Overall height, in.	52.4	53.6	57.9
Front track, in.	57.1	57.1	57.1
Rear track, in.	57.3	57.3	57.3
Turn diameter, ft.	31.5	31.5	31.5
Curb weight, lbs.	1933	2039	2130[1]
Cargo vol., cu. ft.	25.0	12.2	60.3
Fuel capacity, gal.	11.9	11.9	11.9
Seating capacity	5	5	5
Front headroom, in.	38.2	38.5	39.4
Front shoulder room, in.	53.5	53.5	53.5
Front legroom, max., in.	43.3	43.1	41.2
Rear headroom, in.	36.6	37.4	38.0
Rear shoulder room, in.	52.3	53.0	53.5
Rear legroom, min., in.	30.4	32.0	33.3

1. *2366 lbs., 4WD wagon*

Powertrain layout: transverse front engine/front-wheel drive or automatically engaged 4WD.

Engines

	ohc I-4	ohc I-4	ohc I-4
Size, liters/cu. in.	1.5/91	1.5/91	1.6/97
Fuel delivery	TBI[1]	TBI[1]	PFI
Horsepower @ rpm	70 @ 5500	92 @ 6000	108 @ 6000
Torque (lbs./ft.) @ rpm	83 @ 3000	89 @ 4500	100 @ 5000
Availability	S[2]	S[3]	S[4]

EPA city/highway mpg

4-speed OD manual	34/38		
5-speed OD manual		31/36	28/33
4-speed OD automatic		27/32	

1. *Two injectors*   2. *Civic Hatchback*   3. *DX, LX, 2WD wagon*   4. *Si, 4WD wagon*

3-door hatchbacks, 4-door sedans and 5-door wagons; Honda's automatically engaged 4WD system is available only on the wagon. The Si hatchback, which sat out the 1988 model year, returns with the 16-valve 1.6-liter 4-cylinder engine that's used in the 2-seat CRX Si. For '89, the 16-valve engine's horsepower and torque are up by three, to 108 and 100 respectively. The Si also has sport seats, gas-filled shock absorbers and a power sunroof. It's the only Civic with 14-inch tires (others have 13s) and the only hatchback equipped with a tachometer. The Si is available with a 5-speed manual only. The base Civic hatchback comes with a 70-horsepower 1.5-liter four and 4-speed manual transmission only. A 92-horsepower 1.5-liter four is standard on the DX hatchbacks, DX and LX sedans, and the 2WD wagon, teamed with either a 5-speed manual or a 4-speed overdrive automatic. The 4WD wagon uses the same 108-horsepower 1.6-liter four as the Civic Si.

For

- Fuel economy   • Handling
- 4WD traction (4WD wagon)   • Visibility

Against

- Performance (automatic transmission)
- Cargo room (hatchbacks)

Summary

Smoothness characterizes these subcompacts, which blend function, precision and thoughtful design. Civics aren't fast—though the new Si should be quick—but their spunky engines rev willingly and, when equipped with the fine 5-speed, they're downright fun to drive. With automatic transmission, the engine has trouble mustering enough power for safe passing, and the transmission is slow to downshift unless you floor the throttle. Steering response is quick at speeds above 5 mph, but parking a Civic with manual steering can be a chore. Roadholding is quite good, and you'll find yourself slipping through city traffic and zipping around country bends. The firm suspensions can be jarring over rough pavement, though, and road and engine noise are frequent companions in the hatchbacks. We're happy to see the return of the Si, which in its previous incarnation was a delightful mix of sportiness and economy. Smoothness is also the byword in the spacious and intelligently designed cabin, where the controls work with little friction. Four adults, if they're of average height, ride in comfort in the sedan; rear headroom in the hatchback is limited. The view from the supportive seats is panoramic and the quality of construction is outstanding for the class, though some interior materials on base and DX versions are budget-grade. You'll pay for all this goodness in steep prices, but shop carefully and you'll likely find some dealers willing to part with base and DX hatchbacks and sedans at reasonable sums.

Prices

Honda Civic	Retail Price	Dealer Invoice	Low Price
3-door hatchback, 4-speed	$6385	$5746	$6035
DX 3-door hatchback, 5-speed	8445	7178	8095
DX 3-door hatchback, automatic	9295	7900	8945
Si 3-door hatchback, 5-speed	9980	8483	9630
DX 4-door notchback, 5-speed	9190	7811	8990
DX 4-door notchback, automatic	10090	8576	9890
LX 4-door notchback, 5-speed	10150	8627	9950
LX 4-door notchback, automatic	10720	9112	10520
5-door wagon, 5-speed	10125	8606	9775
5-door wagon, automatic	11140	9469	10790
4WD 5-door wagon, 6-speed	12210	10378	11860
4WD 5-door wagon, automatic	12810	10888	12460
Destination charge	245	245	245

Standard Equipment:

1.5-liter SOHC 16-valve PFI 4-cylinder engine, 4-speed manual transmission, reclining front bucket seats, left remote mirror, 50/50 folding rear seatbacks, trip odometer, coolant temperature gauge, rear defroster, remote fuel door and hatch releases, bodyside moldings, 165/70SR13 tires. **DX adds:** 5-speed

> **KEY: ohv** = overhead valve; **ohc** = overhead cam; **dohc** = double overhead cam; **I** = inline cylinders; **V** = cylinders in V configuration; **flat** = horizontally opposed cylinders; **bbl.** = barrel (carburetor); **PFI** = port (multi-point) fuel injection; **TBI** = throttle-body (single-point) fuel injection; **rpm** = revolutions per minute; **OD** = overdrive transmission; **S** = standard; **O** = optional; **NA** = not available.

Prices are accurate at time of printing; subject to manufacturer's change.

manual or 4-speed automatic transmission, power steering (on 4-door with automatic transmission), rear wiper/washer (3-door), tilt steering column (3-door), tinted glass, remote hatch release (3-door), intermittent wipers (3-door), 175/70R13 tires; Si has 1.6-liter engine, uprated suspension, sport seats, power sunroof, tachometer, remote mirrors, 185/60R14 tires. **LX** adds to DX: power steering, tachometer, power mirrors, tilt steering column, power windows and locks, one-piece folding rear seatback, rear head restraints, remote trunklid release. **2WD wagon** has tilt steering column, 60/40 folding rear seatbacks, rear head restraints, rear wiper/washer, intermittent wipers, tinted glass, remote liftgate release, 175/70R13 tires. **4WD wagon** adds: 1.6-liter engine, 6-speed manual or 4-speed automatic transmission, storage drawer under passenger seat, 165SR13 tires. Si has higher-output engine, uprated suspension, sport seats, power sunroof.

OPTIONS are available as dealer-installed accessories.

Honda CRX

Honda CRX Si

What's New

Like the related Civics, Honda's 2-seat CRX coupes were redesigned for '88 and return almost unchanged for '89. The redesign brought more horsepower, all-independent suspension and a longer, wider body. Three front-drive models continue. The HF (high fuel economy) uses an 8-valve, 63-horsepower 1.5-liter 4-cylinder engine. The mid-range CRX gets a 16-valve, 92-horsepower 1.5 and is the only model available with automatic transmission. The CRX Si has a 16-valve 1.6-liter four that gets three additional horsepower this year, to 108, and three more pounds/feet of torque, to 100. The Si also has a stiffer suspension and 14-inch tires, versus 13s on other models. CRX and Si models come with a 2-piece cargo cover.

For

- Performance (manual transmission)   • Fuel economy
- Handling and maneuverability   • Instruments/controls

Against

- Performance (automatic transmission)   • Ride   • Noise

Summary

If a practical, sporty, low-cost, economical car is what you seek, and you don't need more than two seats, there's no better choice. The Si has true sports car moves. It takes

---

KEY: ohv = overhead valve; **ohc** = overhead cam; **dohc** = double overhead cam; **I** = inline cylinders; **V** = cylinders in V configuration; **flat** = horizontally opposed cylinders; **bbl.** = barrel (carburetor); **PFI** = port (multi-point) fuel injection; **TBI** = throttle-body (single-point) fuel injection; **rpm** = revolutions per minute; **OD** = overdrive transmission; **S** = standard; **O** = optional; **NA** = not available.

---

some restraint to keep the front tires from spinning in hard acceleration, but once you hone your technique, the slick 5-speed, precise steering and sharp handling are extremely rewarding. The mid-level CRX is only slightly less feisty than the Si, but benefits from a ride that's not as brutally stiff. In fact, the Si's rock-hard suspension makes it more attractive as a weekend fun car than an everyday commuter. The HF gets great fuel economy, but if mileage is your goal, there are more practical designs. Automatic transmission plain wastes the CRX's spirit. All these 2-seat Hondas have controls that are models of clarity and simplicity. Their seats are low and thinly padded, but offer ample support without lots of fancy adjustments. No CRX is quiet, a penalty of the hatchback body style and minimal sound insulation. A vertical glass panel in the rear hatch improves visibility for parking, but the view aft is still not great. There are compromises, but the CRX has many charms. Check with your insurance agent before succumbing, however. Rates for 2-seaters can be extremely high.

Specifications

	3-door hatchback
Wheelbase, in.	90.6
Overall length, in.	147.8
Overall width, in.	65.7
Overall height, in.	50.0
Front track, in.	57.1
Rear track, in.	57.3
Turn diameter, ft.	29.5
Curb weight, lbs.	1819[1]
Cargo vol., cu. ft.	23.2
Fuel capacity, gal.	11.9[2]
Seating capacity	2
Front headroom, in.	37.0
Front shoulder room, in.	53.5
Front legroom, max., in.	43.9
Rear headroom, in.	—
Rear shoulder room, in.	—
Rear legroom, min., in.	—

1. CRX HF; 1992 lbs., CRX; 2017 lbs. CRX Si.   2. 10.6 gals. CRX HF

Powertrain layout: transverse front engine/front-wheel drive.

Engines	ohc I-4	ohc I-4	ohc I-4
Size, liters/cu. in.	1.5/91	1.5/91	1.6/97
Fuel delivery	PFI	TBI	PFI
Horsepower @ rpm	62 @ 4500	92 @ 6000	108 @ 6000
Torque (lbs./ft.) @ rpm	90 @ 2000	89 @ 4500	100 @ 5000
Availability	S[1]	S[2]	S[3]
EPA city/highway mpg			
5-speed OD manual	45/52	34/41	28/33
4-speed OD automatic		29/36	

1. CRX HF   2. CRX   3. CRX Si

Prices

Honda CRX	Retail Price	Dealer Invoice	Low Price
HF 3-door hatchback, 5-speed	$8895	$7560	$8395
3-door hatchback, 5-speed	9310	7913	8810
3-door hatchback, automatic	9880	8398	9380
Si 3-door hatchback, 5-speed	10930	9290	10430
Destination charge	245	245	245

Standard Equipment:

HF: 1.5-liter 8-valve PFI 4-cylinder engine, 5-speed manual transmission, reclining front bucket seats, left remote mirror, tinted glass, tachometer,

---

Prices are accurate at time of printing; subject to manufacturer's change.

coolant temperature gauge, trip odometer, intermittent wipers, rear defogger, remote fuel door and hatch releases, bodyside moldings, 165/70R13 tires. **CRX** adds: 1.5-liter SOHC 16-valve PFI engine (92 bhp), tilt steering column, dual remote mirrors, digital clock, cargo cover. **Si** adds: 105-bhp PFI engine, uprated suspension, rear wiper/washer, 185/60HR14 tires on alloy wheels.

OPTIONS are available as dealer-installed accessories.

Honda Prelude

Honda Prelude Si

What's New

Honda introduced its third-generation Prelude last year and brings it back unchanged for 1989. All-new sheetmetal on a 6.5-inch-longer body, 230 additional pounds of curb weight and the world's first production 4-wheel steering (4WS) option were hallmarks of the redesign. The base Prelude S continues with a 104-horsepower 2.0-liter overhead-cam 4-cylinder engine with 12 valves and dual side-draft carburetors. The Si has a 135-horsepower 2.0-liter four with 16 valves, dual overhead camshafts and multi-point fuel injection. It also gets 14-inch wheels in place of the S's 13s. A 5-speed manual transmission is standard. Optional is a 4-speed automatic with a driver-selective "sport" mode for improved acceleration. The 4WS Prelude beat Mazda's 626 Turbo 4WS to the market by mere months. Honda's 4WS is a mechanical system that links front and rear steering boxes. It steers the rear wheels a few degrees left or right depending on the angle of the steering wheel and is designed to improve handling and maneuverability. The turning diameter is 31.5 feet for the 4WS model and 34.8 feet for other Preludes.

For

● Performance ● Handling/roadholding ● Braking

Against

● Passenger room
● Driveability (automatic transmission) ● Price (Si 4WS)

Summary

A top-notch car became subtly better with the '88 redesign, but the Prelude is dangerously close to being priced out of its league, which is still sporty coupe rather than sports and GT car. The flagship Si 4WS (which includes alloy wheels and bronze-tinted glass) goes for more than $19,000 with automatic, while a Prelude S is some $4000 cheaper. We'll pass on the fuel-injected engine and 4WS to save that kind of money. Any Prelude is an agile coupe with a pinch more rear-seat space than many of the breed. Their feel is typically Honda, and that means an efficiency of design and a pleasing mechanical smoothness. However, ordering an automatic transmission robs the car of much of its verve, and the combination of a taut suspension and firm seats can make for jarring passage over broken pavement. Plus, you might find a little too much plastic in the cabin, considering the price. As for the 4WS, it does enhance low-speed maneuverability: the car seems to pivot on an axis. It does not seem to aid overall handling and in high-speed turns can induce an initial flick of the tail that's disconcerting. In Europe and Japan, 70 percent of Preludes are sold with 4WS; the figure is around 2 percent in the U.S. We're unconvinced of its advantage, especially considering its cost, complexity and added weight.

Specifications

	2-door notchback
Wheelbase, in.	101.0
Overall length, in.	175.6
Overall width, in.	67.3
Overall height, in.	51.0
Front track, in.	58.3
Rear track, in.	57.9
Turn diameter, ft.	34.8[1]
Curb weight, lbs.	2522
Cargo vol., cu. ft.	11.2
Fuel capacity, gal.	15.9
Seating capacity	4
Front headroom, in.	36.9
Front shoulder room, in.	53.1
Front legroom, max., in.	43.1
Rear headroom, in.	34.1
Rear shoulder room, in.	51.1
Rear legroom, min., in.	27.1

1. 31.4 ft. on 4WS

Powertrain layout: transverse front engine/front-wheel drive.

Engines

	ohc I-4	dohc I-4
Size, liters/cu. in.	2.0/119	2.0/119
Fuel delivery	2 × 1 bbl.	PFI
Horsepower @ rpm	104 @ 5800	135 @ 6200
Torque (lbs./ft.) @ rpm	111 @ 4000	127 @ 4500
Availability	S	S[1]

EPA city/highway mpg

5-speed OD manual	23/28	23/26
4-speed OD automatic	20/26	21/26

1. Si

Prices

Honda Prelude	Retail Price	Dealer Invoice	Low Price
S 2-door notchback, 5-speed	$13945	$11713	$13445
S 2-door notchback, automatic	14670	12322	14170
Si 2-door notchback, 5-speed	16965	14250	16465
Si 2-door notchback, automatic	17690	14859	17190
Si 4WS 2-door notchback, 5-speed	18450	15498	17950
Si 4WS 2-door notchback, automatic	19175	16107	18675
Destination charge	245	245	245

Standard Equipment:

S: 2.0-liter 12-valve carbureted 4-cylinder engine, 5-speed manual or 4-speed automatic transmission, power steering, power 4-wheel disc brakes, cloth reclining front bucket seats, cloth trim, tilt steering column, console, AM/FM

Prices are accurate at time of printing; subject to manufacturer's change.

ST ET cassette, power antenna, door-mounted passive lap and shoulder belts, tachometer, coolant temperature gauge, trip odometer, intermittent wipers, 185/70HR13 tires. **Si** adds: DOHC 16-valve PFI engine, air conditioning, cruise control, adjustable lumbar support and side bolsters, diversity antenna, 195/60HR14 tires. **Si 4WS** adds: 4-wheel steering, power door locks, bronze tinted glass, alloy wheels.

OPTIONS are available as dealer-installed accessories.

Hyundai Excel

1988 Hyundai Excel GS 3-door

What's New

The major change for the Excel front-drive subcompact this year is a longer warranty that provides 3-year/36,000-mile "bumper-to-bumper" coverage. Last year's warranty covered the entire car for 12 months/12,500 miles and the powertrain for 3 years/36,000 miles. A 3-year/unlimited-miles corrosion warranty is unchanged. Excel, sold in the U.S. since 1986, is the best-selling single imported model. It is based on the design for a previous generation Mitsubishi Mirage (also sold through Dodge and Plymouth dealers as the Colt), but the exterior and interior styling are Hyundai's. Mitsubishi owns part of Hyundai and supplies engines, transmissions and other mechanical components to Hyundai, and allows Hyundai to build some components under license. Hyundai builds 3- and 5-door hatchbacks from this design that are sold in the U.S. by Mitsubishi as the Precis. Excel uses a Mitsubishi-designed 1.5-liter 4-cylinder engine with a 2-barrel carburetor and 68 horsepower. Transmission choices are 4- and 5-speed manuals and a 3-speed automatic. Three body styles are available: 3- and 5-door hatchbacks and a 4-door notchback sedan.

For

• Fuel economy • Maneuverability • Value

Against

• Performance • Interior room • Noise

Summary

Hyundai's extraordinary success with this car is mainly because of its lower price, though astute marketing also has

---

KEY: ohv = overhead valve; **ohc** = overhead cam; **dohc** = double overhead cam; **I** = inline cylinders; **V** = cylinders in V configuration; **flat** = horizontally opposed cylinders; **bbl.** = barrel (carburetor); **PFI** = port (multi-point) fuel injection; **TBI** = throttle-body (single-point) fuel injection; **rpm** = revolutions per minute; **OD** = overdrive transmission; **S** = standard; **O** = optional; **NA** = not available.

---

helped. About the only car that matches Excel in price and value per dollar is the Dodge Omni/Plymouth Horizon; comparably equipped Japanese subcompacts are far more expensive than Excel. Beware that Hyundai dealers capitalize on Excel's popularity by price gouging, padding the low sticker prices with "protection packages" and bogus charges such as "ocean freight" and a nonexistent "import tariff." The average out-the-door price of a 1988 Excel was around $8300. These cars are fine values at Hyundai's suggested retail prices, but not nearly as good if a dealer adds $1000-2000 in accessories and additional profit. When the price gets above $9000, you're into the same range as a Mazda 323, Toyota Tercel or even a basic Honda Civic. We would pay a little more to get one of those instead of an Excel, even if it meant sacrificing some convenience features. Excel's shortcomings include sluggish acceleration from an engine that generates more noise than power, especially with automatic transmission, a cramped rear seat, and mediocre handling and roadholding ability. In addition, recent insurance industry statistics give Excel a poor rating for occupant protection. Low suggested retail prices still make this car quite attractive to those on a tight budget, but Hyundai dealers are under no obligation to sell Excels at those prices. We suggest you keep this car on your shopping list, but be sure to look elsewhere to compare actual selling prices.

Specifications

	3-door hatchback	4-door notchback	5-door hatchback
Wheelbase, in.	93.7	93.7	93.7
Overall length, in.	161.0	168.0	161.0
Overall width, in.	63.1	63.1	63.1
Overall height, in.	54.1	54.1	54.1
Front track, in.	54.1	54.1	54.1
Rear track, in.	52.8	52.8	52.8
Turn diameter, ft.	33.8	33.8	33.8
Curb weight, lbs.	2156	2156	2178
Cargo vol., cu. ft.	26.6	11.2	26.0
Fuel capacity, gal.	10.6[1]	10.6[1]	10.6[1]
Seating capacity	5	5	5
Front headroom, in.	37.5	37.5	37.5
Front shoulder room, in.	52.1	52.1	52.1
Front legroom, max., in.	40.9	40.9	40.9
Rear headroom, in.	36.9	36.9	36.9
Rear shoulder room, in.	51.6	51.6	51.6
Rear legroom, min., in.	32.4	32.4	32.4

1. 13.2 gals. on GLS with automatic

Powertrain layout: transverse front engine/front-wheel drive.

Engines

	ohc I-4
Size, liters/cu. in.	1.5/90
Fuel delivery	2 bbl.
Horsepower @ rpm	68 @ 5500
Torque (lbs./ft.) @ rpm	82 @ 3500
Availability	S

EPA city/highway mpg

4-speed manual	27/33
5-speed OD manual	28/37
3-speed automatic	27/31

Prices

Hyundai Excel	Retail Price	Dealer Invoice	Low Price
3-door hatchback, 4-speed	$5499	—	—
GL 3-door hatchback, 5-speed	6699	—	—

Prices are accurate at time of printing; subject to manufacturer's change.

CONSUMER GUIDE®

	Retail Price	Dealer Invoice	Low Price
GL 3-door hatchback, automatic	7184	—	—
GS 3-door hatchback, 5-speed	7699	—	—
GS 3-door hatchback, automatic	8184	—	—
GL 5-door hatchback, 5-speed	6949	—	—
GL 5-door hatchback, automatic ...	7434	—	—
GLS 5-door hatchback, 5-speed	7599	—	—
GLS 5-door hatchback, automatic	8084	—	—
4-door notchback, 4-speed	6199	—	—
GL 4-door notchback, 5-speed	7149	—	—
GL 4-door notchback, automatic	7634	—	—
GLS 4-door notchback, 5-speed	7749	—	—
GLS 4-door notchback, automatic	8234	—	—
Destination charge	255	255	255

Dealer invoice and low price not available at time of publication.

Standard Equipment:

1.5-liter 2bbl. 4-cylinder engine, 4-speed manual transmission, power brakes, cloth and vinyl reclining front bucket seats, variable intermittent wipers, coolant temperature gauge, graphic display, rear defogger, locking fuel filler door, split folding rear seat, rear-seat heat ducts, 155/80R13 all-season SBR tires with full-size spare. **GL** adds: 5-speed manual or 3-speed automatic transmission, cloth trim, dual remote mirrors, analog clock, tinted glass, rear wiper/washer (hatchbacks), remote fuel door and hatch/trunk releases, full center console, bodyside moldings, wheel covers. **GLS** adds: cloth headliner, windshield sunshade band, digital clock, driver's seat height and lumbar support adjustments, tachometer, underseat tray (4- and 5-doors), full cloth seats, AM/FM ST ET cassette, upgraded carpeting, visor mirror. **GS** adds: front sport seats, P175/70R13 tires on alloy wheels.

Optional Equipment:

Air conditioning	735	—	—
5 175/70R13 tires on alloy wheels	325	—	—
Excel Option Pkg.	175	—	—
Right outside mirror, tinted glass, bodyside molding.			
Passive restraint system	75	—	—
Power steering	260	—	—
AM/FM ST cassette (std. GLS)	295	—	—
Panasonic AM/FM ST ET cassette, GL ...	430	—	—
GLS	135	—	—
Power sunroof	395	—	—
Two-tone paint	125	—	—

Hyundai Sonata

Hyundai Sonata

What's New

Hyundai, already the fourth largest importer behind Toyota, Honda and Nissan while selling just one model, adds a larger car this year to its U.S. lineup. Sonata is a new front-wheel-drive sedan scheduled to go on sale in November to give Hyundai a competitor against compact and mid-size cars. Like the subcompact Excel, Hyundai's only U.S. model till now, it is built in Korea. At 184.3 inches overall,

Sonata is nearly two feet longer than Excel and Hyundai is calling it a mid-size car. Wheelbase is 104.3 inches, one to two inches longer than compacts such as the Honda Accord, Toyota Camry and Chevrolet Corsica, but an inch or two shorter than domestic intermediates such as the Ford Taurus and Chevrolet Celebrity. At 68.9 inches wide, Sonata also falls between contemporary compacts and intermediates in that dimension. Base curb weight is about 2750 pounds. Available only as a 4-door notchback sedan, Sonata was styled by Italian designer Giorgio Giugiaro, who also did the Excel. Unlike the Excel, which relies mainly on Mitsubishi mechanical components, Sonata's engineering is mostly by Hyundai's own staff. An exception is the engine, which is licensed by Mitsubishi. The U.S. Sonata will use a fuel-injected 4-cylinder engine in the 2.0- to 2.4-liter range; this engine likely will be related to the 2.4-liter engine used in the 1985-87 Mitsubishi Galant sedan and the current Mitsubishi Wagon/Van. An electronically controlled 4-speed overdrive automatic transmission will be offered. Hyundai hadn't announced prices for the Sonata, but most guesses place the base price in the $10,000 range, which would undercut Japanese compacts such as the Accord and Camry, and domestic intermediates such as the Taurus and Celebrity. We haven't driven Sonata, so we can't comment on its performance.

Prices not available at time of publication.

Isuzu I-Mark

Isuzu I-Mark RS 3-door

What's New

Isuzu's front-drive subcompact gets a new double-overhead-cam, 16-valve 1.6-liter engine that produces 125 horsepower. The potent new engine comes only with a 5-speed manual transmission in the sporty RS versions of I-Mark, which debuted last spring with a suspension tuned by Lotus, the British sports car company now owned by General Motors. Lotus helped select suspension components and the 185/60R14 tires for the RS 3-door hatchback and 4-door sedan, both of which wear "Handling by Lotus" badges. RS models are available in white, red or grey monochromatic exterior colors. Optional for the RS is a Recaro Package that includes Recaro front bucket seats, a leather-wrapped steering wheel and a removable glass sunroof. The I-Mark LS 4-door also has the Lotus-tuned suspension, but uses a 110-horsepower, turbocharged 1.5-liter 4-cylinder engine, also available only with a 5-speed manual.

Prices are accurate at time of printing; subject to manufacturer's change.

Isuzu

Base I-Mark S and higher-priced XS hatchbacks and sedans return for 1989 with a 70-horsepower, carbureted 1.5-liter 4-cylinder engine, available with either a 5-speed manual or 3-speed automatic transmission. GM owns part of Isuzu and sells a version of the I-Mark through Chevrolet dealers as the Geo Spectrum. While Isuzu is pushing ahead with a performance image this year for I-Mark, Chevrolet has dropped the turbocharged engine from Spectrum and will not get the new twin-cam engine for 1989. A new Sunsport option available for the I-Mark S 3-door includes an electric canvas sunroof with built-in air deflector, body color mirrors, wheel covers and front bumper, and cloth sport seats.

For

- ●Fuel economy ●Handling (RS, LS)
- ●Performance (RS, LS)

Against

- ●Performance (S, XS) ●Noise

Summary

We haven't driven the RS with its new double-overhead-cam engine or tried the Lotus suspension, but we are familiar with I-Mark's turbocharged engine and the base 70-horsepower engine. Both generate too much noise for peaceful motoring, though the turbocharged engine also generates enough power to justify the racket it makes. The 70-horsepower engine is quite economical, but lags well behind the 1.5-liter engines offered by Honda and Toyota in acceleration and refinement. With the optional automatic transmission, Isuzu's carbureted 1.5 sounds hoarse and harried just trying to keep up with traffic. In most other areas, I-Mark has ordinary credentials. The sporty LS and RS models are exceptions; their firm suspensions and wide tires give them much more tenacious handling ability than the other models. However, if we wanted a sporty car, we'd probably look somewhere else first. All told, we aren't enthused about this car.

Specifications

	3-door hatchback	4-door notchback
Wheelbase, in.	94.5	94.5
Overall length, in.	157.9	160.7
Overall width, in.	63.6	63.6
Overall height, in.	54.1	54.1
Front track, in.	54.7	54.7
Rear track, in.	54.3	54.3
Turn diameter, ft.	32.8	32.8
Curb weight, lbs.	1984	2011
Cargo vol., cu. ft.	29.7	11.4
Fuel capacity, gal.	11.1	11.1
Seating capacity	5	5
Front headroom, in.	38.0	37.7
Front shoulder room, in.	52.8	52.8
Front legroom, max., in.	41.7	41.7
Rear headroom, in.	37.1	37.6
Rear shoulder room, in.	52.8	52.8
Rear legroom, min., in.	33.3	33.5

Powertrain layout: transverse front engine/front-wheel drive.

Engines

	ohc I-4	Turbo ohc I-4	dohc I-4
Size, liters/cu. in.	1.5/90	1.5/90	1.6/97
Fuel delivery	2 bbl.	PFI	PFI

	ohc I-4	Turbo ohc I-4	dohc I-4
Horsepower @ rpm	70 @ 5400	110 @ 5400	125 @ 6800
Torque (lbs./ft.) @ rpm	87 @ 3400	120 @ 3500	102 @ 5400
Availability	S[1]	S[2]	S[3]
EPA city/highway mpg			
5-speed OD manual	37/41	26/34	24/32
3-speed automatic	31/33		

1. S, XS   2. LS   3. RS

Prices

Isuzu I-Mark (1988 prices)	Retail Price	Dealer Invoice	Low Price
S 4-door notchback, 5-speed	$8009	$7209	$7609
S 4-door notchback, automatic	8439	7579	8009
S 3-door hatchback, 5-speed	7659	6894	7277
S 3-door hatchback, automatic	8089	7264	7677
XS 4-door notchback, automatic	9649	8391	9020
XS 3-door hatchback, 5-speed	9029	7856	8443
XS 3-door hatchback, automatic	9459	8226	8843
Turbo 4-door notchback, 5-speed	11189	9735	10462
Turbo 3-door hatchback, 5-speed	9829	8552	9191
Destination charge	259	259	259

Standard Equipment:

S: 1.5-liter 2bbl. 4-cylinder engine, 5-speed manual or 3-speed automatic transmission, power brakes, reclining front bucket seats, cloth upholstery, carpet, assist grips, cargo area cover (3-door), rear defogger, tinted glass, door map pockets, right visor mirror, wide bodyside moldings, split fold-down rear seat (3-door), telescopic steering column, tachometer, headlamps-on chime, all-season SBR tires (P155/80R13 on 4-door, P175/70R13 on 3-door). **XS** adds: remote fuel filler and trunk/hatch releases, fog lamps, dual power mirrors, trip odometer, intermittent wipers, rear wiper/washer (3-door), digital clock, luggage compartment lamp, upgraded carpeting, soft-grip steering wheel, alloy wheels. **Turbo** adds: 1.5-liter PFI turbocharged engine, power steering, AM/FM ST w/cassette and EQ, sunroof, sport suspension, P185/60R14 performance SBR tires.

Optional Equipment:

Air conditioning	680	578	629
Cruise control, XS	195	160	178
Carpet mats	50	35	43
AM/FM cassette, exc. Turbo	185	130	158
AM/FM ET cassette, XS	410	287	349
Sunroof, XS	300	240	270
Power steering, S	220	187	204

Isuzu Impulse

What's New

The rear-drive Impulse sport coupe carries on for its final year with refinements and minor equipment changes, including new body-colored bumpers. Last year Impulse gained new suspension components and tires selected by Lotus, the British car company now owned by General Motors. Like this year's I-Mark, Impulse carries "Handling by Lotus" badges to note the British firm's influence. Base engine remains a 110-horsepower 2.3-liter 4-cylinder engine, while Turbo models have a 140-horsepower, turbocharged 2.0-liter 4-cylinder. Both engines are available with a 5-speed manual or a 4-speed overdrive automatic. Styled in Italy and built in Japan, Impulse debuted in the U.S. in 1983.

Prices are accurate at time of printing; subject to manufacturer's change.

1988 Isuzu Impulse

For

● Performance (turbo) ●Handling

Against

● Ride ●Noise (turbo)

Summary

This slow-selling coupe has humble origins—its rear-drive chassis is distantly related to the one used for the discontinued Chevrolet Chevette—but appealing styling that still looks contemporary. It took a few years for Isuzu to get it together, but Impulse now has impressive performance credentials to go with its good looks. The potent turbocharged engine provides fine performance across a broad range of engine speeds, and turbo boost comes on smoothly rather than abruptly. However, the engine sounds strained and thrashy above 3000 rpm, which discourages winding the engine to higher speeds to tap its full power potential. The Lotus-tuned suspension provides impressive cornering ability, though that comes at the expense of ride comfort. The firm suspension feels jittery on rough pavement and the rear tires lose traction in bumpy turns. Inside, the low roof leaves little head room for six-footers and there's only token rear seat room, though that's typical of sporty coupes. Gadgets abound in the Impulse's interior and you certainly get your money's worth in convenience features: Air conditioning, power locks and windows, power mirrors, and a stereo radio with cassette player are standard. Impulse has never been a big seller, but that could actually work in its favor in the future. Its heritage includes a General Motors chassis, Italian styling, Japanese manufacturing and British suspension engineering. Coupled with low volume, that can add up to collectible car status.

Specifications

	3-door hatchback
Wheelbase, in.	96.1
Overall length, in.	172.6
Overall width, in.	65.2
Overall height, in.	51.4
Front track, in.	53.3
Rear track, in.	53.9
Turn diameter, ft.	31.5
Curb weight, lbs.	2727
Cargo vol., cu. ft.	29.4
Fuel capacity, gal.	15.1
Seating capacity	4
Front headroom, in.	36.9

	3-door hatchback
Front shoulder room, in.	54.5
Front legroom, max., in.	41.9
Rear headroom, in.	35.8
Rear shoulder room, in.	54.0
Rear legroom, min. in.	28.1

Powertrain layout: longitudinal front engine/rear-wheel drive.

Engines	ohc I-4	Turbo ohc I-4
Size, liters/cu. in.	2.3/137	2.0/122
Fuel delivery	PFI	PFI
Horsepower @ rpm	110 @ 5000	140 @ 5400
Torque (lbs./ft.) @ rpm	127 @ 3000	166 @ 3000
Availability	S	S
EPA city/highway mpg		
5-speed OD manual	20/26	21/27
4-speed OD automatic	20/26	20/26

Prices not available at time of publication.

Isuzu Trooper II

1988 Isuzu Trooper II 4-door

What's New

The 2-door model gets replaced by a new 2-door built on a shorter wheelbase, and a V-6 engine purchased from General Motors becomes optional this year for Trooper II, Isuzu's compact 4-wheel-drive vehicle. The wheelbase and overall length of the new 2-door weren't available, but they will be shorter than the 104.3-inch wheelbase and 175.2-inch overall length of last year's model. The new 2-door is scheduled to arrive by January. A 4-door model built on the 104.3-inch wheelbase returns from last year. (The number of doors listed for these models here excludes the dual swing-out rear doors.) A GM 2.8-liter V-6 engine used in the Chevrolet S10 Blazer, S10 pickup and several GM car lines

---

KEY: ohv = overhead valve; **ohc** = overhead cam; **dohc** = double overhead cam; **I** = inline cylinders; **V** = cylinders in V configuration; **flat** = horizontally opposed cylinders; **bbl.** = barrel (carburetor); **PFI** = port (multi-point) fuel injection; **TBI** = throttle-body (single-point) fuel injection; **rpm** = revolutions per minute; **OD** = overdrive transmission; **S** = standard; **O** = optional; **NA** = not available.

Prices are accurate at time of printing; subject to manufacturer's change.

will be offered in both Trooper body styles. Horsepower is rated at 125 by GM. The standard engine remains a 120-horsepower 2.6-liter 4-cylinder, available with either a 5-speed manual or 4-speed automatic. Later in the year, Isuzu plans to introduce a new 4-wheel-drive sport utility, the Amigo. Isuzu hadn't released information on the Amigo, except that it is built on a modified truck chassis, is smaller than the Trooper and will be offered as a convertible, which makes it sound similar to the Suzuki Samurai. Isuzu, however, insists the Amigo will not be a Samurai imitator, and will attract different kinds of buyers.

For

- 4WD traction •Passenger room •Cargo room

Against

- Ride •Noise •Entry/exit

Summary

Trooper's tall, boxy design gives this 4x4 plenty of passenger room and cargo space. In fact, the 4-door version has more room than nearly every rival, save Mitsubishi's new 5-door model. What this vehicle has lacked in the past is snappy performance. The base 4-cylinder engine is adequate, but lacks the low-end torque that 3500-pound vehicles of this type need. We're not sure the GM-built V-6 will be that big an improvement, though we haven't driven a Trooper with that engine. Note than GM now offers a larger V-6 as an option above the 2.8 on its 4x4s; previously, Jeep used the 2.8 but dropped it once it developed a more powerful 4.0-liter inline 6-cylinder. Even without a brawny engine, Trooper acquits itself well in slippery conditions and off-road. However, the lack of shift-on-the-fly for the 4WD system can be downright inconvenient, especially when you need to frequently change in or out of 4WD. The vehicle has to be stopped to change in or out of 4WD. If you want a rugged 4x4 for off-road work, then Trooper II is still one of the more reasonably priced vehicles around. If you want a more luxurious 4x4 mainly for highway use, then it's better to look elsewhere.

1988 Specifications

	2-door wagon	4-door wagon
Wheelbase, in.	104.3	104.3
Overall length, in.	175.2	175.2
Overall width, in.	65.0	65.0
Overall height, in.	71.7	71.7
Front track, in.	54.7	54.7
Rear track, in.	55.1	55.1
Turn diameter, ft.	35.4	35.4
Curb weight, lbs.	3366	3421
Cargo vol., cu. ft.	79.0	79.0
Fuel capacity, gal.	21.9	21.9
Seating capacity	5	5
Front headroom, in.	40.2	40.2
Front shoulder room, in.	53.8	53.8
Front legroom, max., in.	39.9	39.9

KEY: ohv = overhead valve; **ohc** = overhead cam; **dohc** = double overhead cam; **I** = inline cylinders; **V** = cylinders in V configuration; **flat** = horizontally opposed cylinders; **bbl.** = barrel (carburetor); **PFI** = port (multi-point) fuel injection; **TBI** = throttle-body (single-point) fuel injection; **rpm** = revolutions per minute; **OD** = overdrive transmission; **S** = standard; **O** = optional; **NA** = not available.

	2-door wagon	4-door wagon
Rear headroom, in.	39.4	39.4
Rear shoulder room, in.	53.3	53.3
Rear legroom, min., in.	36.1	36.1

Powertrain layout: longitudinal front engine/rear-wheel and on-demand 4WD.

Engines

	ohc I-4
Size, liters/cu. in.	2.6/156
Fuel delivery	PFI
Horsepower @ rpm	120 @ 4600
Torque (lbs./ft.) @ rpm	146 @ 2600
Availability	S

EPA city/highway mpg

5-speed OD manual	16/19
4-speed OD automatic	16/18

Prices

Isuzu Trooper II (1988 prices)	Retail Price	Dealer Invoice	Low Price
S 3-door wagon, 5-speed	$11909	$10600	$11255
S 5-door wagon, 5-speed	12639	11249	11944
XS 3-door wagon, 5-speed	12909	11490	12200
XS 5-door wagon, 5-speed	13439	11961	12700
XS 5-door wagon, automatic	14439	12851	13645
LS 5-door wagon, 5-speed	14799	13172	13986
LS 5-door wagon, automatic	15799	14062	14931
Destination charge	259	259	259

Standard Equipment:

2.6-liter PFI 4-cylinder engine, 5-speed manual or 4-speed automatic transmission, 2-speed transfer case, automatic locking front hubs (with 5-speed; manual with automatic transmission), power steering, power 4-wheel disc brakes, skid plates, front bucket seats with height adjustments, tinted glass, dual outside mirrors, tachometer, oil pressure and coolant temperature gauges, 235/75R15 tires with full-size spare. **LS** adds: captain's chairs, cruise control, rear defogger and wiper/washer, rear privacy glass, cloth upholstery.

Optional Equipment:

Air conditioning	850	NA	NA
Spare tire cover, S & XS	50	30	40
Rear defogger & wiper/washer, S	200	170	185
Rear heater	300	255	278
Carpet mats	50	35	43
AM/FM ST	185	130	158
AM/FM ST ET cassette w/2 speakers	385	270	328
w/4 speakers	475	335	405
Roof rack	142	100	121
Rear seat, S & XS	400	343	372
LS	500	428	464
Wind deflector	50	35	43

Jaguar XJ-S and XJ6

What's New

A factory-built XJ-S convertible went on sale in September, and both the new convertible and the carryover 2-door XJ-S coupe have anti-lock brakes as standard equipment. The convertible is the first ragtop XJ-S built by Jaguar. Last year, Jaguar offered a special-order convertible built in the U.S. by Hess & Eisenhardt, a custom car builder. Jaguar also offered the XJ-SC, a 2-seat cabriolet with removable Targa

Jaguar XJ-S convertible

Specifications

Specifications	XJ-S 2-door notchback	XJ-S 2-door conv.	XJ6 4-door notchback
Wheelbase, in.	102.0	102.0	113.0
Overall length, in.	191.7	191.7	196.4
Overall width, in.	70.6	70.6	78.9
Overall height, in.	47.0	47.8	54.3
Front track, in.	58.6	58.6	59.1
Rear track, in.	59.2	59.2	59.0
Turn diameter, ft.	39.4	39.4	40.8
Curb weight, lbs.	4015	4190	3903
Cargo vol., cu. ft.	10.6	NA	12.7
Fuel capacity, gal.	24.0	24.0	23.2
Seating capacity	4	2	5
Front headroom, in.	36.1	36.1	36.6
Front shoulder room, in.	55.5	NA	57.5
Front legroom, max., in.	41.3	41.3	41.7
Rear headroom, in.	33.4	—	36.5
Rear shoulder room, in.	52.0	—	57.6
Rear legroom, min., in.	23.4	—	33.1

Powertrain layout: longitudinal front engine/rear-wheel drive.

Engines

Engines	dohc I-6	ohc V-12
Size, liters/cu. in.	3.6/219	5.3/326
Fuel delivery	PFI	PFI
Horsepower @ rpm	195 @ 5000	262 @ 5000
Torque (lbs./ft.) @ rpm	232 @ 4000	290 @ 3000
Availability	S[1]	S[2]

EPA city/highway mpg

3-speed automatic		13/17
4-speed OD automatic	17/24	

1. XJ6   2. XJ-S

Prices

Jaguar	Retail Price	Dealer Invoice	Low Price
XJ6 4-door notchback	$43500	—	—
XJ6 Vanden Plas 4-door notchback	47500	—	—
XJ-S 2-door notchback	47000	—	—
XJ-S convertible	56000	—	—
Destination charge	450	450	450

Dealer invoice and low price not available at time of publication.

Standard Equipment:

XJ6: 3.6-liter DOHC 24-valve PFI 6-cylinder engine, 4-speed automatic transmission, power steering, 4-wheel disc brakes, anti-lock braking system, self-leveling suspension, automatic climate control, power front bucket seats with leather facings, power windows, power locks with infrared remote, analog tachometer and speedometer, LED gauges (coolant temperature, oil pressure, voltmeter), heated power mirrors, motorized front shoulder belts with manual lap belts outboard rear lap/shoulder belts, intermittent wipers, trip computer, rear defogger, rear fog lights, tinted glass, AM/FM ST ET cassette with channel 19 CB monitor, folding rear armrest, seatback pockets, console with cassette storage, map light, Pirelli P5 205/70R15 tires on alloy wheels. **Vanden Plas** adds: limited-slip differential, headlamp washers, heated front seats, seatback walnut picnic tables, full leather seat trim, rear storage armrest, rear reading lamps. **XJ-S:** 5.3-liter PFI V-12, 3-speed automatic transmission, 4-wheel disc brakes, anti-lock braking system, automatic climate control, full leather interior, heated reclining front bucket seats with power lumbar, motorized front shoulder belts and manual lap belts (convertible has active 3-point belts), outboard rear lap/shoulder belts, power windows and locks, heated power mirrors, cruise control, trip computer, tinted glass, intermittent wipers, coolant temperature and oil pressure gauges, AM/FM ST ET cassette with power antenna, rear defogger, leather-wrapped steering wheel, front fog lights, Pirelli P600 235/60VR15 tires on alloy wheels.

---

roof panels and rear roof section. Both have been succeeded by the new factory convertible, a 2-seater with a power folding top and heated glass rear window. A luggage shelf behind the seats has a lockable storage compartment that supplements the trunk. The anti-lock brakes standard on the convertible and the 2+2 coupe were developed by Jaguar and the Alfred Teves Company. Both XJ-S models have new electrically heated seats with power lumbar adjustment, a new steering wheel and new alloy wheels wrapped in Pirelli P600 235/60VR15 tires. Power comes from a 262-horsepower 5.3-liter V-12 engine linked to a 3-speed automatic transmission. The only change to the XJ6 sedan is a shorter final drive ratio for quicker acceleration. The ratio changes from 2.88:1 to 3.58:1 as Jaguar adopts the same final drive used in European models. Jaguar says the change makes little difference in fuel economy, so the XJ6 still avoids the gas-guzzler tax. XJ6 was redesigned for an early 1988 introduction, succeeding a model with the same name that survived for 20 years in the luxury car market. Standard equipment on the base XJ6 and plusher Vanden Plas includes anti-lock brakes (by Bosch rather than Teves) and motorized automatic front shoulder belts. The XJ6 uses a dual-cam, 3.6-liter inline 6-cylinder engine that last year received a boost in compression ratio from 8.2:1 to 9.6:1, and 14 more horsepower (to 195). A 4-speed overdrive automatic transmission is standard.

For

• Ride   • Anti-lock brakes   • Refinement

Against

• Fuel economy   • Interior room (XJ-S)   • Price

Summary

The XJ6 is a fashionable executive sedan and the XJ-S an attractive 2-door with timeless styling. The forte for either is high-speed cruising, though they're just as nice in town. The much-newer XJ6, the bigger seller of the two, has an all-independent suspension that irons out bumps, soaks up dips and fills in patches with ease. Braking also is first rate thanks to the across-the-board anti-lock equipment. With either an XJ6 or an XJ-S you're getting about two tons of mass, so despite the muscular engines, you get acceleration that's brisk more than quick. While Jaguar has dodged the guzzler tax, mileage is mediocre; we averaged 15.5 mpg in mostly urban driving with a 1988 XJ6. The XJ6 is too small inside for being so large outside (its trunk isn't vast, either), while the XJ-S is even cozier inside. The seats are superbly comfortable and the interior furnishings are impressive—polished wood trim, aromatic leather and abundant chrome. Jaguar workmanship has come a long way in recent years but, judging by our most recent XJ6 test car, remains below BMW and Mercedes standards.

Prices are accurate at time of printing; subject to manufacturer's change.

Jeep Cherokee/ Wagoneer

Jeep Cherokee Limited 3-door

What's New

Optional anti-lock brakes are available for '89 on Cherokee and Wagoneer models with the 6-cylinder engine, automatic transmission and the more sophisticated of Jeep's two 4-wheel drive systems, Selec-Trac. Jeep says these are the first light trucks with an anti-lock system that works on all four wheels; other light-truck anti-lock systems work only on the rear wheels. Another distinction is that the Jeep system, which was designed by Bendix, works in both 2WD and 4WD; most anti-lock systems on 4x4s work in 2WD only. Among other changes to Jeep's sport-utility wagons, power steering is now standard on the base Cherokee and a 20-gallon fuel tank replaces a 13.5-gallon tank on all models. Now standard on the top-of-the-line Wagoneer and Cherokee Limited models is a remote-control entry system. Dual power outside mirrors, fog lamps and a tachometer are now standard on Wagoneer Limited. Cherokee and Wagoneer are basically trim and equipment variations on the same 3- and 5-door bodies. The base Cherokee has a 2.5-liter 4-cylinder engine and 2WD. Standard on the Cherokee 4x4 is Command-Trac, a part-time 4WD system (for use on wet or slippery surfaces only) with shift-on-the-fly capability. A Cherokee Sport 3-door, added to the line in early 1988, is available in 2WD or Command-Trac 4WD and only with the 4.0-liter inline 6-cylinder engine and 5-speed manual transmission. The plush Limited models come standard with the 4.0-liter six, a 4-speed automatic transmission and Selec-Trac 4WD, which can be used on smooth, dry pavement. Jeeps are now sold through the Jeep-Eagle division of Chrysler Motors. Chrysler acquired Jeep when it purchased AMC in 1987.

For

- 4WD traction • Performance (6-cylinder)
- Passenger room • Cargo space • Anti-lock brakes

Against

- Fuel economy • Performance (4-cylinder) • Price

Summary

We rate these the best compact 4x4s because of their convenient 4WD systems, abundant passenger and cargo room, towing ability, off-road capabilities and civilized on-road manners. Though it uses quite a lot of fuel, the 4.0 six puts them among the most powerful vehicles in this class. Now, 4-wheel anti-lock brakes make them the only ones with that added measure of control under braking. We prefer the 6-cylinder because under a heavy load, the base 4-cylinder is weak; it's feeble with automatic transmission. Cherokee and Wagoneer are spacious inside for their modest exterior dimensions. Head room is adequate all around and the back seat accommodates three adults. Folding the rear seatback gives you great luggage space for the modest overall length. Limited versions are the Cadillacs of compact 4x4s, with prices to match. An alternative is a more modestly equipped Cherokee. It'll have fewer comfort and convenience features, but still be a well-rounded vehicle that doesn't trade an unreasonable degree of on-road control or comfort for its impressive off-road capability.

Specifications

	3-door wagon	5-door wagon
Wheelbase, in.	101.4	101.4
Overall length, in.	165.3	165.3
Overall width, in.	70.5	70.5
Overall height, in.	64.1	64.1
Front track, in.	58.0	58.0
Rear track, in.	58.0	58.0
Turn diameter, ft.	36.1	36.1
Curb weight, lbs.	2915	2958
Cargo vol., cu. ft.	71.2	71.2
Fuel capacity, gal.	20.0	20.0
Seating capacity	5	5
Front headroom, in.	38.3	38.3
Front shoulder room, in.	55.3	55.3
Front legroom, max., in.	39.9	39.9
Rear headroom, in.	38.0	38.0
Rear shoulder room, in.	55.3	55.3
Rear legroom, min., in.	35.3	35.3

Powertrain layout: longitudinal front engine/rear-wheel drive or on-demand 4-wheel drive.

Engines

	ohv I-4	ohv I-6
Size, liters/cu. in.	2.5/150	4.0/242
Fuel delivery	TBI	PFI
Horsepower @ rpm	121 @ 5000	173 @ 4500
Torque (lbs./ft.) @ rpm	141 @ 3500	220 @ 2500
Availability	S	O

EPA city/highway mpg

5-speed OD manual 2WD	21/24	17/23
4-speed OD automatic 2WD	18/24	16/21
4-speed OD automatic 4WD	18/23	16/20
5-speed OD manual 4WD	19/24	17/22

Prices

Jeep Cherokee	Retail Price	Dealer Invoice	Low Price
3-door wagon, 2WD	$12160	$10895	$11528
5-door wagon, 2WD	12795	11455	12125
3-door wagon, 4WD	13654	12132	12893
Limited 3-door wagon, 4WD	23130	20549	21840
5-door wagon, 4WD	14200	12691	13446
Limited 5-door wagon, 4WD	24058	21366	22712
Destination charge	429	429	429

Prices are accurate at time of printing; subject to manufacturer's change.

Standard Equipment:

2.5-liter TBI 4-cylinder engine, 5-speed manual transmission, Command-Trac part-time 4WD (4WD models) power steering and brakes, vinyl front bucket seats, AM ET, tinted glass, 195/75R15 all-season SBR tires. **Limited adds:** 4.0-liter PFI 6-cylinder engine, Selec-Trac full-time 4WD, air conditioning, retractable cargo area cover, cruise control, deep-tinted glass, upgraded sound insulation, floormats, fog lights, Gauge Group with tachometer, Light Group, power mirrors, power windows and locks, keyless entry system, AM/FM ST ET cassette with Accusound by Jensen speakers, front vent windows, rear quarter vent windows (3-door), roof rack, tilt steering column, intermittent wipers, rear defogger and wiper/washer, carpet, cargo tiedowns, console with armrest, rear heat ducts, upgraded interior and exterior trim, spare tire cover, leather-wrapped steering wheel, front bumper guards, 225/70R15 Goodyear Eagle GT+4 tires on alloy wheels.

Optional Equipment:

	Retail Price	Dealer Invoice	Low Price
4.0-liter PFI 6-cylinder engine	613	521	564
Anti-lock braking system	852	724	784
4-speed automatic transmission	757	643	696
Selec-Trac, 4WD	398	338	366
Requires 4.0 and automatic transmission.			
Rear Trac-Lok differential	287	244	264
Requires 4.0 and conventional spare tire.			
Air conditioning	845	718	777
Pioneer Pkg.	1254	1066	1154
Carpet, cargo tiedowns, dome/map light, Gauge Group (oil pressure and coolant temperature, voltmeter, trip odometer, LCD clock, low fuel), rocker/reclining bucket seats with head restraints, spare tire cover, Light Group, AM/FM ST ET, rear quarter vent windows (3-door), bright bumpers, front air dam, fender flares, upgraded interior and exterior trim.			
Laredo Pkg.	3025	2571	2783
Pioneer interior items plus wing-back seats, upgraded door panels and carpet, cargo area cover, console with armrest, cargo area skid strips, door edge gaurds, added sound insulation, floormats, roof rack, leather-wrapped steering wheel, tachometer, power mirrors, intermittent wipers, rear defogger, 215/75R15 OWL tires on alloy wheels.			
Sport Pkg., 3-door	973	827	895
4.0-liter engine, carpet, bodyside stripes, exterior moldings, 225/75R15 OWL tires on alloy wheels.			
Option Group 1, base 3-door	269	229	247
Base 5-door	1517	1289	1396
3-door: carpet, spare tire cover, 205/75R15 tires, wheel trim rings. 5-door adds: air conditioning, console with armrest, AM/FM ST ET, roof rack, rear wiper/washer.			
Sport Pkg. Option Group, 3-door	1269	1079	1167
Sport Pkg., console with armrest, Gauge Group with tachometer, AM/FM ST ET, spare tire cover.			
Pioneer Pkg. Option Group 1, 3-door	2507	2131	2306
Pioneer Pkg., air conditioning, console with armrest, upgraded sound insulation, Protection Group, roof rack, tilt steering column, remote mirrors, intermittent wipers, rear defogger.			
Pioneer Pkg. Option Group 2, 3-door	3165	2690	2912
Group 1 plus cruise control, power windows and locks, keyless entry, AM/FM ST ET cassette.			
Laredo Pkg. Option Group	3739	3178	3440
Laredo Pkg., cruise control, fog lamps, power windows and locks, keyless entry, AM/FM ST ET cassette.			
Fabric rocker/recliner seats, base & Sport	138	117	127
Carpet, base	211	179	194
Console with armrest	122	104	112
Requires carpet.			
Cruise control	224	190	206
Requires Visibility Group; not available with 4-cylinder engine and manual transmission.			
Deep tinted glass, base & Sport 3-door	310	264	285
Pioneer	147	125	135
Rear defogger	163	139	150
HD alternator & battery	136	116	125
w/air conditioning	73	62	67
Extra Quiet Insulation Pkg., Pioneer	128	109	118
Front fog lamps	111	94	102
Gauge Group, Sport	160	136	147
Tachometer, LCD clock, coolant temperature and oil pressure, low fuel, trip odometer.			

	Retail Price	Dealer Invoice	Low Price
Tachometer, Pioneer	64	54	59
Remote mirrors, base, Sport & Pioneer	77	65	71
Power Windows & Locks Group, Pioneer & Laredo 3-door	431	366	397
Pioneer & Laredo 5-door	578	491	532
Power windows and locks, keyless entry.			
Power front seats, Laredo	420	357	386
Protection Group, base, Sport & Pioneer	201	171	185
Skid Plate Group, 4WD	145	123	133
AM/FM ST ET, base & Sport	193	164	178
AM/FM ST ET cassette, base & Sport	396	337	364
Pioneer & Laredo	203	173	187
Six premium speakers, Pioneer & Laredo	176	150	162
Requires Power Windows & Locks Group and stereo radio.			
Roof rack, base, Sport & Pioneer	140	119	129
Rear quarter vent windows, base & Sport	163	139	150
Outside spare tire carrier, exc. Ltd	175	149	161
Spare tire cover, base & Sport	46	39	42
Leather-wrapped steering wheel, Sport	48	41	44
Manual sunroof, Pioneer & Laredo	361	307	332
Requires roof rack.			
Off Road Suspension Pkg., 4WD			
Base w/4.0 & automatic	992	843	913
Base w/5-speed	950	808	874
Sport w/automatic	599	509	551
Sport w/5-speed	557	473	512
Pioneer w/4.0 & automatic	891	757	820
Pioneer w/5-speed	849	722	781
Laredo w/4.0 & automatic	624	530	574
Laredo w/5-speed	582	495	535
High-pressure gas shocks, 225/75R15 OWL tires with conventional spare, styled steel wheels (alloy on Laredo), front and rear tow hooks, Skid Plate Pkg., HD cooling.			
Tilt steering column	128	109	118
Requires Visibility Group.			
Trailer Towing Pkg. A (2000 lbs.)	121	103	111
Trailer Towing Pkg. B (5000 lbs.)	362	308	333
Front vent windows, base, Sport & Pioneer			
Visibility Group, base, Sport & Pioneer	289	246	266
Remote mirrors, intermittent wipers, rear defogger.			
Rear wiper/washer, base & Sport	148	126	136
California emissions pkg.	126	107	116

Jeep Wagoneer Limited

	Retail Price	Dealer Invoice	Low Price
5-door 4WD wagon	$23127	$20597	$21862
Destination charge	429	429	429

Standard Equipment:

4.0-liter PFI 6-cylinder engine, 4-speed automatic transmission, Selec-Trac full-time 4WD, power steering and brakes, air conditioning, power windows and locks, keyless entry system, power front seats, AM/FM ST ET, retractable cargo cover, roof rack, cruise control, tilt steering column, tinted glass, rear defogger and wiper/washer, intermittent wipers, Light Group, cargo tiedowns, LCD clock, coolant temperature and oil pressure gauges, voltmeter, trip odometer, leather-wrapped steering wheel, cloth/leather upholstery, P205/75R15 WSW self-sealing tires on alloy wheels.

Optional Equipment:

	Retail Price	Dealer Invoice	Low Price
Anti-lock braking system	852	724	784
California emissions pkg.	126	107	116
Trac-Lok rear differential	287	244	264
AM/FM ST ET cassette	203	173	187

KEY: ohv = overhead valve; **ohc** = overhead cam; **dohc** = double overhead cam; **I** = inline cylinders; **V** = cylinders in V configuration; **flat** = horizontally opposed cylinders; **bbl.** = barrel (carburetor); **PFI** = port (multi-point) fuel injection; **TBI** = throttle-body (single-point) fuel injection; **rpm** = revolutions per minute; **OD** = overdrive transmission; **S** = standard; **O** = optional; **NA** = not available.

Prices are accurate at time of printing; subject to manufacturer's change.

	Retail Price	Dealer Invoice	Low Price
Six premium speakers & power antenna ..	176	150	162
Skid Plate Group	145	123	133
Manual sunroof	361	307	332
Trailer Towing Pkg. A (2000 lbs.)	121	103	111
Trailer Towing Pkg. B (5000 lbs)	362	308	333
205/70R15 BSW tires	NC	NC	NC
Conventional spare tire, WSW self-sealing .	203	173	187
BSW non-self-sealing	163	139	150

Jeep Wrangler

Jeep Wrangler Islander

What's New

A tropical-theme model, the Islander, joins the 4-wheel-drive Wrangler line. Basically a graphics package, Islander features exteriors like "Malibu yellow" and "Pacific blue," offset by a large sunset-orange globe decal. It comes standard with a soft top and silver-painted steel wheels. Another "theme" Wrangler, the Sahara, was introduced during the '88 model year. It comes in either "khaki metallic" or "coffee" exterior colors with khaki-colored wheels, khaki soft top (a tan hard top is optional) and water-resistant khaki "Trailcloth" interior fabric. Sahara carries its theme a bit further than Islander with fog lamps, gas shock absorbers, wheel flares and integrated body side steps as standard. Dealer-installed Sahara accessories include a brush/grille guard, soft top boot and a bug screen kit. Also introduced in mid-1988 was the entry-level Wrangler S, which comes only with a soft top and the fuel-injected 2.5-liter 4-cylinder engine and 5-speed manual transmission that are standard on all Wranglers. Base, Sport Decor and Laredo are the other trim levels. A hard top and full metal doors are standard on the Laredo, optional on the others. A carbureted 4.2-liter 6-cylinder and 3-speed automatic transmission (available only with the six) are optional. All models come with a part-time 4WD system (not for use on dry pavement) with full shift-on-the-fly capability. Wrangler can be equipped to tow a 2000-pound trailer.

> **KEY: ohv** = overhead valve; **ohc** = overhead cam; **dohc** = double overhead cam; **I** = inline cylinders; **V** = cylinders in V configuration; **flat** = horizontally opposed cylinders; **bbl.** = barrel (carburetor); **PFI** = port (multi-point) fuel injection; **TBI** = throttle-body (single-point) fuel injection; **rpm** = revolutions per minute; **OD** = overdrive transmission; **S** = standard; **O** = optional; **NA** = not available.

For

- 4WD traction
- Maneuverability
- Visibility

Against

- Ride and handling
- Fuel economy
- Cargo room

Summary

The all-new Wrangler debuted in 1987, but Jeep's marketing strategy dictated that it retain the personality and appeal of its CJ predecessor. Thus, you get a rugged, military-style vehicle with a stiff, jarring suspension and a high center of gravity that makes it feel tipsy in corners. You get only adequate acceleration from either available engine and mediocre mileage. It's a tall step up over the door sills into the interior, which has a cramped rear seat and tiny cargo area; with the rear seat tilted out of the way cargo room is adequate. While its overall behavior and performance surpasses that of the CJ and the less expensive Suzuki Samurai, we'd buy a Wrangler only if we frequently went off road or slogged through deep snow. There, it feels more in its element, with fine maneuverability and reassuring traction from the convenient 4WD system. We realize these sporty-looking vehicles are jam packed with "character," and so emotion often prevails over matters of practicality at buying time. Still, consider seriously how you'd use a Wrangler and if the compromises in on-road ride and handling are worth it. Finally, check with your insurance agent; some companies charge higher rates for these specialty vehicles.

Specifications

	2-door w/hard top	2-door w/soft top
Wheelbase, in.	93.4	93.4
Overall length, in.	152.0	152.0
Overall width, in.	66.0	66.0
Overall height, in.	69.3	72.0
Front track, in.	58.0	58.0
Rear track, in.	58.0	58.0
Turn diameter, ft.	33.7	33.7
Curb weight, lbs.	2902	NA
Cargo vol., cu. ft.	43.2	43.2
Fuel capacity, gal.	15.0	15.0
Seating capacity	4	4
Front headroom, in.	40.2	41.4
Front shoulder room, in.	53.1	53.1
Front legroom, max., in.	39.5	39.5
Rear headroom, in.	40.5	40.3
Rear shoulder room, in.	56.3	56.3
Rear legroom, min., in.	35.0	35.0

Powertrain layout: longitudinal front engine/rear-wheel drive or on-demand 4-wheel drive.

Engines

	ohv I-4	ohv I-6
Size, liters/cu. in.	2.5/150	4.2/256
Fuel delivery	TBI	2 bbl.
Horsepower @ rpm	117 @ 5000	112 @ 3000
Torque (lbs./ft.) @ rpm	135 @ 3500	210 @ 2000
Availability	S	O

EPA city/highway mpg

5-speed OD manual	18/20	16/20
3-speed automatic		15/16

Prices are accurate at time of printing; subject to manufacturer's change.

CONSUMER GUIDE®

Prices

Jeep Wrangler

	Retail Price	Dealer Invoice	Low Price
Wrangler S	$8995	$8510	$8753
Wranger Base	10929	9763	10346
Wrangler Islander	11628	10378	11003
Wrangler Sahara	12760	11374	12067
Wrangler Laredo	14774	13146	13960
Destination charge	429	429	429

Standard Equipment:

S: 2.5-liter TBI 4-cylinder engine, 5-speed manual transmission, part-time 4-wheel drive with 2-speed transfer case, power brakes, vinyl front bucket seats, tachometer, coolant temperature and oil pressure gauges, voltmeter, trip odometer, padded roll bar, soft top, tinted windshield, fuel tank skid plate, vinyl spare tire cover, 4 215/75R15 all-terrain SBR tires. **Base** adds: folding rear seat, AM radio, right outside mirror. **Islander** adds: charcoal carpet, upholstery, soft top and spare tire cover, door pockets, charcoal fender flares. **Sahara** adds to base: center console, door pouches, leather-wrapped steering wheel, khaki upholstery, soft top and spare tire cover, fender flares. **Laredo** adds: 4.2-liter 2bbl. 6-cylinder engine, power steering, upgraded sound insulation, AM/FM ST ET, floormats, carpeted lower door panels with pockets, front bumper extensions with tow hooks, chrome front bumper and rear bumperette, mud guards, side steps, deep-tinted glass, hardtop with full metal doors, courtesy lights, intermittent wipers, console, tilt steering column, fog lights, spare wheel lock, off-road gas shocks, alloy wheels.

Optional Equipment:

	Retail	Dealer	Low
4.2-liter 6-cylinder engine (NA base; std. Laredo)	430	366	396
3-speed automatic transmission (NA base) .	497	422	457
Requires 4.2 and tilt steering column.			
Power steering, S, base & Islander	303	258	279
Air conditioning (NA base)	887	754	816
Rear Trac-Lok differential	281	239	259
Trailcloth fabric seats, Islander	108	92	99
Hardtop, S & base	596	507	548
w/deep-tinted glass, Islander & Sahara .	748	636	688
California emissions pkg.	129	110	119
Base Option Group	NA	NA	NA
Carpet, Convenience Group, extra-capacity fuel tank, power steering, tilt steering column, 215/75R15 OWL tires.			
Islander Option Group	NA	NA	NA
Convenience Group, extra-capacity fuel tank, power steering, tilt steering column, AM/FM ST ET.			
Option Group, Sahara	NA	NA	NA
Laredo	NA	NA	NA
Extra-capacity fuel tank, tilt steering column.			
Metallic paint, base, Islander & Laredo	175	149	161
Carpet, S & base	138	117	127
Convenience Group, base & Islander	187	159	172
w/tilt steering column	125	106	115
Courtesy lights, intermittent wipers, glovebox lock, engine compartment light, center console.			
Cruise control, Sahara & Laredo	225	191	207
Requires 4.2, automatic transmission and tilt steering column.			
Rear defogger for hardtop (NA base)	166	141	153
Deep tinted glass, Islander & Sahara	152	129	140
20-gallon fuel tank (NA base)	63	54	58
HD alternator & battery group (NA base) .	136	116	125
Floormats, base, Islander & Sahara	33	28	30
Requires carpet.			
Off Road Pkg.			
Base & Islander w/215 black tires	308	262	283
Base & Islander w/215 or 225 OWL tires .	196	167	180
Islander w/215 OWL tires	325	276	299
AM radio, S	101	86	93
AM/FM ST ET, S	273	232	251
Base & Islander	172	146	158

	Retail Price	Dealer Invoice	Low Price
AM/FM ST ET cassette, base	540	459	497
Base & Islander	439	373	404
Sahara & Laredo	267	227	246
Folding rear seat, S	446	379	410
Tilt steering column, base & Islander	190	162	175
Sahara & Laredo	126	107	116
5 215/75R15 OWL tires, base & Sahara	231	196	213
5 225/75R15 OWL tires, base & Sahara	423	360	389
Islander	322	274	296
Laredo	193	164	178
Alloy wheels, base, Islander & Sahara	324	275	298

Lincoln Continental

Lincoln Continental Signature Series

What's New

The all-new 1988 Continental was the first Lincoln with front-wheel drive and fewer than eight cylinders. For '89, it becomes the first U.S. car with air bags standard for both the driver and front-seat passenger. The driver's-side air bag is contained in the steering wheel hub; the passenger's is loaded above the glove compartment in a revamped dashboard. Available only as a 4-door, 6-passenger notch-back, Continental is 4.4 inches longer but 170 pounds lighter and has more interior room than the rear-drive car it replaced. Standard and Signature Series trim levels are offered. Ford's 3.8-liter V-6 and 4-speed overdrive automatic transmission is the only powertrain. Power ratings are unchanged, but Lincoln says engine response is better thanks to higher wide-open-throttle shift speeds and a revised final-drive ratio. Continental's chassis has a computer-controlled damping system designed to maintain a smooth ride by adjusting in fractions of a second to changes in the road surface. Its power steering has speed-sensitive and variable-effort assist. Brakes are 4-wheel discs with anti-lock control. Lincoln says Continental's electronic gauges have enhanced contrast this year for improved readability. The door trim panels are also revised and front seat-track travel is increased. Also, ordering the optional Insta-Clear heated windshield brings a heavy-duty battery.

For

● Air bags ● Anti-lock brakes ● Ride ● Interior room
● Quietness

Against

● Performance ● Instrumentation

Prices are accurate at time of printing; subject to manufacturer's change.

Lincoln

Summary

Ford helped fuel a seller's market for the new Continental by keeping initial production low, but it's no secret the car was a sales success on its merits as well. The styling is contemporary and aerodynamic, yet it's still a large car that's unmistakably American. With its fake wood accents and cushy seats, the roomy interior has all the amenities buyers of domestic luxury cars expect. In addition, the trunk is spacious. Lincoln makes good use of high technology, including anti-lock brakes and the dual air bags, two safety features we applaud. The car cruises quietly and its sophisticated suspension provides a comfortable ride, even if it does get a touch floaty over high-speed dips. Ford insiders say engineers have also been working to quell that touch of high-speed float. Front-wheel drive didn't seem to put off an audience used to traditional rear-drive premium sedans. Neither did the V-6, though we consider it Continental's Achilles' heel. Performance is adequate, but a car in this class needs more power. Ford evidently agrees, and is reportedly working on a 4.6-liter V-8 for use in the 1990 Continental. Another of our complaints was that the digital instrumentation conveyed too little information and washed out in daylight. We'll await a road test of an '89 to see if the new dashboard remedies that problem.

Specifications

	4-door notchback
Wheelbase, in.	109.0
Overall length, in.	205.1
Overall width, in.	72.7
Overall height, in.	55.6
Front track, in.	62.3
Rear track, in.	61.1
Turn diameter, ft.	38.0
Curb weight, lbs.	3628
Cargo vol., cu. ft.	19.0
Fuel capacity, gal.	18.6
Seating capacity	6
Front headroom, in.	38.7
Front shoulder room, in.	57.5
Front legroom, max., in.	41.7
Rear headroom, in.	38.4
Rear shoulder room, in.	57.5
Rear legroom, min., in.	39.2

Powertrain layout: transverse front engine/front-wheel drive.

Engines

	ohv V-6
Size, liters/cu. in.	3.8/232
Fuel delivery	PFI
Horsepower @ rpm	140 @ 3800
Torque (lbs./ft.) @ rpm	215 @ 2200
Availability	S

EPA city/highway mpg

4-speed OD automatic	19/28

Prices

Lincoln Continental	Retail Price	Dealer Invoice	Low Price
4-door notchback	$27468	$23305	$25587
Signature 4-door notchback	29334	24873	27304
Destination charge	524	524	524

Standard Equipment:

3.8-liter PFI V-6 engine, 4-speed automatic transmission, speed-sensitive power steering, power 4-wheel disc brakes, anti-lock braking system, dual front airbags, automatic climate control air conditioning, 50/50 front seats with recliners, leather upholstery (cloth is available at no cost), folding front and rear armrests, rear lap/shoulder belts, cruise control, automatic parking brake release, AM/FM cassette, tinted glass, heated power mirrors, rear defogger, remote fuel door and decklid releases, power windows and locks, intermittent wipers, tilt steering column, right visor mirror, electronic instruments with coolant temperature, oil pressure and voltage gauges, trip computer, service interval reminder, digital clock, vinyl bodyside moldings, bright rocker panel moldings, cornering lamps, 205/70R15 tires. **Signature** adds: power recliners, power passenger seat, power decklid pulldown, Autolamp on/off/delay, automatic headlamp dimmer, upgraded upholstery, lighted visor mirrors, alloy wheels.

Optional Equipment:

	Retail Price	Dealer Invoice	Low Price
Alloy wheels	478	401	440
Keyless illuminated entry	209	175	192
Anti-theft alarm	200	168	184
Power glass moonroof	1319	1108	1213
Memory seat w/power lumbar	301	253	277
Leather-wrapped steering wheel	115	96	106
Ford JBL Audio System	525	441	483
CD player	617	519	568
Insta-Clear heated windshield	253	213	233
Comfort/Convenience Group	819	688	753

Power decklid pulldown, power passenger seat, lighted visor mirrors, automatic headlamp dimmer, Autolamp on/off/delay, rear floormats, power passenger recliner.

Overhead Console Group	226	190	208

Digital compass, automatic day/night mirror.

Cloth seat trim	NC	NC	NC
California emissions pkg.	99	83	91

Lincoln Mark VII

Lincoln Mark VII LSC

What's New

Lincoln's premium coupe is carried over virtually unchanged after getting a more powerful engine, a better stereo and minor restyling for 1988. The only '89 alteration is the addition of an engine-management computer malfunction warning light to cars sold in all 50 states; the light previously was standard in California only. Two Mark VII models are offered at the same price: the sporty LSC and the luxury Bill Blass Designer Series. Both are rear-drive, 5-seat, 2-door coupes. They retain the 225-horsepower 5.0-liter V-8 acquired in '88. A 4-speed overdrive automatic

Prices are accurate at time of printing; subject to manufacturer's change

transmission and anti-lock brakes are standard on both. The LSC has 225/60R16 black-sidewall high-performance tires on alloy wheels, quick-ratio power steering, a handling suspension and leather upholstery. The Designer Series gets 215/70R15 whitewalls, regular power steering, added chrome exterior trim and a choice of leather or cloth seating. Standard equipment on both includes automatic climate control, a self-leveling suspension, remote decklid release and power decklid pulldown and 6-way power front seats. A power glass moonroof is among the options.

For

- Performance • Anti-lock brakes • Quietness
- Handling/roadholding (LSC) • Luxury appointments

Against

- Fuel economy • Rear seat room • Trunk space

Summary

There's plenty to recommend the Mark VII, especially in sporty LSC guise. Its V-8 delivers satisfying performance in a refined, well-behaved manner. A firm suspension and wide tires give the LSC the commendable handling and roadholding of a sporty European coupe. Traditional Lincoln buyers will likely be turned off by the LSC's firm ride, but we prefer a little harshness to excessive bouncing and poor high-speed ride control. The standard anti-lock brakes have impressive stopping ability, while the plush interior furnishings and lavish standard equipment make for comfortable, pleasant motoring. Drawbacks include a rear seat that has ample leg room for adults, but not much head room and a driveline hump that discourages squeezing three people into the back. The trunk's center well is shallow and narrow, and the spare tire eats up some luggage space. We've averaged as much as 26 mpg on the highway, but around town we usually record mileage in the 15-17 mpg range. On balance, though, the Mark VII LSC has much to offer, including a reasonable price for the luxury-coupe class. We're not saying it's an inexpensive car, but it's much cheaper than rivals such as the BMW 635CSi and Jaguar XJ-S. Its strongest domestic competitor is the Cadillac Eldorado with the optional touring suspension. The chief difference is the Eldo's front-drive layout, more formal styling, and slightly higher price.

Specifications

	2-door notchback
Wheelbase, in.	108.5
Overall length, in.	202.8
Overall width, in.	70.9
Overall height, in.	54.2
Front track, in.	58.4
Rear track, in.	59.0
Turn diameter, ft.	40.0
Curb weight, lbs.	3722
Cargo vol., cu. ft.	14.2
Fuel capacity, gal.	22.1
Seating capacity	5
Front headroom, in.	37.8
Front shoulder room, in.	56.0
Front legroom, max., in.	42.0
Rear headroom, in.	37.1
Rear shoulder room, in.	57.8
Rear legroom, min., in.	36.9

Powertrain layout: longitudinal front engine/rear-wheel drive.

Engines

	ohv V-8
Size, liters/cu. in.	5.0/302
Fuel delivery	PFI
Horsepower @ rpm	225 @ 4000
Torque (lbs./ft.) @ rpm	300 @ 3200
Availability	S
EPA city/highway mpg	
4-speed OD automatic	17/24

Prices

Lincoln Mark VII	Retail Price	Dealer Invoice	Low Price
Bill Blass 2-door notchback	$27218	$23087	$25053
LSC 2-door notchback	27218	23087	25053
Destination charge	524	524	524

Standard Equipment:

LSC: 5.0-liter PFI V-8, 4-speed automatic transmission, power steering, power 4-wheel disc brakes, anti-lock braking system, electronic air suspension with automatic level control, automatic climate control, overhead console with warning lights and reading lamps, power decklid release, defroster group, power windows and door locks, remote fuel door release, tinted glass, automatic headlamp dimmer, Autolamp on/off/delay, illuminated entry, analog instruments including tachometer and coolant temperature gauge, heated power mirrors, AM/FM ST ET cassette, power seats, cruise control, tilt steering column, intermittent wipers, full-length console with lockable compartment, leather interior trim including steering wheel, shift knob, and console, handling suspension, P225/60R16 tires on aluminum wheels. **Bill Blass** has electronic instrument cluster, prairie mist metallic clearcoat paint, bodyside and decklid paint stripes, choice of leather, UltraSuede or cloth/leather seat trim, P215/70R15 tires on wire-spoke aluminum wheels.

Optional Equipment:

Traction-Lok axle	101	85	93
Anti-theft alarm	200	168	184
Power glass moonroof	1319	1108	1213
Automatic day/night mirror	89	75	82
Ford JBL Audio System	525	441	483
California emissions pkg.	99	83	91

Lincoln Town Car

What's New

A Gucci Designer Series arrives for 1989 to top the Cartier Designer and Signature series as the ultimate expression of Lincoln's best-selling model, the Town Car. The full-size, rear-drive luxury car returns functionally unchanged. All models are 4-door sedans with 6-passenger seating. Power is from a 150-horsepower fuel-injected 5.0-liter V-8 engine mated to a 4-speed overdrive automatic transmission. Standard power equipment includes windows, mirrors and "twin-comfort lounge" seats; cruise control and an automatic climate system are also standard. Town Car starts with the base sedan, which gets a new Frenched rear win-

> **KEY: ohv** = overhead valve; **ohc** = overhead cam; **dohc** = double overhead cam; **I** = inline cylinders; **V** = cylinders in V configuration; **flat** = horizontally opposed cylinders; **bbl.** = barrel (carburetor); **PFI** = port (multi-point) fuel injection; **TBI** = throttle-body (single-point) fuel injection; **rpm** = revolutions per minute; **OD** = overdrive transmission; **S** = standard; **O** = optional; **NA** = not available.

Prices are accurate at time of printing; subject to manufacturer's change

Lincoln

Lincoln Town Car Signature Series

dow for its standard full vinyl roof this year. The Signature Series upgrades interior appointments and supplants the base model's full vinyl top with a coach roof. The Cartier Designer Series adds platinum bodyside moldings, a new-for-'89 dual-shade paint treatment and other amenities. The new Gucci series features a full Cambria top and color-keyed bodyside accent stripes. An engine-computer management system malfunction warning light has been added to all models. A restyled but still-rear-drive Town Car is due for 1990.

For

- Luxury appointments • Passenger room
- Trunk space • Driveability • Quietness

Against

- Fuel economy • Size and weight
- Handling/roadholding • Wind noise

Summary

Town Car's direct competitor is the Cadillac Brougham, and we prefer the Lincoln to the Cadillac, though neither full-size rear-driver makes our hearts sing. Both feel clumsy in tight maneuvers and their overall handling pales in comparison to European luxury sedans. The Town Car's V-8 is more powerful and responsive than the Brougham's and seems to work better with its 4-speed overdrive automatic transmission. Both have expansive interiors and ample luggage space. But neither offers anti-lock brakes, an important safety feature that's either standard or optional on many smaller, less expensive sedans. Nearly all road and engine noise is sealed out by lavish amounts of sound deadening materials, but Town Car's tall, brick-like styling results in excessive wind noise at highway speeds compared to newer, more aerodynamic designs. We put the Town Car at the top of the short list of traditional full-size, rear-drive American luxury cars, but we think buyers spending this much should also look at some newer, more roadworthy cars. They'll get more substance for their money.

KEY: ohv = overhead valve; **ohc** = overhead cam; **dohc** = double overhead cam; **I** = inline cylinders; **V** = cylinders in V configuration; **flat** = horizontally opposed cylinders; **bbl.** = barrel (carburetor); **PFI** = port (multi-point) fuel injection; **TBI** = throttle-body (single-point) fuel injection; **rpm** = revolutions per minute; **OD** = overdrive transmission; **S** = standard; **O** = optional; **NA** = not available.

Specifications

	4-door notchback
Wheelbase, in.	117.3
Overall length, in.	219.0
Overall width, in.	78.1
Overall height, in.	55.9
Front track, in.	62.2
Rear track, in.	62.0
Turn diameter, ft.	40.0
Curb weight, lbs.	4051
Cargo vol., cu. ft.	22.4
Fuel capacity, gal.	18.0
Seating capacity	6
Front headroom, in.	39.0
Front shoulder room, in.	60.7
Front legroom, max., in.	43.5
Rear headroom, in.	38.2
Rear shoulder room, in.	60.7
Rear legroom, min., in.	42.1

Powertrain layout: longitudinal front engine/rear-wheel drive.

Engines

	ohv V-8
Size, liters/cu. in.	5.0/302
Fuel delivery	PFI
Horsepower @ rpm	150 @ 3200
Torque (lbs./ft.) @ rpm	270 @ 2000
Availability	S

EPA city/highway mpg
4-speed OD automatic	17/24

Prices

Lincoln Town Car	Retail Price	Dealer Invoice	Low Price
4-door notchback	$25205	$21395	$23300
Signature 4-door notchback	28206	23916	26061
Cartier 4-door notchback	29352	24879	27116
Gucci 4-door notchback	NA	NA	NA
Destination charge	524	524	524

Standard Equipment:

5.0-liter PFI V-8 engine, 4-speed automatic transmission, power steering and brakes, power vent and door windows, power locks, automatic climate control, tilt steering column, automatic parking brake release, tinted glass, front bumper guards, cornering lamps, coach lamps, power mirrors, variable intermittent wipers, cruise control, full vinyl roof with padded rear pillars, Twin-Comfort lounge front seats with six-way power driver's seat and dual fold-down center armrests, folding rear armrest, AM/FM ST ET cassette, Premium Sound System, passenger assist handles, electronic warning chimes, full interior courtesy lighting, rear defogger, P215/70R15 tires w/full-size spare. **Signature** adds: 3.27 axle ratio, coach roof, manual front passenger seatback recliner, pleat-pillow-style upholstery, seatback assist straps and map pockets, door and quarter trim woodtone accents. **Cartier** adds: two-tone paint, full textured-vinyl roof, platinum bodyside moldings, turbine-spoke aluminum wheels, leather-wrapped steering wheel. **Gucci** adds: full cambria roof.

Optional Equipment:

Traction-Lok axle	101	85	93
Wire wheel covers, base	341	286	314
Signature & Cartier (credit)	(137)	(115)	(115)
Lacy spoke alloy wheels, base	478	401	440
Signature & Cartier (credit)	(NC)	(NC)	(NC)
Wire spoke alloy wheels, base	873	733	803
Signature & Cartier	395	332	363
Electronic instrument panel	822	691	756

Prices are accurate at time of printing; subject to manufacturer's change

	Retail Price	Dealer Invoice	Low Price
Keyless illuminated entry	209	175	192
Anti-theft alarm	200	168	184
Power glass moonroof	1319	1108	1213
Comfort/Convenience Group	694	583	638

Power decklid pulldown, lighted visor mirrors, automatic headlamp dimmer, Autolamp on/off/delay, power seats, rear floormats.

Leather-wrapped steering wheel	115	96	106
Automatic day/night mirror	89	75	82
Automatic load leveling	202	170	186
Bodyside molding	70	59	64
Carriage roof, Signature	710	596	653
Others	1069	898	983
Valino luxury coach roof	359	302	330
Ford JBL Audio System	525	441	483
CD player	617	519	568
Leather seat trim, Signature	469	394	431
Others	531	446	489
California emissions pkg	99	83	91
Dual exhaust	83	70	76
Class III trailer tow pkg.	546	458	502

3.55 Traction-Lok axle, dual exhaust, trailer towing suspension and wiring harness, power steering and transmission oil coolers, HD U-joint and rear brakes, HD radiator, automatic load leveling, HD turn signals and flasher. For 5000-lb. capacity/750 lb. tongue load.

Mazda MPV

Mazda MPV

What's New

Called MPV, for "Multi-Purpose Vehicle," Mazda's belated entry in the minivan segment went on sale in October as a rear-drive passenger model with seats for five or seven and as a 2-seat cargo model. A 4-wheel-drive passenger version is due next spring. Though built in Japan, the MPV was designed strictly for North America, which is its only market for now. It's packaged for the same car-like comfort and driving feel that have helped make the pioneering front-drive Chrysler minivans so popular. It's somewhat shorter, higher, wider and heavier than the standard-size Caravan/Voyager. Where the 4-door MPV parts company with other minivans is in having a conventional side-opening right-rear door instead of a sliding door. Mazda says this reduces weight, increases available interior width and makes for a more rigid structure. Standard power for rear-drive MPVs is a new 121-horsepower 2.6-liter inline 4-cylinder engine with multi-point fuel injection and three valves per cylinder—two intake and one exhaust. Optional with rear drive and standard with the planned 4WD is a 150-horsepower version of the Mazda 929's 3.0-liter V-6. Both engines mate with a standard 5-speed overdrive manual transmission; a 4-speed overdrive automatic is optional. Mazda says it chose rear-drive configuration for its superior handling and towing capabilities compared to front drive, and to facilitate engineering of the 4WD model. The MPV's standard towing capacity is 2300 pounds in 4-cylinder form, 2600 pounds with the V-6. The V-6 model can be upped to a creditable 4300 pounds with an optional trailering package.

For

- Performance (V-6)   • Passenger room   • Cargo space
- Maneuverability

Against

- Performance (4-cylinder)   • Noise

Summary

Though Mazda stuck with rear-wheel drive, the MPV is the most car-like minivan since the Caravan/Voyager. Entering the cabin requires a slightly higher step-up than in the Caravan/Voyager, but there's a similarly car-like driving position and, if anything, more fore/aft seat travel. The MPV's middle seat has a reclining backrest and fore/aft movement; the recliner allows you to convert the middle and rear seats (where fitted) into a makeshift bed. Seven-passenger seating is arrayed 2/2/3, and kids can walk or adults can wriggle from front to rear (or vice versa) without leaving the vehicle. Quick-release latches facilitate removing the middle seat, but the rear seat is bolted down. The side-opening right-rear door is easier to open and close than most minivans' sliding type, and may prove less prone to rattles. Our MPV driving has so far been confined to a short time with a V-6/automatic on city streets, but a few things stand out. Snappy acceleration for one. Mazda claims 12 seconds in the 0-60 mph run with automatic, and 11 seconds for the V-6/5-speed. Add two seconds to each figure for 4-cylinder models. Maneuverability seems better than average too—a rear-drive benefit. This combines with a smaller exterior package to make parking the MPV easier than a Ford Aerostar or a Grand Caravan/Voyager. With its higher build and center of gravity, the MPV corners with more apparent body roll than the Chrysler products, but doesn't feel at all tippy unless you're going too fast. Noise suppression isn't the best. The V-6 is subdued except when pushed, but wind noise is prominent despite the low-drag styling. We think most MPV buyers will be happier with the more muscular V-6, especially if they plan on any towing. In all, Mazda's new Multi-Purpose Vehicle strikes us as pleasant, practical and thoroughly professional. It lives up to its name by combining station wagon comfort and driving ease with minivan versatility in a compact package, and thus stacks up as about the only direct alternative for those attracted to the Chrysler products.

Specifications

	4-door van
Wheelbase, in.	110.4
Overall length, in.	175.8
Overall width, in.	71.9
Overall height, in.	68.1
Front track, in.	60.0
Rear track, in.	60.6
Turn diameter, ft.	36.1
Curb weight, lbs.	3463
Cargo vol., cu. ft.	37.5
Fuel capacity, gal.	15.9[1]
Seating capacity	7

Prices are accurate at time of printing; subject to manufacturer's change

Mazda

	4-door van
Front headroom, in.	40.0
Front shoulder room, in.	57.5
Front legroom, max., in.	40.6
Rear headroom, in.	39.0
Rear shoulder room, in.	57.5
Rear legroom, min., in.	34.8

1. 19.6 with V-6

Powertrain layout: longitudinal front engine/rear-wheel drive.

Engines

	ohc I-4	ohc V-6
Size, liters/cu. in.	2.6/159	3.0/180
Fuel delivery	PFI	PFI
Horsepower @ rpm	121 @ 4600	150 @ 5000
Torque (lbs./ft.) @ rpm	149 @ 3500	165 @ 4000
Availability	S	O
EPA city/highway mpg		
5-speed OD manual	20/24	18/23
4-speed OD automatic	19/24	17/22

Prices

Mazda MPV	Retail Price	Dealer Invoice	Low Price
5-passenger, 2.6L, 5-speed	$12909	—	—
5-passenger, 2.6L, automatic	13609	—	—
7-passenger, 2.6L, 5-speed	13759	—	—
7-passenger, 2.6L, automatic	14459	—	—
7-passenger, 3.0L, 5-speed	14359	—	—
7-passenger, 3.0L, automatic	15109	—	—
Destination charge	269	269	269

Dealer invoice and low price not available at time of publication.

Standard Equipment:

2.6-liter PFI 4-cylinder or 3.0-liter PFI V-6 engine, 5-speed manual transmission, power steering and brakes, cloth reclining front bucket seats, 3-passenger middle seat (5-passenger), 2-passenger middle and 3-passenger rear seats (7-passenger), power mirrors, tachometer, coolant temperature gauge, trip odometer, tilt steering column, intermittent wipers, illuminated entry, rear wiper/washer, tinted glass, AM/FM ST ET cassette, lighted visor mirrors, floormats, rear defogger, remote fuel door release, 205/70R14 all-season tires.

Options prices not available at time of publication.

Mazda RX-7

What's New

A facelifted RX-7 is due in January, but until then Mazda dealers will be selling leftover 1988 models of its 2-seat, rotary-powered sports car. Mazda hasn't released information about the forthcoming facelift, so we can't either. The second-generation RX-7 debuted for 1986 and sales have slid the past couple of years, partly because of higher prices and, Mazda claims, because high insurance rates on 2-seaters are scaring away buyers. The lineup at the end of 1988 included base SE models in 2-seat and 2+2 configurations; performance-oriented GTU and Turbo 2-seaters; luxury 2-seat and 2+2 GXL models; and a 2-seat convertible, added to the roster last year. In addition, a 10th anniversary model based on the Turbo went on sale in May. All are powered by a 1.3-liter, twin-rotor Wankel engine. The base engine makes 146 horsepower and the turbo engine 182.

1988 Mazda RX-7 SE

For

- Performance ● Handling/roadholding
- Anti-lock brakes option

Against

- Fuel economy ● Interior room ● Ride

Summary

Tenacious handling and terrific acceleration put the RX-7 Turbo among the elite of mass-produced sports car performers. The turbocharger has minimal lag and boost comes up around 2300 rpm, triggering a smooth rush of power. Naturally aspirated RX-7s aren't slouches either; while they lack the Sunday punch of the turbo, they display the same free-revving, willing nature. With either engine, the 5-speed manual transmission makes it easier to tap into the abundant horsepower developed at higher engine speeds. Fuel economy isn't one of RX-7's strong points: We averaged 16.5 mpg in a Turbo and 18.5 mpg in a convertible. RX-7 has quick reflexes, good grip in fast turns and fine stability at high speed. Stopping ability is sure and drama-free on all models, and an anti-lock feature is optional on the higher-priced GXL and Turbo models. Now for the bad news. The ride is punishingly harsh, especially on the Turbo, and it can compromise control if bumps or ruts surprise you in the middle of a fast turn. It's so stiff on the Turbo over rocky pavement that you'll seek out a smoother path. The fat, low-profile tires roar constantly and boom over bumps, plus they have mediocre traction in the wet. Inside, taller people will likely wish for more foot room and a higher roof. The low seating also hurts entry/exit and promotes a closed-in feeling. Most of these criticisms can be leveled at other sports cars, so RX-7 isn't any worse than its rivals. In many ways it's better, so we still rate this car highly. If you're interested in a convertible, the RX-7 is one of the nicest open-air cars available.

1988 Specifications	3-door hatchback	2+2 3-door hatchback	2-door conv.
Wheelbase, in.	95.7	95.7	95.7
Overall length, in.	168.9	168.9	168.9
Overall width, in.	66.5	66.5	66.5
Overall height, in.	49.8	49.8	49.8
Front track, in.	57.1	57.1	57.1
Rear track, in.	56.7	56.7	56.7
Turn diameter, ft.	32.2	32.2	32.2
Curb weight, lbs.	2625[1]	2645	3003
Cargo vol., cu. ft.	19.5	20.9	4.1
Fuel capacity, gal.	16.6	16.6	16.6
Seating capacity	2	4	2
Front headroom, in.	37.2	37.2	36.5

Prices are accurate at time of printing; subject to manufacturer's change

	3-door hatchback	2+2 3-door hatchback	2-door conv.
Front shoulder room, in.	52.8	52.8	52.8
Front legroom, max., in.	43.7	43.7	43.7
Rear headroom, in.	—	33.0	—
Rear shoulder room, in.	—	NA	—
Rear legroom, min., in.	—	NA	—

1. 2850 lbs., Turbo

Powertrain layout: longitudinal front engine/rear wheel drive.

Engines	2-rotor Wankel	Turbo 2-rotor Wankel
Size, liters/cu. in.	1.3/80	1.3/80
Fuel delivery	PFI	PFI
Horsepower @ rpm	146 @ 6500	182 @ 6500
Torque (lbs./ft.) @ rpm	138 @ 3500	183 @ 3500
Availability	S	S[1]
EPA city/highway mpg		
5-speed OD manual	17/24	17/23
4-speed OD automatic	17/23	

1. Turbo

Prices

Mazda RX-7 (1988 Prices)	Retail Price	Dealer Invoice	Low Price
SE 3-door hatchback	$16150	$14030	$14940
SE 2+2 3-door hatchback	16650	14460	15405
GTU 3-door hatchback	18150	15569	16710
GXL 3-door hatchback	20050	16983	18367
GXL 2+2 3-door hatchback	20550	17403	18617
Turbo 3-door hatchback	22750	19251	20851
2-door convertible	21550	18243	19747
2-door convertible w/Option Pkg.	24050	20343	22047
10th Anniversary 3-door hatchback	24650	20897	22624
Destination charge	269	269	269

Standard Equipment:

SE: 1.3-liter PFI rotary engine, 5-speed manual transmission, power steering, power 4-wheel disc brakes, reclining front bucket seats, tachometer, trip odometer, coolant temperature and oil pressure gauges, voltmeter, remote liftgate and fuel door releases, AM/FM ST ET cassette, power antenna, digital clock, theft-deterrent system, 185/70HR14 SBR tires on alloy wheels. **GTU** adds: limited-slip differential, aerodynamic body addenda, sport-tuned suspension, sport seats, power mirrors. **GXL** adds to base: Auto Adjusting Suspension, air conditioning, cruise control, graphic EQ, power windows, power sunroof, rear wiper/washer, leather-wrapped steering wheel. **Turbo** adds to GTU: power sunroof, air conditioning, cruise control, power windows, graphic EQ, leather-wrapped steering wheel, 205/55VR16 tires. **Convertible** has GXL equipment plus upgraded stereo. **Convertible Option Package** adds leather trim, compact disc player, headrest speakers. **10th Anniversary** has Turbo equipment plus monochromatic exterior treatment, bronze tinted glass, headlamp washer.

Optional Equipment:

4-speed automatic transmission, SE & GXL	670	570	620
Anti-lock brakes, GXL & Turbo	1400	1190	1295
Air conditioning (std. GXL & Turbo)	859	688	774
Leather seats, GXL 2-seater	850	680	765
GXL 2+2	1075	860	968
Turbo	1000	800	900
Power sunroof, SE & GTU	595	506	551
Compact Disc player, GXL & Turbo	875	705	790
Cruise control, conv.	239	179	209
Graphic EQ, SE & GTU	149	112	131
Armrest	53	40	47
Glass breakage detector, 10th Anniversary	88	70	79

Mazda 323 GTX

What's New

Mazda has trimmed its front-drive 323 subcompact line to five models for 1989. Gone are last year's 5-door station wagon, the turbocharged GT 4-door and the base 4-door. Returning unchanged are base, SE and GTX Turbo 3-door hatchbacks, and SE and LX 4-door sedans. All have front-wheel drive except the GTX hatchback, which arrived in mid-1988 with full-time 4-wheel-drive. While there are no changes this fall, Mazda is working on a new "budget" sports car. Set for spring release as an early 1990 model, it will be a front-drive convertible built on a modified 323 chassis, like the Mercury Capri (also planned as a 1990 model). The new 2-seat Mazda will be built in Japan, not Australia, where Capri will be built. Power will likely be supplied by a derivative of the 323's 1.6-liter overhead-cam 4-cylinder. A turbo model is a strong possibility. Mazda's returning 323 lineup has two engines and three transmissions. All except the GTX use an 82-horsepower 1.6; the GTX's turbocharged 1.6 has dual overhead cams and 132 horsepower. The base 3-door hatchback comes only with a 4-speed manual; the SE 3-door and both 4-door sedans come with either a 5-speed manual or 4-speed overdrive automatic; the GTX comes only with a 5-speed manual.

For

- Fuel economy
- Cargo space
- 4WD traction (GTX)
- Performance (GTX)

Against

- Noise
- Seat comfort

Summary

There's nothing here that's unusual or trend-setting, except for the turbocharged, 4WD GTX, but there's still plenty to recommend about the 323. They're reasonably priced, well-built, economical subcompacts that are protected by a 3-year/50,000-mile "bumper-to-bumper" warranty and 5-

KEY: ohv = overhead valve; **ohc** = overhead cam; **dohc** = double overhead cam; **I** = inline cylinders; **V** = cylinders in V configuration; **flat** = horizontally opposed cylinders; **bbl.** = barrel (carburetor); **PFI** = port (multi-point) fuel injection; **TBI** = throttle-body (single-point) fuel injection; **rpm** = revolutions per minute; **OD** = overdrive transmission; **S** = standard; **O** = optional; **NA** = not available.

Prices are accurate at time of printing; subject to manufacturer's change

Mazda

year/unlimited-miles corrosion warranty. Their 82-horse-power 4-cylinder engine provides sufficient power and good fuel economy. You'll get the best performance and highest mileage with the 5-speed manual transmission, but the optional automatic transmission has an overdrive fourth gear, which helps highway fuel economy. Both the hatchback and the sedan have fairly roomy interiors, and both suffer from too much engine and road noise, plus flat, poorly padded seats. The hatchback benefits from a folding rear seat for increased cargo room, while the 4-door's deep, cavernous trunk holds a surprising amount of luggage. Suggested retail prices start at $7399 for the SE hatchback and $8299 for the SE 4-door. Air conditioning, automatic transmission, power steering and a stereo radio add about $2000 to those prices, so figure on from $8400 to almost $10,500 for a 323 with those features. Shop around, you may find a Mazda dealer who's selling 323s for less than suggested retail.

Specifications

	3-door hatchback	4-door notchback
Wheelbase, in.	94.5	94.5
Overall length, in.	161.8	169.7
Overall width, in.	64.8	64.8
Overall height, in.	54.7	54.7
Front track, in.	54.7	54.7
Rear track, in.	55.7	55.7
Turn diameter, ft.	30.8	30.8
Curb weight, lbs.	2100[1]	2175
Cargo vol., cu. ft.	10.5	14.7
Fuel capacity, gal.	12.7[2]	12.7
Seating capacity	5	5
Front headroom, in.	38.3	38.3
Front shoulder room, in.	52.8	52.8
Front legroom, max., in.	41.5	41.5
Rear headroom, in.	37.0	37.4
Rear shoulder room, in.	52.8	52.8
Rear legroom, min., in.	34.7	34.7

1. 2600 lbs., GTX   2. 13.2 gals., GTX

Powertrain layout: transverse front engine/front-wheel drive (permanent 4WD on GTX).

Engines

	ohc I-4	Turbo dohc I-4
Size, liters/cu. in.	1.6/97	1.6/97
Fuel delivery	PFI	PFI
Horsepower @ rpm	82 @ 5000	132 @ 6000
Torque (lbs./ft.) @ rpm	92 @ 2500	136 @ 3000
Availability	S	S[1]

EPA city/highway mpg

4-speed OD manual	26/30	
5-speed OD manual	28/33	21/24
4-speed OD automatic	24/30	

1. GTX

Prices

Mazda 323	Retail Price	Dealer Invoice	Low Price
3-door hatchback	$6299	$5865	$6150
SE 3-door hatchback	7399	6892	7146
SE 4-door notchback	8299	7554	7927
LX 4-door notchback	9499	8349	8924
GTX 3-door hatchback	12999	11264	12132
Destination charge	269	269	269

Standard Equipment:

1.6-liter PFI 4-cylinder engine, 4-speed manual transmission, power brakes, vinyl reclining front bucket seats, one-piece folding rear seatback, trip odometer, coolant temperature gauge, intermittent wipers, cargo cover, rear defogger, tinted glass, map lights, left remote mirror, bodyside moldings, 155SR13 tires. **SE** adds: 5-speed manual transmission, cloth upholstery, 50/50 folding rear seatbacks, remote fuel door and liftgate/decklid releases, day/night mirror, center console. **LX** adds: power steering, folding rear armrest, illuminated entry, power mirrors, tilt steering column, tachometer, variable intermittent wipers. **GTX** adds: turbocharged DOHC 16-valve engine, permanent 4-wheel drive, 4-wheel disc brakes, sport seats, 185/60R14 tires on alloy wheels.

Optional Equipment:

	Retail Price	Dealer Invoice	Low Price
4-speed automatic transmission (NA GTX)	700	630	665
Power steering, base & SE	250	213	232
Air conditioning	760	616	688
Manual sunroof, LX & GTX	350	298	324
Alloy wheels, LX	400	320	360
Cruise control, 4-doors	200	160	180
Power windows & locks, LX & GTX	300	240	270
AM/FM ST ET cassette, 4-doors & GTX	450	342	396
Floormats	55	39	47
Armrest	59	47	53
Rear spoiler, SE hatchback	199	159	179

Mazda 626/MX-6

Mazda MX-6 LX

What's New

Slow sales of the 4-wheel-steering (4WS) 626 4-door has resulted in the 4WS being moved to the sporty MX-6 2-door instead, the coupe being seen as more appropriate for such a specialized feature. The move also gives Mazda a more direct competitor to Honda's 4WS Prelude. The 626 Turbo 4WS 4-door was the most expensive model in last year's 626/MX-6 lineup, ending 1988 with a base price of nearly $18,000. Besides the new home for the 4-wheel steering feature, the major change for these front-drive compacts is that the price of the optional anti-lock brakes (ABS) has been cut from $1400 to $1000. ABS is available only on turbocharged 626 and MX-6 models. The 626 again comes in 4-door and 5-door Touring Sedan styling, but the 145-horsepower turbocharged 2.2-liter engine is available only on the Touring Sedan this year. The revised 626 lineup looks like this: DX and LX 4-door sedans, and LX and Turbo Touring Sedans. The DX and LX use a 110-horsepower 2.2-liter four. The MX-6 coupe shares the 626's platform and engines, but has different styling and is built in Japan and at Mazda's Flat Rock, Mich., assembly plant. All 626s are imported from Japan. Ford's similar Probe hatchback coupe also is built at Flat Rock. The DX and LX coupes use the 110-horsepower engine, while the MX-6 GT and GT 4WS use the 145-horsepower turbo engine. A 5-speed manual transmission is standard on all models and a 4-speed automatic is optional for all except the MX-6 GT 4WS.

Prices are accurate at time of printing; subject to manufacturer's change

CONSUMER GUIDE®

For
- ●Performance (turbo) ●Anti-lock brakes option
- ●Handling ●Passenger room

Against
- ●Driveability (automatic transmission)
- ●Wind and road noise

Summary

The best news about the 626 and MX-6 is that anti-lock brakes, a safety feature we recommend, is less expensive this year. You have to buy one of the expensive turbo-charged models to get ABS, though these are still among the lowest-priced cars that offer ABS. We tested a 1988 626 Turbo 4WS and found the rear-wheel steering entertaining, but not very useful. You feel it most in parking and low-speed turns, when it shortens the turning diameter. Otherwise, we see little value to this expensive technical innovation. Our advice is to skip 4WS and spend the money on anti-lock brakes instead. We're most familiar with the 626 LX 4-door, since one is in our long-term test fleet. This has proven to be a comfortable, capable compact sedan with ample room for four people, a nicely furnished interior and pretty good handling for a family car. However, the automatic transmission diminishes the driving pleasure because it balks at downshifting to furnish a little more power. That forces us to floor the throttle, which results in a tardy burst of power. The 626 rides more firmly than most family cars, which we don't mind, except there's too little wheel travel and absorbency at the rear, so the car bangs and clunks over sharp bumps and railroad tracks. Despite its aerodynamic design, the 626 suffers a lot of wind noise around the side windows on the highway and too much road noise for peaceful cruising. On the plus side, we have had decent fuel economy and trouble-free service from this car.

Specifications

	MX-6 2-door notchback	626 4-door notchback	626 5-door hatchback
Wheelbase, in.	99.0	101.4	101.4
Overall length, in.	177.0	179.3	179.3
Overall width, in.	66.5	66.5	66.5
Overall height, in.	53.5	55.5	54.1
Front track, in.	57.3	57.3	57.3
Rear track, in.	57.7	57.7	57.7
Turn diameter, ft.	35.3[1]	36.0	36.0
Curb weight, lbs.	2535[2]	2590	2680
Cargo vol., cu. ft.	15.4	15.9	22.0
Fuel capacity, gal.	15.9	15.9	15.9
Seating capacity	4	5	5
Front headroom, in.	38.4	39.0	38.7
Front shoulder room, in.	54.9	54.9	54.9
Front legroom, max., in.	43.6	43.7	43.6
Rear headroom, in.	37.8	37.8	37.2
Rear shoulder room, in.	53.3	54.9	54.9
Rear legroom, min., in.	31.8	36.6	32.9

1. 31.5 with 4WS 2. 2705 lbs., GT; 2888 lbs., GT 4WS

Powertrain layout: Transverse front engine/front-wheel drive.

Engines

	ohc I-4	Turbo ohc I-4
Size, liters/cu. in.	2.2/133	2.2/133
Fuel delivery	PFI	PFI
Horsepower @ rpm	110 @ 4700	145 @ 4300

	ohc I-4	Turbo ohc I-4
Torque (lbs./ft.) @ rpm	130 @ 3000	190 @ 3500
Availability	S	S[1]
EPA city/highway mpg		
5-speed OD manual	24/31	21/28
4-speed OD automatic	22/28	19/25

1. 626 Turbo, MX-6 GT

Prices

Mazda 626/MX-6

	Retail Price	Dealer Invoice	Low Price
626 DX 4-door notchback	$11299	$9869	$10584
626 LX 4-door notchback	13199	11380	12290
626 LX 5-door Touring Sedan	13399	11551	12475
626 Turbo 5-door Touring Sedan	15049	12962	14006
MX-6 DX 2-door notchback	11399	9841	10620
MX-6 LX 2-door notchback	13299	11333	12316
MX-6 GT Turbo 2-door notchback	15499	13192	14346
Destination charge	269	269	269

Standard Equipment:

DX: 2.2-liter PFI 4-cylinder engine, 5-speed manual transmission, power steering and brakes, cloth reclining front bucket seats, tachometer, coolant temperature gauge, trip odometer, tilt steering column, tinted glass, intermittent wipers, digital clock, center console, rear defogger, remote fuel door and trunk/liftgate releases, remote mirrors, 185/70R14 SBR tires. **LX** adds: power windows and locks, cruise control, AM/FM ST ET cassette, power antenna, map lights, variable intermittent wipers; Touring Sedan has removable shelf panel and rear wiper/washer. **Turbo** models add: 2.2-liter intercooled turbocharged engine, 4-wheel disc brakes, Auto Adjusting Suspension (exc. MX-6 4WS), graphic EQ (MX-6), 195/60HR15 tires on alloy wheels.

Optional Equipment:

4-speed automatic transmission	720	634	677
Anti-lock brakes, Turbos	1000	850	925
Air conditioning	800	642	721
Sunroof (NA DX, MX-6 GT)	555	444	500
Alloy wheels, exc. Turbo	400	320	360
Cruise control, DX	220	176	198
AM/FM ST ET cassette, DX	450	342	396
Graphic EQ	149	112	131
Floormats	58	41	50
Armrest w/lid (NA DX)	57	44	51
Rear spoiler, MX-6 LX	375	300	338

Mazda 929

What's New

A 4-speed overdrive automatic is the only transmission available this year on the 929 luxury sedan, which gains a power moonroof and a power driver's seat as standard equipment. A 5-speed manual transmission was a no-cost option last year, but has been dropped due to lack of interest. The moonroof and power driver's seat were options last year. Lackluster U.S. response to the rear-drive 929, introduced a year ago, is behind feverish efforts at giving

> **KEY: ohv** = overhead valve; **ohc** = overhead cam; **dohc** = double overhead cam; **I** = inline cylinders; **V** = cylinders in V configuration; **flat** = horizontally opposed cylinders; **bbl.** = barrel (carburetor); **PFI** = port (multi-point) fuel injection; **TBI** = throttle-body (single-point) fuel injection; **rpm** = revolutions per minute; **OD** = overdrive transmission; **S** = standard; **O** = optional; **NA** = not available.

Prices are accurate at time of printing; subject to manufacturer's change

Mazda 929

the staid luxury sedan some needed styling zip. A facelift has been rumored for as early as mid-1989, but we've been told not to expect it before model year 1990. Its 158-horsepower 3.0-liter V-6 engine is unchanged. Anti-lock brakes (ABS) are optional: $1000 as an individual item and $1355 with the electronic Auto Adjusting Suspension. Last year, ABS alone was $1400 and the auto suspension/ABS package was $1800.

For

- Performance • Anti-lock brakes option
- Ride • Room/comfort

Against

- Driver seating • Fuel economy

Summary

Generic, even dull styling may indeed be the cause of lukewarm 929 sales, but we think there's a pretty good luxury sedan here at a reasonable price. There's nothing outstanding to rave about, yet the 929 does its job done competently and quietly. In fact, it feels about as bland as it looks. That's not all bad, especially if you want your driving to be peaceful and undemanding. The 929 has a polished, well-balanced powertrain that delivers sufficient power to keep pace with traffic without strain. The electronically controlled automatic transmission changes gears smoothly and doesn't hunt in and out of overdrive on the expressway. The suspension copes well with sharp bumps and handily smothers rough patches, while the all-season tires provide good control and grip in turns. Strong braking power can be augmented with the anti-lock option, which is more reasonably priced this year. A spacious interior provides ample room for up to five adults, though with the optional leather upholstery the wide, flat driver's seat has no lateral support, so positioning and comfort are a compromise even with the standard power adjustments. A functional dashboard has a refreshing lack of gimmickry and visibility is good to all directions. In all, the 929 is a substantial, comfortable and competent car with a decidedly conservative nature. With anti-lock brakes and even a compact disc player added, the price is still on the right side of $25,000, or some $5000 less than a top-line Acura Legend LS.

Specifications

	4-door notchback
Wheelbase, in.	106.7
Overall length, in.	193.1
Overall width, in.	66.9
Overall height, in.	54.5
Front track, in.	56.9
Rear track, in.	57.5
Turn diameter, ft.	35.4
Curb weight, lbs.	3373
Cargo vol., cu. ft.	15.1
Fuel capacity, gal.	18.5
Seating capacity	5
Front headroom, in.	37.8
Front shoulder room, in.	55.2
Front legroom, max., in.	43.3
Rear headroom, in.	37.4
Rear shoulder room, in.	55.2
Rear legroom, min., in.	37.0

Powertrain layout: longitudinal front engine/rear-wheel drive.

Engines

	ohc V-6
Size, liters/cu. in.	3.0/180
Fuel delivery	PFI
Horsepower @ rpm	158 @ 4500
Torque (lbs./ft.) @ rpm	170 @ 4000
Availability	S

EPA city/highway mpg

4-speed OD automatic	19/23

Prices

Mazda 929	Retail Price	Dealer Invoice	Low Price
4-door notchback	$21920	$18329	$20125
Destination charge	269	269	269

Standard Equipment:

3.0-liter PFI V-6 engine, 4-speed automatic transmission, power steering, power 4-wheel disc brakes, automatic air conditioning, cloth reclining front bucket seats, power driver's seat, power windows, locks and mirrors, power moonroof, rear armrest, tachometer, coolant temperature gauge, voltmeter, trip odometer, intermittent wipers, AM/FM ST ET cassette with EQ, P195/65R15 tires on alloy wheels.

Optional Equipment:

Anti-lock brakes	1000	850	925
Incl. Auto Adjusting Suspension	1355	1152	1254
Leather power seats	880	730	805
Digital instruments	530	451	491
Cold Pkg.	250	208	229
All-season tires, HD battery, semi-concealed wipers, heated driver's seat, larger washer fluid reservoir.			
Floormats	79	55	67
Armrest w/lid	79	55	67

Mercedes-Benz S-Class

What's New

A passenger-side air bag is a new option for all S-Class cars except the 2-seat 560SL. A driver's-side air bag has been standard equipment for the S-Class since 1986. The passenger-side air bag is mounted in the space usually occupied by the glovebox. It has nearly three times the volume of the driver's-side air bag because of the additional space it must fill; it has two inflator units instead of

Mercedes-Benz 300SEL

	560SL 2-door conv.	560SEC 2-door hatchback	300SE 4-door notchback	SEL 4-door notchback
Front track, in.	57.7	61.5	61.5	61.5
Rear track, in.	57.7	60.4	60.4	60.4
Turn diameter, ft.	34.4	38.1	30.0	40.6
Curb weight, lbs.	3705	3915	3730	3770[1]
Cargo vol., cu. ft.	0.0	11.9	15.2	15.2
Fuel capacity, gal.	25.5	27.1	23.6	27.1
Seating capacity	2	4	5	5
Front headroom, in.	36.5	37.8	38.5	38.6
Front shoulder room, in.	51.6	57.2	56.2	56.2
Front legroom, max., in.	42.4	41.9	41.9	41.9
Rear headroom, in.	—	36.7	36.5	37.2
Rear shoulder room, in.	—	54.2	55.7	55.7
Rear legroom, min., in.	—	30.6	34.4	39.6

1. 300SEL; 3885 lbs.; 420SEL; 4080 lbs.; 560SEL.

Powertrain layout: longitudinal front engine/rear-wheel drive.

Engines

	ohc I-6	ohc V-8	ohc V-8
Size, liters/cu. in.	3.0/181	4.2/256	5.6/338
Fuel delivery	PFI	PFI	PFI
Horsepower @ rpm	177 @ 5700	201 @ 5200	238 @[1] 4800
Torque (lbs./ft.) @ rpm	188 @ 4400	228 @ 3600	287 @[1] 3500
Availability	S	S	S

EPA city/highway mpg

4-speed automatic	17/19	15/18	14/16

1. 227 horsepower and 279 lbs./ft. torque, 560SL.

one so it can fully inflate in the same amount of time. Models with the passenger-side air bag get a lockable storage box in the center console in place of the glovebox. With the addition of the 300SE model last spring, there are six S-Class models. The 300SE sedan rides a 115.6-inch wheelbase and has a 177-horsepower 3.0-liter inline 6-cylinder engine. The 300SEL sedan uses the same engine, but rides a 121.1-inch wheelbase. The 420SEL and 560SEL sedans have the long wheelbase, and 4.2-liter and 5.6-liter V-8 engines respectively. The 5.6-liter V-8 also powers the 560SEC 4-seat coupe and 560SL roadster. In fashion news, S-Class models gain a new soft leather upholstery as standard equipment; "crystal silver" replaces "astral silver" as an exterior color choice, "garnet red" has been added as a new choice, and "champagne metallic" has been dropped; and the number of colors for bumpers and protective side moldings has increased from four to 12. All models also have revised first and second gear ratios (slightly shorter in first, slightly taller in second). The 300SE gains the 10-speaker, 100-watt stereo system found in the other S-Class sedans and the 560SEC.

For

- Anti-lock brakes   • Air bags   • Performance (V-8s)
- Ride   • Handling   • Resale value

Against

- Fuel economy   • Price

Summary

All the S-Class cars are built with the finest materials and excellent craftsmanship, and all offer commendable ride and handling, anti-lock brakes and a driver's-side air bag as standard. The choices among these cars are how many seats you want, how much performance you need and how much you're willing to pay. The 6-cylinder 300SE and 300SEL are basically "price leaders" for this line, offering all the features of the V-8-powered cars but with less horsepower at reduced costs. Our only complaint with the 6-cylinder models is that their acceleration isn't always sparkling, a penalty of too much weight for the available power. They still accelerate briskly and cruise at high speeds with minimal effort, but you can get similar performance for a lot less money. What you can't get for a lot less money is Mercedes' impeccable workmanship and exceptional engineering: These cars just work better than most others. The V-8 models are the most enjoyable.

Specifications	560SL 2-door conv.	560SEC 2-door hatchback	300SE 4-door notchback	SEL 4-door notchback
Wheelbase, in.	96.7	112.2	115.6	121.1
Overall length, in.	180.3	199.2	202.6	208.1
Overall width, in.	70.5	72.0	71.7	71.7
Overall height, in.	51.1	55.0	56.6	56.7

Prices

Mercedes-Benz S-Class	Retail Price	Dealer Invoice	Low Price
300SE 4-door notchback	$51400	—	—
300SEL 4-door notchback	55100	—	—
420SEL 4-door notchback	61210	—	—
560SEL 4-door notchback	72280	—	—
560SEC 2-door notchback	79840	—	—
560SL 2-door convertible	64230	—	—
Destination charge	250	250	250
Gas Guzzler Tax, 300SE, SEL	650	650	650
420SEL	1050	1050	1050
560SL	1300	1300	1300
560SEL, SEC	1500	1500	1500

Dealer invoice and low price not available at time of publication.

Standard Equipment:

300SE, 300SEL, 420SEL: 3.0-liter PFI 6-cylinder or 4.2-liter PFI V-8 engine, 4-speed automatic transmission, power steering, anti-lock braking system, 4-wheel disc brakes, Supplemental Restraint System, outboard rear lap/shoulder belts, anti-theft alarm, central locking, power windows, air conditioning, automatic climate control, AM/FM ST ET cassette, leather power front seats with 2-position memory, power telescopic steering column, leather-wrapped steering wheel and shift handle, rear defogger, cruise control, headlight wipers and washers, heated power mirrors, outside temperature indicator, tachometer, coolant temperature and oil pressure gauges, lighted visor mirrors, 205/65VR15 SBR tires on alloy wheels. **560SEL and 560SEC** add: 5.6-liter PFI V-8, automatic rear level control, limited-slip differential. **560SL** has removable hardtop and folding convertible top.

OPTIONS prices not available at time of publication.

KEY: ohv = overhead valve; **ohc** = overhead cam; **dohc** = double overhead cam; **I** = inline cylinders; **V** = cylinders in V configuration; **flat** = horizontally opposed cylinders; **bbl.** = barrel (carburetor); **PFI** = port (multi-point) fuel injection; **TBI** = throttle-body (single-point) fuel injection; **rpm** = revolutions per minute; **OD** = overdrive transmission; **S** = standard; **O** = optional; **NA** = not available.

Prices are accurate at time of printing; subject to manufacturer's change

Mercedes-Benz 190

Mercedes-Benz 190E 2.6

What's New

The 2.3-liter 4-cylinder gas engine has been dropped from the 190 line, leaving a 158-horsepower 2.6-liter inline 6-cylinder as the only gas engine for Mercedes' compact sedan, which also gets a mild facelift this year. Returning from last year's lineup is the 190D 2.5, Mercedes' only diesel-powered North American model. The 190D 2.5, which isn't available in California, has a new glow-plug system and a revised exhaust gas recirculation system for lower emissions. Its 2.5-liter diesel 5-cylinder is rated at 90 horsepower, three less than last year. Anti-lock brakes are now standard on the 190D 2.5, making ABS standard on all of Mercedes' North American models. Both 190 models gain wide bodyside moldings, a deeper front air dam and restyled bumpers. Bumpers and lower side moldings now come in a choice of 12 colors instead of four. Inside, redesigned seats have softer upholstery and more side support, plus Mercedes claims nearly an inch more rear knee room. Height adjustments on the center pillars for the front seat shoulder belts have been added. Mercedes' Supplemental Restraint System, which includes a driver's-side air bag, is standard.

For

- Anti-lock brakes • Air bag
- Performance (except diesel) • Ride • Handling

Against

- Price • Rear seat room • Trunk space
- Performance (diesel)

Summary

As the price of Mercedes' smallest sedan has crept up, sales have gone down, showing that even the high end of the car market is price sensitive. Mindful of that, Mercedes has promised that the price of the 190E 2.6 will be "somewhat lower" for 1989 when it is announced. Base price of the 190E 2.6 was $33,500 in the East and Midwest and $33,940 in the West at the end of the 1988 model year. BMW has reduced prices on its 3-Series cars for 1989, so one of Mercedes' main rivals also is feeling the pinch. The 190 sedans have the same high-level quality and engineering as the larger Mercedes models, but much less sheetmetal and interior room. The rear seat is downright cramped (though this year's redesigned seats will add a little more space) and only two adults can fit in back without bumping

elbows and knees; the trunk provides modest luggage space. High price and cramped quarters aside, the 190E 2.6 offers satisfying acceleration, capable handling, a taut and stable ride, excellent stopping ability from its anti-lock brakes, and the added protection of a standard driver's-side air bag. Throw in a long list of comfort and convenience features, and this is a lot of car in a compact package. Too bad there isn't more interior room for a sedan that's this pricey.

Specifications

	4-door notchback
Wheelbase, in.	104.9
Overall length, in.	175.1
Overall width, in.	66.5
Overall height, in.	54.7
Front track, in.	56.6
Rear track, in.	55.8
Turn diameter, ft.	34.8
Curb weight, lbs.	2955
Cargo vol., cu. ft.	11.7
Fuel capacity, gal.	16.5
Seating capacity	5
Front headroom, in.	38.0
Front shoulder room, in.	53.5
Front legroom, max., in.	41.9
Rear headroom, in.	36.7
Rear shoulder room, in.	53.2
Rear legroom, min., in.	30.9

Powertrain layout: longitudinal front engine/rear-wheel drive.

Engines	ohc I-6	Diesel ohc I-5
Size, liters/cu. in.	2.6/159	2.5/152
Fuel delivery	PFI	PFI
Horsepower @ rpm	158 @ 5800	90 @ 4600
Torque (lbs./ft.) @ rpm	162 @ 4600	117 @ 2800
Availability	S	S
EPA city/highway mpg		
5-speed OD manual		19/27
4-speed automatic	20/23	28/33

Prices

Mercedes-Benz 190	Retail Price	Dealer Invoice	Low Price
190D 2.5 4-door notchback, automatic	$30980	—	—
190E 2.6 4-door notchback, 5-speed	31590	—	—
190E 2.6 4-door notchback, automatic	32500	—	—
Destination charge	250	250	250

Dealer invoice and low price not available at time of publication.

Standard Equipment:

2.6-liter PFI 6-cylinder engine (2.5-liter diesel on 190D 2.5), 5-speed manual or 4-speed automatic transmission, power steering, anti-lock braking system, 4-wheel disc brakes, Supplemental Restraint System, outboard rear lap/shoulder belts, automatic climate control, power windows and locks, cruise control, intermittent wipers, vinyl reclining front bucket seats, heated power mirrors, AM/FM ST ET cassette, tachometer, coolant temperature and oil pressure gauges, trip odometer, lighted visor mirrors, wide bodyside moldings, 185/65R15 (V-rated on 190E 2.6) SBR tires on alloy wheels.

OPTIONS prices not available at time of publication.

Prices are accurate at time of printing; subject to manufacturer's change

Mercedes-Benz 260/300

Mercedes-Benz 300TE

What's New

A passenger-side air bag is a new option as the biggest change for Mercedes' mid-size sedans. As with the S-Class cars, the passenger-side air bag is mounted in the dashboard where the glove box is usually located. All Mercedes sold in North America come with a driver's-side air bag standard. This year's lineup duplicates the 1988 roster: 260E and 300E sedans, 300CE coupe and 300TE station wagon. A 158-horsepower 2.6-liter inline 6-cylinder is standard on the 260E. The others use a 3.0-liter six making 177 horsepower. The 5-speed manual transmission standard last year on the sedans has been dropped, so all models in this series have a 4-speed automatic transmission. The compact 190E 2.6 is the only Mercedes model with manual transmission for 1989. All models also get the heated windshield washer reservoir lines and spray nozzles that debuted last year in the S-Class. The 300CE, new for '88, gains eight more color choices for the bumpers and protective side moldings, for a total of 12.

For

● Anti-lock brakes   ● Air bags
● Performance   ● Handling   ● Ride

Against

● Price   ● Fuel economy

Summary

This is Mercedes' best-selling line, accounting for nearly 40 percent of its U.S. sales. Far roomier than the compact 190 yet not as costly as the S-Class cars, the 260/300 line is the most practical. All models have ample room for five people except the 300CE, which seats four. Acceleration is quite brisk with the 3.0-liter six, and nearly as brisk with the 2.6. The automatic transmission doesn't always react promptly when you need a quick burst of power for passing, but driveability is otherwise flawless with either engine. All models in this line also have a taut, composed ride, direct and responsive steering, and athletic handling ability that devours most curves with aplomb. The firm suspension, however, transmits a lot of road feel to the driver and the firm power steering can't be turned with one finger. The interior design nearly duplicates what you find in the S-Class, so anyone familiar with Mercedes' functional control layout will feel right at home. With anti-lock brakes, a driver's-side air bag and a power sunroof as standard, there are only a handful of comfort and convenience options to pad the prices. Most shoppers will find the base prices out of reach anyway. If price isn't a problem for you, these cars belong at the top of your shopping list. If you have trouble fitting $40,000 or so into your budget, we suggest you lower your sights and look at the Acura Legend instead.

Specifications

	300CE 2-door notchback	260E, 300E 4-door notchback	300TE 5-door wagon
Wheelbase, in.	106.9	110.2	110.2
Overall length, in.	183.9	187.2	188.2
Overall width, in.	68.5	68.5	68.5
Overall height, in.	55.5	56.9	59.8
Front track, in.	58.9	58.9	58.9
Rear track, in.	58.6	58.6	58.6
Turn diameter, ft.	35.8	36.7	36.7
Curb weight, lbs.	3310	3210	3530
Cargo vol., cu. ft.	14.4	14.6	76.8
Fuel capacity, gal.	20.9	20.9	21.4
Seating capacity	4	5	5
Front headroom, in.	36.0	36.9	37.4
Front shoulder room, in.	55.7	55.9	55.9
Front legroom, max., in.	41.9	41.7	41.7
Rear headroom, in.	35.5	36.9	36.8
Rear shoulder room, in.	50.2	55.7	55.6
Rear legroom, min., in.	29.6	33.5	33.9

Powertrain layout: longitudinal front engine/rear-wheel drive.

Engines

	ohc I-6	ohc I-6
Size, liters/cu. in.	2.6/159	3.0/181
Fuel delivery	PFI	PFI
Horsepower @ rpm	158 @ 5800	177 @ 5700
Torque (lbs./ft.) @ rpm	162 @ 4600	188 @ 4400
Availability	S	S

EPA city/highway mpg

4-speed automatic	20/24	18/22

Prices

Mercedes-Benz 260/300	Retail Price	Dealer Invoice	Low Price
260E 4-door notchback	$39200	—	—
300E 4-door notchback	44850	—	—
300CE 2-door notchback	53880	—	—
300TE 5-door wagon	48210	—	—
Destination charge	250	—	—
Gas Guzzler Tax, 300TE	650	650	650

Dealer invoice and low price not available at time of publication.

Standard Equipment:

260E: 2.6-liter PFI 6-cylinder engine, 4-speed automatic transmission, power steering, anti-lock braking system, 4-wheel disc brakes, Supplemental Restraint System, cruise control, rear headrests, outboard rear lap/shoulder belts, heated power mirrors, automatic climate control, power windows and locks, rear defogger, tachometer, coolant temperature and oil pressure

> **KEY: ohv** = overhead valve; **ohc** = overhead cam; **dohc** = double overhead cam; **I** = inline cylinders; **V** = cylinders in V configuration; **flat** = horizontally opposed cylinders; **bbl.** = barrel (carburetor); **PFI** = port (multi-point) fuel injection; **TBI** = throttle-body (single-point) fuel injection; **rpm** = revolutions per minute; **OD** = overdrive transmission; **S** = standard; **O** = optional; **NA** = not available.

Prices are accurate at time of printing; subject to manufacturer's change

Mercedes-Benz 300E

gauges, trip odometer, intermittent wipers, heated windshield washer fluid reservoir and nozzles, 195/65VR15 SBR tires; power sunroof is available at no charge. **300E** adds: 3.0-liter engine, headlamp wipers and washers, anti-theft alarm system, power telescopic steering column, power front seat, outside temperature indicator (300CE has leather upholstery; velour is available at no charge). **300TE wagon** adds: automatic level control, roof rack, rear wiper/washer.

OPTIONS prices not available at time of publication.

Mercury Cougar

Mercury Cougar XR7

What's New

Cougar shares with the Ford Thunderbird an all-new for 1989 mid-size coupe design. Both cars, which will go on sale Dec. 26, are bigger on the inside and smaller on the outside—but about 400 pounds heavier—than the 6-year-old design they replace. They gain an all-independent suspension and retain their rear-drive layout. Cougar's formal rear roofline continues to set it apart from the fastback Thunderbird. Compared to its predecessor, Cougar is 3.4 inches shorter, but its wheelbase is 8.8 inches longer. Every cabin measurement is increased, including seven additional inches of rear-seat hip room and four more inches of rear shoulder room. Two Cougar models are offered. The base LS retains a 3.8-liter V-6 and a 4-speed overdrive automatic transmission as its only powertrain. Ford says new hydraulic engine mounts allowed it to lighten the 3.8 by removing its vibration-reducing internal balance shaft. No longer offered is the 5.0-liter V-8 that was standard last year on the XR7 and optional on the LS. Now the XR7 has a supercharged and intercooled version of the 3.8 V-6; a 5-speed manual transmission is standard and the 4-speed

automatic is optional. Visually, the XR7 is set apart from the LS by alloy wheels and monochromatic coloring. XR7s come standard with 4-wheel disc brakes and an anti-lock system—a setup that's optional on the LS in place of its disc/drum brakes. Wheels on base Cougars increase from 14- to 15-inch diameter for '89; XR7s retain last year's 16-inch performance tires. Standard on the supercharged car is adjustable shock absorber damping that lets the driver choose a soft or a firm ride. Speed-sensitive variable-assist power steering, air conditioning and power windows are standard on all Cougars. LS models' digital instrumentation includes a tachometer; XR7s get analog gauges that add a supercharger boost gauge. XR7 has power front seats with adjustable backrest wings and inflatable lumbar bolsters. All models get automatic front shoulder belts, with manual-locking lap belts, and 3-point lap-shoulder belts for outboard rear-seat passengers. XR7s also get split, folding rear seatbacks that open to the trunk for added cargo capacity. Prices won't be announced until shortly before Cougar goes on sale.

For

● Performance (XR7)   ● Handling/roadholding
● Anti-lock brakes

Against

● Performance (LS)

Summary

Cougar's squared-off roofline gives its rear seat passengers a hair more head room and nearly one inch more rear-seat leg room than the Thunderbird. And Cougar is slightly less aerodynamic than its Ford sibling. Otherwise, the two are functionally identical. Thus, the driving impressions we garnered from short test-track rides in Thunderbirds hold for Cougar as well. Our initial impression is of an outstanding personal-luxury coupe design. It looks distinctive and has a spacious interior that was missing in its predecessors. Road manners seem above average, with a laudable combination of a comfortable ride and capable handling. At Ford's Michigan proving grounds, the supercharged engine seemed to lack much power below about 3000 rpm, but delivered a solid kick and superb response above that. The automatic shifts well, but compared to the Mazda-built 5-speed manual, doesn't allow the degree of engine-speed control that best exploits the supercharged 3.8. With its higher horsepower, wider tires and sportier appointments, the XR7 had much better test-track manners than the LS, though we may prefer the base car's ride and handling on public roads, where bumps and potholes abound. On the

Mercury Cougar LS

Prices are accurate at time of printing; subject to manufacturer's change

downside, the anti-lock brakes on the pre-production proto-types we drove were hard to modulate and grabbed quite late. Inside, the some dashboard buttons looked and felt too inexpensive for this class. Otherwise, we were highly impressed. We look forward to a full test drive.

Specifications

	2-door notchback
Wheelbase, in.	113.0
Overall length, in.	198.7
Overall width, in.	72.7
Overall height, in.	52.7
Front track, in.	61.4
Rear track, in.	60.2
Turn diameter, ft.	35.6
Curb weight, lbs.	3553
Cargo vol., cu. ft.	14.7
Fuel capacity, gal.	19.0
Seating capacity	5
Front headroom, in.	38.1
Front shoulder room, in.	59.1
Front legroom, max., in.	42.5
Rear headroom, in.	37.6
Rear shoulder room, in.	59.1
Rear legroom, min., in.	36.7

Powertrain layout: longitudinal front engine/rear-wheel drive.

Engines	ohv V-6	Supercharged ohv V-6
Size, liters/cu. in.	3.8/232	3.8/232
Fuel delivery	PFI	PFI
Horsepower @ rpm	140 @ 3800	210 @ 4000
Torque (lbs./ft.) @ rpm	215 @ 2400	315 @ 2600
Availability	S	S[1]

EPA city/highway mpg		
5-speed OD manual		NA
4-speed OD automatic	NA	NA

1. XR7

Equipment Summary

Mercury Cougar
(prices not available at time of publication):

Standard Equipment:

LS: 3.8-liter PFI V-6, 4-speed automatic transmission, power steering and brakes, air conditioning, cloth reclining front bucket seats, tinted glass, intermittent wipers, electronic instruments (tachometer, coolant temperature, oil pressure, voltmeter, trip computer, service interval reminder), power windows and mirrors, visor mirrors, AM/FM ST ET, motorized front shoulder belts with manual lap belts, outboard rear lap/shoulder belts, 205/70R15 tires. **XR7** adds: supercharged engine with dual exhaust, 5-speed manual transmission, 4-wheel disc brakes, anti-lock braking system, handling suspension, Traction-Lok axle, sport seats with power lumbar and side bolsters, analog instruments (tachometer, coolant temperature, oil pressure, boost/vacuum), maintenance monitor, 225/60VR16 tires on alloy wheels.

Mercury Grand Marquis

What's New

Mercury celebrates its 50th Anniversary with a special edition Grand Marquis for '89, but otherwise makes few changes to its full-size, rear-drive 4-door sedans and 5-door wagons. The Grand Marquis is a slightly plusher ver-

Mercury Grand Marquis 4-door

sion of the Ford LTD Crown Victoria. It uses the same 5.0-liter fuel-injected V-8 and 4-speed overdrive automatic transmission as its sole powertrain. Grand Marquis gained new front and rear styling and lost the 2-door body style in 1988. The carried-over sedan seats six; the wagon has 8-passenger capacity with the optional dual rear-facing seats. Both are available in base GS and upscale LS trim; the 50th Anniversary Edition will come as a sedan only when it arrives at mid-year as a limited-volume entry with unique bodyside and hood stripes, lacy spoke aluminum wheels with gold center ornaments, a premium sound system, luxury cloth and leather trim, 6-way power front seats, and a leather-wrapped 4-spoke steering wheel. It deletes the drip moldings standard on other models. For '89, clear-coat metallic paint is a new option, cast aluminum wheels are a required option on all LS models and bumper guards, previously standard on sedans, are now optional.

For
- Performance
- Driveability
- Passenger room
- Cargo space
- Quietness
- Trailer towing ability

Against
- Fuel economy
- Size and weight
- Maneuverability

Summary

Among the dwindling number of full-size, rear drive sedans and wagons, the Grand Marquis and LTD Crown Victoria are our favorites, beating GM's models mainly through better engine performance and driveability. Ford's V-8 has slightly more torque than GM's, plus Ford's overdrive automatic transmission is better behaved. The transmission stays out of overdrive until you reach 40-45 mph, which reduces the amount of gear shifting and improves engine response in urban traffic. In most other areas, the GM cars are comparable to the Grand Marquis and Crown Victoria, so you get true 6-passenger seating, ample cargo room and the ability to tow hefty trailers (up to 5000 pounds on a properly equipped Grand Marquis). What you don't get are good fuel economy, maneuverability and sporty handling. These heavy, full-size cars are cumbersome to maneuver and feel loose and floaty on the open road. In their favor,

> **KEY: ohv** = overhead valve; **ohc** = overhead cam; **dohc** = double overhead cam; **I** = inline cylinders; **V** = cylinders in V configuration; **flat** = horizontally opposed cylinders; **bbl.** = barrel (carburetor); **PFI** = port (multi-point) fuel injection; **TBI** = throttle-body (single-point) fuel injection; **rpm** = revolutions per minute; **OD** = overdrive transmission; **S** = standard; **O** = optional; **NA** = not available.

Prices are accurate at time of printing; subject to manufacturer's change

Mercury

they have among the best ratings from the insurance industry for occupant protection. Also, there's not much short of a truck or van that can haul so many people and so much cargo, plus tow a trailer at the same time. There are a lot of more modern mid-size cars that cost nearly as much, yet don't have as much room or brawn. And where else can big-car buyers go for traditional American styling?

Specifications

	4-door notchback	5-door wagon
Wheelbase, in.	114.3	114.3
Overall length, in.	213.5	218.3
Overall width, in.	77.5	79.3
Overall height, in.	55.4	57.0
Front track, in.	62.2	62.2
Rear track, in.	62.0	62.0
Turn diameter, ft.	39.1	39.1
Curb weight, lbs.	3828	4019
Cargo vol., cu. ft.	22.4	88.0
Fuel capacity, gal.	18.0	18.0
Seating capacity	6	8
Front headroom, in.	37.9	38.8
Front shoulder room, in.	61.6	61.6
Front legroom, max., in.	43.5	43.5
Rear headroom, in.	37.2	39.1
Rear shoulder room, in.	61.5	61.5
Rear legroom, min., in.	39.3	37.9

Powertrain layout: longitudinal front engine/rear-wheel drive.

Engines

	ohv V-8
Size, liters/cu. in.	5.0/302
Fuel delivery	PFI
Horsepower @ rpm	150 @ 3200
Torque (lbs./ft.) @ rpm	270 @ 2000
Availability	S

EPA city/highway mpg

4-speed OD automatic	17/24

Prices

Mercury Grand Marquis	Retail Price	Dealer Invoice	Low Price
GS 4-door notchback	$16701	$14366	$15534
LS 4-door notchback	17213	14801	16007
GS Colony Park 5-door wagon	17338	14907	16123
LS Colony Park 5-door wagon	17922	15404	16663
Destination charge	480	480	480

Standard Equipment:

GS: 5.0-liter PFI V-8 engine, 4-speed automatic transmission, power steering and brakes, air conditioning, tinted glass, reclining front seats, cloth trim (vinyl on wagon), power windows, intermittent wipers, digital clock, trip odometer, Autolamp system, optical horn, padded half vinyl roof, right visor mirror, AM/FM ST ET, power 3-way tailgate (wagon), lockable storage compartments (wagon), simulated woodgrain exterior applique (wagon), 215/70R15 WSW SBR tires. **LS** adds: rear armrest (4-door), front seatback map pockets, Light Group.

Optional Equipment:

	Retail	Dealer	Low
Preferred Pkg. 156A, GS 4-door	974	829	896
Preferred Pkg. 157A, GS 4-door	1200	1019	1104
Preferred Pkg. 171A, LS 4-door	1126	957	1036
Preferred Pkg. 172B, LS 4-door	1216	1031	1119
Preferred Pkg. 192A, GS wagon	1143	971	1052
LS wagon	1097	932	1009
Preferred Pkg. 193A, GS wagon	1392	1181	1281
LS wagon	1346	1142	1238
Conventional spare tire	73	62	67
Automatic climate control	211	179	194
w/any Preferred Pkg.	66	56	61
Rear defogger	145	123	133
Insta-Clear heated windshield	250	213	230
HD battery	27	23	25
Power Lock Group	245	208	225
Power locks, power decklid or tailgate release.			
Power decklid release	50	43	46
Power driver's seat	251	214	231
Power front seats	502	427	462
AM/FM ST ET cassette	137	116	126
High Level Audio System	472	401	434
w/Pkg. 157A	335	286	308
w/Pkg. 172A or 193A	167	142	154
AM/FM ST ET delete (credit)	(206)	(175)	(175)
Premium sound system	168	143	155
Power antenna	76	64	70
Cornering lamps	68	58	63
Bodyside moldings	66	56	61
Two-tone paint	159	135	146
Formal coach roof	665	563	612
Clearcoat paint	226	192	208
Leather-wrapped steering wheel	59	50	54
Floormats	43	37	40

Mercury Sable

Mercury Sable LS wagon

What's New

Mercury gives its slightly plusher version of the Ford Taurus some new interior and exterior trim for '89, and also takes steps to make it quieter and smoother running. The aerodynamic noses of the front-drive 4-door sedan and 5-door wagon are subtly reworked to include new headlamps, parking and turn signal lamps. The "grille" between the headlights is now illuminated its full width and there's a new bumper. Sedans get a revised taillamp lens design. Inside, new door-trim panels include larger speakers that replace instrument-panel speakers. And controls for the optional power windows are now floodlit on the driver's side. Returning as the standard engine is a 3.0-liter V-6, but Mercury has given it revised dampers and isolators and air-induction noise controls as part of its program to reduce noise, vibration and harshness. The program extends to the optional 3.8-liter V-6, which also gets sequential fuel injection for '89. Both engines are rated at 140 horsepower, but the 3.8 makes more torque, 215 pounds/feet to the 3.0's

Prices are accurate at time of printing; subject to manufacturer's change.

150

CONSUMER GUIDE®

160. Mercury says revised door weatherstripping will reduce wind noise and that the brake pedal now has improved feel. The standard 14-inch wheel covers are of a new design; 15-inch alloy wheels remain optional.

For

●Performance (3.8 V-6) ●Handling/roadholding ●Ride
●Passenger room ●Cargo space

Against

●Fuel economy ●Electronic instruments

Summary

The standard 3.0-liter V-6 is adequate for Sable, but since these cars debuted for 1986 we've felt that more power was desirable. The 3.8 provides it, with greater torque for brisker acceleration from a standing start and stronger passing response. Neither engine is particularly fuel efficient, however. Expect under 20 mpg in city driving and around 23-25 mpg on the highway. Like the similar Ford Taurus, Sable has uncommonly good handling for a domestic family car. The well-designed suspension promotes spirited cornering with minimal body lean, while the responsive power steering is firm and accurate. Even the standard tires combine good grip in turns with reassuring wet-road traction. Sable's suspension is a little softer than Taurus's, so the Mercury rides a little better over rough pavement, though it's still firmer than what many American-car buyers are accustomed to. Interior space and furnishings are nearly identical to those found in Taurus. Among our few complaints: the optional electronic instrument cluster is hard to read in sunlight and offers few advantages over the standard analog cluster.

Specifications

	4-door notchback	5-door wagon
Wheelbase, in.	106.0	106.0
Overall length, in.	190.9	191.9
Overall width, in.	70.8	70.7
Overall height, in.	54.3	55.1
Front track, in.	61.6	61.6
Rear track, in.	60.5	59.9
Turn diameter, ft.	39.8	39.8
Curb weight, lbs.	3054	3228
Cargo vol., cu. ft.	18.5	81.0
Fuel capacity, gal.	16.0	16.0
Seating capacity	6	8
Front headroom, in.	38.3	38.6
Front shoulder room, in.	57.5	57.5
Front legroom, max., in.	41.7	41.7
Rear headroom, in.	37.6	38.4
Rear shoulder room, in.	57.5	57.5
Rear legroom, min., in.	37.1	36.6

Powertrain layout: transverse front engine/front-wheel drive.

Engines	ohv V-6	ohv V-6
Size, liters/cu. in.	3.0/182	3.8/232
Fuel delivery	PFI	PFI
Horsepower @ rpm	140 @ 4800	140 @ 3800
Torque (lbs./ft.) @ rpm	160 @ 3000	215 @ 2200
Availability	S	O

EPA city/highway mpg

4-speed OD automatic	21/29	19/28

Prices

Mercury Sable	Retail Price	Dealer Invoice	Low Price
GS 4-door notchback	$14101	$12134	$13118
LS 4-door notchback	15094	12978	14036
GS 5-door wagon	14814	12731	13768
LS 5-door wagon	15872	13639	14756
Destination charge	450	450	450

Standard Equipment:

GS: 3.0-liter PFI V-6 engine, 4-speed automatic transmission, power steering and brakes, air conditioning, 50/50 cloth reclining front seats with armrests, 60/40 split rear seatback (wagons), outboard rear lap/shoulder belts, tinted glass, digital clock, intermittent wipers, tachometer, coolant temperature gauge, trip odometer, optical horn, tiedown hooks (wagon), front cornering lamps, power mirrors, luggage rack (wagon), AM/FM ST ET, front door map pockets, rear armrest (except wagons), covered package tray storage bin (4-doors), visor mirrors, 205/70R14 all-season SBR tires. **LS** adds: power windows, automatic parking brake release, warning lights (low fuel, washer fluid and oil, lamp outage, door ajar), Light Group, bodyside cladding, remote fuel door and liftgate releases, upgraded upholstery, power front lumbar support adjustments, front seatback map pockets, lighted visor mirrors.

Optional Equipment:

3.8-liter V-6	400	340	368
Automatic air conditioning	183	155	168
Air conditioning delete, GS (credit)	(788)	(670)	(670)
Preferred Pkg. 450A	547	464	503
Preferred Pkg. 451A	1018	865	937
Preferred Pkg. 460A	462	393	425
Preferred Pkg. 461A	978	830	900
Preferred Pkg. 462A	2054	1745	1890
205/70R14 WSW tires	82	69	75
205/65R15 tires	65	55	60
205/65R15 WSW tires	146	124	134
Conventional spare tire	73	62	67
Alloy wheels	172	146	158
Polycast wheels	123	105	113
Autolamp system	73	62	67
Automatic parking brake release	12	10	11
Lighted visor mirrors, GS	100	85	92
Electronic instrument cluster	351	299	323
Insta-Clear heated windshield	250	213	230
Keyless entry	202	172	186
Power moonroof	741	630	682
Rear defogger	145	123	133
Cruise control	182	154	167
Tilt steering column	124	106	114
Light Group, GS	59	50	54
HD battery	27	23	25
HD suspension	26	22	24
Extended-range fuel tank	46	39	42
California emissions pkg.	100	85	92
Power Lock Group, GS	287	244	264
LS	195	166	179
Power locks, remote fuel door and decklid/liftgate releases.			
Power driver's seat	251	214	231
Dual power seats	502	427	462
w/pkg. containing power driver's seat	251	214	231
Power windows, GS	296	252	272
AM/FM ST ET delete (credit)	(206)	(175)	(175)
AM/FM ST ET cassette	137	116	126

> **KEY: ohv** = overhead valve; **ohc** = overhead cam; **dohc** = double overhead cam; **I** = inline cylinders; **V** = cylinders in V configuration; **flat** = horizontally opposed cylinders; **bbl.** = barrel (carburetor); **PFI** = port (multi-point) fuel injection; **TBI** = throttle-body (single-point) fuel injection; **rpm** = revolutions per minute; **OD** = overdrive transmission; **S** = standard; **O** = optional; **NA** = not available.

Prices are accurate at time of printing; subject to manufacturer's change.

	Retail Price	Dealer Invoice	Low Price
Premium sound system	168	143	155
High Level Audio System	472	401	434
w/Pkg. 450 or 451	335	285	308
w/Pkg. 461	167	142	154
Power antenna	76	64	70
Clearcoat paint	183	155	168
Paint stripe	61	52	56
Floormats	43	36	40
Leather-wrapped steering wheel	59	50	54
Leather seat trim	415	352	382
All-vinyl trim	37	31	34
Picnic tray, wagons	66	56	61
Rear-facing third seat, wagons	155	132	143
Liftgate wiper/washer, wagons	126	107	116
Cargo area cover, wagons	66	56	61
Engine block heater	18	16	17

Mercury Topaz

Mercury Topaz LS 4-door

What's New

Topaz coasts into 1989 little changed from the freshened appearance and horsepower boost it received for '88. The compact, front-drive 2-door sedan comes in GS and sporty XR5 trim, while the 4-door sedan is available in GS, LS and sporty LTS guise. GS and LS models are available with an All Wheel Drive option, a part-time 4-wheel-drive system not for use on dry pavement. A cargo tiedown net is now standard in the luggage compartment of LS, XR5 and LTS models. The driver's-side air bag supplemental restraint system, an $815 option on GS and LS models last year, is now available on LTS versions. Motorized automatic front shoulder belts are standard on all Topazes. The base 2.3-liter 4-cylinder engine, standard in Topaz GS and LS, has 98 horsepower. The high-output 2.3-liter four, standard in the XR5, LTS and All Wheel Drive Topaz, is rated at 100 horsepower. Front-drive models have a 5-speed manual transmission standard with both engines and a 3-speed automatic optional. Automatic transmission is mandatory with All Wheel Drive models.

For

● 4WD traction (All Wheel Drive option) ● Air bag option

Against

● Engine noise ● Driveability

Summary

Since Topaz is nearly identical to the Ford Tempo except for styling, see our comments under that entry as well. Freshened styling in '88 improved Topaz's exterior appearance, and a redesign inside gave it a more functional dashboard, but changes under the hood weren't so effective. Performance improved, but only to where it's now adequate. It's not in the same class as, say, the Honda Accord. Part of the problem is that either 2.3 continues to generate more noise than useful power. Also, the 3-speed automatic transmission changes up to high gear too quickly to maintain good acceleration, and then it balks at downshifting for prompt passing response. Topaz can be nicely equipped for a reasonable price. But there are other compact sedans with more interior room, plus more overall refinement (the Accord and Toyota Camry to name two), though at higher cost. The All Wheel Drive option is reasonably priced as well, and gives Topaz great traction in rain and snow. The added weight of 4WD, however, slows acceleration and raises fuel consumption.

Specifications

	2-door notchback	4-door notchback
Wheelbase, in.	99.9	99.9
Overall length, in.	176.7	177.0
Overall width, in.	68.3	66.8
Overall height, in.	52.7	52.8
Front track, in.	54.9	54.9
Rear track, in.	57.6	57.6
Turn diameter, ft.	38.7	38.7
Curb weight, lbs.	2567	2606
Cargo vol., cu. ft.	13.2	12.9
Fuel capacity, gal.	15.4[1]	15.4[1]
Seating capacity	5	5
Front headroom, in.	37.5	37.5
Front shoulder room, in.	53.2	53.2
Front legroom, max., in.	41.5	41.5
Rear headroom, in.	36.9	36.9
Rear shoulder room, in.	53.3	53.4
Rear legroom, min., in.	36.0	36.0

1. 14.7 with 4WD

Powertrain layout: transverse front engine/rear-wheel drive.

Engines	ohv I-4	ohv I-4
Size, liters/cu. in.	2.3/141	2.3/141
Fuel delivery	PFI	PFI
Horsepower @ rpm	98 @ 4400	100 @ 4400
Torque (lbs./ft.) @ rpm	124 @ 2200	130 @ 2600
Availability	S	S[1]

EPA city/highway mpg		
5-speed OD manual	23/31	23/31
3-speed automatic	22/26	22/26

1. XR5, LTS, 4WD

Prices

Mercury Topaz	Retail Price	Dealer Invoice	Low Price
GS 2-door notchback	$9577	$8626	$9102
GS 4-door notchback	9734	8766	9250
XR5 2-door notchback	10498	9446	9972
LS 4-door notchback	11030	9920	10475
LTS 4-door notchback	11980	10765	11373
Destination charge	425	425	425

Prices are accurate at time of printing; subject to manufacturer's change.

Standard Equipment:

GS: 2.3-liter PFI 4-cylinder engine, 5-speed manual transmission, power steering and brakes, reclining cloth and vinyl front bucket seats, motorized front shoulder belts, manual lap belts, tachometer, coolant temperature gauge, trip odometer, tinted glass, intermittent wipers, AM/FM ST ET, diagnostic alert module, door pockets, map lights, 185/70R14 tires. **LS** adds: tilt steering column, power windows and locks, cruise control, all-cloth upholstery, rear defogger, remote decklid and fuel door releases, Light Group, cargo net, headlamps-on tone, front armrest, console cassette storage, touring suspension, performance tires. **XR5** adds to GS: high-output engine, 3.73 final drive ratio, performance suspension, leather-wrapped steering wheel, tilt steering column, sport seats with power lumbar, Light Group, cargo net, front armrest, remote decklid and fuel door releases, performance tires on alloy wheels. **LTS** adds to LS: high-output engine, 3.73 final drive ratio, performance suspension, air conditioning, cruise control, sport seats with power lumbar, cargo net, performance tires on alloy wheels.

Optional Equipment:

	Retail Price	Dealer Invoice	Low Price
3-speed automatic transmission	515	437	474
All Wheel Drive, GS	1441	1229	1326
GS w/Pkg. 363A	927	788	853
LTS	1352	1132	1244
LS	1424	1215	1310
LS w/Pkg. 365A	915	777	842
Driver's side air bag, GS	815	692	750
LS & LTS	622	529	572
Air conditioning	788	670	725
Special Value Pkg. 361A, GS	436	371	401
Special Value Pkg. 363A, GS	751	639	691
Supplemental Restraint Pkg. 362A, GS 4-door	1124	954	1034
Special Value Pkg. 371A, XR5	427	363	393
Special Value Pkg. 365A, LS	NC	NC	NC
Supplemental Restraint Pkg. 366A, LS . . .	896	761	824
Supplemental Restraint Pkg. 376A, LTS . .	622	529	572
185/70R14 WSW tires	82	69	75
185/70R14 performance tires	82	69	75
Comfort/Convenience Group, GS	179	152	165
Folding center armrest, Light Group, remote fuel door and decklid releases.			
Rear defogger	145	123	133
Power Lock Group, GS 2-door	237	201	218
XR5	156	133	144
GS 4-door	288	245	265
GS 2-door w/Comfort/Convenience	156	133	144
GS 4-door w/Comfort/Convenience	207	176	190
Power locks, remote fuel door and decklid releases.			
Clearcoat metallic paint	91	78	84
Premium sound system	138	117	127
Speed control	182	154	167
Tilt steering column	124	106	114
Polycast wheels	178	151	164
Decklid luggage rack	115	97	106
Power windows, 4-doors	296	252	272
AM/FM ST ET cassette, GS	137	116	126
AM/FM ST ET cassette, GS (credit)	(245)	(208)	(208)
AM/FM cassette delete, XR5, LS, LTS . . .	(382)	(324)	(324)
Power driver's seat	251	214	231
Locking spoke wheel covers	212	180	195
California emissions pkg.	100	85	92
Engine block heater	20	17	18
Vinyl seat trim	37	31	34

Mercury Tracer

What's New

Introduced in March 1987 as an '88 model, Tracer is a front-drive subcompact built in Hermosillo, Mexico, from the design for the Mazda 323. It's an upscale replacement for the Lynx, which was Mercury's version of the Ford Escort. It comes in 3-and 5-door hatchback and 5-door wagon body

Mercury Tracer 3-door

styles. For 1989, the 323 comes in 3-door hatchback and 4-door notchback styling; Mazda has dropped its version of the wagon. All models have folding rear seatbacks that enlarge the cargo hold. The sole '89 change is the addition of yellow paint to mark service points in the engine compartment. Tracer's multi-point fuel-injected 1.6-liter 4-cylinder engine and front-drive chassis are lifted from the Mazda. Mercury revised Mazda's transmission gearing and replaced some suspension pieces for its versions, but it neglected to pick up the 323's 4-speed overdrive automatic transmission, opting instead for an optional 3-speed automatic. A 5-speed manual is standard. Mercury markets Tracer as a fully-equipped subcompact, listing only a handful of individual options and option groups.

For

●Fuel economy   ●Handling/roadholding   ●Cargo room

Against

●Noise   ●Driveability (automatic transmission)

Summary

Mazda is partly owned by Ford Motor Company and the 323's design inspired similar Ford-badged models in Australia and Canada before the Tracer landed in the U.S. Though Tracer is based on the 323, it lacks some of the variations that make Mazda's subcompact special, namely a 4-wheel-drive model, an available turbocharged engine with dual overhead cams and an optional 4-speed overdrive automatic transmission. Of the three, Tracer needs the 4-speed automatic the most. Its 3-speed automatic has the engine running at a frantic 3300 rpm at 60 mph. With the 5-speed manual, it runs a more relaxed and hushed 2600 rpm at 60. An overdrive top gear added to the automatic would cut engine speed, reduce noise and improve fuel economy. As it is, Tracer is still pretty economical with automatic (a 27-mpg average in our last test). However, you'll get better fuel economy and improved driveability with the 5-speed. Tracer is a generally enjoyable subcompact that handles competently, feels surefooted and rides fairly well for a small car. The rear seat is a little cramped,

KEY: ohv = overhead valve; **ohc** = overhead cam; **dohc** = double overhead cam; **I** = inline cylinders; **V** = cylinders in V configuration; **flat** = horizontally opposed cylinders; **bbl.** = barrel (carburetor); **PFI** = port (multi-point) fuel injection; **TBI** = throttle-body (single-point) fuel injection; **rpm** = revolutions per minute; **OD** = overdrive transmission; **S** = standard; **O** = optional; **NA** = not available.

Prices are accurate at time of printing; subject to manufacturer's change.

but all models have folding rear seatbacks that create a generous-sized cargo area. In addition, Tracer is solidly built, reasonably priced and comes with Ford's competitive warranties. Overall, not a bad little car.

Specifications

	3-door hatchback	5-door hatchback	5-door wagon
Wheelbase, in.	94.7	94.7	94.7
Overall length, in.	162.0	162.0	169.7
Overall width, in.	65.2	65.2	65.2
Overall height, in.	53.0	53.0	53.7
Front track, in.	54.9	54.9	54.9
Rear track, in.	56.0	56.0	56.0
Turn diameter, ft.	30.8	30.8	30.8
Curb weight, lbs.	2158	2185	2233
Cargo vol., cu. ft.	28.9	28.9	57.6
Fuel capacity, gal.	11.9	11.9	11.9
Seating capacity	5	5	5
Front headroom, in.	38.3	38.3	38.2
Front shoulder room, in.	51.9	51.9	51.9
Front legroom, max., in.	41.5	41.5	41.5
Rear headroom, in.	37.0	37.0	38.1
Rear shoulder room, in.	51.9	51.9	51.9
Rear legroom, min., in.	34.7	34.7	34.7

Powertrain layout: transverse front engine/front-wheel drive.

Engines

	ohc I-4
Size, liters/cu. in.	1.6/97
Fuel delivery	PFI
Horsepower @ rpm	82 @ 5000
Torque (lbs./ft.) @ rpm	92 @ 2500
Availability	S

EPA city/highway mpg

5-speed OD manual	26/29
3-speed automatic	26/29

Prices

Mercury Tracer	Retail Price	Dealer Invoice	Low Price
3-door hatchback	$8556	$7714	$8135
5-door hatchback	9242	8324	8783
5-door wagon	9726	8755	9241
Destination charge (1988)	415	415	415

Standard Equipment:

1.6-liter PFI 4-cylinder engine, 5-speed manual transmission, power brakes, urethane lower body protection, front mudguards, rear stone guards, bodyside moldings, composite halogen headlamps, tachometer, trip odometer, coolant temperature gauge, electronic diagnostic warning system, digital clock, side window demisters, rear seat heat ducts, tinted glass, cloth upholstery, cut-pile carpeting, reclining front bucket seats, driver's seat height and lumbar support adjustments, split folding rear seatback with headrests and recliners, AM/FM ST ET, door map pockets, rear defogger, dual power mirrors, headlamps-on reminder, optical horn, ignition and door lock illumination, lighted ashtray, cigar lighter, cargo area lamp, dual map lights, instrument panel storage tray, full center console with storage tray, lockable glovebox, pullout tray under front passenger seat, remote fuel door

> **KEY: ohv** = overhead valve; **ohc** = overhead cam; **dohc** = double overhead cam; **I** = inline cylinders; **V** = cylinders in V configuration; **flat** = horizontally opposed cylinders; **bbl.** = barrel (carburetor); **PFI** = port (multi-point) fuel injection; **TBI** = throttle-body (single-point) fuel injection; **rpm** = revolutions per minute; **OD** = overdrive transmission; **S** = standard; **O** = optional; **NA** = not available.

release, remote hatch release (hatchbacks), rear passenger walk-in device (3-door), liftgate outside release handle with interior latch (wagon), front and rear passenger grab handles, right visor mirror, cargo area cover, carpeted lower door panels, P175/70R13 tires on steel wheels, full wheel covers.

Optional Equipment (1988 prices):	Retail Price	Dealer Invoice	Low Price
3-speed automatic transmission	415	352	384
Air conditioning	688	585	637
AM/FM ST delete (credit)	(206)	(175)	(175)
Preferred Equipment Pkg. 511B, 3-door	235	200	218
Preferred Equipment Pkg. 512B, 3-door	554	471	513
5-door & wagon	318	271	295
Power steering, cruise control, AM/FM ST cassette.			
Preferred Equipment Pkg. 513B, 3-door	1424	1211	1318
5-door	1199	1011	1105
Wagon	1006	855	931
Pkg. 512B plus air conditioning, aluminum wheels.			

Merkur Scorpio and XR4Ti

1988 Merkur Scorpio

What's New

Scorpio and XR4Ti are rear-drive hatchbacks imported from Ford of Germany and sold through Lincoln-Mercury dealers with Merkur franchises. Merkur (pronounced mare-COOR) is the German word for Mercury. Scorpio is a luxury 5-door with a 2.9-liter V-6 engine and standard anti-lock disc brakes. It came to the U.S. in May 1987 as an '88 model after being named Europe's 1986 "Car of the Year." XR4Ti is a smaller 3-door sport sedan with a turbocharged 2.3-liter 4-cylinder. It kicked off the Merkur program in 1985 and is known to European buyers as the Sierra. Additions to the Scorpio for '89 include emergency tensioning retractors for the front seat belts. Heated front seats are a new 1989 option. Changes carried over from mid-1988 include gas struts in place of the hood prop rod, a power recliner added to the 4-way power adjustments for the front seats, and an "Off" setting for the automatic climate control system. Scorpio has 60/40 split rear seatbacks that fold down for extra cargo room and have separate power recliners for each side. The XR4Ti gets some new exterior touches for 1989, but most of what's new was added as a running change during the 1988 model year. Cosworth-style cast aluminum wheels replaced thick-spoked alloy wheels, and speedometer graphics were changed from an 85-mph top speed to 150 mph. Cruise control also was added as an option. New exterior appearance touches for '89 are double accent tape stripes and black bodyside and bumper inserts in place of bright inserts.

For

● Anti-lock brakes (Scorpio) ● Cargo space
● Handling/roadholding ● Ride ● Passenger room

Prices are accurate at time of printing; subject to manufacturer's change.

Against

●Fuel economy ●Instruments/controls

Summary

Ford has been advertising the Scorpio and XR4Ti without using the Merkur name, though it continues to appear on the cars and on dealers' signs. That's an indication of the muddled marketing that has contributed to lackluster sales for these two credible "captive imports." Sales of the XR4Ti, a roadworthy and spacious sports sedan, suffered in part because Lincoln-Mercury dealers, accustomed to conservative, big-car customers, seemed ill at ease with the avant-garde-looking fastback. We like the XR4Ti best in 5-speed manual form, though the turbo 2.3 is not nearly as refined as engines in rival sports sedans. Our major complaint with the Scorpio also is under the hood, where its V-6 provides only adequate performance for this price class. Scorpio is otherwise a fine luxury tourer, with a cavernous and well-appointed cabin and confidence-inspiring road manners. Ford has stepped up its Scorpio promotions, and it's paying off. Scorpio is outselling the XR4Ti for the first time, and based on the first half of 1988, Scorpio sales figure to reach around 8000 for the year, a jump of 3000 from 1987. Ford now indexes Scorpio's resale value to that of the Mercedes-Benz 190E and pays any difference in cash to Scorpio owners who trade for a new Lincoln, Mercury or Merkur. A 32-valve 4.6-liter V-8 is reportedly being tested for use in the Scorpio, among other cars, after 1990. XR4Ti sales, meanwhile, are on a pace to total about 5000 for 1988, capping a steady decline from a high of 14,000 in 1986. Ford is reportedly considering a V-6 engine for this car in 1990.

Specifications

	XR4Ti 3-door hatchback	Scorpio 5-door hatchback
Wheelbase, in.	102.7	108.7
Overall length, in.	178.4	186.4
Overall width, in.	68.0	69.5
Overall height, in.	53.8	54.6
Front track, in.	57.2	58.1
Rear track, in.	57.8	58.1
Turn diameter, ft.	35.4	34.1
Curb weight, lbs.	2920	3230
Cargo vol., cu. ft.	35.5	37.2
Fuel capacity, gal.	15.0	16.9
Seating capacity	5	5
Front headroom, in.	38.5	37.0
Front shoulder room, in.	53.9	56.4
Front legroom, max., in.	41.0	41.3
Rear headroom, in.	37.7	37.3
Rear shoulder room, in.	54.1	56.4
Rear legroom, min., in.	34.4	38.6

Powertrain layout: longitudinal front engine/rear-wheel drive.

1988 Merkur XR4Ti

Merkur

Engines

	ohv V-6	Turbo ohc I-4
Size, liters/cu. in.	2.9/177	2.3/140
Fuel delivery	PFI	PFI
Horsepower @ rpm	144 @ 5500	175 @ 5000[1]
Torque (lbs./ft.) @ rpm	162 @ 3000	200 @ 3000[1]
Availability	S[2]	S[3]

EPA city/highway mpg

5-speed OD manual	17/23	19/25
3-speed automatic		18/21
4-speed OD automatic	17/23	

1. 145 horsepower and 180 lbs./ft. torque with automatic transmission 2. Scorpio 3. XR4Ti

Prices

Merkur Scorpio (1988 Prices)

	Retail Price	Dealer Invoice	Low Price
5-door hatchback	$24090	$20697	$22392
Destination charge	142	142	142

Standard Equipment:

2.9-liter PFI V-6 engine, 5-speed manual transmission, power steering, power 4-wheel disc brakes, anti-lock braking system, power windows, central locking including hatch lid, automatic temperature control air conditioning, front bucket seats with 16-way adjustments, 60/40 split folding rear seatback with electric recliners, cruise control, tilt/telescopic steering column, heated power mirrors, remote fuel door release, intermittent wipers, rear wiper/washer, AM/FM ST ET cassette, overhead console with digital clock, calendar, stopwatch and outside thermometer, automatic side window demisters, P205/60HR15 Pirelli P6 tires on cast aluminum wheels.

Optional Equipment:

4-speed automatic transmission	550	468	510
Power moonroof	1055	897	975
Touring Pkg.	2465	2096	2280

Power moonroof with shade, leather seats, trip computer.

Merkur XR4Ti (1988 prices)

3-door hatchback	$19065	$16966	$18016
Destination charge	142	142	142

Standard Equipment:

2.3-liter PFI turbocharged 4-cylinder engine, 5-speed manual transmission, power steering, power brakes, rear defogger, bronze tinted glass, dual-note horn, cruise control, intermittent wipers, rear wiper/washer, halogen foglamps and headlamps, dual heated power mirrors, bodyside molding, metallic paint, flip-out rear quarter windows, air conditioning, folding rear center armrest, full cut-pile carpeting, electronic multi-function clock, console with cassette storage and armrest, cloth door panel trim, door map pockets, lighted lockable glovebox, Graphic Information Module (door ajar, low air temperature and bulb outage monitor), three passenger assist handles, side window demisters, tachometer, coolant temperature and turbo boost gauges, warning lights (oil pressure, oil level, low fuel, low coolant, low washer fluid, front brake pad wear), footwell courtesy lights with time delay, cargo area tiedowns and net, front seatback map pockets, lighted right visor mirror, rear package tray with storage box, AM/FM ST ET cassette, cloth seat trim, dual sport seats with stepless recliners, driver's seat height adjuster, asymmetrically split folding rear seatback, 195/60HR15 Pirelli P6 tires on aluminum wheels.

Optional Equipment:

3-speed automatic transmission	427	363	395
Heated front seats	225	192	209
Gray leather interior	890	757	824
Tilt/slide screened moonroof	549	466	505

Prices are accurate at time of printing; subject to manufacturer's change.

Mitsubishi Eclipse

1990 Mitsubishi Eclipse GS

What's New

This new sporty coupe is the first car built by Diamond-Star Motors, a 50-50 partnership between Chrysler and Japan's Mitsubishi Motors. Chrysler-Plymouth dealers will sell a nearly identical version as the Plymouth Laser. Both are built at the new, highly automated Diamond-Star plant in Normal, Ill., and will go on sale in January as 1990 models. Mitsubishi designed the interior and supplies the powertrains. The 3-door hatchback body is the result of a styling collaboration between Mitsubishi and Chrysler. Eclipse and Laser share engines and general content. They differ only slightly in trim, with the biggest difference in the top-rung turbocharged models. Turbo Eclipses have a rear-deck spoiler and vaned alloy wheels, turbo Lasers have no spoiler and lattice-style alloys. Their front-drive platform was taken from the Mitsubishi Galant compact sedan, but their wheelbase is five inches shorter at 97 inches. At 170 inches overall, Eclipse is seven inches shorter than the Ford Probe, one of the principal targets. Three engines are offered. A 90-horsepower 1.8-liter 4-cylinder powers base Eclipses and upgraded GS versions. The Eclipse GS DOHC 16 Valve takes its name from its twin-cam 135-horsepower 2.0-liter four. A turbocharged and intercooled 2.0 DOHC four rated at 190 horsepower forms the basis for the GS Turbo. All but the turbo are available with an optional 4-speed automatic transmission. Four-wheel disc brakes are standard on all models; Mitsubishi says an anti-lock system is in the works for 1991. A 4-wheel-drive turbo model is to be introduced next fall.

For

- Performance (2.0 and 2.0 turbo) ● Braking
- Handling/roadholding

Against

- Performance (automatic transmission)
- Rear-seat room

Summary

Prices had not been announced at publication time, but Mitsubishi says the base Eclipse will start at $10,000 or less, while a fully optioned turbo will go for around $15,000. That would undercut such competitors as the Probe, Toyota Celica and Acura Integra—cars the Eclipse equals or exceeds on most points by which sporty coupes are measured. The best-balanced Eclipse is the GS DOHC 16 Valve equipped with the slick-shifting 5-speed. It has significantly more muscle than the 1.8-liter version, but does without the

feistiness under full power that can make the Turbo a demanding car to drive fast. Its firm ride is only marginally stiffer than the base car's, but wider, lower-profile tires puts its handling among the best in the class. The power steering is a touch light for some drivers. Eclipse's interior is a model of sporty-car efficiency, though the back seat is for children only. Outward visibility is a weak point, and the heavy hatchlid and high cargo liftover are a nuisance. Still, Eclipse appears to be a strong competitor right out of the box.

Mitsubishi Galant

Mitsubishi Galant LS

What's New

On sale since May, Galant is a front-drive compact sedan that replaced the smaller Tredia to give Mitsubishi a new rival for cars such as the Honda Accord, Toyota Camry and Mazda 626. Galant was Japan's Car of the Year, though it won that honor with features such as 4-wheel-drive and 4-wheel steering, which aren't offered here. Only a 4-door body style is available, but it comes three ways: base Galant, luxury LS and sporty GS. The base and LS use a 102-horsepower 2.0-liter 4-cylinder engine. A 5-speed manual transmission is standard on the base Galant and a 4-speed overdrive automatic is standard on the LS, optional on the base. A 135-horsepower dual-cam, 16-valve 2.0 powers the GS, which comes only with manual transmission. However, due to popular demand, automatic transmission will be offered on the GS late in the '89 model year. The GS also has 4-wheel disc brakes, high-performance tires and Mitsubishi's Active-Electronically Controlled Suspension as standard equipment. Anti-lock brakes are optional for the GS. Power steering, reclining front bucket seats, tilt steering column, intermittent wipers, a rear defogger and a tachometer are standard on all models. The LS and GS also come with power locks, windows and mirrors. Galant and the subcompact Mirage have a 7-year/100,000-mile rust warranty; all other Mitsubishis have a 5-year/unlimited-mileage rust warranty.

For

- Performance (GS) ● Anti-lock brakes (GS) ● Ride
- Handling/roadholding ● Passenger room
- Cargo space

Against

- Driveability (automatic transmission)
- Road noise (GS)

Summary

The new Galant represents a major step upward for Mitsubishi into the hotly competitive compact field. Our conclu-

---

Prices are accurate at time of printing; subject to manufacturer's change.

CONSUMER GUIDE®

sion is that you should add this car to your shopping list. It's roomy, comfortable, solid and competent. The 102-horsepower engine delivers adequate acceleration, though with automatic transmission it has a hard time coping with hills and fast-paced expressway traffic. The transmission is often slow to downshift for passing and seems to be always changing gears in hilly terrain. The 135-horsepower engine in the GS furnishes much more exhilarating acceleration, but currently comes only with manual transmission, which will limit its appeal. The front-drive Galant has steady, predictable handling ability and a stable, absorbent ride. Here, too, the GS outshines the others, thanks to its firmer suspension and larger tires, plus it's the only Galant available with anti-lock brakes. However, the base and LS aren't slouches on curvy roads, and we're not convinced the electronically controlled suspension on the GS is the right choice for low-cost, long-term reliability. Inside, there's ample room for four adults, plenty of cargo room, and well-designed controls that have easy operation and precision feel. Galant isn't a revolutionary compact sedan, but it's good in many ways and competitively priced.

Specifications

	4-door notchback
Wheelbase, in.	102.4
Overall length, in.	183.9
Overall width, in.	66.7
Overall height, in.	53.5
Front track, in.	57.5
Rear track, in.	57.5
Turn diameter, ft.	34.8
Curb weight, lbs.	2601
Cargo vol., cu. ft.	12.3
Fuel capacity, gal.	15.9
Seating capacity	5
Front headroom, in.	38.6
Front shoulder room, in.	54.7
Front legroom, max., in.	41.9
Rear headroom, in.	37.4
Rear shoulder room, in.	54.4
Rear legroom, min., in.	36.0

Powertrain layout: transverse front engine/front-wheel drive.

Engines

	ohc I-4	dohc I-4
Size, liters/cu. in.	2.0/122	2.0/122
Fuel delivery	PFI	PFI
Horsepower @ rpm	102 @ 5000	135 @ 6000
Torque (lbs./ft.) @ rpm	116 @ 4500	125 @ 5000
Availability	S[1]	S[2]

EPA city/highway mpg

5-speed OD manual	23/30	21/27
4-speed OD automatic	22/28	

1. Base, LS   2. GS

Prices

Mitsubishi Galant	Retail Price	Dealer Invoice	Low Price
4-door notchback, 5-speed	$10971	—	—
4-door notchback, automatic	11819	—	—
LS 4-door notchback, automatic	13579	—	—
GS 4-door notchback, 5-speed	15269	—	—
Destination charge (1988)	225	225	225

Dealer invoice and low price not available at time of publication.

Standard Equipment:

2.0-liter PFI 4-cylinder engine, 5-speed manual transmission, power steering and brakes, cloth reclining front bucket seats, outboard rear lap/shoulder belts, tachometer and coolant temperature gauge, dual trip odometers, intermittent wipers, optical horn, automatic-off headlamps, digital clock, tinted glass, remote mirrors, tilt steering column, center console with covered storage, cup and coin holders, remote fuel door and decklid releases, right visor mirror, rear defogger, 185/70SR14 all-season SBR tires. **LS** adds: 4-speed automatic transmission with power and economy modes, variable intermittent wipers, cruise control, power windows and locks, power mirrors, front spot lamps, driver's seat height and lumbar support adjustments, velour upholstery, rear center armrest with trunk-through, front seatback pockets, storage tray under passenger seat, bilevel console, AM/FM ST ET cassette with power antenna, rear heat ducts. **GS** adds to base: DOHC 16-valve engine, 4-wheel disc brakes, variable-assist steering, electronically controlled suspension, electronic time and alarm system, leather-wrapped steering wheel and shift knob, switchable green or orange instrument lighting, cruise control, power windows and locks, power mirrors, front and rear spot lamps, driver's seat height adjustment, contoured rear seat, sport cloth upholstery, power antenna, front mud guards, rear heat ducts, 195/60HR15 SBR performance tires on alloy wheels.

Optional Equipment:

	Retail Price	Dealer Invoice	Low Price
Anti-lock braking system, GS	1495	—	—
Air conditioning	790	—	—
AM/FM ST ET cassette, base	552	—	—
w/EQ, LS & GS	316	—	—
Power sunroof, LS & GS	685	—	—
Wheel covers, base	78	—	—
Floormats	67	—	—
Mud guards, base & LS	98	—	—

Mitsubishi Mirage

Mitsubishi Mirage 4-door

What's New

A redesigned Mirage went on sale in June as a more upscale subcompact than Mitsubishi's Korean-built Precis. The previous Mirage had a strong mechanical kinship with Precis, but the new Mirage doesn't. Two body styles and two engines are offered for the front-drive Mirage. A base 3-door hatchback comes with a fuel-injected, 81-horsepower 1.5-liter 4-cylinder engine and 3-speed automatic

KEY: ohv = overhead valve; **ohc** = overhead cam; **dohc** = double overhead cam; **I** = inline cylinders; **V** = cylinders in V configuration; **flat** = horizontally opposed cylinders; **bbl.** = barrel (carburetor); **PFI** = port (multi-point) fuel injection; **TBI** = throttle-body (single-point) fuel injection; **rpm** = revolutions per minute; **OD** = overdrive transmission; **S** = standard; **O** = optional; **NA** = not available.

Prices are accurate at time of printing; subject to manufacturer's change.

Mitsubishi

transmission. A 4-door notchback sedan comes in base and more lavish LS trim levels with the same engine, but a choice of 5-speed manual or 3-speed automatic transmissions. The hot number of the line is the Mirage Turbo hatchback, powered by a 135-horsepower, dual-cam 1.6-liter four and available only with a 5-speed manual. Neither engine is available in Precis. The 4-door has a longer wheelbase (96.7 inches versus 93.9 for the hatchback) and at 170.1 inches overall is nearly a foot longer. A split folding rear seatback is standard on the base and Turbo hatchbacks, and the LS sedan. Dodge and Plymouth dealers sell a similar version of the hatchback as the Colt, while Eagle dealers have a restyled version of the 4-door called the Summit. Chrysler Corp. owns 25 percent of Mitsubishi. All Mirages are now imported from Japan, but eventually the 4-door model and Summit will be built at the Diamond-Star plant in Illinois, where the Eclipse and Plymouth Laser also will be built. Mirage and the compact Galant have a 7-year/100,000-mile rust warranty; all other Mitsubishis have a 5-year/unlimited-mileage rust warranty.

For

- ●Performance (Turbo) ●Fuel economy ●Handling

Against

- ●Performance (automatic transmission)
- ●Cargo room (hatchbacks)

Summary

Mirage is a comfortable, generally competent subcompact in base and LS form, though it doesn't plow much new ground or set standards among small cars. The Turbo hatchback, on the other hand, delivers exciting performance. Turbo lag is minimal, so there's plenty of power when you want it and it's provided over a broad range of engine speeds. The power also is delivered with smoothness and refinement, while the manual shift linkage is light and precise. Most Mirages, however, will be sold with the milder 81-horsepower engine. It runs smoothly and doesn't suffer any flat spots, but with automatic transmission it struggles up hills and doesn't have much extra power for passing quickly. Fuel economy will be good with automatic, and great with the overdrive manual transmission, which also helps extract better performance. The 4-door offers decent passenger room for a small sedan, and on the LS version, the convenience of split, fold-down rear seatbacks for extra cargo room. The base and LS Mirages trail the trend-setting Honda Civic in performance, but offer comparable interior space and longer factory warranties. The fleet Mirage Turbo plain outruns most rivals.

Mitsubishi Mirage Turbo

Specifications

	3-door hatchback	4-door notchback
Wheelbase, in.	93.9	96.7
Overall length, in.	158.7	170.1
Overall width, in.	65.7	65.7
Overall height, in.	52.0	52.8
Front track, in.	56.3	56.3
Rear track, in.	56.3	56.3
Turn diameter, ft.	30.2	30.8
Curb weight, lbs.	2238	2271
Cargo vol., cu. ft.	11.5	10.3
Fuel capacity, gal.	13.2	13.2
Seating capacity	5	5
Front headroom, in.	38.3	39.1
Front shoulder room, in.	53.5	53.5
Front legroom, max., in.	41.9	41.9
Rear headroom, in.	36.9	37.5
Rear shoulder room, in.	50.7	53.1
Rear legroom, min., in.	32.5	34.4

Powertrain layout: transverse front engine/front-wheel drive.

Engines	ohc I-4	Turbo dohc I-4
Size, liters/cu. in.	1.5/90	1.6/97
Fuel delivery	PFI	PFI
Horsepower @ rpm	81 @ 5500	135 @ 6000
Torque (lbs./ft.) @ rpm	91 @ 3000	141 @ 3000
Availability	S	S

EPA city/highway mpg

5-speed OD manual	29/35	23/29
3-speed automatic	27/29	

Prices

Mitsubishi Mirage	Retail Price	Dealer Invoice	Low Price
3-door hatchback, automatic	$9159	—	—
4-door notchback, 5-speed	8859	—	—
4-door notchback, automatic	9329	—	—
LS 4-door notchback, 5-speed	10209	—	—
LS 4-door notchback, automatic	10699	—	—
Turbo 3-door hatchback, 5-speed	11969	—	—
Destination charge (1988)	225	225	225

Dealer invoice and low price not available at time of publication.

Standard Equipment:

1.5-liter PFI 4-cylinder engine, 3-speed automatic transmission (hatchback; 5-speed manual on 4-door), power brakes, reclining front bucket seats, cloth upholstery (vinyl with cloth inserts on 4-door), coolant temperature gauge, trip odometer, outboard rear lap/shoulder belts, rear defogger, console with storage bin, split folding rear seat (hatchback), remote hatch release, tinted glass, optical horn, 155/80SR13 all-season tires. **LS** adds: 5-speed manual or 3-speed automatic transmission, power steering, velour upholstery, split folding rear seat, tachometer, digital clock, intermittent wipers, remote fuel door and trunk releases, right outside mirror, 175/70SR13 tires. **Turbo** adds to base: 1.6-liter PFI DOHC PFI turbocharged engine, 5-speed manual transmission, power steering, power 4-wheel disc brakes, sport suspension, tachometer, sport seats, digital clock, anti-lift windshield wipers, intermittent wipers, rear wiper/washer, tilt/telescopic steering column, air dam and sill extensions, rear spoiler, remote fuel door release, 195/60HR14 performance tires on alloy wheels.

Optional Equipment:

Air conditioning	744	—	—
Power steering, base	262	—	—
4-door requires automatic transmission.			

Prices are accurate at time of printing; subject to manufacturer's change.

	Retail Price	Dealer Invoice	Low Price
AM/FM ST ET, base	353	—	—
AM/FM ST ET cassette	552	—	—
Digital clock, base	56	—	—
Rear wiper/washer, 3-doors	100	—	—
Floormats	63	—	—
Mud guards	67	—	—
Trim rings, base 4-door	58	—	—

Mitsubishi Montero

Mitsubishi Montero V6 5-door

What's New

Two more doors on the outside and two more cylinders under the hood are the major changes for Montero, Mitsubishi's compact 4-wheel-drive vehicle. A new 5-door body style has a longer wheelbase and overall length than the 3-door, which continues for 1989, while a new 3.0-liter V-6 is standard on the 5-door and optional on the 3-door. It is the same V-6 previously available in the Dodge Caravan/Plymouth Voyager minivans. Dodge sells a similar version of this vehicle as the Raider, but for 1989 Mitsubishi's U.S. partner only gets the V-6, not the 5-door body style. With the 5-door, Montero grows by two feet overall (from 157.3 inches to 181.7), while the wheelbase is 13.6 inches longer at 106.1. V-6 models also gain a wider rear track, 55.7 inches instead of 54.1. Coil springs replace leaf springs at the rear on all models. Most of the added length goes into the rear seat, where leg room is significantly improved, but cargo volume also benefits, growing from 62.3 cubic feet to 94.9. The 143-horsepower V-6 is available with a 5-speed manual or 4-speed automatic transmission. A 109-horsepower 2.6-liter 4-cylinder engine remains standard on the 3-door, and comes only with manual transmission this year. The fuel tank on the 5-door holds 24.3 gallons, while on the V-6 3-door it holds 19.8. A 15.8-gallon tank comes with the 4-cylinder engine. An on-demand, part-time 4WD system (not for use on dry pavement) with automatic locking front hubs is standard. The vehicle has to be stopped to engage or disengage 4WD, but once engaged, it allows intermittent 4WD operation. Models include base SP and upgraded Sport 3-doors, and base and plusher LS 5-door versions. An electric sunroof is optional on the LS, while a new Power Package (power windows and locks, plus cruise control) is standard on the LS, and optional on the base 5-door and Sport 3-door.

For

- Passenger room
- Cargo space
- Ride (5-door)
- Workmanship

Against

- Performance (4-cylinder)
- Driving position

Summary

The new dimensions give Montero the longest wheelbase and overall length among compact 4x4s. The only rivals available as 5-door models are the popular Jeep Cherokee/Wagoneer and the Isuzu Trooper II, though Chevrolet and Ford are planning 5-door models. After some 600 miles with the top-line LS 5-door, we think the big new V-6 Montero equals the Jeep Cherokee in some respects and even beats it in others. In performance, it doesn't quite match the Cherokee's optional 4.0-liter straight six. The Montero 4-door is heavy for the V-6's torque. With the 4-speed automatic that most buyers will choose, it quickly loses steam up steep hills—and that's with just the driver aboard, which makes us wonder about a full load. The V-6 is still better than the base 4-cylinder, with fine acceleration on the flat, easy and fairly quiet 65-mph cruising, and a responsive automatic transmission in change-of-pace traffic. Economy is nothing to write the EPA about—our near-new LS returned 17.4 mpg—though it's not bad for the class. Against the 3-door Montero, the 5-door's extra weight and wheelbase mean a much smoother ride and a far roomier, more useful interior (we especially like the easy-folding back seat). You sit higher than in a Cherokee, but the driver's seat doesn't go back that far and the steering wheel is rather high. Otherwise, you could almost be driving a passenger car. Workmanship on this LS easily beat most Cherokees we've tested, and overall refinement makes this a better all-around vehicle than Isuzu's similar 5-door Trooper II. Montero still trails Cherokee in the 4WD department; Jeep offers two 4WD systems, one full-time and one part-time, both with shift-on-the-fly. Mitsubishi offers one part-time system, without true shift-on-the-fly. Montero's price seems right, however. Our loaded LS lacked only the optional power sunroof yet stickered for under $20,000. The base 5-door goes for just over $17,000 but is still well equipped. All of which makes Mitsubishi's big new all-drive wagon worth a look, especially when a similarly equipped Cherokee can easily run $23,000 or more.

Specifications

	3-door wagon	5-door wagon
Wheelbase, in.	92.5	106.1
Overall length, in.	154.9	183.1
Overall width, in.	66.1	66.1
Overall height, in.	72.4	74.4
Front track, in.	55.1	55.1
Rear track, in.	54.1	55.7
Turn diameter, ft.	34.1	38.7
Curb weight, lbs.	3115	3781
Cargo vol., cu. ft.	62.3	94.9
Fuel capacity, gal.	15.8[1]	24.3
Seating capacity	5	5
Front headroom, in.	40.7	40.5
Front shoulder room, in.	55.1	55.1
Front legroom, max., in.	39.6	39.6

KEY: ohv = overhead valve; **ohc** = overhead cam; **dohc** = double overhead cam; **I** = inline cylinders; **V** = cylinders in V configuration; **flat** = horizontally opposed cylinders; **bbl.** = barrel (carburetor); **PFI** = port (multi-point) fuel injection; **TBI** = throttle-body (single-point) fuel injection; **rpm** = revolutions per minute; **OD** = overdrive transmission; **S** = standard; **O** = optional; **NA** = not available.

Prices are accurate at time of printing; subject to manufacturer's change.

Mitsubishi

	3-door wagon	5-door wagon
Rear headroom, in.	33.4	40.5
Rear shoulder room, in.	57.1	57.1
Rear legroom, min., in.	34.4	34,6

1. 19.8 with V-6

Powertrain layout: longitudinal front engine/rear-wheel drive or on-demand 4WD.

Engines

	ohc I-4	ohc V-6
Size, liters/cu. in.	2.6/156	3.0/181
Fuel delivery	2 bbl.	PFI
Horsepower @ rpm	109 @ 5000	143 @ 5000
Torque (lbs./ft.) @ rpm	142 @ 3000	168 @ 2500
Availability	S	S

EPA city/highway mpg

5-speed OD manual	16/19	15/18
4-speed OD automatic		17/17

Prices

Mitsubishi Montero	Retail Price	Dealer Invoice	Low Price
SP 3-door wagon, 5-speed	$12299	—	—
SP 3-door wagon, V-6, 5-speed	13949	—	—
SP 3-door wagon, V-6, automatic	14639	—	—
Sport 3-door wagon, V-6, automatic	15399	—	—
5-door wagon, V-6, 5-speed	17099	—	—
5-door wagon, V-6, automatic	17789	—	—
LS 5-door wagon, V-6, automatic	18389	—	—
Destination charge (1988)	225	225	225

Dealer invoice and low price not available at time of publication.

Standard Equipment:

SP: 2.6-liter 2bbl. 4-cylinder engine, 5-speed manual transmission, part-time 4-wheel drive with 2-speed transfer case, power steering and brakes, reclining front bucket seats, tilt steering column, door pockets, center console with storage, remote fuel door and liftgate releases, right visor mirror, intermittent wipers, rear defogger, tachometer, coolant temperature and oil pressure gauges, voltmeter, inclinometer, skid plates (front end, transfer case, fuel tank), tow hooks, mud flaps, tinted glass, opening front vent windows, dual outside mirrors, 225/75R15 M + S SBR tires, full-size spare with lock. **SP V-6** adds: 3.0-liter PFI V-6, 5-speed manual or 4- speed automatic transmission, 235/75R15 all-season SBR tires. **Sport** adds: limited-slip differential, suspension driver's seat, digital clock, AM/FM ST ET cassette, two-tone paint, headlamp washers, rear wiper/washer. **5-door** adds: folding rear seat with head restraints, rear heat ducts, bodyside molding, raccoon paint treatment. **LS** adds: power windows and locks, cruise control, reclining rear seats.

Mitsubishi Montero V6 5-door

Optional Equipment:

	Retail Price	Dealer Invoice	Low Price
Air conditioning	810	—	—
AM/FM ST ET, SP	324	—	—
AM/FM ST ET cassette, SP	503	—	—
w/EQ, Sport & LS	321	—	—
Digital clock, SP	56	—	—
Rear wiper/washer, SP	166	—	—
Alloy wheels, V-6 models	464	—	—
Power Pkg. (NA SP)	644	—	—
Power windows and locks, cruise control.			
Power sunroof, LS	685	—	—
Floormats, SP & Sport	73	—	—
Others	79	—	—
Spare tire cover	35	—	—

Mitsubishi Precis

Mitsubishi Precis 3-door

What's New

An audible wear indicator warning has been added to the front disc brakes on all Precis models, and all models except the base hatchback gain a new center console, center armrest and velour upholstery. Precis, built in Korea by Hyundai (which is partly owned by Mitsubishi), is based on an older design for the subcompact Mirage, not the current version. The nearly identical Hyundai Excel is built from the same design as Precis; both use a Mitsubishi-designed front-drive chassis and wear Hyundai-designed sheetmetal. Importing Precis from Korea, where wages and production costs are lower than in Japan, allows Mitsubishi to offer a less expensive alternative to its new Mirage. Precis comes as a 3- or 5-door hatchback with a 68-horsepower, carbureted 1.5-liter 4-cylinder engine. A 4-speed manual transmission is standard on the base 3-door, while a 5-speed manual is standard on the RS 3-door. LS versions have the 5-speed manual standard and a 3-speed automatic optional. Precis' warranties are shorter than those on other Mitsubishi vehicles: 12 months/12,500 miles on the entire car; 3 years/36,000 miles on the powertrain; and 3 years/unlimited miles on rust.

For

•Fuel economy •Maneuverability •Value

Against

•Performance •Interior room •Noise

Summary

If low price and high fuel economy are important to you, this is a good place to start your shopping. Precis, a kissing

Prices are accurate at time of printing; subject to manufacturer's change.

cousin to the Hyundai Excel, offers both. You'll get the most economy with one of the manual transmissions, especially the overdrive 5-speed. The 3-speed automatic returns the worst mileage and the worst performance. Acceleration is weak with automatic and passing response lackluster. Performance improves to adequate levels with manual transmission, but models we've driven suffered from flat spots that slowed progress. Besides their additional comfort and convenience features, the top-of-the-line LS models also have wider tires than the base and RS: 175/70R13 instead of skinny 155/80R13. The wider tires don't turn the LS versions into cornering fools, yet their larger footprint does provide a little more grip, which these cars can certainly use. A soft suspension allows plenty of body lean in turns and considerable understeer (resistance to turning). Inside, there are a lot of standard features for the money, but not a lot of room because of the short wheelbase. By comparison, the new Mirage 4-door has a longer wheelbase for greater interior room, a spunkier engine and improved handling ability. However, the Mirage costs quite a bit more, as do most other Japanese subcompacts. That makes Precis more attractive from a budget standpoint. Also check out the Dodge Omni/Plymouth Horizon twins, which come with more power and longer powertrain and rust warranties at a similar price.

Specifications

	3-door hatchback	5-door hatchback
Wheelbase, in.	93.7	93.7
Overall length, in.	160.9	160.9
Overall width, in.	63.1	63.1
Overall height, in.	54.1	54.1
Front track, in.	54.1	54.1
Rear track, in.	52.8	52.8
Turn diameter, ft.	33.8	33.8
Curb weight, lbs.	2161	2216
Cargo vol., cu. ft.	26.6	26.0
Fuel capacity, gal.	10.6[1]	10.6[1]
Seating capacity	5	5
Front headroom, in.	37.5	37.5
Front shoulder room, in.	52.1	52.1
Front legroom, max., in.	40.9	40.9
Rear headroom, in.	36.9	36.9
Rear shoulder room, in.	52.0	52.0
Rear legroom, min., in.	32.4	32.4

1. 13.2 on LS with automatic transmission

Powertrain layout: transverse front engine/front-wheel drive.

Mitsubishi Precis 5-door

Engines

	ohc I-4
Size, liters/cu. in.	1.5/90
Fuel delivery	2 bbl.
Horsepower @ rpm	68 @ 5500
Torque (lbs./ft.) @ rpm	82 @ 3500
Availability	S

EPA city/highway mpg

4-speed manual	27/33
5-speed OD manual	28/37
3-speed automatic	27/31

Prices

Mitsubishi Precis	Retail Price	Dealer Invoice	Low Price
3-door hatchback, 4-speed	$5499	—	—
RS 3-door hatchback, 5-speed	6699	—	—
LS 3-door hatchback, 5-speed	7349	—	—
LS 3-door hatchback, automatic	7839	—	—
LS 5-door hatchback, 5-speed	7599	—	—
LS 5-door hatchback, automatic	8089	—	—
Destination charge (1988)	225	225	225

Dealer invoice and low price not available at time of publication.

Standard Equipment:

1.5-liter 2bbl. 4-cylinder engine, 4-speed manual transmission, power brakes, vinyl reclining front bucket seats, folding rear seatbacks, cargo cover, variable intermittent wipers, rear defogger, rear heat ducts, trip odometer, coolant temperature gauge, low fuel and door/hatch ajar warning lights, locking fuel door, 155/80R13 all-season tires with full-size spare. **RS** adds: 5-speed manual transmission, cloth trim, upgraded door panels with map pockets, console, analog clock, wide bodyside moldings, remote fuel door and hatch releases, rear wiper/washer, dual remote mirrors, tinted glass. **LS** adds: 5-speed manual or 3-speed automatic transmission, tachometer, digital clock, upgraded steering wheel, storage tray under front passenger seat (5-door), right visor mirror, AM/FM ST ET cassette, dark upper windshield band, roll-down rear windows (5-door), wheel covers, 175/70R13 all-season tires.

Optional Equipment:

Air conditioning (NA base)	735	—	—
AM/FM ST ET cassette, base & RS	295	—	—
High Power, LS	135	—	—
High Power, RS	565	—	—
Passive restraint, LS 5-door	75	—	—
Alloy Wheel Pkg., LS	295	—	—
Power steering, LS	260	—	—
Power sunroof, LS	395	—	—
Floormats	51	—	—

Mitsubishi Sigma

What's New

Formerly known as Galant, and for most of 1988 as the Galant Sigma, this front-drive sedan now comes calling merely as Sigma to avoid confusion with the new Galant compact. Though Sigma has the same 102.4-inch wheel-

> **KEY: ohv** = overhead valve; **ohc** = overhead cam; **dohc** = double overhead cam; **I** = inline cylinders; **V** = cylinders in V configuration; **flat** = horizontally opposed cylinders; **bbl.** = barrel (carburetor); **PFI** = port (multi-point) fuel injection; **TBI** = throttle-body (single-point) fuel injection; **rpm** = revolutions per minute; **OD** = overdrive transmission; **S** = standard; **O** = optional; **NA** = not available.

Prices are accurate at time of printing; subject to manufacturer's change.

Mitsubishi

Mitsubishi Sigma

base as Galant and is just two inches longer at 185.8 inches overall, it has a higher mission in life, as well as a higher price: it's Mitsubishi's challenger to Japanese luxury sedans such as the Nissan Maxima and Toyota Cressida. Last year Sigma received a facelift and a 142-horsepower 3.0-liter V-6 as standard equipment in place of a 110-horsepower 4-cylinder. For 1989, anti-lock brakes (also new last year) are available as an individual option as well as part of the Eurotech Package, which includes Mitsubishi's Electronically Controlled Suspension (ECS). ECS lets the driver choose from three manual suspension modes, or it can be set to automatically adjust firmness and ride height. Other changes for 1989 are: an auto-down feature for the driver's power window; doors can be locked from the outside without holding the door handle; the standard theft-deterrent system can be activated by key-locking the doors; a new delay feature allows closing the power sunroof 30 seconds after the ignition has been turned off; and underhood service items are color-coded to make them easier to find.

For

- Performance   • Anti-lock brakes option
- Handling/roadholding

Against

- Electronically controlled suspension   • Interior room

Summary

Mitsubishi's 3.0-liter V-6 makes Sigma much more potent than it was with the old 4-cylinder engine, and the refined engine works in concert with the overdrive automatic transmission to deliver a smooth power flow. We're also glad to see the anti-lock brakes as a separate option instead of being part of the expensive Eurotech Package, which includes the electronically controlled suspension. We don't notice enough benefit from ECS to make it worthwhile. Anti-lock brakes, however, are well worth the extra money for the improved stopping ability they provide in emergencies. Last year's restyling did nothing to give Sigma's cramped interior more space, so there's still too little room front and rear. The driver's seat doesn't go back far enough for tall people to have a comfortable reach to the wheel and pedals, plus the optional power sunroof leaves inadequate

head room for most. Rear seat passengers are squeezed unless the front seats are moved well forward. Though short on room, Sigma offers good performance and optional anti-lock braking at reasonable cost.

Specifications

	4-door notchback
Wheelbase, in.	102.4
Overall length, in.	185.8
Overall width, in.	66.7
Overall height, in.	51.6
Front track, in.	56.9
Rear track, in.	55.7
Turn diameter, ft.	34.8
Curb weight, lbs.	3075
Cargo vol., cu. ft.	12.4
Fuel capacity, gal.	15.9
Seating capacity	5
Front headroom, in.	37.5
Front shoulder room, in.	53.5
Front legroom, max., in.	40.3
Rear headroom, in.	36.7
Rear shoulder room, in.	53.2
Rear legroom, min., in.	36.4

Powertrain layout: transverse front engine/front-wheel drive.

Engines

	ohc V-6
Size, liters/cu. in.	3.0/181
Fuel delivery	PFI
Horsepower @ rpm	142 @ 5000
Torque (lbs./ft.) @ rpm	168 @ 2500
Availability	S

EPA city/highway mpg

4-speed OD automatic	18/22

Prices

Mitsubishi Sigma	Retail Price	Dealer Invoice	Low Price
4-door notchback	$17069	—	—
Destination charge (1988)	225	225	225

Dealer invoice and low price not available at time of publication.

Standard Equipment:

3.0-liter PFI V-6 engine, 4-speed automatic transmission, power steering and brakes, reclining front bucket seats, 8-way adjustable driver's seat, velour trim, underseat tray, seatback pockets, folding rear seatbacks, rear armrest, door pockets, rear heat ducts, power windows and locks, heated power mirrors, tinted glass with dark upper band, tilt steering column, rear defogger, speed-sensitive variable intermittent wipers, digital clock, trip odometer, tachometer, coolant temperature gauge, voltmeter, cruise control, low fuel and washer fluid lights, AM/FM ST ET cassette, power antenna, theft deterrent system, remote fuel door and decklid releases, 195/60HR15 tires on alloy wheels.

Optional Equipment:

Anti-lock brakes	1495	—	—
Eurotech Pkg.	2042	—	—
Electronically controlled suspension, anti-lock brakes.			
Power sunroof	685	—	—
Leather seats	816	—	—
AM/FM ST ET cassette w/EQ	361	—	—
Floormats	80	—	—

KEY: ohv = overhead valve; **ohc** = overhead cam; **dohc** = double overhead cam; **I** = inline cylinders; **V** = cylinders in V configuration; **flat** = horizontally opposed cylinders; **bbl.** = barrel (carburetor); **PFI** = port (multi-point) fuel injection; **TBI** = throttle-body (single-point) fuel injection; **rpm** = revolutions per minute; **OD** = overdrive transmission; **S** = standard; **O** = optional; **NA** = not available.

Prices are accurate at time of printing; subject to manufacturer's change.

CONSUMER GUIDE®

Mitsubishi Starion

Mitsubishi Starion ESI-R

What's New

The rear-drive, turbocharged Starion sports car is back for its final season with a handful of standard equipment changes and only one model. Last year's base ESI model and optional automatic transmission are gone, leaving only an ESI-R model with a 5-speed manual transmission. Starion is the oldest model in Mitsubishi's lineup and the only one still around from the company's U.S. sales launch for 1983. A new Starion is due as a 1990 model. Chrysler dealers sell a nearly identical model as the Conquest TSi. Starion comes with a turbocharged/intercooled 2.6-liter 4-cylinder engine that makes 188 horsepower. Changes this year include 3-point outboard rear seat belts, an auto-down feature for the driver's power window, a heavy-duty rear defroster and a revised theft-deterrent system that can be activated by key locking the doors. Standard equipment includes motorized front shoulder belts, anti-lock rear brakes, a limited-slip rear differential, automatic air conditioning, graphic equalizer and steering-wheel-mounted stereo controls. An optional Sports Handling Package adds wider wheels and tires, and manually adjustable shock absorbers.

For

- Performance • Anti-lock rear brakes
- Handling/roadholding

Against

- Interior room • Fuel economy • Noise • Ride

Summary

Starion is a Japanese muscle car that feels and sounds like it has six or eight cylinders, not just four. Its turbocharged 4-cylinder provides impressive acceleration with little turbo lag. The engine gets loud and coarse at higher speeds, however, plus there's a lot of suspension and road noise. Wide tires and a stiff suspension give Starion impressive cornering ability, but bumpy roads will make you wish for more compliance. Worse, the huge performance tires slip easily in rain and snow, resulting in limited traction despite the standard limited-slip rear differential. Other high-performance rear-drive cars suffer similar problems, so Starion is no worse. We would prefer 4-wheel anti-lock brakes, but an anti-lock feature for the rears only is better than none. As with other cars of this type, most of the interior room is devoted to the front seats, leaving only token space for the rear. Cargo room is adequate, plus the rear seatbacks fold down for extra room. Overall, Starion packs a strong punch and comes loaded with gadgets, at a price lower than Japanese rivals such as the Nissan 300ZX and Toyota Supra. However, with base prices now over $20,000, we can't get too excited when the Ford Mustang GT is some $4000 less fully loaded.

Specifications

	3-door hatchback
Wheelbase, in.	95.9
Overall length, in.	173.2
Overall width, in.	68.3
Overall height, in.	50.2
Front track, in.	57.7
Rear track, in.	57.3
Turn diameter, ft.	31.5
Curb weight, lbs.	3036
Cargo vol., cu. ft.	19.0
Fuel capacity, gal.	19.8
Seating capacity	4
Front headroom, in.	36.6
Front shoulder room, in.	52.4
Front legroom, max., in.	40.8
Rear headroom, in.	35.4
Rear shoulder room, in.	51.2
Rear legroom, min., in.	29.1

Powertrain layout: longitudinal front engine/rear-wheel drive.

Engines

	Turbo ohc I-4
Size, liters/cu. in.	2.6/156
Fuel delivery	TBI
Horsepower @ rpm	188 @ 5000
Torque (lbs./ft.) @ rpm	234 @ 2500
Availability	S

EPA city/highway mpg

5-speed OD manual	18/22

Prices

Mitsubishi Starion	Retail Price	Dealer Invoice	Low Price
ESI-R 3-door hatchback	$19859	—	—
Destination charge (1988)	225	225	225

Dealer invoice and low price not available at time of publication.

Standard Equipment:

2.6-liter TBI turbocharged, intercooled 4-cylinder engine, 5-speed manual transmission, power steering, power 4-wheel disc brakes, anti-lock rear brakes, limited-slip differential, automatic climate control, power windows and locks, heated power mirrors, cruise control, automatic-off headlamps, velour reclining front bucket seats, motorized front shoulder belts with manual lap belts, outboard rear lap/shoulder belts, split folding rear seat, door pockets, tilt steering column, bronze-tinted glass, rear defogger and wiper/washer, leather-wrapped steering wheel, removable cargo cover, remote fuel door and liftgate releases, speed-sensitive intermittent wipers, right visor mirror, console with armrest and storage, tachometer, coolant temperature and oil pressure gauges, voltmeter, boost gauge, trip odometer, digital clock, AM/FM ST ET cassette w/EQ, power antenna, audio controls in steering wheel hub, theft protection device, 205/55VR16 front and 225/50VR16 rear tires on alloy wheels.

Prices are accurate at time of printing; subject to manufacturer's change.

Mitsubishi

225/50VR16 front and 245/45VR16 rear tires on wider alloy wheels, adjustable shock absorbers.

Mitsubishi Wagon/Van

Mitsubishi Wagon

What's New

Mitsubishi's compact van gets several minor standard-equipment revisions and new blackout exterior trim around the windows that gives it a raccoon-like appearance. Mitsubishi calls the 7-seat passenger version Wagon and the 2-seat cargo version Van. The conversion model of the Van has been dropped for 1989, as has the 2-tone paint option for the Wagon. New standard features include a high-mounted stop lamp, heavy-duty rear defroster with a new timer, tie-down hooks in the cargo area, a new on-board engine diagnostics system, and one-touch front door locks. The top-line LS Wagon also gains power mirrors and an auto-down feature for the driver's power window. Mitsubishi's minivan has a rear-drive, mid-engine design that's similar to Nissan's and Toyota's vans. A 107-horsepower 2.4-liter 4-cylinder engine is mounted behind the front axle and between the front seats. All models come with a 4-speed overdrive automatic transmission. Passenger models have reclining front bucket seats, two captain's chairs in the middle that recline, swivel and slide fore and aft, and a 3-place rear bench seat with reclining seatback. The rear bench can also be removed for additional cargo room. Passenger and cargo models have a sliding side door and rear liftgate.

For

• Passenger room • Cargo room • Maneuverability

Against

• Ride • Driver seating

Summary

Except for the new Mazda MPV, the Japanese minivans are peas in a pod. Mitsubishi, Nissan and Toyota use similar mid-engine layouts on a rear-drive chassis (Toyota also offers 4-wheel-drive) with a short wheelbase and considerable body overhang, resulting in a bouncy, pitchy, un-

comfortable ride under most conditions. Mitsubishi uses a firm suspension that has stiff reactions to bumps and allows excessive bouncing. In its favor, the Wagon handles better than the Toyota Van, feels more stable and has less body lean in turns. The short wheelbase and tiny 29.5-foot turn circle aid maneuverability, a plus for parking and city driving. However, the front doors are located right over the front wheels, and it's a tall step up into the interior through narrow openings. Once inside, the driver faces a cramped, bus-like driving position, and with the engine between the front seats, the driver and front passenger can't move to the rear of the Wagon from the inside. On the plus side, interior space is competitive with most other compact vans and Mitsubishi offers one of the most flexible seating arrangements on wheels. There's little cargo space behind the rear bench seat, but you can pivot the whole seat forward for plenty of room, or remove it. Mitsubishi gets credit for coming up with a flexible seating arrangement, but we're still convinced that the front-drive Dodge Caravan/Plymouth Voyager are better choices for passenger use.

Specifications

	4-door van
Wheelbase, in.	88.0
Overall length, in.	175.2
Overall width, in.	66.5
Overall height, in.	71.3
Front track, in.	56.9
Rear track, in.	54.3
Turn diameter, ft.	29.5
Curb weight, lbs.	3285
Cargo vol., cu. ft.	161.6
Fuel capacity, gal.	14.2
Seating capacity	7
Front headroom, in.	38.6
Front shoulder room, in.	57.3
Front legroom, max., in.	40.0
Rear headroom, in.	39.0
Rear shoulder room, in.	60.6
Rear legroom, min., in.	41.1

Powertrain layout: longitudinal mid engine/rear-wheel drive.

Engines

	ohc I-4
Size, liters/cu. in.	2.4/143
Fuel delivery	PFI
Horsepower @ rpm	107 @ 5000
Torque (lbs./ft.) @ rpm	132 @ 3500
Availability	S

EPA city/highway mpg

4-speed OD automatic	18/21

Prices

Mitsubishi Wagon	Retail Price	Dealer Invoice	Low Price
4-door van	$14929	—	—
4-door van w/LS Pkg.	16579	—	—
Destination charge (1988)	225	225	225

Dealer invoice and low price not available at time of publication.

Standard Equipment:

2.4-liter PFI 4-cylinder engine, 4-speed automatic transmission, power brakes, carpeting, console with storage compartment, rear defogger, remote

Prices are accurate at time of printing; subject to manufacturer's change.

CONSUMER GUIDE®

fuel door release, tinted glass, rear heater, dual outside mirrors, velour upholstery, two recline/swivel/slide middle seats, third fold-down, removable bench seat, tilt steering column, rear side storage bins, variable intermittent wipers, P205/75R14 tires. **LS Pkg.** adds: power steering, digital clock, cruise control, power windows and door locks, bronze tinted glass, bodyside molding, AM/FM ST ET cassette.

Optional Equipment:	Retail Price	Dealer Invoice	Low Price
Dual air conditioning	1426	—	—
AM/FM ST ET cassette, base	511	—	—
Power sunroof	685	—	—
Alloy wheels, LS	335	—	—
Power steering, base	268	—	—
Power mirrors, base	65	—	—
Digital clock, base	56	—	—
Floormats	52	—	—

Nissan Maxima

Nissan Maxima GXE

What's New

Redesigned for 1989, Maxima's boxy styling is discarded in favor of all-new aero sheetmetal and some advances under the skin. Last year's 5-door wagon has been dropped, so all '89s are 4-door sedans. Maxima retains its predecessor's front-drive layout, but grows enough to climb from the compact- to the mid-size class. The wheelbase is stretched by four inches, the new body is longer by six inches and wider by three. Every interior dimension is up as well, though curb weight increases only 46 pounds. Carried over is the 3.0-liter V-6 that's essentially the same engine used in the 300ZX sports car. A new intake manifold design and other alterations raise its horsepower by three, to 160. Torque also is up, from 168 pounds-feet at 3600 rpm to 181 at 3200. Two trim levels are offered: luxury GXE, which comes only with a 4-speed automatic transmission; and sporty SE, with a choice of 5-speed manual or automatic. SE's standard 4-wheel disc brakes have an anti-lock option; GXEs have only a front disc/rear drum setup. An Electronics Package optional for the GXE includes a new heads-up display that repeats the dashboard's digital speedometer reading by reflecting it onto the lower-left corner of the windshield, in the driver's field of vision. The package also includes a keyless entry system, automatic temperature control and Sonar Suspension II, a revision of the system introduced last year. A sonar sensor under the front bumper "reads" the road surface and electronically adjusts shock-absorber damping to try to maintain a supple ride. Standard on SE and optional on GXE is a new Bose music system acousti-cally matched to the interior. Motorized front shoulder belts are standard. Nissan has increased its basic warranty for 1989 from 12 months/12,500 miles to 3 years/36,000 miles, including the powertrain warranty.

For
- Performance •Anti-lock brakes (SE)
- Handling/roadholding (SE)

Against
- Ride (SE) •Noise
- Driveability (automatic transmission)

Summary

Nissan designed the new Maxima mainly for the U.S. market and it's the largest car Nissan has ever sold here. Nissan says it expects sales to split evenly between the GXE and the SE, and that 40 percent of SEs will have manual transmission. It also says it hopes to price the cars near 1988 levels, which means $18,000 to around $21,000 (1989 prices hadn't been announced when this issue went to press). At that they are a good value, based on our test drives of prototypes over public roads. The SE with a 5-speed is a credible sports sedan. It feels agile and sure-footed, has good interior room, seems well built and enjoys the security of optional anti-lock brakes. A harsh ride over rough pavement is its only major flaw. We think its performance is far superior to the previous model's, making it a realistic alternative for the sport-sedan buyer who's been priced out of the BMW market, especially since it's less expensive than the Acura Legend Sedan. The GXE is softer by nature, with a ride that's almost too soft, suffering floating and pitching when driven hard on twisty roads. We'd expect it to be comfortable and competent in most city and highway travel, however. Some flaws: The automatic transmission is slow to downshift for passing, and the shortage of rear-seat head room for riders over 5-foot-7 is a serious drawback in a mid-size sedan. The best thing about the GXE's optional heads-up display is that it does not compromise the car's function; it does seem to make speedometer-watching easier. A switch turns it off, leaving you with only the rather poorly designed digital gauges in the instrument pod. Cabin design is generally first rate, though, and these cars strike us as thoroughly modern inside and out, a welcome departure from their glitzy and angular predecessors.

Specifications	4-door notchback
Wheelbase, in.	104.3
Overall length, in.	187.6
Overall width, in.	69.3
Overall height, in.	55.1
Front track, in.	59.4
Rear track, in.	58.7
Turn diameter, ft.	36.7
Curb weight, lbs.	3086
Cargo vol., cu. ft.	14.5

KEY: ohv = overhead valve; **ohc** = overhead cam; **dohc** = double overhead cam; **I** = inline cylinders; **V** = cylinders in V configuration; **flat** = horizontally opposed cylinders; **bbl.** = barrel (carburetor); **PFI** = port (multi-point) fuel injection; **TBI** = throttle-body (single-point) fuel injection; **rpm** = revolutions per minute; **OD** = overdrive transmission; **S** = standard; **O** = optional; **NA** = not available.

Prices are accurate at time of printing; subject to manufacturer's change.

Nissan Maxima SE

Nissan Pathfinder SE

	4-door notchback
Fuel capacity, gal.	18.5
Seating capacity	5
Front headroom, in.	39.5
Front shoulder room, in.	NA
Front legroom, max., in.	43.7
Rear headroom, in.	36.9
Rear shoulder room, in.	NA
Rear legroom, min., in.	33.2

Powertrain layout: transverse front engine/front-wheel drive.

Engines

	ohc V-6
Size, liters/cu. in.	3.0/181
Fuel delivery	PFI
Horsepower @ rpm	160 @ 5200
Torque (lbs./ft.) @ rpm	181 @ 3200
Availability	S

EPA city/highway mpg

5-speed OD manual	20/26
4-speed OD automatic	19/26

Prices not available at time of publication.

Nissan Pathfinder

What's New

A 2-wheel-drive model is scheduled to be added in January, but until then, Nissan's compact sporty utility vehicle is a virtual carryover from last year. All Pathfinders now come with an on-demand, part-time 4-wheel-drive system standard. You have to stop and then back up slightly to disengage 4WD, though you can shift from 2WD to 4WD High below 25 mph. The new 2WD model will serve as a price leader for Pathfinder; rivals such as the Chevrolet S10 Blazer/GMC S15 Jimmy, Ford Bronco II and Jeep Cherokee also offer 2WD price-leader models. A 145-horsepower 3.0-liter V-6 became standard last year, when Nissan trimmed the model choices from three to two, dropping the base S and leaving only XE and SE models. Both are available with a 5-speed manual or 4-speed overdrive automatic transmission. Pathfinder comes only in a 3-door wagon body style. It's covered by Nissan's new 36-month/36,000-mile basic warranty (including the powertrain), which replaces the previous 12-month/12,500-mile plan.

For

● 4WD traction ● Passenger room ● Cargo space

Against

● Fuel economy ● Performance ● Noise

Summary

Pathfinder is sluggish off the line, though there's decent power between 30 and 60 mph to help you keep pace with traffic. At 3700 pounds or more, these sport-utility vehicles certainly aren't overpowered. The V-6 sounds agricultural and unrefined in this form, more like Nissan's 4-cylinder truck engine than a sports-car-derived engine. It's surprisingly harsh as well, feeling rough and reluctant to rev above about 3000 rpm. Pathfinder's suspension absorbs big bumps fairly well, though we're not convinced this is the best riding compact sport utility vehicle, as Nissan claims. With the ride height typical of the breed, Pathfinder has lots of body roll and a "tippy" feel in tight turns; it also tends to get blown about by winds. It's a high step up into the cabin and getting into the rear seat requires some contortions, but the long wheelbase means there's plenty of leg room in the rear, so front-seaters don't have to give up any of their space. The rear seat isn't really wide enough to hold three adults without crowding, but the SE's standard split rear backrests can be reclined slightly. Cargo volume is good, the load floor is wide and long, and, on the SE, those split-folding rear seatbacks lay flat if you tilt the cushions forward. We'd like more off-the-line punch and a less truck-like driving feel, but with its spacious interior and overall versatility, Pathfinder represents a sensible compromise between on-road wagon and off-road warrior.

Specifications

	3-door wagon
Wheelbase, in.	104.3
Overall length, in.	171.9
Overall width, in.	66.5
Overall height, in.	66.7
Front track, in.	56.1
Rear track, in.	54.5
Turn diameter, ft.	35.4
Curb weight, lbs.	3735
Cargo vol., cu. ft.	71.3
Fuel capacity, gal.	21.1
Seating capacity	5
Front headroom, in.	39.0
Front shoulder room, in.	54.7

Prices are accurate at time of printing; subject to manufacturer's change.

	3-door wagon
Front legroom, max., in.	41.9
Rear headroom, in.	36.8
Rear shoulder room, in.	54.7
Rear legroom, min., in.	41.0

Powertrain layout: longitudinal front engine/rear-wheel drive or on-demand 4WD.

Engines

	ohc V-6
Size, liters/cu. in.	3.0/181
Fuel delivery	TBI
Horsepower @ rpm	145 @ 4800
Torque (lbs./ft.) @ rpm	166 @ 2800
Availability	S

EPA city/highway mpg

5-speed OD manual	15/18
4-speed OD automatic	14/16

Prices

Nissan Pathfinder (1988 prices)	Retail Price	Dealer Invoice	Low Price
XE 3-door wagon, 5-speed	$15399	13326	14363
XE 3-door wagon, automatic	16549	14321	15435
SE 3-door wagon, 5-speed	17499	15144	16322
SE 3-door wagon, automatic	18399	15923	17161
Destination charge	250	250	250

Standard Equipment:

XE: 3.0-liter PFI V-6 engine, 5-speed manual or 4-speed automatic transmission, part-time 4-wheel drive with automatic locking front hubs, power steering and brakes, cloth reclining front bucket seats, split folding rear seat, tinted glass, dual outside mirrors, tachometer, coolant temperature gauge, trip odometer, digital clock, cruise control (with automatic transmission only), remote fuel door release, memory tilt steering column, rear defogger, console, skid plates, front and rear tow hooks, 215/75R15 all-season SBR tires. **SE adds:** power windows and locks, cruise control, power mirrors, AM/FM cassette with diversity antenna, rear side privacy glass, driver's seat height and lumbar support adjustments, reclining rear seatback, variable intermittent wipers, rear wiper/washer, right visor mirror, voltmeter, 235/75R15 tires on chromed wheels.

Optional Equipment:

Air conditioning	795	652	724
Sport Pkg., XE	900	738	819
SE	2000	1640	1820

XE: outside spare tire carrier, spare tire cover and wheel locks, limited-slip differential, 235/75R15 tires on chrome-spoke wheels. SE adds: flip-up glass sunroof, adjustable suspension, rear disc brakes, fender flares, 31×10.5R15 tires on alloy wheels.

Two-tone paint	300	246	273
Vehicle security system	159	131	145
Graphics	100	82	91
Fog lights	145	119	132
Roof luggage rack	155	127	141

Nissan Pulsar NX

What's New

A new cylinder head for the base 1.6-liter 4-cylinder engine boosts horsepower from 69 to 90 as the major change for Pulsar, Nissan's front-drive sporty coupe built off the Sentra platform. The new cylinder head has three valves per cylin-

Nissan Pulsar NX SE

der—two intake and one exhaust—instead of two. The 1.6-liter engine, standard in the XE version of Pulsar, retains its single-point fuel-injection system. The 1.6 is available with a 5-speed manual or 3-speed automatic transmission. The SE model returns with a double-overhead cam, 125-horsepower 1.8-liter engine with four valves per cylinder. A 5-speed manual or a 4-speed overdrive automatic is available on the SE. Both models come in 3-door hatchback styling with a removable hatch lid and roof panels. The optional fiberglass canopy roof called Sportbak that gave Pulsar the look of a sporty station wagon has been dropped. A folding 2-place rear seat is standard, giving Pulsar nominal seating for four. For 1989, Pulsar is covered by Nissan's new 36-month/36,000-mile basic warranty on the entire car. It replaces a 12-month, 12,500-mile warranty.

For

- Performance (SE) ● Handling/roadholding
- Driveability (5-speed)

Against

- Visibility ● Noise

Summary

Nissan's created a niche within the sporty-coupe market with the versatile Pulsar NX. Removable hatchlid and T-tops give Pulsar an open-air edge over non-convertible rivals. We haven't yet sampled the 90-horsepower 1.6-liter engine, but the 125-horsepower 1.8 coupled to a 5-speed manual gives pleasing if unspectacular performance. A firm and supportive front bucket places the driver within perfect reach of the smooth-working clutch and gearshift. The control layout is excellent. Handling is above average, and the responsive engine allows you to power through corners. A compliant ride is the rule, though the rear end can slam over really sharp brows in the road. Visibility is hampered by the low seating position and by busy bodywork to the rear and over the shoulders. The 2-place rear seat is fine for children and acceptable for adults only on short trips. Although interior and exterior assembly are tight, there's lots of wind and road noise even before you

> **KEY: ohv** = overhead valve; **ohc** = overhead cam; **dohc** = double overhead cam; **I** = inline cylinders; **V** = cylinders in V configuration; **flat** = horizontally opposed cylinders; **bbl.** = barrel (carburetor); **PFI** = port (multi-point) fuel injection; **TBI** = throttle-body (single-point) fuel injection; **rpm** = revolutions per minute; **OD** = overdrive transmission; **S** = standard; **O** = optional; **NA** = not available.

Prices are accurate at time of printing; subject to manufacturer's change.

Nissan

start removing roof panels. Popping the T-tops is simple; taking off the rear hatch is more of a chore, requiring removal of four screws, four bolts and the struts that hold up the hatch lid. Nissan provides a wrench and screwdriver, but you'll need two people to lift off the hatch, plus a place to store it, since it doesn't fit in the car (leaving you vulnerable to unexpected rain showers). While this isn't as convenient as a convertible, it's still quite versatile. Overall, Pulsar NX is a solidly built, fun-to-drive junior sports car with some unique attributes.

Specifications

	3-door hatchback
Wheelbase, in.	95.7
Overall length, in.	166.5
Overall width, in.	65.7
Overall height, in.	50.8
Front track, in.	56.7
Rear track, in.	56.7
Turn diameter, ft.	33.5
Curb weight, lbs.	2388
Cargo vol., cu. ft.	7.0
Fuel capacity, gal.	13.2
Seating capacity	4
Front headroom, in.	38.0
Front shoulder room, in.	NA
Front legroom, max., in.	44.2
Rear headroom, in.	33.9
Rear shoulder room, in.	NA
Rear legroom, min., in.	31.1

Powertrain layout: transverse front engine/front-wheel drive.

Engines

	ohc I-4	dohc I-4
Size, liters/cu. in.	1.6/97	1.8/110
Fuel delivery	TBI	PFI
Horsepower @ rpm	90 @ 6000	125 @ 6400
Torque (lbs./ft.) @ rpm	96 @ 3200	112 @ 4800
Availability	S	S[1]

EPA city/highway mpg

5-speed OD manual	26/34	23/29
3-speed automatic	24/29	
4-speed OD automatic		21/28

1. SE

Prices

Nissan Pulsar NX	Retail Price	Dealer Invoice	Low Price
XE 3-door hatchback, 5-speed	$11749	—	—
XE 3-door hatchback, automatic	12269	—	—
SE 3-door hatchback, 5-speed	12999	—	—
SE 3-door hatchback, automatic	13754	—	—

Dealer invoice, low price and destination charge not available at time of publication.

Standard Equipment:

XE: 1.6-liter TBI 4-cylinder engine, 5-speed manual or 3-speed automatic transmission, power steering and brakes, T-bar roof, removable hatchback lid, variable intermittent wipers, tachometer, coolant temperature gauge, trip odometer, AM/FM ST ET with diversity antenna, power mirrors, tilt steering column, center console, remote fuel door and hatch releases, reclining front cloth bucket seats, 185/70R13 tires. **SE** adds: 1.8-liter DOHC 16-valve PFI engine, 5-speed manual or 4-speed automatic transmission, tweed-type upholstery, 195/60R14 tires on alloy wheels.

Optional Equipment:	Retail Price	Dealer Invoice	Low Price
Air conditioning	795	—	—
Security system/alloy wheel locks, SE	199	—	—
Fog lights, SE	145	—	—
Sport graphics	105	—	—

Nissan Sentra

Nissan Sentra Standard 2-door

What's New

Nissan's most popular U.S. model line, the subcompact Sentra, gains 21 horsepower, while the 4-wheel-drive station wagon gets a permanently engaged 4WD system. Sentra's 1.6-liter 4-cylinder engine has a new cylinder head with three valves per cylinder instead of two, and horsepower jumps from 69 to 90. The engine still has the single-point fuel injection that became standard last year. The 4WD wagon trades its on-demand, part-time 4WD system for a new permanently engaged, full-time system that uses a viscous coupling to split torque between the front and rear wheels as needed for sufficient traction. Nissan now calls it the All-Wheel Drive wagon. The 4WD wagon comes only with a manual transmission; all other Sentras, which have front-wheel drive, are available with manual or a 3-speed automatic. In addition to the 5-door wagon, Sentra body styles include 2- and 4-door sedans, and a 3-door hatchback Sport Coupe; the 3-door hatchback sedan has been dropped. All have new grilles and taillamps this year, the first appearance change since Sentra was redesigned for the 1987 model year. Sentra gets a 36-month/36,000-mile basic warranty for 1989, in place of the 12-month/12,500-mile coverage Nissan used previously. Some Sentras are built at Nissan's Tennessee assembly plant and some are imported from Japan.

For

- Fuel economy
- 4WD traction (wagon)
- Handling/roadholding (SE)

Against

- Performance
- Passenger room
- Noise

Summary

Sentra doesn't offer extraordinary performance or have exceptionally low prices, but it does stand as a pretty good value among subcompacts. Sentra rates highly among cars under $20,000 in J.D. Power & Associates customer satisfaction surveys, so apparently it satisfies most owners. We haven't yet driven the 90-horsepower 1.6, but the 21 additional horsepower over last year's engine will be a welcome and needed boost. Our 1988 automatic-transmission GXE felt underpowered, especially with three or four aboard and the air conditioner on. We averaged about 27 mpg, however, so it was at least economical. Acceleration and fuel economy are both better with manual transmission. The sporty SE coupe has larger tires and a firmer suspension, so it rates a couple of notches above other Sentras in handling. Otherwise, ride, room, comfort and road manners are middle-of-the-subcompact-pack: not up there with the Honda Civic or Volkswagen Golf, but comparable to Ford Escort or Hyundai Excel. We rate the Standard 2-door sedan and E models as the best values among Sentras. XE, GXE and SE models offer more comfort and convenience features, but they're no longer "low-cost transportation" compared to such rivals as the Excel and Dodge Omni/Plymouth Horizon. The Sentra Standard and E models are indeed basic transportation, but they should be reliable transportation as well.

Specifications

Specifications	2-door notchback	4-door notchback	3-door Sport Coupe	5-door wagon
Wheelbase, in.	95.7	95.7	95.7	95.7
Overall length, in.	168.7	168.7	166.5	172.2
Overall width, in.	64.6	64.6	65.6	64.6
Overall height, in.	54.3	54.3	52.2	54.3
Front track, in.	56.3	56.3	56.5	56.3
Rear track, in.	56.3	56.3	56.5	56.3
Turn diameter, ft.	30.2	30.2	30.2	30.2
Curb weight, lbs.	2200	2231	2258	2304
Cargo vol., cu. ft.	12.0	12.0	16.0	24.0
Fuel capacity, gal.	13.2	13.2	13.2	13.2
Seating capacity	5	5	4	5
Front headroom, in.	38.2	38.2	37.0	38.2
Front shoulder room, in.	52.1	52.1	52.3	52.1
Front legroom, max., in.	41.8	41.8	41.6	41.8
Rear headroom, in.	36.8	36.8	29.2	39.3
Rear shoulder room, in.	51.3	52.1	50.3	52.1
Rear legroom, min., in.	31.4	31.4	NA	31.4

Powertrain layout: transverse front engine/front-wheel drive or permanent 4WD (wagon).

Engines	ohc I-4
Size, liters/cu. in.	1.6/97
Fuel delivery	TBI
Horsepower @ rpm	90 @ 6000
Torque (lbs./ft.) @ rpm	96 @ 3200
Availability	S

EPA city/highway mpg	
5-speed OD manual	28/36
3-speed automatic	26/30

Prices

Nissan Sentra	Retail Price	Dealer Invoice	Low Price
Standard 2-door notchback, 4-speed	$6849	—	—
E 2-door notchback, 5-speed	7999	—	—
E 2-door notchback, automatic	8869	—	—
XE 2-door notchback, 5-speed	9649	—	—
XE 2-door notchback, automatic	10544	—	—
E 4-door notchback, 5-speed	8549	—	—
E 4-door notchback, automatic	9419	—	—
XE 4-door notchback, 5-speed	10299	—	—
XE 4-door notchback, automatic	11194	—	—
E 5-door wagon, 5-speed	9224	—	—
E 5-door wagon, automatic	10094	—	—
XE 5-door wagon, 5-speed	10724	—	—
XE 5-door wagon, automatic	11619	—	—
XE 4WD 5-door wagon, 5-speed	11524	—	—
XE 4WD 5-door wagon, automatic	12419	—	—
XE 3-door coupe, 5-speed	10674	—	—
XE 3-door coupe, automatic	11569	—	—
SE 3-door coupe, 5-speed	11924	—	—
SE 3-door coupe, automatic	12444	—	—

Dealer invoice, low price and destination charge not available at time of publication.

Standard Equipment:

Standard: 1.6-liter TBI 4-cylinder engine, 4-speed manual transmission, power brakes, coolant temperature gauge, rear defogger, taillamp/brake failure warning, headlamps-on tone, reclining front bucket seats, center console, 155R13 all-season SBR tires. **E** adds: 5-speed manual or 3-speed automatic transmission, power steering and tilt steering column (with automatic transmission only), tinted glass, dual outside mirrors, black bodyside moldings, trip odometer, intermittent wipers. **XE** adds: dual remote mirrors, cloth upholstery, tilt steering column, tachometer (coupe), remote decklid/hatch releases, AM/FM ST (with automatic transmission only), power steering, rear wiper/washer (coupe), 175/70R13 all-season tires. **GXE** adds: wide bodyside moldings, upgraded cloth upholstery, AM/FM ST ET with diversity antenna, variable intermittent wipers, driver's seat lumbar support, tachometer, visor mirrors, large console with lid, digital clock. **SE** adds to XE coupe: pop-up sunroof, multiple driver's seat adjustments, visor mirrors, digital clock, large console with lid, AM/FM ST ET with diversity antenna, variable intermittent wipers, 185/60R14 tires on alloy wheels. **All Wheel Drive** wagon has permanent, full-time 4-wheel drive.

Optional Equipment:

	Retail Price	Dealer Invoice	Low Price
Air conditioning	795	—	—
Removable glass sunroof, XE coupe	450	—	—
Two-tone paint, SE coupe	300	—	—
Metallic paint, Standard	100	—	—

Nissan Stanza

What's New

The Stanza 5-door station wagon has been dropped, but Nissan plans to bring in a new wagon next spring under a new model name. The remaining body style is a compact-sized 4-door notchback sedan available in base E and more expensive GXE trim. A new grille, taillamps and interior fabrics are the only changes for 1989. While the wagon was available with front- or 4-wheel drive, the sedan comes only with front-drive. Stanza uses a 97-horsepower 2.0-liter 4-cylinder engine, hitched to either a 5-speed manual or 4-

---

KEY: ohv = overhead valve; **ohc** = overhead cam; **dohc** = double overhead cam; **I** = inline cylinders; **V** = cylinders in V configuration; **flat** = horizontally opposed cylinders; **bbl.** = barrel (carburetor); **PFI** = port (multi-point) fuel injection; **TBI** = throttle-body (single-point) fuel injection; **rpm** = revolutions per minute; **OD** = overdrive transmission; **S** = standard; **O** = optional; **NA** = not available.

Prices are accurate at time of printing; subject to manufacturer's change

Nissan

Nissan Stanza GXE

speed overdrive automatic transmission. All Nissans are covered by a 36-month/36,000-mile basic warranty for 1989. It replaces a 12-month/12,500-mile warranty.

For

● Fuel economy   ● Passenger room

Against

● Performance

Summary

Since an all-new wagon is on the way, we'll confine our remarks to the Stanza sedan. It's a pleasant, well-equipped compact priced below such Japanese rivals as the Honda Accord and Toyota Camry. Its main shortcoming is a lack of muscle: 97 horsepower for 2800 pounds. You frequently have to floor the throttle pedal to keep pace with traffic. The automatic transmission further handcuffs performance by changing quickly to the higher gears and by a reluctance to downshift promptly for passing. The car is at least economical; we averaged 27 mpg in our last test, which included a lot of city miles. Body lean is well controlled in turns but the power steering is light and the tires let you know early that they're not up to hard cornering. The penalty for the soft ride is a little too much bouncing over freeway dips. The driving position is workable, with a comfortable seat, clear and sensibly designed controls that are easy to reach, and good outward visibility. Interior room is fine for four adults and the trunk has more usable space than its 12-cubic foot volume would indicate, plus the rear seatbacks fold for extra cargo room. Stanza is a comfortable, well-built compact that simply needs more power.

Specifications

	4-door notchback
Wheelbase, in.	100.4
Overall length, in.	177.8
Overall width, in.	66.5
Overall height, in.	54.9

KEY: **ohv** = overhead valve; **ohc** = overhead cam; **dohc** = double overhead cam; **I** = inline cylinders; **V** = cylinders in V configuration; **flat** = horizontally opposed cylinders; **bbl.** = barrel (carburetor); **PFI** = port (multi-point) fuel injection; **TBI** = throttle-body (single-point) fuel injection; **rpm** = revolutions per minute; **OD** = overdrive transmission; **S** = standard; **O** = optional; **NA** = not available.

	4-door notchback
Front track, in.	55.7
Rear track, in.	55.7
Turn diameter, ft.	32.2
Curb weight, lbs.	2770
Cargo vol., cu. ft.	12.0
Fuel capacity, gal.	16.1
Seating capacity	5
Front headroom, in.	38.9
Front shoulder room, in.	NA
Front legroom, max., in.	42.0
Rear headroom, in.	37.4
Rear shoulder room, in.	NA
Rear legroom, min., in.	33.0

Powertrain layout: transverse front engine/front-wheel drive.

Engines

	ohc I-4
Size, liters/cu. in.	2.0/120
Fuel delivery	PFI
Horsepower @ rpm	97 @ 5200
Torque (lbs./ft.) @ rpm	114 @ 2800
Availability	S

EPA city/highway mpg

5-speed OD manual	22/28
4-speed OD automatic	21/27

Prices

Nissan Stanza	Retail Price	Dealer Invoice	Low Price
E 4-door notchback, 5-speed	$11849	—	—
E 4-door notchback, automatic	12599	—	—
GXE 4-door notchback, 5-speed	13799	—	—
GXE 4-door notchback, automatic	14549	—	—

Dealer invoice, low price and destination charge not available at time of publication.

Standard Equipment:

2.0-liter PFI 4-cylinder engine, 5-speed manual or 4-speed automatic transmission, power steering and brakes, tinted glass, reclining front bucket seats, coolant temperature gauge, dual trip odometers, tilt steering column, intermittent wipers, rear defogger, 185/70R14 tires. **GXE** adds: power mirrors, tinted upper windshield band, upgraded upholstery and carpet, cruise control, AM/FM cassette with diversity power antenna, variable intermittent wipers, tachometer, oil pressure and voltage gauges, driver's seat height and lumbar support adjustments, split folding rear seatbacks, visor mirrors.

Optional Equipment:

Air conditioning	795	—	—
Alloy wheels, GXE	420	—	—
Value Option Pkg., GXE	999	—	—
Air conditioning, power glass sunroof, vehicle security system, pinstripe.			
Two-tone paint, GXE	300	—	—

Nissan Van

What's New

Nissan's compact Van gets only minor equipment revisions this year. Returning models are base XE and plusher GXE models; the XE comes with a 5-speed manual or 4-speed overdrive automatic, while the GXE comes only with the automatic. The Nissan Van's mechanical layout is similar to

Prices are accurate at time of printing; subject to manufacturer's change.

Nissan Van GXE

them up. These Japanese mini-vans have quality assembly and low-speed maneuverability on their side, but for daily passenger duty, we prefer the front-engine/front-drive design of the Dodge Caravan and Plymouth Voyager.

Specifications

	4-door van
Wheelbase, in.	92.5
Overall length, in.	178.0
Overall width, in.	66.5
Overall height, in.	72.4
Front track, in.	56.3
Rear track, in.	55.1
Turn diameter, ft.	30.2
Curb weight, lbs.	3330
Cargo vol., cu. ft.	157.6
Fuel capacity, gal.	17.7
Seating capacity	7
Front headroom, in.	39.0
Front shoulder room, in.	57.1
Front legroom, max., in.	39.8
Rear headroom, in.	39.0
Rear shoulder room, in.	57.7
Rear legroom, min., in.	31.3

Powertrain layout: longitudinal mid-engine/rear-wheel drive.

Engines

	ohc I-4
Size, liters/cu. in.	2.4/146
Fuel delivery	TBI
Horsepower @ rpm	106 @ 4800
Torque (lbs./ft.) @ rpm	137 @ 2000
Availability	S

EPA city/highway mpg

5-speed OD manual	18/22
4-speed OD automatic	18/21

that of the compact vans from Mitsubishi and Toyota. The rear wheels are driven by a 4-cylinder engine mounted behind the front axle and between the front seats. Nissan uses a 106-horsepower 2.4-liter 4-cylinder engine borrowed from its compact pickup trucks. Seats for seven are standard, arranged in a 2-2-3 configuration from front to rear. The XE has a 2-place removable bench seat in the center, while the GXE substitutes removable, swiveling captain's chairs. The 3-place rear bench seat that's standard on both models also is removable. Nissan's Van has a sliding side door and a split rear liftgate that allows opening the window separately. Front and rear air conditioning units are standard. The Van, like all Nissans, gets a new 36-month/36,000-mile basic warranty for 1989. It replaces the previous 12-month/12,500-mile coverage.

For

•Passenger room •Cargo space •Maneuverability

Against

•Ride •Entry/exit

Summary

The Nissan Van and the other Japanese mini-vans have spacious, nicely done interiors, but their mid-engine designs dictate some serious compromises. Getting in and out of the front seats is made difficult by the need to climb over the front wheel wells, and once in, the engine blocks access to the rear compartment. Plus, to service the hard-to-reach engine, you first have to detach and lift cumbersome interior panels. Still, a couple of things make the Nissan Van a little different than the others. Its liftgate has a window that can be opened separately, and Nissan seems to have devoted more of its interior to passengers, at the expense of cargo room. There's more leg room for those in the middle and rear seats than in the Mitsubishi and Toyota vans, plus a slight improvement in head room (even with the optional dual sunroofs). There's precious little luggage space at the rear when all seats are in place, so you'll have to reduce passenger capacity to carry a lot of cargo. On the road, the Nissan Van gets adequate acceleration from the 2.4-liter engine, though it takes a heavy throttle foot to climb even small hills. Nissan's Van isn't as bouncy as the Mitsubishi Wagon on the road, but the suspension has less absorbency, so it bangs over bumps rather than soaking

Prices

Nissan Van (1988 prices)	Retail Price	Dealer Invoice	Low Price
XE 4-door van, 5-speed	$14799	12807	13803
XE 4-door van, automatic	15744	13624	14684
GXE 4-door van, automatic	17099	14797	15948
Destination charge	250	250	250

Standard Equipment:

2.4-liter TBI 4-cylinder engine, 5-speed manual or 4-speed automatic transmission, power steering, power brakes, front and rear air conditioning, 7-passenger seating including cloth reclining front bucket seats with driver's seat lumbar support, reclining rear bench seats, AM/FM ST ET with clock, cruise control (with automatic transmission only), tinted glass, front and rear heaters, dual outside mirrors with power right, tilt/swing-up steering column, tachometer, coolant temperature gauge, trip odometer, headlamps-on warning, variable intermittent wipers, full wheel covers, P195/75R14 tires. **GXE** adds: second-row captain's chairs, power windows and locks, privacy glass, dual power mirrors, AM/FM ST ET cassette, rear wiper/washer.

Optional Equipment:

Dual sunroofs & alloy wheels, GXE	1300	1066	1183
Includes flip-up front sunroof, sliding rear sunroof and 205/70R14 tires.			
Two-tone paint	300	246	273
Roof luggage rack	155	127	141

Prices are accurate at time of printing; subject to manufacturer's change.

Nissan 240SX

Nissan 240SX SE 3-door

What's New

An all-new sports coupe debuts this fall as a replacement for the 200SX. The new model, called 240SX to denote its larger engine, offers among its options an innovative heads-up instrument display. Like its predecessor, the 240SX has rear-wheel drive and is available as a 2-door XE Coupe and a 3-door hatchback, the SE Fastback. Both ride a 97.4-inch wheelbase, nearly two inches longer than the 200SX. Overall length is up by 3.6 inches to 178. The only engine offered is a 2.4-liter 4-cylinder with three valves per cylinder—two intake, one exhaust. Nissan rates it at 140 horsepower and 152 pounds/feet torque. That's 41 more horsepower and 36 more pounds/feet torque than the 2.0 four standard in '88, but 25 fewer horsepower and 16 fewer pounds/feet torque than last year's optional 3.0-liter V-6. Transmission choices are a 5-speed manual and 4-speed automatic. Four-wheel disc brakes are standard; the SE offers optional anti-lock brakes. Last year's independent rear suspension is replaced by an all-new multi-link, independent design. The optional heads-up display (HUD) is available only on the XE coupe. It projects the digital speedometer reading onto the lower left corner of the windshield, where the driver should be able to read it without taking his or her eyes from the road. Nissan's 1989 Maxima also offers a HUD, and Oldsmobile will offer one on the Cutlass Supreme as a limited-production option in spring 1989. Nissan has replaced its previous 12-month/12,500-mile warranty with a 36-month/36,000 plan for 1989.

For

- Handling/roadholding
- Ergonomics
- Anti-lock brakes (SE)
- Driveability

Against

- Ride
- Air conditioner
- Rear-seat room
- Cargo space (XE Coupe)

Summary

Nissan hopes sporty-coupe shoppers will recall the original 240Z when they consider the 240SX. Introduced as a 1970 model, the 240Z became a legend for its budget performance and smart styling. Nissan says one of its chief goals with the 240SX is to offer a sleek sporty coupe that's not expensive to own. To this end, it did without a sophisticated engine and took pains to design simple-to-repair exterior and chassis pieces. An SE with popular options such as air conditioning, sunroof, and alloy wheels probably will list for $15,000-16,000, with another $1000 or so for anti-lock brakes. Based on test drives of early production models, those prices would make the 240SX a decent value, despite some drawbacks. Those we drove suffered a rough, unforgiving ride, though handling was balanced and predictable, and high-speed stability was outstanding. An SE, even with the 5-speed, felt short of power, though. Nissan says it wanted to give the 240SX good low-end and midrange performance, and did not want the added cost, weight and insurance penalties of a turbocharger or a 16-valve design. Decent low-end and midrange power the 240SX has, but the engine doesn't feel particularly eager to rev and seems to run out of steam at about 4400 rpm, just where smaller 16-valvers from Honda and Toyota get their second wind. Around-town performance should be perfectly adequate, better even than some of those 16-valvers that need lots of revs to deliver power. Still, the hot styling calls out for more spirit under the hood. The cockpit is well-suited to spirited driving, however, with clear gauges, gimmick-free controls and supportive, comfortable seats. The cramped back seat is suitable only for children, however, and the air conditioner feels weak in hot, muggy conditions. But while the 240SX doesn't hit the high notes of some rivals, its overall competence and pleasing design are qualities we dare not dismiss.

Specifications

	2-door notchback	3-door hatchback
Wheelbase, in.	97.4	97.4
Overall length, in.	178.0	178.0
Overall width, in.	66.5	66.5
Overall height, in.	50.8	50.8
Front track, in.	57.7	57.7
Rear track, in.	57.5	57.5
Turn diameter, ft.	30.8	30.8
Curb weight, lbs.	2657	2690
Cargo vol., cu. ft.	8.6	14.2
Fuel capacity, gal.	15.9	15.9
Seating capacity	4	4
Front headroom, in.	37.8	37.8
Front shoulder room, in.	52.0	52.0
Front legroom, max., in.	42.0	42.0
Rear headroom, in.	34.5	33.3
Rear shoulder room, in.	52.0	51.8
Rear legroom, min., in.	23.8	23.8

Powertrain layout: longitudinal front engine/rear-wheel drive.

Engines

	ohc I-4
Size, liters/cu. in.	2.4/146
Fuel delivery	PFI
Horsepower @ rpm	140 @ 5600
Torque (lbs./ft.) @ rpm	152 @ 4400
Availability	S

EPA city/highway mpg

5-speed OD manual	20/26
4-speed OD automatic	20/25

Prices

Nissan 240SX	Retail Price	Dealer Invoice	Low Price
XE 2-door notchback, 5-speed	$12999	—	—
XE 2-door notchback, automatic	13759	—	—
SE 3-door hatchback, 5-speed	13199	—	—
SE 3-door hatchback, automatic	13959	—	—

Dealer invoice, low price and destination charge not available at time of publication.

Prices are accurate at time of printing; subject to manufacturer's change.

Standard Equipment:

XE: 2.4-liter PFI 4-cylinder engine, 5-speed manual or 4-speed automatic transmission, power steering, power 4-wheel disc brakes, reclining front bucket seats with driver's side lumbar support adjustment, motorized front shoulder belts with manual lap belts, outboard rear lap/shoulder belts, tilt steering column, tachometer, coolant temperature gauge, trip odometer, digital clock, variable intermittent wipers, tinted glass, dual remote mirrors, AM/FM ST ET with diversity antenna, rear defogger, door pockets, remote fuel door and decklid/liftgate releases, 195/60R15 all-season M + S SBR tires.

Optional Equipment:

Anti-lock braking system, SE	1400	—	—
Requires Sport Pkg. & air conditioning.			
Air conditioning	795	—	—
Power Convenience Group, XE	999	—	—
Active speakers, head-up display, cruise control, power windows and locks, power mirrors, radio upgrade with cassette, map and footwell courtesy lights.			
Power Convenience Group, SE	799	—	—
Cruise control, power windows and locks, power mirrors, radio upgrade with cassette, rear wiper/washer.			
Sport Pkg., SE	799	—	—
Cruise control, sport suspension, front air dam, rear spoiler, leather-wrapped steering wheel and shift knob, upgraded upholstery.			
Power Convenience Group & Sport Pkg., SE	1399	—	—
Two-tone paint	300	—	—
Power glass sunroof, XE	800	—	—
Removable glass sunroof, SE	450	—	—
XE requires Power Convenience Group; SE requires Power Convenience Group or Sport Pkg.			

Nissan 300ZX

Nissan 300ZX Turbo

What's New

Nissan's rear-drive sports car is carried over unchanged in anticipation of an all-new model due in the spring of 1989. Nissan hasn't released information on the new Z-car, except to acknowledge that while the current one has grown into somewhat of a hefty grand tourer, the new car will be oriented more toward outright performance. The 1990 model will have more horsepower than the current edition and may feature a high-performance version with two turbochargers. For 1989, the base engine stays a 165-horsepower 3.0-liter V-6; the optional 205-horsepower turbocharged 3.0 also is carried over. The naturally aspirated 300ZX is available in 2+2 or 2-seat versions, while the 300ZX Turbo only comes in 2-seat guise. The current design was introduced for the 1984 model year. For '89, the 300ZX is covered by Nissan's new basic 36-month/36,000-mile warranty, which replaces the previous 12-month/12,500-mile coverage.

For

● Performance   ● Handling/roadholding   ● Driver seating

Against

● Fuel economy   ● Entry/exit   ● Interior room

Summary

The 300ZX is competitive in price and performance with its nearest rival, the Toyota Supra. Both offer base models with naturally aspirated 6-cylinder engines and high-performance editions with turbocharged engines. The Nissan has consistently outsold the Toyota, despite what we see as several Supra advantages. Supra is available with anti-lock brakes, it rides better and the interior is more modern and harmonious. A Mazda RX-7 also gives the 300ZX a run for its money, and offers a turbo model with optional anti-lock brakes. And don't overlook the rear-drive Mitsubishi Starion/Chrysler Conquest as a less costly alternative. We agree with critics that the 300ZX looks and feels somewhat dated. But don't take that to imply that it doesn't have performance and handling far beyond that of most cars; it does, especially in turbo trim. It's just that the 300ZX goes about its business with a nod toward comfort and convenience features that others in its class might overlook in their quest for fast 0-60 mph times and sky-high cornering limits. That doesn't make it bad, just different. Rumors about the bare-knuckles nature of the all-new 1990 Z-car, however, indicate that even Nissan believes it's time for a change.

Specifications

	3-door hatchback	3-door 2+2 hatchback
Wheelbase, in.	91.3	99.2
Overall length, in.	170.7	178.5
Overall width, in.	67.9	67.9
Overall height, in.	49.7	49.7
Front track, in.	55.7	55.7
Rear track, in.	56.5	56.5
Turn diameter, ft.	32.2	34.8
Curb weight, lbs.	3139	3265
Cargo vol., cu. ft.	14.7	20.3
Fuel capacity, gal.	19.0	19.0
Seating capacity	2	4
Front headroom, in.	36.6	37.2
Front shoulder room, in.	54.0	54.2
Front legroom, max., in.	43.6	43.6
Rear headroom, in.	—	34.3
Rear shoulder room, in.	—	NA
Rear legroom, min., in.	—	25.3

Powertrain layout: longitudinal front engine/rear-wheel drive.

Engines	ohc V-6	Turbo ohc V-6
Size, liters/cu. in.	3.0/181	3.0/181
Fuel delivery	PFI	PFI
Horsepower @ rpm	165 @ 5200	205 @ 5300

> **KEY: ohv** = overhead valve; **ohc** = overhead cam; **dohc** = double overhead cam; **I** = inline cylinders; **V** = cylinders in V configuration; **flat** = horizontally opposed cylinders; **bbl.** = barrel (carburetor); **PFI** = port (multi-point) fuel injection; **TBI** = throttle-body (single-point) fuel injection; **rpm** = revolutions per minute; **OD** = overdrive transmission; **S** = standard; **O** = optional; **NA** = not available.

Prices are accurate at time of printing; subject to manufacturer's change.

	ohc V-6	Turbo ohc V-6
Torque (lbs./ft.) @ rpm	173 @ 4000	227 @ 3600
Availability	S	S[1]
EPA city/highway mpg		
5-speed OD manual	17/25	17/25
4-speed OD automatic	18/26	17/24
1. Turbo		

Prices

Nissan 300ZX	Retail Price	Dealer Invoice	Low Price
GS 3-door hatchback, 5-speed	$22299	—	—
GS 3-door hatchback, automatic	23049	—	—
Turbo 3-door hatchback, 5-speed	24699	—	—
Turbo 3-door hatchback, automatic	25499	—	—
GS 2+2 3-door hatchback, 5-speed	23499	—	—
GS 2+2 3-door hatchback, automatic	24199	—	—

Dealer invoice, low price and destination charge not available at time of publication.

Standard Equipment:

GS: 3.0-liter PFI V-6 engine, 5-speed manual or 4-speed automatic transmission, 4-wheel power disc brakes, power steering, air conditioning, AM/FM ST ET cassette, cloth reclining front bucket seats with 8-way adjustments, front seats, tilt steering column, theft-deterrent system, power windows and locks, tachometer, trip odometer, coolant temperature and oil pressure gauges, tinted glass, 215/60R15 tires on alloy wheels. **Turbo** adds: turbocharged engine, adjustable shock absorbers, uprated brakes, headlight washers, turbo boost gauge, 225/50VR16 tires.

Optional Equipment:

Electronic Equipment Pkg.	1375	—	—

Premium stereo with EQ, automatic temperature control, audio and cruise control switches in steering wheel hub, power driver's seat, heated mirrors, illuminated entry, left visor mirror.

Leather Trim Pkg., 2-seat	1055	—	—
2+2s	1215	—	—

Partial leather seating surfaces, imitation leather door panel inserts, cargo area cover, bronze-tinted glass; requires Electronic Equipment Pkg.

Digital instrumentation pkg	710	—	—

Digital instruments, trip computer, Auto Check System, oil temperature gauge; requires Electronic Equipment and Leather Pkg.

Pearlglow paint	350	—	—

Oldsmobile Custom Cruiser

What's New

The full-size, rear-drive Custom Cruiser wagon gains 3-point shoulder belts for the outboard rear seating positions, but otherwise is a carryover. Custom Cruiser is built from the same design as the Buick LeSabre/Electra Estate, Chevrolet Caprice and Pontiac Safari wagons. All use a 140-horsepower, carbureted 5.0-liter V-8 engine and 4-speed overdrive automatic transmission. With a folding, rear-facing third seat standard, there's seating for eight. A 5000-pound towing package is optional on Custom Cruiser, whose design dates to the 1977 model year.

For

- Room/comfort • Cargo room
- Trailer towing capacity

Oldsmobile Custom Cruiser

Against

- Fuel economy • Size and weight
- Handling/maneuverability

Summary

Full-size rear-drive wagons such as the Custom Cruiser have become scarce, yet they still have enough of a following to justify continuing their production. It's easy to see why when you look at the plus points: Room for eight people, generous cargo capacity and 5000-pound trailer towing capability. You'll have to look at vans to beat those figures, and vans just aren't everyone's cup of tea. The Cruiser's pluses come from its size and brawn, but so do its minuses. The V-8 works up a heavy thirst hauling around two tons of curb weight, and that thirst will eventually doom these vehicles if the price of gasoline soars again. They can be cumbersome to drive in urban areas because of their ponderous handling, so they're much more at home cruising on the open road, where they have room to stretch their legs. We prefer something smaller and more agile, but since station wagons are supposed to be roomy beasts of burden, there are still good reasons for buying one of these. In their favor, they aren't much more expensive than most front-drive, mid-size wagons (such as the Oldsmobile Cutlass Cruiser) that have less cargo volume and lower trailer-towing limits.

Specifications

	5-door wagon
Wheelbase, in.	116.0
Overall length, in.	220.3
Overall width, in.	79.8
Overall height, in.	58.5
Front track, in.	62.1
Rear track, in.	64.1
Turn diameter, ft.	39.2
Curb weight, lbs.	4221
Cargo vol., cu. ft.	87.2
Fuel capacity, gal.	22.0
Seating capacity	8
Front headroom, in.	39.6
Front shoulder room, in.	60.6
Front legroom, max., in.	42.2
Rear headroom, in.	39.3
Rear shoulder room, in.	60.5
Rear legroom, min., in.	37.2

Powertrain layout: longitudinal front engine/rear-wheel drive.

Prices are accurate at time of printing; subject to manufacturer's change.

Engines

	ohv V-8
Size, liters/cu. in.	5.0/307
Fuel delivery	4 bbl.
Horsepower @ rpm	140 @ 3200
Torque (lbs./ft.) @ rpm	225 @ 2000
Availability	S
EPA city/highway mpg	
4-speed OD automatic	17/24

Prices

Oldsmobile Custom Cruiser	Retail Price	Dealer Invoice	Low Price
5-door wagon	$16795	$14494	$15645
Destination charge	505	505	505

Standard Equipment:

5.0-liter 4bbl. V-8, 4-speed automatic transmission, power steering and brakes, power tailgate window, air conditioning, tinted glass, left remote mirror, trip odometer, AM/FM ST ET, right visor mirror, chime tones, rear storage compartment lock, 55/45 bench seat, outboard rear lap/shoulder belts, rear-facing third seat, P225/75R15 tires.

Optional Equipment:

Option Pkg. 1SB	1072	911	986
Intermittent wipers, cruise control, tilt steering column, wire wheel covers, door edge guards, floormats, power mirrors, passenger recliner, power antenna, Reminder Pkg., rear defogger, accent stripe.			
Option Pkg. 1SC	2493	2119	2294
Pkg. 1SB plus power windows and locks, luggage rack, bodyside moldings, lighted visor mirrors, power driver's seat, AM/FM ST ET cassette, side paneling.			
Option Pkg. 1SD	3095	2631	2847
Pkg. 1SC plus automatic climate control, dome/reading lamps, Illumination Pkg., cornering lamps, exterior lamp monitor, automatic day/ night mirror, power passenger seat.			
Passenger recliner	45	38	41
Power locks	245	208	225
Power windows	285	242	262
Automatic leveling	175	149	161
Limited-slip differential	100	85	92
Engine block heater	18	15	17
Instrument panel cluster	66	56	61
Trailering Pkg.	96	82	88
California emissions pkg.	100	85	92

Oldsmobile Cutlass Calais

What's New

A new 3.3-liter V-6 arrives as the largest engine available in Calais, Oldsmobile's front-drive compact, and all models get a facelift and standard 3-point rear shoulder belts. The new V-6, called "3300" by GM, is rated at 160 horsepower and 185 pounds/feet of torque. It replaces last year's 3.0-liter V-6, which was rated at 125 horsepower and 150 pounds/feet torque. Late in the model year, a high-output 2.3-liter 4-cylinder Quad 4 engine is scheduled to become optional in the sporty Calais International Series model. The HO Quad 4 produces 185 horsepower, 25 more than the base Quad 4; torque is the same for both versions at 160 pounds/feet. A 2.5-liter 4-cylinder remains the base engine for Calais, but horsepower has been bumped to 110 from 98

Oldsmobile Cutlass Calais SL 4-door

thanks to cylinder-head revisions and a larger throttle body for its single-point fuel injection system. The 2.5-liter 4-cylinder is standard on all models except the International Series, which comes with the Quad 4. Both 4-cylinder engines come with either a 5-speed manual or 3-speed automatic. The 3.3 V-6 is optional on the Calais S and SL models and comes only with automatic. Oldsmobile has dropped the Firenza subcompact, so Calais is the smallest, least expensive model in its '89 lineup. A new entry-level Calais, the Value Leader (VL) is available only with the 2.5-liter four and limited options to keep the price down. All models have new grilles, front and rear fascias, exterior trim and taillamps, plus the International Series has new ground-effects front and side body trim, monochromatic paint and fog lamps. The "I Series" goes on sale about mid-year and rides on new 205/55R16 tires. A power sunroof is a new option. All Calais models are available in 2-door coupe or 4-door sedan styling. Calais is built from the same design as the Buick Skylark and Pontiac Grand Am.

For

● Performance (V-6, Quad 4) ● Visibility

Against

● Noise (Quad 4) ● Rear seat room ● Trunk space

Summary

Calais' base 2.5-liter four delivers adequate acceleration and decent fuel economy, but insufficient amounts of either to rate as the best choice. We haven't been able to measure fuel economy on the new 3.3-liter V-6, but rate this engine as the quietest and most refined of the three. It also has the most low-end torque, making it the best with automatic transmission. Surprisingly, the V-6 doesn't provide a real strong surge of power initially or instant passing response. We think most of that is due to the gear ratios for the 3-speed automatic, which are chosen for economy rather than performance. The Quad 4 has a much different character; it develops its power at much higher engine speeds, making it better suited for a manual transmission. Few Calais buyers want manual shift, however, and the one available isn't the smoothest or easiest to operate. The

---

KEY: ohv = overhead valve; **ohc** = overhead cam; **dohc** = double overhead cam; **I** = inline cylinders; **V** = cylinders in V configuration; **flat** = horizontally opposed cylinders; **bbl.** = barrel (carburetor); **PFI** = port (multi-point) fuel injection; **TBI** = throttle-body (single-point) fuel injection; **rpm** = revolutions per minute; **OD** = overdrive transmission; **S** = standard; **O** = optional; **NA** = not available.

Prices are accurate at time of printing; subject to manufacturer's change.

Oldsmobile

Quad 4 is much noisier than the V-6. We suggest you try the V-6 and Quad 4 to see which one better fits your needs. Like its cousins at Buick and Pontiac, Calais suffers from too little interior and cargo space compared to Japanese compacts such as the Honda Accord and Toyota Camry. While we can't get too excited about this car, it does offer more engine choices than most other compacts.

Specifications

	2-door notchback	4-door notchback
Wheelbase, in.	103.4	103.4
Overall length, in.	178.8	178.8
Overall width, in.	66.7	66.7
Overall height, in.	52.4	52.4
Front track, in.	55.6	55.6
Rear track, in.	55.2	55.2
Turn diameter, ft.	35.4	35.4
Curb weight, lbs.	2512	2573
Cargo vol., cu. ft.	13.2	13.2
Fuel capacity, gal.	13.6	13.6
Seating capacity	5	5
Front headroom, in.	37.7	37.7
Front shoulder room, in.	53.9	53.9
Front legroom, max., in.	42.9	42.9
Rear headroom, in.	37.1	37.1
Rear shoulder room, in.	55.1	53.5
Rear legroom, min., in.	34.3	34.3

Powertrain layout: transverse front engine/front-wheel drive.

Engines

	ohv I-4	dohc I-4	ohv V-6
Size, liters/cu. in.	2.5/151	2.3/138	3.3/204
Fuel delivery	TBI	PFI	PFI
Horsepower @ rpm	110 @ 5200	150 @ 5200	160 @ 5200
Torque (lbs./ft.) @ rpm	135 @ 3200	160 @ 4000	185 @ 3200
Availability	S	O[1]	O

EPA city/highway mpg

5-speed OD manual	23/33	24/35	
3-speed automatic	23/30	23/32	20/27

1. Standard, I-Series

Prices

Oldsmobile Cutlass Calais	Retail Price	Dealer Invoice	Low Price
2-door notchback	$9995	$9225	$9610
4-door notchback	9995	9225	9610
S 2-door notchback	10895	9729	10312
S 4-door notchback	10995	9819	10407
SL 2-door notchback	11895	10622	11259
SL 4-door notchback	11995	10712	11354
I Series 2-door notchback	14395	12855	13625
I Series 4-door notchback	14495	12944	13720
Destination charge	425	425	425

Standard Equipment:

2.5-liter TBI 4-cylinder engine, 5-speed manual transmission, power steering and brakes, cloth front bucket seats (reclining on 2-door), AM/FM ST ET,

---

> **KEY: ohv** = overhead valve; **ohc** = overhead cam; **dohc** = double overhead cam; **I** = inline cylinders; **V** = cylinders in V configuration; **flat** = horizontally opposed cylinders; **bbl.** = barrel (carburetor); **PFI** = port (multi-point) fuel injection; **TBI** = throttle-body (single-point) fuel injection; **rpm** = revolutions per minute; **OD** = overdrive transmission; **S** = standard; **O** = optional; **NA** = not available.

---

trip odometer, dual outside mirrors, tinted glass, automatic front seatbelts, outboard rear lap/shoulder belts, 185/80R13 all-season SBR tires. **S** adds: full-length console with armrest and storage bin, left remote mirror, reclining front seatbacks. **SL** adds: Convenience Group (reading lights, lighted right visor mirror, misc. lights), 4-way driver's seat, split folding rear seat, upgraded steering wheel, two-tone paint. **International Series** adds: 2.3-liter DOHC PFI Quad 4 engine, FE3 suspension, air conditioning, door pockets, floormats, fog lamps, rocker panel extensions and wheel flares, Driver Information System (trip computer, service reminder), rallye instruments (tachometer, coolant temperature, oil pressure, voltmeter), AM/FM ST ET cassette, tilt steering column, leather-wrapped steering wheel and shift handle, power decklid release, intermittent wipers, 205/55R16 tires on alloy wheels.

Optional Equipment:

	Retail Price	Dealer Invoice	Low Price
2.3-liter Quad 4, S & SL	660	561	607
3.3-liter V-6, S & SL	710	604	653
3-speed automatic transmission, base	380	323	350
Others	490	417	451
Column shift, SL 4-door (credit)	(110)	(94)	(94)
Option Pkg. 1SB, base 2-door	338	287	311
Base 4-door	388	330	357
Tilt steering column, floormats, 4-way driver's seat, power locks.			
Option Pkg. 1SC, base 2-door	513	436	472
Base 4-door	563	479	518
Pkg. 1 plus air conditioning.			
Option Pkg. 1SB, S	484	411	445
Tilt steering column, floormats, intermittent wipers, Convenience Group, air conditioning.			
Option Pkg. 1SC, S 2-door	809	688	744
S 4-door	934	794	859
Pkg. 1SB plus power windows and locks, power antenna, 4-way driver's seat, remote mirrors, power decklid release.			
Option Pkg. 1SB, SL	633	538	582
Tilt steering column, floormats, intermittent wipers, cruise control, air conditioner, power antenna.			
Option Pkg. 1SC, SL 2-door	1208	1027	1111
SL 4-door	1333	1133	1226
Pkg. 1SB plus power decklid release, power windows and locks, remote mirrors, power driver's seat, Driver Information System.			
Option Pkg. 1SB, I Series 2-door	845	718	777
I-Series 4-door	970	825	892
Power windows and locks, power driver's seat, remote mirrors, cruise control, power antenna.			
Power locks, 2-doors	145	123	133
4-doors	195	166	179
Rear defogger	145	123	133
Engine block heater	18	15	17
185/80R13 WSW tires, base & S	68	58	63
SL (credit)	(26)	(22)	(22)
California emissions pkg.	100	85	92
Accent stripe, S & SL	45	38	41
Touring Car Ride & Handling Pkg., S	491	417	452
SL	397	337	365
Removable sunroof (NA base)	350	298	322

Oldsmobile Cutlass Calais SL 2-door

	Retail Price	Dealer Invoice	Low Price
Power windows (NA base), 2-doors	220	187	202
4-doors	295	251	271
Cruise control (NA base)	175	149	161
14" alloy wheels, S	319	271	293
SL	250	213	230
13" alloy wheels, S	215	183	198
Rallye instruments, S & SL	126	107	116
AM/FM ST ET cassette, S & SL	147	125	135
w/EQ, SL	297	252	273
w/EQ, I Series	150	128	138
Decklid luggage rack (NA base)	115	98	106
Quad 4 Appearance Pkg., S & SL	1130	961	1040

Oldsmobile Cutlass Ciera

Oldsmobile Cutlass Cruiser wagon

What's New

Fresher styling and a new V-6 engine are the major changes for Ciera, Oldsmobile's front-drive intermediate built from the GM A-body design. The most dramatic change is on the 4-door sedan, which gets a new roof design and more steeply sloped rear window, adopting the rounded look the 2-door coupe has sported since late 1986. All models have new grilles, rear end panels, tail lamps and wide body moldings. The new V-6 is GM's 3.3-liter "3300" rated at 160 horsepower and 185 pounds/feet torque. It replaces a 3.8-liter V-6 (150 horsepower, 200 pounds/feet torque) as Ciera's top engine. An optional 125-horsepower 2.8-liter V-6 returns unchanged, while the base 2.5-liter 4-cylinder gains 12 horsepower from improvements to the cylinder head and fuel-injection system. A 3-speed automatic is standard with the 4-cylinder and a 4-speed overdrive automatic is required with the V-6 engines. Ciera, which debuted for 1982, is similar to the Buick Century, Chevrolet Celebrity and Pontiac 6000. All four divisions offer 4-door sedans and 5-door wagons, while only Buick and Olds still offer 2-door coupes. All models have 3-point shoulder belts for the outboard rear seating positions and an electric speedometer standard. A power sunroof is a new option and wheel locks now come with the optional aluminum wheels. All three body styles come in base and SL (called Brougham last year) price levels, while the coupe and sedan also come in International Series trim. The 3300 V-6 is standard on the International Series, optional in the others.

For

- Performance (3.3-liter V-6)   • Passenger room
- Trunk space   • Quietness
- Handling (International Series)

Against

- Performance (4-cylinder)
- Handling (base suspension)

Summary

Ciera is similar to the Buick Century in packaging and price, and has only one mechanical advantage over the Chevy and Pontiac versions of this design, the new 3.3-liter V-6, which isn't offered on the Celebrity or 6000. Our initial impression is that the 3.3 nearly matches the 3.8-liter V-6 it replaces in acceleration, and it should be stingier with fuel. Where the 2.8-liter V-6 is adequate for this car, the 3.3 gives it more pep for brisk takeoffs from low speeds and prompt passing response on the highway. Most Cieras are sold with the soft base suspension, which provides a cushy ride but lacks good cornering grip and high-speed stability. The International Series improves handling with its firmer suspension and wider tires, yet Olds has managed to retain more ride comfort than some of its corporate cousins have in their sporty versions of the GM A-body. The I-Series also has a tastefully done, functional interior. Ciera and the other A-bodies aren't as modern as the Ford Taurus and Mercury Sable, yet they offer comparable passenger and cargo room at competitive prices.

Specifications

	2-door notchback	4-door notchback	5-door wagon
Wheelbase, in.	104.9	104.9	104.9
Overall length, in.	190.3	190.3	194.4
Overall width, in.	69.5	69.5	69.5
Overall height, in.	54.1	54.1	54.5
Front track, in.	58.7	58.7	58.7
Rear track, in.	57.0	57.0	57.0
Turn diameter, ft.	38.1	38.1	38.1
Curb weight, lbs.	2736	2764	2913
Cargo vol., cu. ft.	15.8	15.8	74.4
Fuel capacity, gal.	15.7	15.7	15.7
Seating capacity	6	6	8
Front headroom, in.	38.6	38.6	38.6
Front shoulder room, in.	55.9	55.9	56.2
Front legroom, max., in.	42.1	42.1	42.1
Rear headroom, in.	37.6	38.0	38.9
Rear shoulder room, in.	56.9	55.9	56.2
Rear legroom, min., in.	35.8	35.8	NA

Powertrain layout: transverse front engine/front wheel drive.

Engines

	ohv I-4	ohv V-6	ohv V-6
Size, liters/cu. in.	2.5/151	2.8/173	3.3/204
Fuel delivery	TBI	PFI	PFI
Horsepower @ rpm	110 @ 5200	125 @ 4500	160 @ 5200
Torque (lbs./ft.) @ rpm	135 @ 3200	160 @ 3600	185 3200
Availability	S	O	O[1]

EPA city/highway mpg

3-speed automatic	23/30		
4-speed OD automatic		20/29	20/29

1. Standard, I-Series

Prices

Oldsmobile Cutlass Ciera	Retail Price	Dealer Invoice	Low Price
2-door notchback	$11695	$10327	$11011
4-door notchback	12195	10524	11360
SL 2-door notchback	12695	11210	11953

Prices are accurate at time of printing; subject to manufacturer's change.

Oldsmobile

	Retail Price	Dealer Invoice	Low Price
SL 4-door notchback	13495	11646	12571
I Series 2-door notchback	15995	14124	15060
I Series 4-door notchback	16795	14494	15645
Cruiser 5-door wagon	12995	11215	12105
Cruiser SL 5-door wagon	13995	12078	13037
Destination charge	450	450	450

Standard Equipment:

2.5-liter TBI 4-cylinder engine, 3-speed automatic transmission, power steering and brakes, front bench seat with armrest, left remote and right manual mirrors, outboard rear lap/shoulder belts, AM/FM ST ET, tinted glass, 185/75R14 all-season SBR tires. **SL** adds: 55/45 front seat, Convenience Group (reading lamps, lighted right visor mirror, chime tones, misc. lights), AM/FM ST ET cassette, rear parcel shelf storage bin, power decklid release. **International Series** adds: 3.3-liter PFI V-6, 4-speed automatic transmission, FE3 suspension, front air dam with fog lights, center console with floorshift, Driver Information System (trip computer, service reminder), rallye instruments (tachometer, coolant temperature, oil pressure, voltmeter, trip odometer), tilt steering column, leather-wrapped steering wheel, intermittent wipers, 215/60R14 tires on alloy wheels.

Optional Equipment:

	Retail	Dealer	Low
2.8-liter V-6, exc. I Series	610	519	561
3.3-liter V-6, exc. I Series	710	604	653
4-speed auto trans, exc. I Series	175	149	161
Option Pkg. 1SB, base	631	536	581
Air conditioning, tilt steering column, intermittent wipers, Convenience Group.			
Option Pkg. 1SC, base 2-door	891	757	820
Base 4-door	1026	872	944
Pkg. 1SB plus power windows and locks, floormats, door edge guards, power antenna.			
Option Pkg. 1SD, base 2-door	1252	1064	1152
Base 4-door	1387	1179	1276
Pkg. 1SC plus Molding Pkg., seatback recliners, remote mirrors, power driver's seat.			
Option Pkg. 1SB, SL 2-door	885	752	814
SL 4-door	895	761	823
Air conditioning, tilt steering column, intermittent wipers, floormats, door edge guards, cruise control, power antenna.			
Option Pkg. 1SC, SL 2-door	1250	1063	1150
SL 4-door	1385	1177	1274
Pkg. 1SB plus power windows and locks, seatback recliners, remote mirrors, power driver's seat.			
Option Pkg. 1SB, I Series 2-door	870	740	800
I Series 4-door	995	846	915
Power windows and locks, cruise control, power driver's seat, power antenna.			
Option Pkg. 1SB, base wagon	1163	989	1070
Air conditioning, cruise control, tilt steering column, intermittent wipers, door edge guards, luggage rack, divided bench seat with storage armrest, floormats.			
Option Pkg. 1SC, base wagon	1841	1565	1694
Pkg. 1SB plus power windows and locks, Convenience Group, rear-facing third seat, power antenna.			
Option Pkg. 1SD, base wagon	2202	1872	2026
Pkg. 1SC plus Molding Pkg., seatback recliners, remote mirrors, power driver's seat.			
Option Pkg. 1SB, SL wagon	1590	1352	1463
Air conditioning, tilt steering column, intermittent wipers, cruise control, luggage rack, floormats, door edge guards, power windows and locks, power antenna, power driver's seat, remote mirrors, seatback recliners.			
Divided bench seat, base	183	156	168
Reclining bucket seats, base	347	295	319
45/45 front seat, SL	275	234	253
Seatback recliners (each), base & SL . . .	45	38	41
Power locks, 2-doors	145	123	133
4-doors	195	166	179
Wagons	245	208	225
Power windows, 2-doors	210	179	193
4-doors & wagons	285	242	262

	Retail Price	Dealer Invoice	Low Price
Rear defogger	145	123	133
Power sunroof, SL & I Series	775	659	713
Accent stripe, exc. I Series	45	38	41
FE2 suspension pkg., base & SL	30	26	28
FE3 suspension pkg., base & SL	455	387	419
Engine block heater	18	15	17
Cruise control, base	175	149	161
Wire wheel covers, SL	267	227	246
Super Stock wheels, base	99	84	91
Alloy wheels, base & SL	283	241	260
185/75R14 WSW tires, base & SL	68	58	63
Rallye instruments, base & SL	142	121	131
AM/FM ST ET cassette, base	147	125	135
w/EQ, SL & I Series	150	128	138
HD cooling, w/A/C	70	60	64
w/o A/C	70	60	64
Decklid luggage rack, base & SL	115	98	106
California emissions pkg.	100	85	92
Vinyl woodgrain paneling, wagons	290	247	267
w/Option Pkg	260	221	239

Oldsmobile Cutlass Supreme

Oldsmobile Cutlass Supreme

What's New

Anti-lock brakes are a new option, a larger V-6 engine arrives at mid-year and air conditioning is now standard on Cutlass Supreme for its second season as a front-drive coupe. The front-drive Cutlass Supreme debuted last year with the Buick Regal and Pontiac Grand Prix, intermediates built under the GM10, or W-body, program as replacements for rear-drive cars. The new anti-lock brakes are also optional on the Regal and Grand Prix; 4-wheel disc brakes are standard. At mid-year a 138-horsepower 3.1-liter V-6 will replace a 130-horsepower 2.8 V-6 as the standard engine with automatic transmission. The 2.8 V-6 will still come with the slow-selling 5-speed manual transmission (standard on the International Series, a credit option on the base and SL). The 2.3-liter 4-cylinder Quad 4 engine will not be offered in Cutlass Supreme this year as Oldsmobile hoped. Engineers say work needs to be done to make the Quad 4 quieter and less harsh. Late in the year, a heads-up instrument display that projects images about 10 feet in front of the driver will be offered as a limited-production option. Vacuum fluorescent displays for a digital speedometer, turn signal indicators, low-fuel warning and high-beam indicator are visible through the windshield with the heads-up display, which was installed in 54 replicas of the Cutlass Supreme pace car for the 1988 Indianapolis 500 race, the first time this technology was used in production passenger cars. Several new features debut for the sophomore season: standard 3-point rear shoulder belts and the options list grows with a new power sunroof, remote-control entry system using radio signals, electronic climate control (required with the keyless entry system), compact disc player, 8-speaker sound system, steering-wheel mounted stereo controls and Driver Information System trip computer. Oldsmobile will sell a 4-door sedan built from this design for the 1990 or 1991 model year as a replacement for the Cutlass Ciera 4-door.

For

- Anti-lock brakes • Ride • Passenger room
- Trunk space

Against

- Performance • Engine noise

Summary

We look forward to the arrival of the new 3.1-liter V-6, since the standard 2.8 isn't powerful enough to motivate Cutlass Supreme with anything approaching quickness. Instead, the 2.8 sounds and feels like it's overtaxed by the 3000-pound curb weight. While the 3.1 won't make a huge difference, even a small improvement is welcome. Also welcome is the new anti-lock brake option, which we recommend for the extra safety it can provide. Cutlass Supreme and the similar W-body coupes have gotten off to slow starts in the marketplace, which might be because they're not offered in the more popular 4-door styling. However, it might also be that they just aren't what most buyers are looking for these days in a mid-size car. The good news is that the 1988 models appeared to have fewer flaws than usual for first-year GM models, a sign that the design and assembly quality were right from the start.

Specifications

	2-door notchback
Wheelbase, in.	107.5
Overall length, in.	192.1
Overall width, in.	71.0
Overall height, in.	52.8
Front track, in.	59.5
Rear track, in.	58.0
Turn diameter, ft.	39.0
Curb weight, lbs.	3084
Cargo vol., cu. ft.	15.5
Fuel capacity, gal.	16.6
Seating capacity	6
Front headroom, in.	37.8
Front shoulder room, in.	57.6
Front legroom, max., in.	42.3
Rear headroom, in.	37.1
Rear shoulder room, in.	57.2
Rear legroom, min., in.	34.8

Powertrain layout: transverse front engine/front-wheel drive.

Engines

	ohv V-6	ohv V-6
Size, liters/cu. in.	2.8/173	3.1/189
Fuel delivery	PFI	PFI
Horsepower @ rpm	130 @ 4500	138 @ 4800

	ohv V-6	ohv V-6
Torque (lbs./ft.) @ rpm	170 @ 3600	183 @ 3600
Availability	S	S[1]

EPA city/highway mpg

5-speed OD manual	10/00	
4-speed OD automatic	20/29	NA

1. Mid-year, with automatic transmission

Prices

Oldsmobile Cutlass Supreme	Retail Price	Dealer Invoice	Low Price
2-door notchback	$14295	$12337	$13316
SL 2-door notchback	15195	13113	14154
I Series 2-door notchback	16995	14667	15831
Destination charge	455	455	455

Standard Equipment:

2.8-liter PFI V-6, 4-speed automatic transmission, power steering, power 4-wheel disc brakes, air conditioning, 55/45 front seat with storage armrest, trip odometer, left remote and right manual mirrors, AM/FM ST ET, automatic front seatbelts, outboard rear lap/shoulder belts, tinted glass, 195/75R14 all-season SBR tires. **SL** adds: Convenience Group (reading lamps, right visor mirror, misc. lights), AM/FM ST ET cassette, power decklid release, alloy wheels. **International Series** adds: 5-speed manual transmission, FE3 suspension, fast-ratio steering, Driver Information System (trip computer and service reminder), electronic instruments (tachometer, coolant temperature, oil pressure, voltmeter), power locks with remote control, rocker panel extensions, power front bucket seats, rear bucket seats, tilt steering column, intermittent wipers, 215/60R16 tires on alloy wheels.

Optional Equipment:

4-speed auto trans, I Series	615	523	566
5-speed manual trans, base & SL (credit)	(615)	(523)	(523)
Anti-lock braking system	925	786	851
Option Pkg. 1SB, base	327	278	301
Tilt steering column, intermittent wiper, Convenience Group, bodyside moldings, door edge guards, power antenna, cruise control.			
Option Pkg. 1SC, base	979	832	901
Pkg. 1SB plus power windows and locks, remote lock control, power decklid release, floormats, power driver's seat, power mirrors.			
Option Pkg. 1SB, SL	295	251	271
Cruise control, tilt steering column, floormats, intermittent wipers, bodyside moldings, door edge guards, power antenna.			
Option Pkg. 1SC, SL	1082	920	995
Pkg. 1SB plus power windows and locks, remote lock control, power mirrors, power driver's seat, steering wheel touch control.			
Power locks, base & SL	145	123	133
Power windows	220	187	202
Reclining bucket seats, base & SL	124	105	114
Power sunroof, SL & I Series	650	553	598
Rear defogger	145	123	133
FE3 suspension pkg., base & SL	284	241	261
Engine block heater	18	15	17
Dual engine cooling fans	40	34	37
195/75R14 WSW tires, base	72	61	66
Alloy wheels w/locks, base	215	183	198
Rallye instruments, base & SL	265	225	244
AM/FM ST ET cassette, base	147	125	135
w/EQ, SL & I Series	255	217	235

> **KEY: ohv** = overhead valve; **ohc** = overhead cam; **dohc** = double overhead cam; **I** = inline cylinders; **V** = cylinders in V configuration; **flat** = horizontally opposed cylinders; **bbl.** = barrel (carburetor); **PFI** = port (multi-point) fuel injection; **TBI** = throttle-body (single-point) fuel injection; **rpm** = revolutions per minute; **OD** = overdrive transmission; **S** = standard; **O** = optional; **NA** = not available.

Prices are accurate at time of printing; subject to manufacturer's change.

Oldsmobile

	Retail Price	Dealer Invoice	Low Price
AM/FM ST ET w/CD player, SL & Series . .	399	339	367
Driver Information System, SL	150	128	138
HD cooling	150	128	138
Decklid luggage rack	115	98	106
Custom leather trim, SL	454	386	418
I Series	364	309	335
California emissions pkg	100	85	92

Oldsmobile Ninety-Eight/ Touring Sedan

Oldsmobile Ninety-Eight Regency Brougham

What's New

A driver's-side air bag is a new option for Ninety-Eight, Oldsmobile's front-drive luxury sedan, while automatic front seat belts and 3-point rear shoulder belts are standard this year. The air bag, called Supplemental Inflatable Restraint by Olds, is mounted in the steering wheel hub. It became optional on last year's Delta 88 as the first GM car to offer an air bag since 1976. The new front seat belts can be left buckled all the time for automatic deployment. Steering-wheel controls covering most climate and stereo system functions are a new option on Ninety-Eight and standard on the top-line Touring Sedan, Oldsmobile's European-style luxury sedan. The air bag cannot be ordered with the steering-wheel controls, so it is not listed as an option on the Touring Sedan. An optional compact disc player will be added at mid-year. Drivers now get a manual seat recliner standard, while a new automatic climate control system with full manual override is optional on the base model and standard on the Brougham and Touring Sedan. The Touring Sedan also gets the Driver Information System trip computer, automatic door locks, graphic equalizer and cassette player, and 215/60R16 high-performance, all-season tires as standard. All models use a 165-horsepower 3.8-liter V-6 and 4-speed overdrive automatic. Anti-lock brakes remain standard on Touring Sedan, optional on Ninety-Eights. Ninety-Eight is built front the same design as the Buick Electra/Park Ave-

KEY: ohv = overhead valve; ohc = overhead cam; dohc = double overhead cam; I = inline cylinders; V = cylinders in V configuration; flat = horizontally opposed cylinders; bbl. = barrel (carburetor); PFI = port (multi-point) fuel injection; TBI = throttle-body (single-point) fuel injection; rpm = revolutions per minute; OD = overdrive transmission; S = standard; O = optional; NA = not available.

nue and Cadillac De Ville/Fleetwood. All debuted for 1985, and were followed in later years by the similar Buick LeSabre, Oldsmobile Delta 88 and Pontiac Bonneville.

For

- Performance • Room/comfort • Air bag
- Anti-lock brakes

Against

- Fuel economy • Ride (Touring Sedan)

Summary

Oldsmobile has given its full-size luxury sedan a dual personality: The Ninety-Eight serves the traditional Oldsmobile buyer, while the Touring Sedan is aimed at a younger buyer who might look first at an import. We think the regular Ninety-Eight is more on target than the Touring Sedan, which doesn't quite make the transition from American luxury car to European sport sedan. While it handles better than a Ninety-Eight, it lacks the polished moves of a BMW or Mercedes and suffers from a stiff, jittery ride that gives up too much comfort. Inside, leather upholstery and real wood trim give the Touring Sedan the right materials, but the seats are flat and unsupportive, or far more American than European. We like the Touring Sedan's standard anti-lock brakes and brisk performance, but you can get both on any Ninety-Eight. In addition, you can order an air bag on Ninety-Eights for additional protection for the driver. You can also get the same engine, brakes and air bag on the similar but less expensive Olds 88 Royale.

Specifications

	4-door notchback
Wheelbase, in.	110.8
Overall length, in.	196.4
Overall width, in.	72.4
Overall height, in.	55.1
Front track, in.	60.3
Rear track, in.	59.8
Turn diameter, ft.	39.4
Curb weight, lbs.	3329
Cargo vol., cu. ft.	16.4
Fuel capacity, gal.	18.0
Seating capacity	6
Front headroom, in.	39.3
Front shoulder room, in.	58.9
Front legroom, max., in.	42.4
Rear headroom, in.	38.1
Rear shoulder room, in.	58.8
Rear legroom, min., in.	41.5

Powertrain layout: transverse front engine/front-wheel drive.

Engines

	ohv V-6
Size, liters/cu. in.	3.8/231
Fuel delivery	PFI
Horsepower @ rpm	165 @ 5200
Torque (lbs./ft.) @ rpm	210 @ 2000
Availability	S

EPA city/highway mpg

4-speed OD automatic	19/28

Prices are accurate at time of printing; subject to manufacturer's change.

Prices

Oldsmobile Ninety-Eight	Retail Price	Dealer Invoice	Low Price
4-door notchback	$19295	$16652	$17974
Brougham 4-door notchback	20495	17687	19091
Touring Sedan 4-door notchback	25995	22434	24215
Destination charge	550	550	550

	Retail Price	Dealer Invoice	Low Price
AM/FM ST ET cassette w/EQ, base & Brougham	235	200	216
AM/FM ST ET w/CD player, base & Brougham	359	305	330
Delco/Bose music system, Touring	523	445	481
Instrument panel cluster, base & Brougham	66	56	61
Electronic instruments, Brougham	245	208	225
HD cooling, base & Brougham	66	56	61
Glamour metallic paint, base & Brougham	210	179	193
Custom leather trim, base & Brougham	433	368	398
California emissions pkg.	100	85	92

Standard Equipment:

3.8-liter PFI V-6, 4-speed automatic transmission, power steering and brakes, air conditioning, tinted glass, power windows and locks, 55/45 seat with power driver's side and storage armrest, automatic front seatbelts, outboard rear lap/shoulder belts, remote mirrors, opera lamps, front and rear armrests, right visor mirrors, trip odometer, reading lamp, AM/FM ST ET cassette, automatic load leveling, 205/75R14 all-season SBR tires, wire wheel covers. **Brougham** adds: automatic climate control, cruise control, tilt steering column, cornering lamps, sail panel reading lamps, intermittent wipers, opera lamp, alloy wheels. **Touring Sedan** adds: anti-lock braking system, FE3 suspension, 215/65R15 Goodyear Eagle GT+4 tires on alloy wheels, front console with storage, Driver Information System (trip computer, service reminder), gauge cluster including tachometer, fog lamps, illumination package, power mirrors with heated left, AM/FM ST ET cassette w/EQ, Twilight Sentinel, steering wheel touch controls, power decklid and fuel door releases.

Optional Equipment:

Anti-lock brakes, base & Brougham	925	786	851
Inflatable restraint (airbag), base & Brougham	850	723	782
Option Pkg. 1SB, base	629	535	579

Floormats, intermittent wipers, cruise control, tilt steering column, power decklid release, power antenna, lighted visor mirrors.

Option Pkg. 1SC, base	1721	1463	1583

Pkg. 1SB plus seatback recliners, automatic climate control, power passenger seat, cornering lamps, Driver Information System, Convenience Value Group.

Option Pkg. 1SB, Brougham	684	581	629

Power decklid release and pulldown, floormats, lighted visor mirrors, power antenna, power passenger seat, seatback recliners.

Option Pkg. 1SC, Brougham	1490	1267	1371

Pkg. 1SB plus Driver Information System, Convenience Value Group, steering wheel touch controls, leather-wrapped steering wheel, automatic power locks.

Seatback recliners (each), base & Brougham	45	38	41
Padded vinyl roof, base & Brougham	260	221	239
Power sunroof, Brougham & Touring	1230	1046	1132
Rear defogger, base & Brougham	145	123	133
Accent stripe, base	45	38	41
FE3 suspension pkg., base	289	246	266
Brougham	246	209	226
Engine block heater	18	15	17
Wire wheel covers, Brougham (credit)	(33)	(28)	(28)
Alloy wheels, base	53	45	49
Steering wheel touch controls, base	179	152	165

Oldsmobile Touring Sedan

Oldsmobile Toronado/Trofeo

Oldsmobile Toronado

What's New

A color cathode ray tube (CRT) called the Visual Information Center is optional on the Toronado/Trofeo front-drive luxury coupe this year, and Oldsmobile is stressing how it's different from the one Buick uses on the similar Riviera. Oldsmobile mounts its CRT in the center of the dashboard, same as Buick, but the color graphics are the most striking difference from Buick's monochrome display. Secondly, Olds has left more functions to conventional buttons that flank the 3.7×2.6-inch screen, 12 on the left for the stereo system and climate control, and five on the right for trip-computer and compass functions. The touch-sensitive points on the screen are 1-inch square, larger than on Riviera's CRT, and it isn't necessary to touch certain points on the screen to change functions, only to break the light beams. With the optional cellular telephone, calls can be dialed through the CRT. Also new this year are steering-wheel-mounted controls covering most stereo and climate system functions through six buttons. They're standard on Trofeo and optional on Toronado, and can be ordered with or without the Visual Information Center. The flagship Trofeo also gets anti-lock brakes, automatic door locks and an oil-level sensor as standard equipment. On Toronado, bucket seats and a console, 15-inch aluminum wheels, monochromatic paint, and the oil-level sensor are new standard items. A front bench seat, formerly standard, is now optional on Toronado. Anti-lock brakes remain optional on Toronado. The lone powertrain is a 165-horsepower 3.8-liter V-6 and 4-speed overdrive automatic. Toronado was redesigned and downsized for 1986, along with GM's other front-drive luxury coupes, the Riviera and Cadillac Eldorado. Sales plummeted for all three. A facelift for

Oldsmobile

Eldorado last year added three inches to its overall length, while Riviera gains 11 inches this year in a major restyling. Toronado is scheduled to be restyled for 1990.

For

• Performance • Anti-lock brakes • Quietness

Against

• Visual Information Center • Fuel economy

Summary

The main reason this car hasn't been a hit is that from a distance it can be mistaken for an Olds Calais coupe. Next year's restyling should change that. Until then, a decent car remains hidden under dull sheetmetal. The 3.8-liter V-6 provides brisk acceleration, though it doesn't exactly snap your head back with brute force, and Toronado's front-drive chassis provides competent handling and a liveable ride. The Trofeo's firmer suspension improves cornering ability at the expense of ride comfort, so we recommend you try both to see which will serve you better. Inside, gadgets galore provide the comfort and convenience you expect in this price range, though rear seat room and trunk space are modest. We'll credit Oldsmobile for improving on Buick's touch-sensitive CRT, but we still don't see a great benefit from the Visual Information Center, which complicates rather than simplifies stereo and climate system controls. We expect to see more of these in the future, but we still prefer simple, convenient, conventional controls—and the fewer the better—to electronic toys.

Specifications

	2-door notchback
Wheelbase, in.	108.0
Overall length, in.	187.5
Overall width, in.	70.8
Overall height, in.	53.0
Front track, in.	60.0
Rear track, in.	60.0
Turn diameter, ft.	38.7
Curb weight, lbs.	3361
Cargo vol., cu. ft.	14.1
Fuel capacity, gal.	18.8
Seating capacity	6
Front headroom, in.	37.8
Front shoulder room, in.	58.3
Front legroom, max., in.	43.0
Rear headroom, in.	37.8
Rear shoulder room, in.	57.3
Rear legroom, min., in.	35.7

Powertrain layout: transverse front engine/front-wheel drive.

Engines

	ohv V-6
Size, liters/cu. in.	3.8/231
Fuel delivery	PFI
Horsepower @ rpm	165 @ 5200
Torque (lbs./ft.) @ rpm	210 @ 2000
Availability	S

EPA city/highway mpg

4-speed OD automatic	19/28

KEY: **ohv** = overhead valve; **ohc** = overhead cam; **dohc** = double overhead cam; **I** = inline cylinders; **V** = cylinders in V configuration; **flat** = horizontally opposed cylinders; **bbl.** = barrel (carburetor); **PFI** = port (multi-point) fuel injection; **TBI** = throttle-body (single-point) fuel injection; **rpm** = revolutions per minute; **OD** = overdrive transmission; **S** = standard; **O** = optional; **NA** = not available.

Prices

Oldsmobile Toronado	Retail Price	Dealer Invoice	Low Price
2-door notchback	$21995	$18982	$20488
Trofeo 2-door notchback	24995	21571	23283
Destination charge	550	550	550

Standard Equipment:

3.8-liter PFI V-6 engine, 4-speed automatic transmission, power steering, power 4-wheel disc brakes, air conditioning, tinted glass, power windows and locks, reclining front bucket seats, outboard rear lap/shoulder belts, front and rear armrests, bumper rub strips, Convenience Group (lamps, visor mirror, chime tones), cruise control, electronic gauge cluster including tachometer and Reminder Package, courtesy lamps, dual remote mirrors, automatic leveling system, AM/FM ST ET cassette with power antenna, power driver's seat, tilt steering column, header panel storage unit, intermittent wipers, 205/75R15 tires. **Trofeo** adds: anti-lock braking system, FE3 suspension, front air dam with fog lamps, rocker panel extensions, electrochromic day/night mirror, perforated leather bucket seats, power passenger seat, emergency kit, power fuel door release, floormats, power decklid pulldown, lighted front door locks, lighted visor mirrors, 215/60R14 tires on alloy wheels.

Optional Equipment:

Anti-lock braking system, base	925	786	851
Option Pkg. 1SB, base	587	499	540

Power passenger seat, courtesy and reading lamps, Illumination Pkg., lighted visor mirrors, power mirrors, power decklid pulldown.

Option Pkg. 1SC, base	967	822	890

Pkg. 1SB plus Twilight Sentinel, automatic power locks, automatic day/night mirror, remote fuel door release, steering wheel touch controls.

Divided bench seat, base (credit)	(110)	(94)	(94)
Power sunroof, Trofeo	1230	1046	1132
Accent stripe, base	45	38	41
FE3 suspension pkg., base	126	107	116
Engine block heater	18	15	17
Delco/Bose music system	703	—	—
AM/FM ST ET cassette w/EQ	120	—	—
Mobile telephone	1795	1526	1651
Visual Information Center	1295	1101	1191
HD cooling, base	40	34	37
Glamour metallic paint, base	210	179	193
Custom leather trim, base	384	326	353
California emissions pkg.	100	85	92

Oldsmobile Trofeo

Prices are accurate at time of printing; subject to manufacturer's change.

CONSUMER GUIDE®

Oldsmobile 88 Royale

Oldsmobile 88 Royale 2-door

What's New

The "Delta" name, used since 1965 on Oldsmobile 88 models, has disappeared, so the moniker is now "88 Royale" for the full-size, front-drive family car. Other changes mostly involve the options list: a new fully automatic climate control system and a manual recliner for the driver's seat have been added; later in the year, a compact disc player becomes available. Wheel locks now come with the optional aluminum wheels, while the glass sunroof option has been dropped. As before, two trim levels are offered, base and Brougham, both available on either the 2-door coupe or 4-door sedan. This year if you order the optional air bag, mounted in the steering wheel hub, you also get a leather-wrapped steering wheel. The air bag is available only on the 4-door. Anti-lock brakes also are optional on the 88. All models this year are powered by a 165-horsepower 3.8-liter V-6, which during 1988 replaced a 150-horsepower 3.8 as standard. The front-drive 88 is similar to the Buick LeSabre and Pontiac Bonneville, which were derived from the design for the Buick Electra/Park Avenue and Olds Ninety-Eight.

For

- Performance • Room/comfort • Air bag
- Anti-lock brakes

Against

- Fuel economy

Summary

The 88 Royale is the lowest-priced car available with both anti-lock brakes and an air bag, two safety features that we highly recommend. While the anti-lock brakes can help stop a car safely to avoid accidents, the air bag reduces the driver's chances of being injured in case of an accident. Both of these safety features are expensive: Anti-lock brakes cost $925 and the air bag, available only on the 88 Royale 4-door, is $850. However, they will more than pay for themselves if just once they prevent a serious accident or injury. A fully equipped 88 with those two options probably will run $18,000-20,000 at suggested retail. That's more expensive than rear-drive cars such as the Chevrolet Caprice or Ford LTD Crown Victoria, but there are other things to recommend about the 88, including a roomy interior, a trunk that is at least sufficient for a full-size car, a 165-horse-

power 3.8-liter V-6 engine that delivers satisfying acceleration and, with front-wheel drive, better traction in rain and snow than rear-drive cars. The 88 is built with 2-sided galvanized steel that offers longer corrosion resistance than ordinary steel, eliminating the need for extra-cost rustproofing. The Buick LeSabre and Pontiac Bonneville offer all the same features, except the optional air bag. These three cars are lower-priced versions of the same design used for the more luxurious Oldsmobile Ninety-Eight and Buick Electra. Since all these cars use the same engines, transmissions, brakes and suspensions, you're only paying extra for plusher furnishings in the Ninety-Eight and Electra.

Specifications

	2-door notchback	4-door notchback
Wheelbase, in.	110.8	110.8
Overall length, in.	196.1	196.1
Overall width, in.	72.4	72.4
Overall height, in.	54.7	54.7
Front track, in.	60.3	60.3
Rear track, in.	59.8	59.8
Turn diameter, ft.	39.4	39.4
Curb weight, lbs.	3215	3265
Cargo vol., cu. ft.	16.2	16.3
Fuel capacity, gal.	18.0	18.0
Seating capacity	6	6
Front headroom, in.	38.1	38.9
Front shoulder room, in.	59.3	59.3
Front legroom, max., in.	42.4	42.4
Rear headroom, in.	37.6	38.3
Rear shoulder room, in.	57.8	59.5
Rear legroom, min., in.	37.4	39.1

Powertrain layout: transverse front engine/front-wheel drive.

Engines

	ohv V-6
Size, liters/cu. in.	3.8/231
Fuel delivery	PFI
Horsepower @ rpm	165 @ 5200
Torque (lbs./ft.) @ rpm	210 @ 2000
Availability	S

EPA city/highway mpg

4-speed OD automatic	19/28

Prices

Oldsmobile 88 Royale	Retail Price	Dealer Invoice	Low Price
2-door notchback	$15195	$13113	$14154
4-door notchback	15295	13200	14248
Brougham 2-door notchback	16295	14063	15179
Brougham 4-door notchback	16395	14149	15272
Destination charge	505	505	505

Standard Equipment:

Royale: 3.8-liter PFI V-6 engine, 4-speed automatic transmission, power steering and brakes, air conditioning, tinted glass, headlamps-on warning, left remote mirror, AM/FM ST ET, bench seat with center armrests, automatic front lap/shoulder belts, rear lap/shoulder belts, 205/75R14 all-season tires. **Brougham** adds: Convenience Group (lamps, right visor mirror, chime tones), 55/45 front seat with storage armrest, power decklid release.

Optional Equipment:

Inflatable Restraint System, 4-doors	850	723	782
Anti-lock brakes	925	786	851

Prices are accurate at time of printing; subject to manufacturer's change.

	Retail Price	Dealer Invoice	Low Price
Option Pkg. 1SB, base	559	475	514
Split bench seat, intermittent wipers, cruise control, tilt steering column, Convenience Group.			
Option Pkg. 1SC, base 2-door	1374	1168	1264
Base 4-door	1509	1283	1388
Pkg. 1SB plus power windows and locks, floormats, door edge guards, power antenna, seatback recliners, power driver's seat.			
Option Pkg. 1SD, base 2-door	1823	1550	1677
Base 4-door	1958	1664	1801
Pkg. 1SC plus power decklid release, Reminder Pkg., Convenience Value Group.			
Option Pkg. 1SB, Brougham 2-door	1162	988	1069
Brougham 4-door	1297	1102	1193
Power windows and locks, intermittent wipers, cruise control, tilt steering column, floormats, door edge guards, power antenna, Reminder Pkg., power driver's seat.			
Option Pkg. 1SC, Brougham 2-door	1806	1535	1662
Brougham 4-door	1941	1650	1786
Pkg. 1SB plus seatback recliners, automatic climate control, opera lamps, Convenience Value Group.			
Power passenger seat, Brougham	240	204	221
Seatback recliners (each)	45	38	41
Power locks, 2-doors	145	123	133
4-doors	195	166	179
Power windows, 2-doors	210	179	193
4-doors	285	242	262
Rear defogger	145	123	133
Vinyl roof, Brougham 4-door	200	170	184
Accent stripe	45	38	41
FE2 suspension pkg.	271	230	249
FE3 suspension pkg.	729	620	671
Engine block heater	18	15	17
Wire wheel discs w/WSW tires	275	234	253
15" alloy wheels	318	270	293
WSW tires	76	65	70
AM/FM ST ET cassette	122	104	112
w/EQ, Brougham	357	303	328
AM/FM ST ET w/CD player, Brougham . . .	481	409	443
Instrument panel cluster	66	56	61
High-capacity cooling	66	56	61
Custom leather trim, Brougham	433	368	398
California emissions pkg	100	85	92

Peugeot 405

What's New

The 405, Peugeot's newest offering for the U.S market, was voted the 1987 "Car of the Year" in Europe by the widest margin in history: 54 out of 57 judges from 17 countries. It is a front-drive 4-door sedan with a wedge-shaped profile that compares size-wise with the BMW 3-series, Mercedes 190, and Audi 80/90. The 405 rides a 105.1-inch wheelbase and measures 177.7 inches overall. Three models are offered: DL, S and Mi 16. All come with 4-wheel independent suspension, power disc brakes, power rack-and-pinion steering, adjustable steering column, central locking and automatic air conditioning. The S and the Mi 16 add a power moonroof, infrared remote locking, heated front seats, alloy wheels, cruise control, heated side mirrors and power windows. The Mi 16 is further treated to leather upholstery,

> **KEY: ohv** = overhead valve; **ohc** = overhead cam; **dohc** = double overhead cam; **I** = inline cylinders; **V** = cylinders in V configuration; **flat** = horizontally opposed cylinders; **bbl.** = barrel (carburetor); **PFI** = port (multi-point) fuel injection; **TBI** = throttle-body (single-point) fuel injection; **rpm** = revolutions per minute; **OD** = overdrive transmission; **S** = standard; **O** = optional; **NA** = not available.

Peugeot 405 Mi 16

wider tires and wheels, and an Alpine anti-theft AM/FM stereo with cassette player. Optional on the Mi 16 are antilock brakes (ABS), and for the S a Luxury Touring Package, which includes ABS plus the leather interior and other Mi 16 goodies. These options, however, will see a delayed introduction. All 405s are powered by a 1.9-liter transversely mounted inline 4-cylinder engine. In the DL and S, the single-overhead cam, 8-valve engine develops 110 horsepower. In the performance-oriented Mi 16, dual-overhead cams and 16 valves increase output to 150 horsepower. A 5-speed manual transmission is standard on all 405s, but only the DL and S can be had with an optional 4-speed overdrive automatic. Peugeot is hoping to sell about 14,000 405s during its introductory year, and only a limited number—about 1000—will be the Mi 16. All 1989 Peugeots come with a 3-year/36,000-mile "bumper-to-bumper" limited warranty and a no-deductible 5-year/ 50,000-mile powertrain warranty. A 3-year membership in the Peugeot Roadside Assistance Plan is also included. It offers 24-hour emergency road service through 14,000 American Automobile Association-approved service centers.

For

● Performance (Mi 16) ● Handling ● Passenger room

Against

● Interior materials

Summary

The only 405 model we've driven is the Mi 16, which we thoroughly enjoyed. Limited availability and the absence of automatic transmission, however, means that few American buyers will enjoy this model. Its eager 16-valve engine pulls quickly and smoothly to higher speeds, where it develops most of its power. Unlike some other multi-valve engines, this one doesn't feel lifeless below 3000 rpm, so you can easily keep pace with traffic with a light throttle foot. The Mi 16 lacks any truly outstanding dynamic qualities, yet it is highly competent in all. The steering is direct and responsive, the handling is surefooted and agile, the stopping ability reassuring. While the ride is firm, in the European mold, it's not so stiff that you'll dread bumpy pavement or rail crossings. It is much firmer than we've come to expect from Peugeot. We can only guess, but DL and S models won't be as frisky because of their less powerful engine, nor as tenacious around corners because of their slightly narrower tires and wheels, but they might have a

Prices are accurate at time of printing; subject to manufacturer's change.

CONSUMER GUIDE®

more absorbent ride. However, all 405 models share the same roomy interior, which offers plenty of room for four adults and sufficient space for their luggage, well-shaped seats, a comfortable driving position, and a generally convenient dashboard design. Power switches for the rear windows are mounted between the front seats in an awkward position, and the interior has too much flimsy-feeling plastic trim, but the cabin is otherwise nicely done. So is the exterior, with its clean lines and subtle aerodynamic styling. We're impressed with the car and we aren't put off by the prices. The Mi 16 is pricey at $20,700 base, but with all models you get a generous amount of standard equipment, leaving few options to increase the cost.

Specifications

	4-door notchback
Wheelbase, in.	105.1
Overall length, in.	177.7
Overall width, in.	67.6
Overall height, in.	55.4
Front track, in.	57.1
Rear track, in.	56.8
Turn diameter, ft.	36.1
Curb weight, lbs.	2460[1]
Cargo vol., cu. ft.	13.7
Fuel capacity, gal.	17.2
Seating capacity	5
Front headroom, in.	36.5
Front shoulder room, in.	NA
Front legroom, max., in.	41.4
Rear headroom, in.	36.9
Rear shoulder room, in.	NA
Rear legroom, min., in.	34.4

1. 2580 lbs., S; 2715 lbs., Mi 16

Powertrain layout: transverse front engine/front-wheel drive.

Engines	ohc I-4	dohc I-4
Size, liters/cu. in.	1.9/116	1.9/116
Fuel delivery	PFI	PFI
Horsepower @ rpm	110 @ 5200	150 @ 6400
Torque (lbs./ft.) @ rpm	120 @ 4250	128 @ 5000
Availability	S[1]	S[2]

EPA city/highway mpg		
5-speed OD manual	20/26	20/27
4-speed OD automatic	20/24	

1. DL and S   2. Mi 16

Prices

Peugeot 405	Retail Price	Dealer Invoice	Low Price
DL 4-door notchback	$14500	$12833	NA
S 4-door notchback	17700	15576	NA
Mi 16 4-door notchback	20700	18113	NA
Destination charge	400	400	400

Low price not available at time of publication.

Standard Equipment:

DL: 1.9-liter PFI 4-cylinder engine, 5-speed manual transmission, power steering, power 4-wheel disc brakes, air conditioning, power locks, velour reclining front bucket seats, tilt steering column, dual remote mirrors, tachometer, coolant temperature and oil level gauges, trip odometer, tinted glass, intermittent wipers, rear defogger, lighted visor mirrors, digital clock, Michelin MXV 185/65R14 tires with full-size spare. **S** adds: power windows,

remote power locks, cruise control, heated front seats, automatic front shoulder belts, outboard rear lap/shoulder belts, center armrest, rear armrest with trunk-through, power sunroof, heated power mirrors, Clarion AM/FM ST ET cassette, alloy wheels. **Mi 16** adds: DOHC 16-valve engine, 3-point front lap/shoulder belts, leather upholstery, leather-wrapped steering wheel, power driver's seat, driver's lumbar and side bolster adjustments, oil temperature gauge, Alpine AM/FM ST ET cassette, 195/60R14 tires on alloy wheels.

Optional Equipment:	Retail Price	Dealer Invoice	Low Price
4-speed auto trans, DL & S	650	573	NA
Anti-lock brakes, Mi 16	NA	NA	NA
Metallic paint	375	330	NA
Leather Pkg., S	1300	1144	NA
Leather sport seats, Alpine AM/FM ST ET cassette.			
Luxury Touring Pkg., S	NA	NA	NA
Leather Pkg. plus anti-lock brakes.			

Peugeot 505

Peugeot 505 STX V6

What's New

Peugeot's large car offering is the 505, which it describes as a "contemporary classic" (it debuted in 1980). Based on a rear-drive 108-inch wheelbase (114.2 for wagons), it is available in eight models: four 4-door sedans and four 5-door station wagons. New for all 505s are a deeper front air dam, lowered side skirts and rear valence, integrated front and rear bumpers, and wider side moldings. The 505 Turbo also sports a trunk-mounted wing. At $26,335, the 505 Turbo sedan is the costliest model in the line, and with 180 horsepower is also the most powerful. Torque is rated at 205 pounds/feet at 2500 rpm, which Peugeot claims is more torque at lower rpm than any car in its class. Two Turbo wagons use the same engine, though it's rated at 160 horsepower (but still with 205 lbs/ft torque, the most for any import wagon, says Peugeot). They are available only with a 4-speed automatic, unlike other 505s, which have a 5-speed manual standard. All seats in the SW8 face forward and even the rearmost passengers receive lap and shoulder belts. A 145-horsepower 2.8-liter V-6 is standard in the S V-6 and STX sedans. The STX has a Torsen limited-slip differential, anti-lock brakes (also on the Turbo sedan), infrared remote-control central locking and anti-theft sound system. A 120-horsepower 2.2-liter four is standard on the S sedan and the DL and SW8 wagons.

Prices are accurate at time of printing; subject to manufacturer's change.

For

- Performance (Turbo, V-6)   ●Anti-lock brakes
- Passenger room/comfort   ●Cargo room (wagons)

Against

- Fuel economy (wagon)   ●Cargo room (sedan)
- Performance (base 4-cylinder)

Summary

Peugeot's 505 series boasts potent engines, anti-lock brakes, roomy interiors, shapely seats that pamper their occupants, and long warranties. Yet, this slow-selling line of European sedans gets little recognition, let alone respect. The Turbo sedan is the quickest, providing impressive sprinting ability once the turbo starts cooking. However, without boost the turbocharged engine feels lifeless. For that reason, the smooth, refined V-6 seems a better choice, offering ample power without the fuss. The base 120-horse-power 4-cylinder engine is just adequate for the sedan, and nearly inadequate for the heavier SW8 wagon, which we tested during 1988. We had to flog the engine to keep pace with traffic and averaged a mediocre 17.6 mpg in a mix of urban and highway driving with the 5-speed manual. The SW8 does indeed hold eight people, but the cramped rear seat is meant for children, not adults, and it's a chore squeezing into the back. While Peugeot lacks the reputation for longevity of European rivals such as Volvo, and dealers are hard to find in some areas, there are enough attractions here to make the 505 worth looking at.

Specifications

	4-door notchback	5-door wagon
Wheelbase, in.	108.0	114.2
Overall length, in.	181.4	194.5
Overall width, in.	67.7	67.7
Overall height, in.	57.0	60.1
Front track, in.	58.7	58.7
Rear track, in.	57.3	57.1
Turn diameter, ft.	35.1	36.7
Curb weight, lbs.	2998	3230
Cargo vol., cu. ft.	13.8	92.5[1]
Fuel capacity, gal.	18.0	18.0
Seating capacity	5	8
Front headroom, in.	37.5	38.1
Front shoulder room, in.	NA	NA
Front legroom, max., in.	40.0	40.0
Rear headroom, in.	36.3	36.3
Rear shoulder room, in.	NA	NA
Rear legroom, min., in.	39.5	37.6

1. 81.1, SW8

Powertrain layout: longitudinal front engine/rear-wheel drive.

Engines

	ohc I-4	Turbo ohc I-4	ohc V-6
Size, liters/cu. in.	2.2/133	2.2/132	2.8/174
Fuel delivery	PFI	PFI	PFI
Horsepower @ rpm	120 @ 5000	180[1] @ 5200	145 @ 5000
Torque (lbs./ft.) @ rpm	131 @ 3500	205 @ 2500	176 @ 2800
Availability	S	S	S
EPA city/highway mpg			
5-speed OD manual	18/24	18/24	18/22
4-speed OD automatic	19/22	18/22	18/22

1. Turbo sedans; 160 horsepower in Turbo wagons

Prices

Peugeot 505

	Retail Price	Dealer Invoice	Low Price
S 4-door notchback	$19295	$16787	$17941
S V-6 4-door notchback	21435	18648	19942
STX V-6 4-door notchback	25895	22270	23983
Turbo 4-door notchback	26335	22648	24392
DL 5-door wagon	17590	15304	16347
SW8 5-door wagon	19995	17396	18596
Turbo 5-door wagon	25540	21964	23652
Turbo SW8 5-door wagon	25695	22098	23797
Destination charge	400	400	400

Standard Equipment:

S 4-door and **DL** wagon: 2.2-liter PFI 4-cylinder engine, 5-speed manual transmission, limited-slip differential, power steering, power 4-wheel disc brakes, automatic climate control, power windows, power locks (remote on 4-door), cruise control, Clarion AM/FM ST ET cassette, lighted visor mirrors, front passive restraints, heated velour reclining front bucket seats, power sunroof (4-door), power mirrors (heated on 4-door), tinted glass, intermittent wipers, rear defogger, rear wiper/washer (wagon), Michelin MXV 185/65HR15 tires. **S V-6** adds: 2.8-liter PFI V-6, alloy wheels, oil level and temperature gauges. **STX** adds: anti-lock braking system, Torsen differential, speed-sensitive power steering, Alpine stereo with TV and weather bands, leather upholstery, leather-wrapped steering wheel, driver's seat height adjustment, rear headrests, fog lights, Michelin MXV 205/60HR15 SBR tires. **Turbo 4-door** has STX equipment plus 2.2-liter turbocharged, intercooled 4-cylinder engine. **Turbo wagon** has 4-speed automatic transmission.

Optional Equipment:

4-speed auto trans (std. Turbo wagons)	650	560	605
Metallic or black lacquered paint	395	340	368
Leather upholstery, Turbo 4-door	960	826	893

Plymouth Acclaim

Plymouth Acclaim LX

What's New

Plymouth's version of Chrysler's new A-car will hit the showrooms in early 1989 as an upscale alternative to the venerable Reliant America. It has the same upright design and front-drive layout as the compact K-car, but Chrysler touts its engineering and performance as far more advanced. The new car's wheelbase is also longer by three

---

Prices are accurate at time of printing; subject to manufacturer's change.

inches and its body by 2.6, though the only significantly larger cabin dimension is an extra 3.1 inches of rear-seat leg space. All Acclaims are 4-door notchback sedans in base, LE or sporty LX trim. Standard in the base and LE is a 2.5-liter 4 cylinder; a turbocharged version is optional. A 5-speed manual transmission is standard, a 3-speed auto matic optional with either 2.5. Exclusive to the LX is a Mitsu-bishi-made 3.0-liter V-6 teamed with Chrysler's new 4-speed overdrive automatic. LXs have firmer suspension settings and 15-inch-diameter alloy wheels versus the other models' 14-inch steel wheels. All Acclaims have front disc/rear drum brakes, power steering, shoulder belts for outboard rear passengers and remote fuel-door and trunk-lid releases. Front bucket seats are standard on all models; a 50/50 split front bench for 6-passenger capacity is op-tional. A 55/45 split, fold-down rear seat is optional on the base model, standard on the other two. All Acclaims have analog instruments, including a tachometer. LXs also get a leather-wrapped steering wheel and cassette stereo sys-tem. A rear-window defroster and tilt steering column are standard on the LE and LX.

Plymouth Acclaim LX

For

● Interior room   ●Cargo space   ●Ride
● Performance (V-6)

Against

● Rear seat comfort   ●Shift action (5-speed)
● Noise (turbo engine)

Summary

Chrysler officials acknowledge that their new A-cars come off a "highly evolved" K-car platform, but they bristle at any suggestion that the Acclaim and similar Dodge Spirit are simply more variations on a tired theme. We agree, based on pre-production examples sampled at Chrysler's Michi-gan proving grounds. The Spirit and Acclaim soundly beat the K-cars in terms of ride, handling, power and quality feel. The ones we sampled didn't have the Honda Accord-like "high-touch" polish that Chrysler officials say was their goal, but they did feel comfortable, competent and competi-tive with their foreign and domestic rivals. Acclaim stresses value and equipment, so it targets a slightly younger family buyer than Spirit. Its personality is softer, but it has the same airy cabin and generously sized trunk. The optional front bench seat isn't uncomfortable, but it lacks the lateral and lower-back support of the buckets on the Acclaim LX. The dashboard has easy-to-read instruments and is flawed only by undersized climate-system buttons. Rear head and knee room are adequate, but the seat cushion is too low and too soft. Performance is best in the V-6 LX, though the 4-speed automatic is slow to downshift for passing. A turbo 2.5 and 3-speed automatic feels marginally quicker, though the engine is loud and coarse when pushed. The base engine provides mediocre pickup. Control in turns is best with the LX's wider tires, though all examples enjoy good overall suspension behavior. The brakes on all the A-cars we sampled had excellent stopping power and pedal feel, though nosedive marred harder stops. The K-car derivation that Acclaim and Spirit most closely re-semble is the mid-size Dodge 600 and Plymouth Caravelle, which Chrysler dubbed the E-body. Both had a 103.3-inch wheelbase, though the A-car's body is shorter by four inches. The 600 and Caravelle are discontinued for 1989, so Acclaim and Spirit are technically their successors, and good ones at that.

Specifications

	4-door notchback
Wheelbase, in.	103.3
Overall length, in.	181.2
Overall width, in.	68.1
Overall height, in.	55.5
Front track, in.	57.5
Rear track, in.	57.2
Turn diameter, ft.	37.6
Curb weight, lbs.	2770
Cargo vol., cu. ft.	14.4
Fuel capacity, gal.	16.0
Seating capacity	6
Front headroom, in.	38.4
Front shoulder room, in.	54.3
Front legroom, max., in.	41.8
Rear headroom, in.	37.9
Rear shoulder room, in.	54.9
Rear legroom, min., in.	38.5

Powertrain layout: transverse front engine/front-wheel drive.

Engines	ohc I-4	Turbo ohc I-4	ohc V-6
Size, liters/cu. in.	2.5/153	2.5/153	3.0/181
Fuel delivery	TBI	PFI	PFI
Horsepower @ rpm	100 @ 4800	150 @ 4800	141 @ 5000
Torque (lbs./ft.) @ rpm	135 @ 2800	180 @ 2000	171 @ 2800
Availability	S	O	S[1]
EPA city/highway mpg			
5-speed OD manual	24/34	20/29	
3-speed automatic	23/29	19/24	
4-speed OD automatic			18/26

1. LX

Equipment Summary

Plymouth Acclaim
(prices not available at time of publication):

Standard Equipment:

2.5-liter TBI 4-cylinder engine, 5-speed manual transmission, power steer-ing, cloth reclining front bucket seats, tachometer, coolant temperature and

---

KEY: ohv = overhead valve; **ohc** = overhead cam; **dohc** = double overhead cam; **I** = inline cylinders; **V** = cylinders in V configuration; **flat** = horizontally opposed cylinders; **bbl.** = barrel (carburetor); **PFI** = port (multi-point) fuel injection; **TBI** = throttle-body (single-point) fuel injection; **rpm** = revolutions per minute; **OD** = overdrive transmission; **S** = standard; **O** = optional; **NA** = not available.

Prices are accurate at time of printing; subject to manufacturer's change.

oil pressure gauges, voltmeter, trip odometer, intermittent wipers, AM/FM ST ET, remote fuel door and decklid releases, 185/70R14 all-season SBR tires. **LE** adds: cruise control, tilt steering column, driver's seat lumbar support adjustment, lighted visor mirrors, rear defogger, message center, 55/45 folding rear seat, armrest with dual cupholders, 195/70R14 tires. **LX** adds: 3.0-liter PFI V-6, 4-speed automatic transmission, sport suspension, decklid luggage rack, trip computer, AM/FM ST ET cassette, leather-wrapped steering wheel, 205/60R15 performance tires on alloy wheels.

quis and Chevrolet Caprice have larger interiors and more cargo space, while front-drive intermediates, such as the new Plymouth Acclaim, offer comparable interior room. We recommend you look at the big Ford sedans or Caprice if you want a full-size car with a V-8, or at a front-drive intermediate if you want a more economical sedan.

Plymouth Gran Fury

Plymouth Gran Fury

What's New

This is the last year for Plymouth's last rear-drive car. No replacement is planned, and the 1989 farewell edition is unchanged. Gran Fury remains a 6-passenger 4-door sedan only. A driver's-side air bag was added late in the 1988 model year and is carried over. Gran Fury was introduced for 1982 as a companion to the similar Dodge Diplomat and Chrysler Fifth Avenue, but its chassis design goes back to the 1976 Plymouth Volare/Dodge Aspen. This 3600-pound sedan falls between medium and full size, and since 1984 has been available only with a 5.2-liter V-8 and a 3-speed automatic transmission. The 318-cubic-inch engine, a Chrysler stalwart for two decades, has a 2-barrel carburetor. Standard equipment includes a tilt steering column and intermittent windshield wipers. Gran Fury has been popular with taxi companies and with police departments. The special police model, with a higher-performance engine and heavy-duty suspension and brakes, will be discontinued along with the standard model.

For

●Driveability ●Air bag

Against

●Fuel economy ●Ride ●Handling/roadholding

Summary

Gran Fury's proven drivetrain is by far its best feature. You get satisfactory performance from the V-8 and crisp shifts from the sturdy 3-speed automatic. Fuel economy is poor, however, and Plymouth recommends expensive premium unleaded fuel. Gran Fury's chassis is inadequate in many ways; the power steering is light and imprecise, the handling is sloppy and the ride loose and bouncy. The full-size, rear-drive Ford LTD Crown Victoria/Mercury Grand Mar-

Specifications

	4-door notchback
Wheelbase, in.	112.7
Overall length, in.	204.6
Overall width, in.	74.2
Overall height, in.	55.3
Front track, in.	60.0
Rear track, in.	59.5
Turn diameter, ft.	40.7
Curb weight, lbs.	3556
Cargo vol., cu. ft.	15.6
Fuel capacity, gal.	18.0
Seating capacity	6
Front headroom, in.	39.3
Front shoulder room, in.	56.0
Front legroom, max., in.	42.5
Rear headroom, in.	37.7
Rear shoulder room, in.	55.9
Rear legroom, min., in.	36.6

Powertrain layout: longitudinal front engine/rear-wheel drive.

Engines

	ohv V-8
Size, liters/cu. in.	5.2/318
Fuel delivery	2 bbl.
Horsepower @ rpm	140 @ 3600
Torque (lbs./ft.) @ rpm	265 @ 1600
Availability	S

EPA city/highway mpg
3-speed automatic	16/22

Prices

Plymouth Gran Fury	Retail Price	Dealer Invoice	Low Price
Salon 4-door notchback	$11995	$10706	$11351
Destination charge	495	495	495

Plymouth Gran Fury

Prices are accurate at time of printing; subject to manufacturer's change.

Standard Equipment:

5 2-liter 2bbl. V-8 engine, 3-speed automatic transmission, power steering and brakes, driver's side air bag, tinted glass, AM ET, cloth and vinyl bench seat with center armrest, ammeter, coolant temperature gauge, headlamps-on tone, 205/75R15 tires.

Optional Equipment:	Retail Price	Dealer Invoice	Low Price
Automatic climate control	855	727	787
Popular Equipment Pkg.	1147	975	1055
Automatic climate control, rear defogger, power mirrors, cruise control, premium wheel covers.			
Luxury Equipment Pkg.	2835	2410	2608
Popular Pkg. plus Light Group, upper door frame molding, power decklid release, power windows and locks, 60/40 cloth seat with recliners, leather-wrapped steering wheel, trunk dress-up, lighted visor mirrors, vinyl roof, wire wheel covers.			
Protection Pkg.	185	157	170
Light Pkg.	133	113	122
Requires Popular Pkg.			
60/40 cloth seat w/recliners	353	300	325
Requires Popular Pkg.			
Rear defogger	149	127	137
California emissions pkg.	102	87	94
Power mirrors	164	139	151
Power locks	201	171	185
Requires power mirrors.			
Power windows	294	250	270
Requires power mirrors and locks.			
AM/FM ST ET cassette	262	223	241
HD suspension	27	23	25
Trunk dress-up	58	49	53
Requires Popular Pkg.			
Conventional spare tire	96	82	88
Vinyl roof	206	175	190
Requires Popular Pkg.			

Plymouth Horizon America

Plymouth Horizon

What's New

America's first domestically built front-drive subcompact continues with only minor changes for its 12th model year. The 5-door hatchback, a twin to the Dodge Omni, acquired the "America" suffix in 1987, when a cost-cutting program consolidated the line into a single trim level with limited options. For '89, the standard 93-horsepower 2.2-liter 4-cylinder engine continues with single-point fuel injection, but gets internal modifications that Chrysler says make it run quieter. Engine-bay items such as the dipstick, power-steering reservoir cap and coolant-overflow cap have been highlighted with bright paint for easier servicing. Standard equipment includes a tachometer, tinted glass and a one-piece fold-down rear seat. A 5-speed manual transmission

is standard; a 3-speed automatic is available as part of an option package that also includes power steering.

For

●Value   ●Fuel economy   ●Performance

Against

●Rear seat room   ●Driving position

Summary

Horizon is an aged subcompact with a cramped interior and an ungainly driving position, but it still stands as excellent value for the money. The 2.2-liter engine delivers potent performance with the standard 5-speed manual transmission and, unlike most rivals, isn't underpowered with the optional automatic transmission. Neither does this engine turn anemic when the air conditioner is on—which is the rule with smaller 4-cylinder engines. Fuel economy isn't bad either. There are many newer subcompacts with more interior room and better ride and handling, but none can match Horizon's performance and standard equipment at such a low price. Though the Hyundai Excel and similar Mitsubishi Precis are price competitive, their engine performance pales by comparison. Chrysler's long warranties covering the powertrain and body rust are an added bonus. The base price has increased $500 for 1989, to $6495, but even with automatic transmission, air conditioning, power steering and a stereo radio the total will still be less than $9000.

Specifications	5-door hatchback
Wheelbase, in.	99.1
Overall length, in.	163.2
Overall width, in.	66.8
Overall height, in.	53.0
Front track, in.	56.1
Rear track, in.	55.7
Turn diameter, ft.	38.1
Curb weight, lbs.	2237
Cargo vol., cu. ft.	33.0
Fuel capacity, gal.	13.0
Seating capacity	5
Front headroom, in.	38.1
Front shoulder room, in.	51.7
Front legroom, max., in.	42.1
Rear headroom, in.	36.9
Rear shoulder room, in.	51.5
Rear legroom, min., in.	33.3

Powertrain layout: transverse front engine/front-wheel drive.

Engines	ohc I-4
Size, liters/cu. in.	2.2/135
Fuel delivery	TBI
Horsepower @ rpm	93 @ 4800
Torque (lbs./ft.) @ rpm	122 @ 3200
Availability	S

KEY: ohv = overhead valve; **ohc** = overhead cam; **dohc** = double overhead cam; **I** = inline cylinders; **V** = cylinders in V configuration; **flat** = horizontally opposed cylinders; **bbl.** = barrel (carburetor); **PFI** = port (multi-point) fuel injection; **TBI** = throttle-body (single-point) fuel injection; **rpm** = revolutions per minute; **OD** = overdrive transmission; **S** = standard; **O** = optional; **NA** = not available.

Prices are accurate at time of printing; subject to manufacturer's change.

EPA city/highway mpg	ohc I-4
5-speed OD manual	26/35
3-speed automatic	24/30

Prices

Plymouth Horizon	Retail Price	Dealer Invoice	Low Price
5-door hatchback	$6595	$6001	$6350
Destination charge	348	348	348

Standard Equipment:

2.2-liter 4-cylinder TBI engine, 5-speed manual transmission, power brakes, rear defogger, rear wiper/washer, trip odometer, tachometer, coolant temperature, oil pressure and voltage gauges, tinted glass, luggage compartment light, black bodyside moldings, left remote mirror, right visor mirror, folding shelf panel, intermittent wipers, cloth and vinyl upholstery, P165/80R13 tires on styled steel wheels.

Optional Equipment:

Basic Pkg	776	660	714
3-speed automatic transmission, power steering.			
Manual Transmission Discount Pkg	705	599	648
Console, power steering, AM & FM ST ET, cloth reclining sport seats, trunk dress-up.			
Automatic Transmission Discount Pkg	1186	1008	1091
Adds 3-speed automatic transmission to Manual Transmission Pkg.			
Air conditioning	694	590	638
Requires Transmission Discount Pkg. and conventional spare tire.			
California emissions pkg	100	85	92
AM & FM ST ET cassette	152	129	140
Requires Transmission Discount Pkg.			
Conventional spare tire	73	62	67
Tinted glass	105	89	97

Plymouth Laser

1990 Plymouth Laser RS Turbo

What's New

The new Laser 2+2 sports coupe is being touted as the "first Plymouth of the 90's." Debuting as a 1990 model, it will start showing up in California showrooms in January, and then spread from there. Young buyers 25 to 35 are the target, especially women (possibly up to 60 percent of sales). Plymouth has slotted the aerodynamically styled Laser as a challenger to the Ford Probe, Toyota Celica, Nissan 240SX, Acura Integra, Mazda MX-6, and even corporate sibling Dodge Daytona. Built in Normal, Ill., at a 50-50 joint venture

> **KEY: ohv** = overhead valve; **ohc** = overhead cam; **dohc** = double overhead cam; **I** = inline cylinders; **V** = cylinders in V configuration; **flat** = horizontally opposed cylinders; **bbl.** = barrel (carburetor); **PFI** = port (multi-point) fuel injection; **TBI** = throttle-body (single-point) fuel injection; **rpm** = revolutions per minute; **OD** = overdrive transmission; **S** = standard; **O** = optional; **NA** = not available.

plant with Mitsubishi, Laser is nearly identical to the Mitsubishi Eclipse, which bows at the same time. Don't be confused by the Laser nameplate that was attached to a 1984-86 Chrysler; that was a Daytona clone, and the new Plymouth Laser is in no way related. Plymouth's version, which rides a 97.2-inch wheelbase and is 170.5 inches long, is based on a Mitsubishi Galant chassis and uses Mitsubishi engines and transmissions. A 1.8-liter 92-horsepower 4-cylinder is standard on both the base model and the top-line RS, but a 2.0-liter 16-valve four is optional on the latter. In normally aspirated form, this dual-overhead-cam unit develops 135 horsepower; intercooling and turbocharging bump horsepower to an impressive 190. A 5-speed manual stick shift is standard, 4-speed automatic optional (except with the turbo motor until June). Base Lasers come with 4-wheel disc brakes, tilt steering column, intermittent wipers, retractable headlamps, dual remote mirrors, tinted glass, tachometer, reclining front bucket seats, console, fold-down rear seat, AM/FM 4-speaker stereo radio, and 185/70R14 all-season tires. RS adds power mirrors, power dome hood (2.0-liter), black roof panel, rear wiper, and a more deluxe interior. RS adds P205/55HR16 rubber with the 2.0 engine and high-speed V-rated tires with the turbo, and a sport suspension.

For

- Performance (2.0 and 2.0 turbo) • Braking
- Handling/roadholding

Against

- Performance (automatic transmission)
- Rear-seat room

Summary

Prices (not announced as of this writing) are expected to range from $10,000 to $15,000, undercutting most competitors. Performance meets or exceeds class standards—with the turbo, Laser thunders from 0-60 mph in 6.9 seconds and tops out at 143 mph, according to Plymouth. The 1.8-liter takes 11.4 seconds to reach 60 mph, the 2.0-liter 8.7; add 2.5 seconds for automatic. Price, performance, and styling should make the Laser a contender in its class. For further comments see the Mitsubishi Eclipse.

Plymouth Reliant America

What's New

The K-car station wagon has been discontinued, leaving 2-and 4-door sedans that are little changed from '88. The optional 2.5-liter 4-cylinder engine gets a slight horsepower boost, to 100 from last year's 96. Bright paint has been added to such underhood items as the dipstick and power-steering reservoir cap to aid in service identification. Chrysler says the front-suspension also has been slightly modified to reduce harshness and improve isolation from road noise. The base engine is a 93-horsepower 2.2-liter four. A 5-speed manual transmission is standard with the 2.2; a 3-speed automatic is optional with the 2.2 and required with the 2.5. Front bucket seats are standard. An optional front bench increases seating capacity to six and is available only with automatic transmission. New options include a 4-speaker stereo system. Reliant and the

---

Prices are accurate at time of printing; subject to manufacturer's change.

Plymouth Reliant 4-door

	2-door notchback	4-door notchback
Front shoulder room, in.	55.0	55.4
Front legroom, max., in.	42.2	42.2
Rear headroom, in.	37.0	37.8
Rear shoulder room, in.	58.8	55.9
Rear legroom, min., in.	35.1	35.4

Powertrain layout: transverse front engine/front-wheel drive.

Engines	ohc I-4	ohc I-4
Size, liters/cu. in.	2.2/135	2.5/153
Fuel delivery	TBI	TBI
Horsepower @ rpm	93 @ 4800	100 @ 2800
Torque (lbs./ft.) @ rpm	122 @ 3200	135 @ 2800
Availability	S	O

EPA city/highway mpg		
5-speed OD manual	25/34	
3-speed automatic	24/30	23/28

identical Dodge Aries, now in their ninth season, were enrolled in Chrysler's "America" program last year. The program is designed to make these front drive compacts more affordable by cutting production costs and by offering only a single trim level with limited options.

For

●Value ●Passenger room ●Fuel economy

Against

●Handling ●Engine noise (2.2)

Summary

Reliant's main weapon is price, and it's a mighty weapon. Start with a base sticker of $7495. Even though it's up $500 from '88, you still get a roomy sedan that holds five adults and a fair amount of luggage. For another $2500 or so you can add the 2.5-liter engine for satisfactory acceleration, air conditioning, power steering, tinted glass, AM/FM stereo radio, a center console, dual outside remote mirrors and a few appearance items, such as sport wheel covers. That's around $10,000, still cheaper than many subcompacts, plus you get Chrysler's competitive warranties against powertrain defects and rust. Of course, K-cars were introduced eight years ago, and though their front-drive platform spawned a host of Chrysler products, it's been surpassed by more refined, newer designs. Don't expect the workmanship or quality of materials to equal that of a Mazda 626 or Toyota Camry, to name two. Don't anticipate the ride and handling of a Honda Accord LXi, either. Still, equipped right, the affordable Aries or Reliant is acceptable for basic family transportation. We recommend the 2.5; it's stronger and quieter than the standard 2.2-liter engine, doesn't get weak-kneed with the air conditioner on, and still returns reasonable fuel economy. The bucket seats are more supportive and comfortable than the front bench. And, driveability is better with the automatic transmission; the 5-speed suffers notchy shift action.

Prices

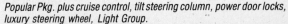

Plymouth Reliant	Retail Price	Dealer Invoice	Low Price
2-door notchback	$7595	$6935	$7365
4-door notchback	7595	6935	7365
Destination charge	454	454	454

Standard Equipment:

2.2-liter TBI 4-cylinder engine, 5-speed manual transmission, power brakes, reclining front bucket seats (bench seat may be substituted at no charge, but requires Popular Equipment Discount Pkg.), cloth and vinyl upholstery, intermittent wipers, optical horn, left remote mirror, right visor mirror with map/reading light, 185/70R13 SBR tires.

Optional Equipment:

2.5-liter engine	279	237	257
Requires Basic or Popular Pkg.			
Power steering	240	204	221
Air conditioning	775	659	713
Requires tinted glass.			
Tinted glass	120	102	110
Basic Equipment Pkg	776	660	714
3-speed automatic transmission, power steering.			
Basic Radio Discount Pkg	929	790	855
Basic Pkg. plus AM & FM ST ET.			
Popular Equipment Pkg., w/bucket seats	1292	1098	1189
w/bench seat	1192	1013	1097
Basic Pkg. (including radio) plus tinted glass, remote mirrors, bodyside tape stripes, added sound insulation, floor console (with bucket seats), trunk dress up, 185/70R14 tires, wheel covers.			
Premium Pkg., 2-door w/bucket seats	1730	1471	1592
2-door w/bench seat	1630	1386	1500
4-door w/bucket seats	1780	1513	1638
4-door w/bench seat	1680	1428	1546
Popular Pkg. plus cruise control, tilt steering column, power door locks, luxury steering wheel, Light Group.			

Specifications

	2-door notchback	4-door notchback
Wheelbase, in.	100.3	100.3
Overall length, in.	178.9	178.9
Overall width, in.	67.9	67.9
Overall height, in.	52.5	52.5
Front track, in.	57.6	57.6
Rear track, in.	57.2	57.2
Turn diameter, ft.	34.8	34.8
Curb weight, lbs.	2317	2323
Cargo vol., cu. ft.	15.0	15.0
Fuel capacity, gal.	14.0	14.0
Seating capacity	6	6
Front headroom, in.	38.2	38.6

Plymouth Reliant 2-door

Prices are accurate at time of printing; subject to manufacturer's change.

	Retail Price	Dealer Invoice	Low Price
500-amp battery, w/2.2	44	37	40
Rear defogger	145	123	133
California emissions pkg	100	85	92
AM & FM ST ET cassette	152	129	140
Requires Popular Pkg.			
Conventional spare tire, 13″	73	62	67
14″	83	71	76
185/70R14 WSW tires	68	58	63
Requires Popular Pkg.			

Plymouth Sundance

Plymouth Sundance 5-door

What's New

Chrysler's new 2.5-liter turbocharged engine replaces a 2.2-liter turbo in the performance model of Plymouth's front-drive subcompact, and all versions get a new grille and taillamps. Sundance, a twin of the Dodge Shadow, comes in 3- and 5-door hatchback body styles. Front-end appearance is altered for '89 thanks to a new grille and aero-style headlamps. New taillamps in a one-piece housing change the look of the rear end slightly. Plymouth says rear-seat knee room has been increased by nearly one inch for '89 through new thin-back front bucket seats. The base engine is a 2.2-liter 4-cylinder. A 100-horsepower 2.5-liter four is optional. This engine, its horsepower boosted from last year's 96, is standard with the sporty Rallye Sport (RS) package, which includes a unique front fascia with fog lamps, two-tone exterior paint and sporty interior touches. Optional with either base or RS trim is a turbo package that includes a turbocharged version of the 2.5 four rated at 150 horsepower. It replaces a 146-horsepower 2.2-liter four as Chrysler's Turbo I engine. The turbo package retains the base car's 185/70R14 tire size, but adds equal-length drive shafts. All engines are available with a 5-speed manual transmission or with a 3-speed automatic.

For

●Performance (turbo) ●Ride/handling ●Cargo room

Against

●Engine noise ●Rear seat room ●Manual shift linkage

Summary

Sundance is second only to Reliant as Plymouth's most popular model, indicating a marketing strategy aimed at young buyers and women has paid off. Plymouth calls Sundance a compact based on interior volume, but we list it as a subcompact based on wheelbase, overall length and the limited amount of rear-seat leg room. There's plenty of head room all around. Luggage space is adequate with the rear seatback up and generous with the seatback folded down. The moderately firm suspension feels a little harsh on bumpy roads, but it's not punishing. There are no suspension or tire options, but Sundance enjoys capable cornering, good road grip and a stable highway ride. Of the three available engines, the one we recommend is the 2.5-liter. It's smoother and quieter than the 2.2 or the turbo 2.5, and it strikes a nice blend of adequate performance and reasonable economy. Avoid the balky 5-speed manual transmission unless you're looking for maximum fuel economy; the optional automatic is easier to live with. Sundance isn't the most polished small hatchback, but it does offer decent performance at an attractively low price, even when fully equipped.

Specifications

	3-door hatchback	5-door hatchback
Wheelbase, in.	97.0	97.0
Overall length, in.	171.7	171.7
Overall width, in.	67.3	67.3
Overall height, in.	52.7	52.7
Front track, in.	57.6	57.6
Rear track, in.	57.2	57.2
Turn diameter, ft.	36.2	36.2
Curb weight, lbs.	2520	2558
Cargo vol., cu. ft.	33.3	33.0
Fuel capacity, gal.	14.0	14.0
Seating capacity	5	5
Front headroom, in.	38.3	38.3
Front shoulder room, in.	54.4	54.7
Front legroom, max., in.	41.5	41.5
Rear headroom, in.	37.4	37.4
Rear shoulder room, in.	52.5	54.5
Rear legroom, min., in.	20.1	20.1

Powertrain layout: transverse front engine/front-wheel drive.

Engines

	ohc I-4	ohc I-4	Turbo ohc I-4
Size, liters/cu. in.	2.2/135	2.5/153	2.5/153
Fuel delivery	TBI	TBI	PFI
Horsepower @ rpm	93 @ 4800	100 @ 4800	150 @ 4800
Torque (lbs./ft.) @ rpm	122 @ 3200	135 @ 2800	180 @ 2000
Availability	S	O	O
EPA city/highway mpg			
5-speed OD manual	24/34	24/34	20/29
3-speed automatic	24/28	23/29	19/24

Prices

Plymouth Sundance	Retail Price	Dealer Invoice	Low Price
3-door hatchback	$8495	$7746	$8121
5-door hatchback	8695	7296	7996
Destination charge	415	415	415

Standard Equipment:

2.2-liter TBI 4-cylinder engine, 5-speed manual transmission, power steering and brakes, mini-console, tinted glass, cloth reclining front bucket seats, split folding rear seatback, headlamps-on tone, tachometer, coolant temper-

Prices are accurate at time of printing; subject to manufacturer's change.

Plymouth Sundance 3-door

Plymouth Grand Voyager LE

ature gauge, voltmeter, trip odometer, optical horn, trip odometer, intermittent wipers, remote liftgate release, remote mirrors, visor mirrors, bodyside moldings, AM/FM ST ET, removable shelf panel, 185/70R14 SBR tires.

Optional Equipment:	Retail Price	Dealer Invoice	Low Price
2.5-liter 4-cylinder engine	287	244	264
Requires Popular or RS Pkg.			
3-speed automatic transmission	552	469	508
Air conditioning	715	608	658
Requires tinted glass.			
Tinted glass	108	92	99
Popular Equipment Pkg., 3-door	362	308	333
5-door	374	318	344
Full console, rear defogger, Light Group, four speakers.			
Deluxe Convenience Pkg	396	337	364
Cruise control, tilt steering column, conventional spare tire, floormats; requires Popular or RS Pkg.			
Power Assist Pkg. I, 3-door w/Popular Pkg. .	178	151	164
5-door w/Popular Pkg.	225	191	207
3- & 5-door w/RS	49	42	45
Power locks and mirrors.			
Power Assist Pkg. II, 3-door	417	354	384
5-door	487	414	448
Power windows and driver's seat; requires Pkg. I.			
RS Pkg., 3-door	1239	1053	1140
5-door	1303	1108	1199
Popular Pkg. plus 2.5-liter engine, performance front seats, fog lights, liftgate luggage rack, power locks, AM/FM ST ET cassette, leather-wrapped steering wheel, two-tone paint, message center, 125-mph speedometer.			
Turbo Sport Pkg., w/Popular Pkg.	950	808	874
w/RS	628	534	578
2.5-liter turbo engine, boost gauge, message center, 185/70R14 SBR performance tires, 125-mph speedometer.			
Pearl coat paint	41	35	38
500-amp battery	45	38	41
Rear defogger	149	127	137
California emissions pkg.	102	87	94
Power locks, 3-door	149	127	137
5-door	201	171	185
Requires Power Assist I or Popular Pkg.			
AM/FM ST ET cassette	207	176	190
w/seek/scan, w/Popular Pkg.	262	223	241
Removable glass sunroof	383	326	352
Requires Popular or RS Pkg.			
Alloy wheels, w/o Deluxe Convenience . . .	270	230	248
w/Deluxe Convenience	397	337	365
Requires Popular or RS Pkg.; not available with 2.2.			

Plymouth Voyager

What's New

A turbocharged engine and a 4-speed overdrive automatic transmission are new for Plymouth's version of Chrysler's highly successful front-drive minivan, a twin of the Dodge Caravan. Standard-wheelbase Voyagers come in base, SE and LE versions that hold five to seven passengers, depending on seating configuration. The extended-length Grand Voyager comes in SE and LE trim and seats up to eight. For '89, power rear-quarter vent windows replace manual-opening vents as an option on SEs and LEs, which also get Voyager's first sunroof option. Leather seating also is a new option, available only on LEs. Voyager's front struts are borrowed this year from the luxury Chrysler New Yorker, a move intended to give the minivan a smoother ride. The new turbo 2.5-liter four is optional only in short-wheelbase SEs and LEs. A naturally aspirated 2.5 is standard in all short-wheelbase versions and in the Grand SE. A Mitsubishi-made V-6 is optional on SE and LE models and on the Grand SE; it's standard on the Grand LE. Either 2.5 may be ordered with a 5-speed manual transmission or a 3-speed automatic. Chrysler's new 4-speed overdrive automatic is standard on Grand LE and is a required option on Voyager LEs and Grand SEs equipped with the V-6 engine. A properly equipped V-6 Voyager can tow up to 2750 pounds. The optional heavy-duty towing package on V-6 Grand Voyagers with the 4-speed automatic can accommodate trailers up to 4000 pounds.

For

• Passenger room   • Cargo space   • Ride/handling

Against

• Fuel economy   • Performance (4-cylinder)

Summary

Chrysler's front-drive minivans continued to sell briskly through 1988, despite price increases of as much as $3000 for Grand Voyagers. Price hikes are more modest for '89, ranging from $425 on the SE to $853 on a Grand Voyager LE. Still, you're looking at around $20,000 for a lavishly equipped Grand LE and $16,000-19,000 for a well-furnished standard-size Voyager or Grand SE. We prefer the V-6 for its clearly superior performance and think most families will be happier with automatic transmission than the somewhat clumsy manual shift linkage. The 2.5 is barely adequate for a lightly loaded Voyager; it's inadequate for one fully loaded. We find Voyager and the iden-

> **KEY: ohv** = overhead valve; **ohc** = overhead cam; **dohc** = double overhead cam; **I** = inline cylinders; **V** = cylinders in V configuration; **flat** = horizontally opposed cylinders; **bbl.** = barrel (carburetor); **PFI** = port (multi-point) fuel injection; **TBI** = throttle-body (single-point) fuel injection; **rpm** = revolutions per minute; **OD** = overdrive transmission; **S** = standard; **O** = optional; **NA** = not available.

Prices are accurate at time of printing; subject to manufacturer's change.

Plymouth

Plymouth Voyager

EPA city/highway mpg	ohc I-4	Turbo ohc I-4	ohc V-6
5-speed OD manual	21/28	18/25	
3-speed automatic	21/23	18/21	18/22
4-speed OD automatic			18/23

1. Standard on Grand Voyager LE

tical Dodge Caravan the easiest compact vans to drive. They have better traction than their rear-drive rivals, better stability on the open road and more car-like handling and ride. The standard-size models nearly match Ford and Chevy rivals in passenger room and cargo space, while the stretched versions improve luggage room dramatically. Despite the high prices, the Caravan/Voyager remain our favorites among compact vans for passenger use.

Prices

Plymouth Voyager

	Retail Price	Dealer Invoice	Low Price
Base SWB 4-door van	$11312	$10105	$10909
SE SWB 4-door van	12039	10744	11592
LE SWB 4-door van	13987	12459	13423
Grand SE 4-door van	13061	11644	12553
Grand LE 4-door van	16362	14549	15656
Destination charge	500	500	500

SWB denotes short-wheelbase models.

Standard Equipment:

2.5-liter TBI engine, 5-speed manual transmission, power steering, power brakes, liftgate wiper/washer, headlamps-on tone, 5-passenger seating, variable intermittent wipers, tinted glass, left remote mirror, AM & FM ST ET, P185/75R14 SBR tires. **SE** adds: highback reclining front seats, rear seat (Grand), front folding center armrests, upgraded door panels, power liftgate release. **LE** adds: front air conditioning, added sound insulation, remote mirrors, bodyside moldings, woodgrain exterior applique, upgraded steering wheel. **Grand LE** adds: 3.0-liter PFI V-6, 4-speed automatic transmission.

Optional Equipment:

2.5-liter turbo engine, SWB	680	578	626
3.0-liter V-6, SE & LE SWB	680	578	626
3-speed automatic transmission (NA 7-pass.)	565	480	520
4-speed automatic, LE SWB & Grand SE .	735	625	676
Requires 3.0 V-6.			
Front air conditioning, base & SE	840	714	773
Rear air conditioning w/heater, Grand . . .	560	476	515
Cloth lowback seats, base	45	38	41
Leather highback seats, LE	671	570	617
Requires Luxury Equipment or LE Decor Pkg.			
7-pass. seating, SWB	389	331	358
8-pass. seating, SE Grand	120	102	110
SE Grand w/Popular Pkg	5	4	5
Converta-Bed, SWB (NA base)	542	461	499
Value Wagon Pkg., base w/5-speed	1045	888	961
Base incl. automatic	1283	1091	1180
SE SWB w/5-speed	994	845	914
SE SWB incl. automatic	1232	1047	1133
Front air conditioning, rear defogger, dual horns, deluxe sound insulation, Light Group, high-back cloth reclining front seats.			
Popular Equipment Discount Pkg., SE SWB .	1661	1412	1528
SE SWB w/Value Wgn Pkg	743	632	684
SE SWB w/7/8-pass	1683	1431	1548
SE SWB w/7/8-pass. & Value Wgn Pkg .	765	650	704
SE Grand	1683	1431	1548
LE SWB	981	834	903
LE SWB w/7-pass, LE Grand	1003	853	923
SE: front air conditioning, forward storage console, overhead console, floormats, gauges, deluxe sound insulation, Light Group, power rear quarter vent windows, conventional spare tire, cruise control, tilt steering column. LE adds: lighted visor mirror, power door locks.			
Luxury Equipment Discount Pkg., SE SWB .	2348	1996	2160
SE SWB w/Value Wgn Pkg	1485	1262	1366
SE SWB w/7/8-pass., SE Grand	2371	2015	2181
SE SWB w/7-pass. & Value Wgn Pkg . .	1507	1281	1386
LE SWB	1603	1363	1475
LE SWB w/7-pass, LE Grand	1626	1382	1496
SE: Popular Pkg. plus lighted visor mirror, power door locks, power mirrors, power front windows, AM & FM ST ET cassette, Eurosport steering wheel. LE adds: power driver's seat.			
Turbo Sport Pkg., SE SWB	1073	912	987
2.5-liter turbo engine, gauges, 205/70R15 all-season tires on alloy wheels.			

Specifications

	4-door van	4-door van
Wheelbase, in.	112.0	119.1
Overall length, in.	179.5	190.5
Overall width, in.	69.6	69.6
Overall height, in.	64.2	65.0
Front track, in.	59.9	59.9
Rear track, in.	62.1	62.1
Turn diameter, ft.	41.0	43.2
Curb weight, lbs.	3003	3304
Cargo vol., cu. ft.	125.0	150.0
Fuel capacity, gal.	15.0[1]	15.0[1]
Seating capacity	7	8
Front headroom, in.	39.0	39.0
Front shoulder room, in.	58.4	58.4
Front legroom, max., in.	38.2	38.2
Rear headroom, in.	37.7	37.6
Rear shoulder room, in.	61.3	61.3
Rear legroom, min., in.	37.7	37.8

1. 20.0 gal opt.

Powertrain layout: transverse front engine/front-wheel drive.

Engines

	ohc I-4	Turbo ohc I-4	ohc V-6
Size, liters/cu. in.	2.5/153	2.5/153	3.0/181
Fuel delivery	TBI	PFI	PFI
Horsepower @ rpm	100 @ 4800	150 @ 4800	141 @ 5000
Torque (lbs./ft.) @ rpm	135 @ 2800	180 @ 2000	171 @ 2800
Availability	S	O	O[1]

KEY: ohv = overhead valve; **ohc** = overhead cam; **dohc** = double overhead cam; **I** = inline cylinders; **V** = cylinders in V configuration; **flat** = horizontally opposed cylinders; **bbl.** = barrel (carburetor); **PFI** = port (multi-point) fuel injection; **TBI** = throttle-body (single-point) fuel injection; **rpm** = revolutions per minute; **OD** = overdrive transmission; **S** = standard; **O** = optional; **NA** = not available.

Prices are accurate at time of printing; subject to manufacturer's change.

CONSUMER GUIDE®

	Retail Price	Dealer Invoice	Low Price
HD Trailer Tow Pkg., SE Grand	392	333	361
HD suspension, 120-amp alternator, conventional spare tire, trailer wiring harness, 205/70R15 all-season tires on styled steel wheels.			
LE Decor Pkg., LF SWB	1449	1232	1333
2.5-liter turbo engine, warm silver fascia, moldings, 7 passenger seating, tape stripes, 205/70R15 all-season tires on alloy wheels.			
Rear defogger	165	140	152
California emissions pkg	99	84	91
Sunscreen glass	406	345	374
Rear heater, SE & LE SWB	329	280	303
120-amp alternator, Grand	62	53	57
Luggage rack, base & SE	144	122	132
Power door locks	203	173	187
Requires Value Wgn or Popular Pkg.			
AM & FM ST ET cassette, base & SE . . .	152	129	140
Requires Value Wagon or Popular Pkg.			
Ultimate Sound stereo	214	182	197
Requires Luxury or LE Decor Pkg.			
AM & FM ST ET delete (credit)	(136)	(116)	(116)
Cruise control, base	207	176	190
Requires tilt steering column and Value Wgn Pkg.			
Tilt steering column, base	122	104	112
Requires cruise control and Value Wgn Pkg.			
HD suspension	68	58	63
Conventional spare tire	104	88	96
Wire wheel covers, LE (Luxury Pkg. req.) .	239	203	220
Sport road wheels (NA base)	415	353	382
Pearl coat paint	46	39	42
Two-tone paint, SE	236	201	217

Pontiac Bonneville

Pontiac Bonneville SSE

What's New

Anti-lock brakes are now optional on the base Bonneville LE and mid-level SE, while the top-line SSE gets a new center console and new front seats. Last year anti-lock brakes were available only as standard equipment on the flagship SSE, a new model for 1988. The new center console on the SSE houses relocated power controls for the reshaped front bucket seats and two auxiliary jacks that can be used for operating radar detectors, portable computers and other electrically powered appliances. Steering-wheel controls for the stereo system, standard last year on the SSE and optional on the LE, have been replaced by a simplified arrangement that covers most stereo and climate system functions. The new steering-wheel controls are standard on the SSE, optional on the LE and SE. The

SSE also gains a subwoofer and larger speakers for its sound system. All models use a 165-horsepower 3.8-liter V-6, which replaced a 150-horsepower version of the same engine during the 1988 model year. Bonneville, available only as a 4-door sedan, debuted for the 1987 model year, using the same basic design as the Buick LeSabre and Oldsmobile 88 Royale. Buick and Oldsmobile also offer 2-door versions.

For

● Performance ● Room/comfort ● Anti-lock brakes ● Handling

Against

● Fuel economy ● Ride (SSE)

Summary

Bonneville is the sportiest of the three GM H-body full-size cars and the Euro-style SSE has the most aggressive styling and performance. Pontiac also has designed the interior to give Bonneville a more contemporary look and feel inside than either the Buick LeSabre and Olds 88. The flagship Bonneville SSE is a little too aggressive for us, with a rear spoiler, body-sill extensions and body-colored wheels, a stiff ride and a deep, muscle-car exhaust note. We'd prefer something more mature than this boy-racer approach, though we still appreciate the SSE for its capable handling, satisfying engine performance and its lengthy standard equipment list, and we applaud Pontiac for keeping this package reasonably priced. Now that anti-lock brakes are optional on the LE and SE, we're more interested in those models. The SE has a less flamboyant exterior, a sport suspension for good handling, and high-performance, all-season tires, plus a lower price than the SSE. The LE has a softer suspension and smaller tires, and an even lower price. Since all three models use the same engine, there's no significant performance different among them. Final drive ratios are different for each model, with the SSE having the shortest (highest numerical) ratio for slightly quicker acceleration. With any of the three, you get a spacious interior, decent cargo volume and an ample selection of convenience equipment.

Specifications

	4-door notchback
Wheelbase, in.	110.8
Overall length, in.	198.7
Overall width, in.	72.1
Overall height, in.	55.5
Front track, in.	60.3
Rear track, in.	59.8
Turn diameter, ft.	39.0
Curb weight, lbs.	3275[1]
Cargo vol., cu. ft.	15.2
Fuel capacity, gal.	18.0
Seating capacity	6
Front headroom, in.	38.9
Front shoulder room, in.	58.9
Front legroom, max., in.	42.4
Rear headroom, in.	38.2
Rear shoulder room, in.	59.1
Rear legroom, min., in.	38.4

1. *3481 lbs., SSE*

Powertrain layout: transverse front engine/front-wheel drive.

Prices are accurate at time of printing; subject to manufacturer's change.

Pontiac

Engines

	ohv V-6
Size, liters/cu. in.	3.8/231
Fuel delivery	PFI
Horsepower @ rpm	165 @ 5200
Torque (lbs./ft.) @ rpm	210 @ 2000
Availability	S

EPA city/highway mpg

4-speed OD automatic	19/28

Prices

Pontiac Bonneville	Retail Price	Dealer Invoice	Low Price
LE 4-door notchback	$14829	$12797	$13813
SE 4-door notchback	17199	14842	16012
SSE 4-door notchback	22899	19762	21331
Destination charge	505	505	505

Standard Equipment:

LE: 3.8-liter PFI V-6, 4-speed automatic transmission, power steering and brakes, air conditioning, tinted glass, cloth bench seat, outboard rear lap/shoulder belts, left remote and right manual mirrors, wide bodyside moldings, AM/FM ST ET, coolant temperature gauge, 205/75R14 tires. **SE adds:** 2.97 axle ratio, 45/55 seat with recliners and storage armrest, cargo security net, intermittent wipers, cruise control, tilt steering column, tachometer and trip odometer, power windows, AM/FM ST ET cassette, Rally Tuned Suspension, 215/65R15 Eagle GT + 4 tires on alloy wheels. **SSE adds:** 3.33 axle ratio, anti-lock braking system, Electronic Ride Control, automatic air conditioning with steering wheel controls, electronic compass, Driver Information System, heated power mirrors, headlamp sentinel, headlamp washers, power seat adjustments (including lumbar support, recliners and head restraints), automatic seatbelts, fog lamps, aero bodyside extensions, AM & FM ST ET cassette with EQ and Touch Control, power antenna, rear armrest with storage, rear defogger, lighted visor mirrors, leather steering wheel trim, illuminated entry, remote decklid release, power door locks, first aid and accessory kits, uprated suspension, 215/60R16 Eagle GT + 4 tires.

Optional Equipment:

	Retail Price	Dealer Invoice	Low Price
Anti-lock brakes, LE & SE	925	505	505
Option Pkg. 1SA, LE	259	220	238
Tilt steering column, intermittent wipers, Lamp Group.			
Option Pkg. 1SB, LE	434	369	399
Pkg. 1SA plus cruise control.			
Option Pkg. 1SC, LE	1325	1126	1219
Pkg. 1SB plus power windows and locks, power driver's seat, power decklid release, lighted right visor mirror.			
Option Pkg. 1SD, LE	1793	1524	1650
Pkg. 1SC plus leather-wrapped steering wheel, illuminated entry, remote fuel door release, power mirrors, power passenger seat.			

	Retail Price	Dealer Invoice	Low Price
Option Pkg. 1SA, SE	274	233	252
Lamp Group, power locks.			
Option Pkg. 1SB, SE	675	574	621
Pkg. 1SA plus power driver's seat, power decklid release, lighted right visor mirror, fog lamps.			
Option Pkg. 1SC, SE	1153	980	1061
Pkg. 1SB plus illuminated entry, remote fuel door release, power mirrors, power passenger seat, Twilight Sentinel.			
Value Option Pkg., LE	424	360	390
205/70R15 tires on alloy wheels, 45/55 seat, AM/FM ST ET cassette.			
Value Option Pkg., LE	327	278	301
45/45 seat, AM/FM ST ET cassette, power antenna.			
Rear defogger, SE & LE	145	123	133
California emissions pkg	100	85	92
Rally gauge cluster, LE	100	85	92
Tachometer and trip odometer.			
Decklid luggage rack, LE & SE	115	98	106
Custom two-tone paint, LE & SE	105	89	97
Power locks, LE & SE	195	166	179
Power windows, LE	355	302	327
AM/FM ST ET cassette, LE & SE	122	104	112
AM/FM ST ET cassette w/EQ, LE w/o VOP	722	614	664
LE w/VOP	600	510	552
LE w/Pkg. 1SD, w/o VOP	672	571	618
LE w/Pkg. 1SD & VOP	550	468	506
SE w/o VOP	672	571	618
SE w/VOP	480	408	442
AM/FM ST ET w/CD & EQ, LE w/o VOP	926	787	852
LE w/VOP	804	683	740
LE w/Pkg. 1SD, w/o VOP	876	745	806
LE w/Pkg. 1SD & VOP	754	641	694
SE w/o VOP	876	745	806
SE w/VOP	684	581	629
SSE	204	173	188
Stereos w/EQ include power antenna.			
Performance Sound System, SE & LE	100	85	92
Requires power windows.			
Power antenna	70	60	64
Seatback recliners, LE w/o Custom Trim	90	77	83
45/55 seat, LE w/o Custom Trim or VOP	133	113	122
45/45 seat w/console, SE	235	200	216
Leather 45/45 seat, SSE	379	322	349
Custom Interior Trim, LE w/o VOP	608	517	559
LE w/VOP	475	404	437
LE w/Pkg. 1SC or 1SD, w/o VOP	253	215	233
LE w/Pkg. 1SC or 1SD & VOP	120	102	110
45/45 seat with recliners, power windows, trunk security net.			
Power glass sunroof, LE & SE	1284	1091	1181
LE & SE w/Option Pkg. (exc. VOP)	1230	1046	1132
205/75R14 WSW tires, LE	76	65	70
205/70R15 tires, LE	48	41	44
205/70R15 WSW tires, LE w/o VOP	114	97	105
LE w/VOP	66	56	61
215/65R15 tires & Y99 suspension, LE w/o VOP	164	139	151
LE w/VOP	116	99	107
All 15-inch tires require alloy wheels on LE.			

Pontiac Bonneville LE

What's New

A turbocharged Firebird goes on sale at mid-year to mark the 20th anniversary of the high-performance Trans Am model. Until then, the major changes are a new theft-deterrent system and rear disc brakes become standard on all four Firebirds: base, Formula, Trans Am and Trans Am GTA. Firebird and the similar Chevrolet Camaro, which have the two highest theft rates among all cars, now come with GM's "pass key" theft-deterrent system that has helped reduce

Prices are accurate at time of printing; subject to manufacturer's change.

Pontiac Firebird Trans Am GTA

Specifications	GTA 2-door notchback	3-door hatchback
Wheelbase, in.	101.0	101.0
Overall length, in.	190.3	190.3
Overall width, in.	72.4	72.4
Overall height, in.	40.8	49.0
Front track, in.	60.7	60.7
Rear track, in.	61.6	61.6
Turn diameter, ft.	36.9	36.9
Curb weight, lbs.	3458	3083[1]
Cargo vol., cu. ft.	12.3	34.8
Fuel capacity, gal.	15.5	15.5
Seating capacity	4	4
Front headroom, in.	37.0	37.0
Front shoulder room, in.	57.7	57.7
Front legroom, max., in.	43.0	43.0
Rear headroom, in.	35.6	35.6
Rear shoulder room, in.	56.3	56.3
Rear legroom, min., in.	28.6	28.6

1. 3318 lbs., Formula; 3337 lbs., Trans Am; 3486 lbs, GTA

Powertrain layout: longitudinal front engine/rear-wheel drive.

Engines	ohv V-6	ohv V-8	ohv V-8	ohv V-8
Size, liters/cu. in.	2.8/173	5.0/305	5.0/305	5.7/350
Fuel delivery	PFI	TBI	PFI	PFI
Horsepower @ rpm	135 @ 4900	170 @ 4000	215 @ 4400	225 @ 4400
Torque (lbs./ft.) @ rpm	160 @ 3900	225 @ 2400	285 @ 3200	330 @ 3200
Availability	S[1]	S[2]	O[3]	S[4]

EPA city/highway mpg

5-speed OD manual	18/27	17/25	17/26	
4-speed OD automatic	19/24	17/26	16/25	14/25

1. Base Firebird   2. Formula, Trans Am   3. Formula, Trans Am, GTA   4. GTA; optional, Formula and Trans Am

Prices

Pontiac Firebird	Retail Price	Dealer Invoice	Low Price
3-door hatchback	$11999	$10715	$11357
Formula 3-door hatchback	13949	12457	13203
Trans Am 3-door hatchback	15999	14287	15143
Trans Am GTA 3-door hatchback	20339	18163	19251
Destination charge	439	439	439

Standard Equipment:

2.8-liter PFI V-6, 5-speed manual transmisson, power steering and brakes, cloth reclining front bucket seats, outboard rear lap/shoulder belts, power hatch pulldown, AM/FM ST ET, gauge cluster including tachometer, Pass Key theft-deterrent system, 215/65R15 tires on alloy wheels. **Trans Am** adds: 5.0-liter TBI V-8, F41 suspension. **Formula** adds: dual exhaust, air conditioning, WS6 performance suspension, 245/50ZR16 tires on alloy wheels. **GTA** adds: 5.7-liter PFI V-8, 4-speed automatic transmission, 4-wheel disc brakes, limited-slip differential, rear defogger, power windows and locks, tinted glass, cruise control, tilt steering column, articulated bucket seats with inflatable lumbar support and thigh bolsters, floormats, upgraded upholstery, cargo cover, AM/FM ST ET cassette with EQ and Touch Control.

> **KEY: ohv** = overhead valve; **ohc** = overhead cam; **dohc** = double overhead cam; **I** = inline cylinders; **V** = cylinders in V configuration; **flat** = horizontally opposed cylinders; **bbl.** = barrel (carburetor); **PFI** = port (multi-point) fuel injection; **TBI** = throttle-body (single-point) fuel injection; **rpm** = revolutions per minute; **OD** = overdrive transmission; **S** = standard; **O** = optional; **NA** = not available.

thefts of Corvettes since 1985. A resistor pellet imbedded in the ignition key must match special coding in the ignition lock, or the starter is temporarily disabled. This theft-deterrent system was standard on the GTA last year. The 4-wheel disc brakes also were standard only on the GTA last year and optional on the Formula and Trans Am. Other across-the-board changes include making rear shoulder belts standard, adopting base coat/clear coat paint finishes, and adding a compact-disc player as a new option (later in the year). The Formula and Trans Am gain air conditioning as standard. New rubber is standard on the GTA with a "Z" speed rating (over 149 mph) instead of a "V" rating (over 130 mph); the size stays the same at 245/50R16. GTA is available as a 3-door hatchback or 2-door notchback (added last year); the others come as a hatchback only. Only about 1500 of the 20th Anniversary Trans Ams will be built, using a turbocharged 3.8-liter V-6 last seen in the Buick Regal Grand National, discontinued a year ago. Pontiac says the turbo V-6 will produce about 250 horsepower and be capable of doing the 0-60 mph sprint in 5.4 seconds or less, and the quarter mile in under 13 seconds. The anniversary model will have a white exterior and tan interior. The anniversary model aside, the engine lineup remains the same as last year's.

For

● Performance (V-8s) ● Handling ● Resale value

Against

● Fuel economy ● Ride ● Interior room

Summary

Since this is so similar to the Camaro, choosing one over the other is largely a matter of personal taste, or whether you want a convertible, which Chevrolet offers and Pontiac doesn't. Before you sign up for 48 or more payments on a Firebird, check with your insurance agent. Even with the new standard theft-deterrent system, these are still going to be costly to insure because they're popular with younger drivers, who tend to get into more trouble on the road than the older crowd. Before you buy, you should also check out the Ford Mustang GT, which offers similar performance at lower cost. With any car of this ilk, performance reigns over comfort and convenience. The tenacious handling you'll enjoy on dry, smooth roads gives way to minimal traction in rain and snow, and the ride is punishing on virtually any surface.

Prices are accurate at time of printing; subject to manufacturer's change.

Pontiac Firebird Formula

Pontiac Grand Am

Pontiac Grand Am SE 2-door

Optional Equipment:	Retail Price	Dealer Invoice	Low Price
Notchback roof, GTA	700	595	644
5.0-liter TBI V-8, base	400	340	368
5.0-liter PFI (TPI) V-8, Formula & T/A . . .	745	633	685
GTA (credit)	(300)	(255)	(255)
Requires limited-slip differential.			
5.7-liter PFI V-8, Formula & T/A	1045	888	961
Requires automatic transmission, engine oil cooler, 4-wheel disc brakes, limited-slip differential and 245/50ZR16 tires.			
5-speed manual transmission, GTA (credit)	(490)	(417)	(417)
4-speed auto trans, base, Formula & T/A .	490	417	451
Air conditioning, base	775	659	713
Limited-slip differential, Formula & T/A . .	100	85	92
Dual exhaust	155	132	143
Engine oil cooler, Formula & T/A	110	94	101
4-wheel disc brakes, Formula	179	152	165
Option Pkg. 1SA, base (credit)	(331)	(281)	(281)
Air conditioning, bodyside moldings, Lamp Group.			
Option Pkg. 1SB, base	137	116	126
Pkg. 1SA plus power windows and locks, 4-way driver's seat, door pockets, cruise control, power decklid release, reading lamps.			
Option Pkg. 1SA, Formula & T/A (credit)	(721)	(613)	(613)
Bodyside moldings, Lamp Group, power windows and locks, door pockets.			
Option Pkg. 1SB, Formula & T/A (credit)	(547)	(465)	(465)
Pkg. 1SA plus 4-way driver's seat, cruise control, power liftgate release, reading lamps, power mirrors.			
Value Option Pkg., base & Formula	802	682	738
T-top roof, AM/FM ST ET cassette.			
Value Option Pkg., T/A	1089	926	1002
T-top roof, Custom Interior Trim, AM/FM ST ET cassette, cargo screen.			
Value Option Pkg., GTA	945	803	869
T-top roof, leather interior.			
Rear defogger, exc. GTA	145	123	133
California emissions pkg.	100	85	92
T-top roof	920	782	846
Two-tone paint, base	150	128	138
Power locks, exc. GTA	145	123	133
Power windows, exc. GTA	240	204	221
AM/FM ST ET cassette, exc. GTA	122	104	112
AM/FM ST ET cassette w/EQ, exc. GTA . .	282	240	259
w/VOP	150	128	138
w/Touch Control, base w/o VOP	447	380	411
Base w/VOP	315	268	290
AM/FM ST ET w/CD & EQ, exc. GTA	447	380	411
w/VOP	394	335	362
GTA	79	67	73
Power antenna, exc. GTA	70	60	64
Cargo security screen, exc. GTA	69	59	63
Luxury Interior Trim, T/A	293	249	270
Luxury seats and door panels, split folding rear seat.			
Leather articulated seats, GTA	375	319	345
245/50ZR16 tires & WS6 suspension, T/A .	385	327	354

What's New

Revamped front and rear styling, more powerful engines and fewer models sum up the changes for Grand Am, the sporty front-drive compact that's Pontiac's best seller. Grand Am's front end now has a swept-back, more rounded look from a new grille and fenders, lower hood, and flush headlamps. The new hood and fenders are made of 2-sided galvanized steel for longer corrosion resistance. At the rear there are new fascias, bumpers and end panels. The base 2.5-liter 4-cylinder engine has a modified cylinder head for better air flow and a larger throttle-body fuel injection unit, resulting in 110 horsepower, 12 more than last year. Last year's base series has been dropped, leaving LE and SE series available in either 2-door coupe or 4-door sedan styling. The 110-horsepower 2.5-liter is standard on the LE and a 165-horsepower turbocharged 2.0-liter four is standard on the performance-oriented SE. The 150-horsepower 2.3-liter Quad 4 dual-cam 4-cylinder returns from last year's debut as the optional engine for all models. Late in the year, a 185-horsepower high-output version of the Quad 4 is supposed to be installed in about 200 SE models as a preview of what will be more widely available for 1990. Grand Am is similar to the Buick Skylark and Oldsmobile Calais, which comprise GM's N-body family of cars. Buick and Olds offer the 2.5-liter four and the 2.3-liter Quad 4, plus a new 3.3-liter V-6 engine in their versions.

For

●Performance (Turbo, Quad 4) ●Visibility

Against

●Noise (Turbo, Quad 4) ●Rear seat room
●Trunk space

Summary

Grand Am is by far the most successful of the three N-body compacts, easily outselling the Skylark and Calais combined. Pontiac has the most focused marketing, aiming Grand Am at younger buyers who want hot performance, or at least a car that looks like it's hot. With the base 2.5-liter engine, Grand Am is strictly lukewarm, even with this year's boost to 110 horsepower. If you're seldom in a hurry, this is the least expensive way to go. The 165-horsepower turbocharged 2.0-liter four and the Quad 4 2.3-liter are huge leaps forward from the base engine, though both generate considerable noise with their abundant power. Of the

Prices are accurate at time of printing; subject to manufacturer's change.

CONSUMER GUIDE®

two high-performance engines, we favor the Quad 4. First, the Quad 4 makes its power from within, relying on a modern, multi-valve design. The turbo engine makes its power from an add-on turbocharger than increases stress, requires more frequent oil changes and can break. The Quad 4's weak point, besides noise, is that it has to be revved a bit to uncork much power. Grand Am's design dictates similarities with the Skylark and Calais, so there's too little interior and trunk space. We don't get too enthused about these cars, but concede that Grand Am is the logical choice of the three for the young and young at heart.

Pontiac Grand Am LE 4-door

Specifications

	2-door notchback	4-door notchback
Wheelbase, in.	103.4	103.4
Overall length, in.	180.1	180.1
Overall width, in.	66.5	66.5
Overall height, in.	52.5	52.5
Front track, in.	55.6	55.6
Rear track, in.	55.2	55.2
Turn diameter, ft.	37.8	37.8
Curb weight, lbs.	2508	2592
Cargo vol., cu. ft.	13.1	13.1
Fuel capacity, gal.	13.6	13.6
Seating capacity	5	5
Front headroom, in.	37.7	37.7
Front shoulder room, in.	52.6	52.6
Front legroom, max., in.	42.9	42.9
Rear headroom, in.	37.1	37.1
Rear shoulder room, in.	55.2	55.2
Rear legroom, min., in.	34.3	34.3

Powertrain layout: transverse front engine/front-wheel drive.

Engines

	ohv I-4	Turbo ohc I-4	dohc I-4
Size, liters/cu. in.	2.5/151	2.0/122	2.3/138
Fuel delivery	TBI	PFI	PFI
Horsepower @ rpm	110 @ 5200	165 @ 5600	150 @ 5200
Torque (lbs./ft.) @ rpm	135 @ 3200	175 @ 4000	160 @ 4000
Availability	S[1]	S[2]	O

EPA city/highway mpg

5-speed OD manual	23/33	21/30	24/35
3-speed automatic	23/30	21/28	23/32

1. LE   2. SE

Prices

Pontiac Grand Am	Retail Price	Dealer Invoice	Low Price
LE 2-door notchback	$10469	$9349	$9909
LE 4-door notchback	10669	9527	10098
SE 2-door notchback	13599	12144	12872
SE 4-door notchback	13799	12323	13061
Destination charge	425	425	425

Standard Equipment:

2.5-liter TBI 4-cylinder engine, 5-speed manual transmission, power steering and brakes, outboard rear lap/shoulder belts, tinted glass, headlamps-on tone, upshift indicator light, dual outside mirrors, AM/FM ST ET, cloth reclining front bucket seats, 185/80R13 all-season SBR tires. **SE** adds: 2.0-liter turbocharged PFI engine, WS6 sport suspension, monotone exterior treatment, sill moldings, fog lamps, power locks, split folding rear seat, tachometer, trip odometer, coolant temperature and oil pressure gauges, voltmeter, 215/60R14 tires on alloy wheels.

Optional Equipment:

	Retail Price	Dealer Invoice	Low Price
2.3-liter DOHC Quad 4 engine, LE	660	561	607
SE (credit)	(108)	(92)	(92)
3-speed automatic transmission	490	417	451
Air conditioning	675	574	621
Option Pkg. 1SA, LE	408	347	375
Air conditioning, console, tilt steering column.			
Option Pkg. 1SB, LE	484	411	445
Pkg. 1SA plus intermittent wipers, cruise control, Lamp Group.			
Option Pkg. 1SC, LE	670	570	616
Pkg. 1SB plus remote fuel door and decklid releases, split folding rear seat.			
Option Pkg. 1SD, LE 2-door	1112	945	1023
LE 4-door	1037	881	954
Pkg. 1SC plus power windows and locks, fog lamps, power driver's seat.			
Option Pkg. 1SA, SE 2-door	431	366	397
SE 4-door	506	430	466
Air conditioning, Lamp Group, power windows.			
Option Pkg. 1SB, SE 2-door	560	476	515
SE 4-door	635	540	584
Pkg. 1SA plus power driver's seat, lighted right visor mirror, power mirrors.			
Value Option Pkg., LE	341	290	314
AM/FM ST ET cassette, 195/70R14 tires on alloy wheels.			
Rear defogger	145	123	133
California emissions pkg	100	85	92
Rally instruments, LE	127	108	117
Decklid luggage rack	115	98	106
Two-tone paint, LE	101	86	93
Power locks, LE 2-door	145	123	133
LE 4-door	195	166	179
Power windows, 2-doors	210	179	193
4-doors	285	242	262
AM/FM ST ET cassette	122	104	112
AM/FM ST ET cassette w/EQ	272	231	250
LE w/VOP	150	128	138
AM/FM ST ET w/CD & EQ	545	463	501
LE w/VOP	423	360	389
Performance Sound System	125	106	115
Requires power windows.			
Articulated sport seats, SE	450	383	414
SE w/Pkg. 1SB	245	208	225
Custom Interior Trim, LE	269	229	247
LE w/Pkg. 1SC or 1SD	119	101	109
Removable glass sunroof	350	298	322
WSW tires, LE	68	58	63
195/70R14 tires, LE	104	88	96

> **KEY: ohv** = overhead valve; **ohc** = overhead cam; **dohc** = double overhead cam; **I** = inline cylinders; **V** = cylinders in V configuration; **flat** = horizontally opposed cylinders; **bbl.** = barrel (carburetor); **PFI** = port (multi-point) fuel injection; **TBI** = throttle-body (single-point) fuel injection; **rpm** = revolutions per minute; **OD** = overdrive transmission; **S** = standard; **O** = optional; **NA** = not available.

Prices are accurate at time of printing; subject to manufacturer's change.

	Retail Price	Dealer Invoice	Low Price
215/60R14 tires, LE	278	236	256
LE w/VOP	174	148	160
w/white letters & WS6 Pkg., LE	370	315	340
w/white letters & WS6 Pkg., LE w/VOP .	266	226	245
60-series tires require alloy wheels.			
14" alloy wheels, LE	215	183	198

Pontiac Grand Prix

Pontiac Grand Prix LE

What's New

Anti-lock brakes are a new option for Grand Prix, a new 3.1-liter V-6 is planned for a mid-year arrival and a 200-horsepower turbocharged model is due late in the year. Grand Prix was redesigned to front-wheel drive last year under GM's W-body program, which includes the Buick Regal and Oldsmobile Cutlass Supreme. All are 2-door coupes. Chevrolet will introduce a 4-door sedan W-body next spring, the Lumina. Eventually, all four divisions will have 2- and 4-door models from this program, though styling will be different for each model. The new anti-lock brake system is optional on all three Grand Prix models: base, LE and SE; 4-wheel disc brakes are standard. Around January, a 140-horsepower 3.1-liter V-6 will replace a 130-horsepower 2.8 V-6 as the standard engine with automatic transmission. The new 3.1 is derived from the 2.8 V-6, which will continue in models with the 5-speed manual transmission. Other changes are that air conditioning is standard on all models, but on the LE and SE it is a new electronically controlled climate system. Rear shoulder belts are also standard. New options include a power sunroof, steering-wheel controls for the stereo system, and a remote keyless entry system that uses radio signals to lock or unlock the doors and trunk from up to 30 feet away. Next spring, the McLaren Turbo Grand Prix is scheduled to be offered in limited numbers (about 2000) with a turbocharged, intercooled version of the 3.1 V-6 developed by the McLaren division of ASC, a specialty-car builder. Pontiac says the engine will produce over 200 horsepower—more than any other front-drive GM car—and have a 4-speed automatic transmission. Standard equipment will include anti-lock brakes, 245/50ZR16 tires on 8-inch wide wheels, power sunroof, tan leather interior and aerodynamic exterior trim. The Turbo will be available in two exterior colors, bright red or metallic black.

For

- Anti-lock brakes
- Ride
- Handling (SE)
- Passenger room
- Trunk space

Against

- Performance
- Engine noise
- Instruments/controls

Summary

Of all the changes for 1989 on Grand Prix, the two that count are the optional anti-lock brakes (ABS) and the 3.1-liter V-6. ABS is an expensive ($925) but worthy safety feature that improves stopping ability in emergencies. The larger V-6 due at mid-year is a welcome addition because the 2.8-liter V-6 just doesn't have enough power to motivate this car swiftly. With automatic transmission you have to floor the throttle and force the engine to rev its heart out in a futile attempt at brisk response. Even this generates more noise than thrust. Grand Prix isn't a slug with the 2.8, but this engine doesn't deliver what the racy exterior promises. We tested a 1988 SE and found its firm suspension provides capable handling, good high-speed stability and a fairly absorbent ride that's more comfortable than we've come to expect from sporty Pontiacs. However, we weren't nearly as impressed with the SE's uncomfortable power front seats and confusing dashboard, which combines digital displays with analog gauges and turns simple functions into complex operations. "Star Trek" fans should love it, but we think Pontiac can learn some lessons from Honda in this area. The sportiest of GM's W-body coupe needs more gumption and fewer gimmicks.

Specifications

	2-door notchback
Wheelbase, in.	107.5
Overall length, in.	193.9
Overall width, in.	70.9
Overall height, in.	53.3
Front track, in.	59.5
Rear track, in.	58.0
Turn diameter, ft.	39.7
Curb weight, lbs.	3167
Cargo vol., cu. ft.	15.0
Fuel capacity, gal.	16.0
Seating capacity	6
Front headroom, in.	37.8

Pontiac Grand Prix SE

KEY: **ohv** = overhead valve; **ohc** = overhead cam; **dohc** = double overhead cam; **I** = inline cylinders; **V** = cylinders in V configuration; **flat** = horizontally opposed cylinders; **bbl.** = barrel (carburetor); **PFI** = port (multi-point) fuel injection; **TBI** = throttle-body (single-point) fuel injection; **rpm** = revolutions per minute; **OD** = overdrive transmission; **S** = standard; **O** = optional; **NA** = not available.

Prices are accurate at time of printing; subject to manufacturer's change.

	2-door notchback
Front shoulder room, in.	57.3
Front legroom, max., in.	42.3
Rear headroom, in.	36.7
Rear shoulder room, in.	57.3
Rear legroom, min., in.	34.8

Powertrain layout: transverse front engine/front-wheel drive.

Engines

	ohv V-6	ohv V-6
Size, liters/cu. in.	2.8/173	3.1/191
Fuel delivery	PFI	PFI
Horsepower @ rpm	130 @ 4500	140 @ 4500
Torque (lbs./ft.) @ rpm	170 @ 3600	185 @ 3600
Availability	S	S[1]

EPA city/highway mpg

5-speed OD manual	18/30	
4-speed OD automatic	20/29	NA

1. Mid-year, with automatic transmission.

Prices

Pontiac Grand Prix	Retail Price	Dealer Invoice	Low Price
2-door notchback	$13899	$11995	$12947
LE 2-door notchback	14849	12815	13832
SE 2-door notchback	15999	13807	14903
Destination charge	455	455	455

Standard Equipment:

2.8-liter PFI V-6 engine, 4-speed automatic transmission (5-speed manual may be substituted for credit), power steering, power 4-wheel disc brakes, air conditioning, cloth split bench seat with folding armrest, AM/FM ST ET, electronic instruments, remote fuel door release, tinted glass, left remote and right manual mirrors, automatic front seatbelts, outboard rear lap/shoulder belts, 195/75R14 tires. **LE** adds: power windows, light group, map pockets, tachometer and trip odometer, 60/40 front seats with recliners, folding rear seatback. **SE** adds: 5-speed manual transmission, cruise control, articulating power front bucket seats, contoured rear seats with head restraints, power mirrors, front floor and rear overhead consoles, analog instruments, front and rear reading lamps, tilt steering column, leather-wrapped steering wheel, fog lamps, intermittent wipers, Rally Tuned Suspension, 215/60R16 Goodyear Eagle GT+4 tires on alloy wheels.

Optional Equipment:

	Retail	Dealer Inv.	Low
4-speed automatic transmission, SE	615	523	566
5-speed manual transmission, base & LE (credit)	(615)	(522)	(522)
Anti-lock brakes	925	786	851
Option Pkg. 1SA, base	183	156	168
Tilt steering column, Lamp Group.			
Option Pkg. 1SB, base	188	160	173
Pkg. 1SA plus intermittent wipers, cruise control.			
Option Pkg. 1SC, base	378	321	348
Pkg. 1SB plus power windows and locks, remote decklid release.			
Option Pkg. 1SA, LE	275	234	253
Tilt steering column, intermittent wipers, cruise control, power locks.			
Option Pkg. 1SB, LE	383	326	352
Pkg. 1SA plus remote decklid release, power driver's seat, lighted right visor mirror.			
Option Pkg. 1SC, LE	594	505	546
Pkg. 1SB plus reading lamps, leather-wrapped steering wheel and shift handle, power mirrors, remote keyless entry.			
Option Pkg. 1SA, SE	260	221	239
Power locks, remote decklid release, power driver's seat.			
Option Pkg. 1SB, SE	563	479	518
Pkg. 1SA plus lighted right visor mirror, remote keyless entry, electronic compass with trip computer and service reminder.			

	Retail Price	Dealer Invoice	Low Price
Value Option Pkg., base	393	334	362
40/60 seat, AM/FM ST ET cassette, gauge pkg., 195/70R15 tires on styled steel wheels.			
Value Option Pkg., LE	375	319	345
AM/FM ST ET cassette, two-tone paint, 195/70R15 tires on alloy wheels.			
Rear defogger	145	123	133
California emissions pkg.	100	85	92
Gauge cluster w/tachometer, base	85	72	78
Decklid luggage rack	115	98	106
Two-tone paint	105	89	97
Power locks	145	123	133
Power glass sunroof, base & LE	675	574	621
LE w/Pkg. 1SC, SE	650	553	598
Power windows, base	220	187	202
AM/FM ST ET cassette	122	104	112
AM/FM ST ET cassette w/EQ, base & LE	447	380	411
Base & LE w/VOP	325	276	299
LE w/Pkg. 1SC, SE	397	337	365
LE w/Pkg. 1SC & VOP	275	234	253
Performance Sound System, LE & SE	125	106	115
Power antenna	70	60	64
40/60 seat, base	133	113	122
Bucket seats, base	193	164	178
Base w/VOP	60	51	55
LE	110	94	101
Leather bucket seats, SE	375	319	345
WSW tires, base & LE	72	61	66
195/70R15 tires, base & LE	48	41	44
215/60R16 tires, base & LE	232	197	213
Base & LE w/VOP	184	156	169
15" styled wheels, base & LE	130	111	120
15" alloy wheels, base & LE	215	183	198
Base w/VOP	85	72	78
Requires 195/70R15 tires.			
16" alloy wheels, base & LE	250	213	230
Base w/VOP	120	102	110
LE w/VOP	35	30	32
Requires 205/60R16 tires.			

Pontiac LeMans

Pontiac LeMans GSE

What's New

Pontiac's Korean-built front-drive subcompact returns unchanged from last year, when it gained a new sport model, the GSE. Later in the 1989 model year, a removable glass sunroof is supposed to be added as an option. Built in Korea by Daewoo (partly owned by GM), LeMans was de-

Prices are accurate at time of printing; subject to manufacturer's change.

Pontiac

signed by GM's Opel division in West Germany. The lineup consists of the Aerocoupe Value Leader 3-door hatchback; plusher LE hatchback; LE and sporty SE 4-door sedans; and the GSE hatchback that was added last spring. All except the GSE use a 74-horsepower 1.6-liter 4-cylinder engine. The GSE comes with a 96-horsepower 2.0-liter four also used in the Pontiac Sunbird, plus larger tires, firmer suspension and racy exterior trim. All models are available with either a 5-speed manual or 3-speed automatic transmission except the Value Leader, which comes only with a 4-speed manual.

For

- Fuel economy   ●Cargo space   ●Performance (GSE)
- Handling

Against

- Performance and driveability (automatic transmission)
- Noise

Summary

Our test of a 1988 LeMans SE showed this is an agile, sure-footed car that's inordinately slow with automatic transmission. Merging into expressway traffic is risky and you have to allow lots of time and space for passing. The engine is loud in hard acceleration and the transmission is slow to downshift for more power. However, our 28 mpg average bordered on exceptional since it was nearly all city driving. A 5-speed we tested was much faster and provided acceptable acceleration, plus slightly better fuel economy. The SE's 175/70R13 all-season tires and firm suspension add up to responsive handling and composed roadholding, but a stiff, jiggly ride on bumpy pavement. LeMans has a 99.2-inch wheelbase, long for a subcompact, so there's more leg room than in some rivals, but the rear doors are narrow for tight access to the back seat. Cargo room is above average for a small car. The trunk on the 4-door has a small opening, but the volume can be increased by folding the rear seatback. The base LeMans is attractively priced, but not as well furnished as a base Hyundai Excel, also made in Korea. A top-line LeMans SE with automatic transmission, power steering and air conditioning runs well over $10,000, which isn't a bargain in this class. LeMans doesn't show much of its German heritage on the road, or much of a cost benefit from being built in Korea.

Specifications

	3-door hatchback	4-door notchback
Wheelbase, in.	99.2	99.2
Overall length, in.	163.7	172.4
Overall width, in.	65.5	65.7
Overall height, in.	53.5	53.7
Front track, in.	55.1	55.1
Rear track, in.	55.4	55.4
Turn diameter, ft.	32.8	32.8
Curb weight, lbs.	2136[1]	2235[2]
Cargo vol., cu. ft.	18.8	18.4
Fuel capacity, gal.	13.2	13.2

> **KEY: ohv** = overhead valve; **ohc** = overhead cam; **dohc** = double overhead cam; **I** = inline cylinders; **V** = cylinders in V configuration; **flat** = horizontally opposed cylinders; **bbl.** = barrel (carburetor); **PFI** = port (multi-point) fuel injection; **TBI** = throttle-body (single-point) fuel injection; **rpm** = revolutions per minute; **OD** = overdrive transmission; **S** = standard; **O** = optional; **NA** = not available.

	3-door hatchback	4-door notchback
Seating capacity	5	5
Front headroom, in.	38.8	38.8
Front shoulder room, in.	53.5	53.5
Front legroom, max., in.	42.0	42.0
Rear headroom, in.	38.0	38.0
Rear shoulder room, in.	53.4	53.4
Rear legroom, min., in.	32.8	32.8

1. 2302 lbs., GSE.   2. 2357 lbs., SE.

Powertrain layout: transverse front engine/front-wheel drive.

Engines

	ohc I-4	ohc I-4
Size, liters/cu. in.	1.6/98	2.0/121
Fuel delivery	TBI	TBI
Horsepower @ rpm	74 @ 5600	96 @ 4800
Torque (lbs./ft.) @ rpm	90 @ 2800	118 @ 3600
Availability	S	S[1]

EPA city/highway mpg

	3-door hatchback	4-door notchback
4-speed OD manual	30/39	
5-speed OD manual	31/40	24/30
3-speed automatic	27/31	22/28

1. GSE

Prices

Pontiac Le Mans	Retail Price	Dealer Invoice	Low Price
Aerocoupe Value Leader 3-door hatchback .	$6399	$5919	$6159
Aerocoupe LE 3-door hatchback	7699	7083	7391
LE 4-door notchback	7999	7359	7679
SE 4-door notchback	9429	8486	8958
Aerocoupe GSE 3-door hatchback	9149	8234	8692
Destination charge	315	315	315

Standard Equipment:

Value Leader: 1.6-liter TBI 4-cylinder engine, 4-speed manual transmission, power brakes, reclining front bucket seats, rear defogger, tachometer, trip odometer, outboard rear lap/shoulder belts, cargo area cover, 175/70SR13 tires with full-size spare. **LE** adds: 5-speed manual transmission, tinted glass, dual remote mirrors, right visor mirror, AM/FM ST ET with clock, swing-out rear side windows (Aerocoupe; roll-down on 4-door). **SE** adds: 2.0-liter engine, sport suspension, sport seats with height adjusters, 60/40 folding rear seat, tilt steering column, fog lamps, 185/60HR14 tires. **GSE** adds: front air dam with fog lamps, sill extensions, rear spoiler, alloy wheels.

Optional Equipment:

3-speed automatic transmission (NA VL)	420	357	386
Air conditioning	660	561	607

Requires power steering; not available on VL).

Pontiac LeMans SE

	Retail Price	Dealer Invoice	Low Price
Power steering, LE & SE	214	182	197
Cruise control	175	149	161
AM/FM ST ET, VL	307	261	282
AM/FM ST ET cassette, VL	429	365	395
Others	122	104	112
Luggage carrier	95	81	87
Removable sunroof	350	298	322

Pontiac Safari

Pontiac Safari

What's New

The full-size, rear-drive Safari wagon is carried over from last year except for the addition of rear shoulder belts as standard equipment. Safari is built from the same design as the Buick LeSabre/Electra Estate, Chevrolet Caprice and Oldsmobile Custom Cruiser wagons and uses the same 140-horsepower 5.0-liter V-8 engine as the others.

For

● Room/comfort ● Cargo room
● Trailer towing capacity

Against

● Fuel economy ● Size and weight
● Handling/maneuverability

Summary

Pontiac does little to promote this full-size, rear-drive wagon since it hardly fits in with the "We Build Excitement" theme, but the division still sells 5000 or so a year. Since this wagon is a spitting image of those sold by Buick, Chevy and Olds, there's nothing to recommend the Safari over the others. Choosing one of the four should come down to which one suits your styling tastes or which dealer offers you the best price. All four offer seats for eight, generous cargo capacity and 5000-pound trailer towing capability. If you want this type of station wagon, then we suggest you also look at the Ford LTD Crown Victoria/Mercury Grand Marquis wagons, the only direct rivals.

Specifications

	5-door wagon
Wheelbase, in.	116.0
Overall length, in.	215.1
Overall width, in.	79.3

	5-door wagon
Overall height, in.	57.4
Front track, in.	62.2
Rear track, in.	64.1
Turn diameter, ft.	39.2
Curb weight, lbs.	4225
Cargo vol., cu. ft.	87.9
Fuel capacity, gal.	22.0
Seating capacity	8
Front headroom, in.	39.6
Front shoulder room, in.	60.8
Front legroom, max., in.	42.2
Rear headroom, in.	39.3
Rear shoulder room, in.	60.9
Rear legroom, min., in.	37.2

Powertrain layout: longitudinal front engine/rear-wheel drive.

Engines

	ohv V-8
Size, liters/cu. in.	5.0/302
Fuel delivery	4 bbl.
Horsepower @ rpm	140 @ 3200
Torque (lbs./ft.) @ rpm	225 @ 2000
Availability	S

EPA city/highway mpg
4-speed OD automatic 17/24

Prices

Pontiac Safari	Retail Price	Dealer Invoice	Low Price
5-door wagon	$15659	$13514	$14587
Destination charge	505	505	505

Standard Equipment:

5.0-liter 4bbl. V-8 engine, 4-speed automatic transmission, power steering and brakes, air conditioning, tinted glass, courtesy lights, left remote and right manual mirrors, bodyside moldings, AM/FM ST ET, cloth split bench seat with center armrest, rear-facing third seat, outboard rear lap/shoulder belts, custom wheel covers, 225/75R15 tires.

Optional Equipment:

Option Pkg. 1SA	402	342	370
Tilt steering column, Lamp Group, cruise control, intermittent wipers.			
Option Pkg. 1SB	1434	1219	1319
Pkg. 1SA plus power windows and locks, power driver's seat, sidewall and tailgate carpet, cornering lamps, remote mirrors, bumper guards, lighted right visor mirror, halogen headlamps.			
HD cooling	40	34	37
Rear defogger	145	123	133
California emissions pkg.	100	85	92
Instrument cluster gauges	71	60	65
Roof luggage rack	155	132	143
Power locks	245	208	225
Power windows	285	242	262
AM/FM ST ET cassette	122	104	112
w/EQ	272	231	250
Power antenna	70	60	64
55/45 seat	133	113	122
Right seatback recliner	45	38	41
Requires 55/45 seat.			
Super-lift shock absorbers	64	54	59
7-lead trailer wiring harness	30	26	28
Wire wheel covers	214	182	197
Simulated woodgrain siding	345	293	317

Prices are accurate at time of printing; subject to manufacturer's change

Pontiac Sunbird

Pontiac Sunbird LE 2-door

What's New

A juggled lineup, a new instrument panel for all models and new front-end styling for some models are the major changes for Sunbird, Pontiac's version of the J-body front-drive subcompact. The J-body family was introduced for 1982 by all five GM divisions, but the slow-selling Cadillac Cimarron and Oldsmobile Firenza have been dropped for 1989, leaving the Buick Skyhawk, Chevrolet Cavalier and Sunbird. Pontiac has dropped the 5-door wagon for 1989 (the 3-door hatchback disappeared last year), leaving a 2-door coupe, 4-door sedan and 2-door convertible. The base sedan from last year also fades away, and a reinstated LE series is now the low-priced Sunbird. The LE coupe and sedan sport new front styling with a lower nose, composite headlamps and twin-outlet grille. The SE coupe and GT coupe and convertible return from last year. The new instrument panel has analog gauges flanked by controls for lights and windshield wiper/washer similar to those used on the Grand Prix. The stereo and heat/vent controls are mounted high on the dash and angled towards the driver. A compact disc player becomes available later in the year. A redesigned center console includes a new horseshoe-shaped parking brake lever. Base engine is a 96-horsepower 2.0-liter 4-cylinder, standard on LE and SE models. A 165-horsepower turbocharged 2.0-liter is standard on the GT models and optional on the SE. The non-turbo engine is available as a credit option on the GTs. A 5-speed manual transmission is standard and a 3-speed automatic optional on all.

For

● Performance (Turbo) ●Fuel economy
● Handling/roadholding (SE, GT)

Against

● Performance (base engine/automatic)
● Rear seat room

Summary

Sunbird differs from its Cavalier and Skyhawk cousins by virtue of its overhead-cam engines (Chevrolet and Buick use overhead-valve engines) and a marketing approach that leans more towards sport and performance than value or luxury. The 96-horsepower base engine provides decent

acceleration and good fuel economy with the 5-speed manual transmission, but loses some of its zest and economy with the optional 3-speed automatic. If rapid transit is your desire, the 165-horsepower turbocharged 4-cylinder is the way to go. It will snap your head back in hard acceleration. Keep both hands on the wheel, however, to control the torque steer, and be prepared to endure a loud, raspy exhaust. Handling and roadholding are best on the GT, but its wide performance tires contribute to a stiff ride. The SE coupe's narrower, all-season tires lack the cornering grip of the GT's tires, yet they improve ride comfort. While the LE and SE look sportier than they really are, they're attractively priced against Japanese subcompacts.

Specifications

	2-door notchback	4-door notchback	2-door conv.
Wheelbase, in.	101.2	101.2	101.2
Overall length, in.	178.2	181.7	178.2
Overall width, in.	66.0	65.0	66.0
Overall height, in.	50.4	53.8	51.9
Front track, in.	55.9	55.6	55.6
Rear track, in.	55.2	55.2	55.2
Turn diameter, ft.	34.7	34.7	34.7
Curb weight, lbs.	2418	2433	2577
Cargo vol., cu. ft.	14.0	15.2	10.4
Fuel capacity, gal.	13.6	13.6	13.6
Seating capacity	5	5	4
Front headroom, in.	37.7	38.6	38.5
Front shoulder room, in.	52.7	53.7	53.7
Front legroom, max., in.	42.9	42.2	42.2
Rear headroom, in.	36.1	37.9	37.4
Rear shoulder room, in.	52.6	53.7	38.0
Rear legroom, min., in.	31.6	34.3	31.1

Powertrain layout: transverse front engine/front-wheel drive.

Engines	ohc I-4	Turbo ohc I-4
Size, liters/cu. in.	2.0/121	2.0/121
Fuel delivery	TBI	PFI
Horsepower @ rpm	96 @ 4800	165 @ 5500
Torque (lbs./ft.) @ rpm	118 @ 3600	175 @ 4000
Availability	S	S[1]
EPA city/highway mpg		
5-speed OD manual	27/36	21/31
3-speed automatic	23/31	20/27

1. GT; optional, SE

Prices

Pontiac Sunbird	Retail Price	Dealer Invoice	Low Price
LE 2-door notchback	$8849	$7902	$8376
LE 4-door notchback	8949	7991	8470
SE 2-door notchback	9099	8125	8612
GT 2-door notchback	11399	10179	10789
GT 2-door convertible	16899	15091	15995
Destination charge	425	425	425

Optional Equipment:

Turbo Pkg., SE	1384	1176	1273
SE w/VOP	1169	994	1075
2.0-liter turbo engine, power steering, Rally Gauge Cluster, engine block heater, WS6 Performance Pkg. (215/60R14 tires on alloy wheels).			
Turbo engine delete (credit), GT	(768)	(653)	(653)
Air conditioning	675	574	621

Prices are accurate at time of printing; subject to manufacturer's change

CONSUMER GUIDE®

	Retail Price	Dealer Invoice	Low Price
Option Pkg. 1SA, LE & SE	233	198	214
SE w/Turbo Pkg.	158	134	145
Tinted glass, power steering, left remote and right manual mirrors.			
Option Pkg. 1SB, LE & SE	330	281	304
SE w/Turbo Pkg.	205	174	189
Pkg. 1SA plus tilt steering column, intermittent wipers, front armrest, floormats.			
Option Pkg. 1SC, SE & LE 2-door	769	654	707
LE 4-door	775	660	713
SE w/Turbo Pkg.	544	462	500
Pkg. 1SB plus rally steering wheel, air conditioning, Lamp Group, cruise control.			
Option Pkg. 1SA, GT notchback	268	228	247
Tinted glass, tilt steering column, intermittent wipers, floormats, air conditioning.			
Option Pkg. 1SB, GT notchback	456	388	420
Pkg. 1SA plus Lamp Group, cruise control, remote decklid release.			
Option Pkg. 1SC, GT notchback	811	689	746
Pkg. 1SB plus power windows and locks, leather-wrapped steering wheel.			
Option Pkg. 1SA, GT conv.	163	139	150
Tilt steering column, intermittent wipers, floormats, air conditioning.			
Option Pkg. 1SB, GT conv.	401	341	369
Pkg. 1SA plus Lamp Group, cruise control, remote decklid release, leather-wrapped steering wheel.			
Value Option Pkg., LE	370	315	340
AM/FM ST ET cassette, rally gauge pkg., 195/70R14 tires on alloy wheels.			
Value Option Pkg., SE	392	333	361
Split folding rear seat, AM/FM ST ET cassette, alloy wheels.			
Rear defogger (NA conv.)	145	123	133
California emissions pkg	100	85	92
Rally gauge cluster, LE	49	42	45
Decklid luggage rack	115	98	106
Two-tone paint, LE	101	86	93
Power locks, 2-doors	145	123	133
LE 4-door	195	166	179
Power windows, 2-doors	210	179	193
LE 4-door	285	242	262
AM/FM ST ET cassette	152	129	140
AM/FM ST ET cassette w/CD player	396	337	364
SE & LE w/VOP	244	207	224
Custom Interior Trim, GT notchback	335	285	308
Upgraded door panels with pockets, split folding rear seat, added sound insulation.			
Custom Interior Trim, GT conv.	142	121	131
Upgraded door panels with pockets.			
Split folding rear seat (NA conv.)	150	128	138
Removable glass sunroof	350	298	322
WSW tires, LE	68	58	63
195/70R14 tires, LE	104	88	96
215/60R14 tires, SE w/o Turbo Pkg.	184	156	169
w/white letters, SE w/o Turbo Pkg.	276	235	254
w/white letters, SE w/Turbo Pkg., GT	102	87	94
Includes WS6 Performance Pkg.; requires alloy wheels.			
13" alloy wheels, LE, SE w/o Turbo Pkg.	215	183	198

Pontiac 6000

What's New

All versions of the 6000 4-door sedans have new exterior features and rear shoulder belts, while the top-line STE comes only as a 4-wheel-drive model for 1989. The sedans get a new roof and rounded rear window, similar to what also appears this year on the Buick Century and Oldsmobile Cutlass Ciera. In addition, the sedans have a new trunk lid and taillamps. The front-drive LE and S/E sedans and 5-door wagons also gain the 6-lamp front treatment (quad headlamps plus integrated fog lamps) first seen on the STE. A body-color grille and moldings give S/E models an even closer resemblance to the STE. The front-drive version of the STE has disappeared, leaving only the All-

Pontiac 6000 LE 4-door

Wheel Drive model that arrived late in the 1988 model year. The 4WD system is permanently engaged, splitting engine torque 60 percent front/40 percent rear. STE comes only as a 4-door with a 140-horsepower 3.1-liter V-6 and 3-speed automatic transmission. Standard equipment also includes anti-lock brakes. The 3.1 V-6 and anti-lock brakes aren't available on other 6000s. STE comes in only two monochromatic paint schemes for 1989: metallic red or blue. A compact-disc player becomes optional later in the year. On the 6000 LE, a larger throttle body for the single-point fuel injection system helps increase horsepower from 98 to 110 in the standard 2.5-liter 4-cylinder engine. A 130-horsepower 2.8-liter V-6 with multi-point injection is standard on the S/E, optional in the LE. Pontiac has dropped the 5-speed manual transmission for the 2.8 V-6 because only a few hundred were sold the past two years. That leaves a 3-speed automatic transmission for the 4-cylinder and the V-6 (sedans only), and a 4-speed overdrive automatic for the V-6 only.

For

- Passenger and cargo room
- Handling (S/E, STE)
- 4WD traction (STE)
- Anti-lock brakes (STE)
- Quietness

Against

- Performance (4-cylinder)
- Handling (LE)

Summary

We still recommend the 6000 and the similar Chevy Celebrity as mid-size cars that offer good value for the money, though they're now seven years old. Their passenger and cargo room are comparable to what you find in the Ford Taurus and Mercury Sable, and the 2.8-liter V-6 engine furnishes sufficient power for adequate performance with a full load. In addition, the 6000 S/E models have firmer suspensions and wider tires that combine for improved handling and roadholding over the base LE versions. Buick and Oldsmobile offer similar versions of this same design, at slightly higher prices, but the only real advantage they have is a new 3.3-liter V-6 that produces 160 horsepower, 30 more than the 2.8 V-6 used by Pontiac and Chevy. The 4WD STE costs a lot more than the other 6000s, but has the equipment to justify its higher price.

KEY: ohv = overhead valve; **ohc** = overhead cam; **dohc** = double overhead cam; **I** = inline cylinders; **V** = cylinders in V configuration; **flat** = horizontally opposed cylinders; **bbl.** = barrel (carburetor); **PFI** = port (multi-point) fuel injection; **TBI** = throttle-body (single-point) fuel injection; **rpm** = revolutions per minute; **OD** = overdrive transmission; **S** = standard; **O** = optional; **NA** = not available.

Prices are accurate at time of printing; subject to manufacturer's change

Specifications

	4-door notchback	5-door wagon
Wheelbase, in.	104.9	104.9
Overall length, in.	188.8	193.2
Overall width, in.	72.0	72.0
Overall height, in.	53.7	54.1
Front track, in.	58.7	58.7
Rear track, in.	57.0	57.0
Turn diameter, ft.	37.0	37.0
Curb weight, lbs.	2760[1]	2897
Cargo vol., cu. ft.	16.2	74.4
Fuel capacity, gal.	15.7	15.7
Seating capacity	6	8
Front headroom, in.	38.6	38.6
Front shoulder room, in.	56.2	56.2
Front legroom, max., in.	42.1	42,1
Rear headroom, in.	38.0	38.9
Rear shoulder room, in.	56.2	56.2
Rear legroom, min., in.	35.8	34.7

1. 3036 lbs., STE

Powertrain layout: transverse front engine/front-wheel drive (permanently engaged 4WD, STE).

Engines

	ohv I-4	ohv V-6	ohv V-6
Size, liters/cu. in.	2.5/151	2.8/173	3.1/191
Fuel delivery	TBI	PFI	PFI
Horsepower @ rpm	110 @ 5200	130 @ 4500	140 @ 4800
Torque (lbs./ft.) @ rpm	135 @ 3200	170 @ 3600	185 @ 3200
Availability	S	S[1]	S[2]

EPA city/highway mpg

3-speed automatic	23/30	20/27	18/23
4-speed OD automatic		20/29	

1. S/E; optional, LE   2. STE

Prices

Pontiac 6000	Retail Price	Dealer Invoice	Low Price
LE 4-door notchback	$11969	$10329	$11149
LE 5-door wagon	13769	11883	12826
S/E 4-door notchback	15399	13289	14344
S/E 5-door wagon	16999	14411	15705
STE All Wheel Drive 4-door notchback	22599	19503	21051
Destination charge	450	450	450

Standard Equipment:

LE: 2.5-liter TBI 4-cylinder engine, 3-speed automatic transmission (4-door); wagon has 2.8-liter PFI V-6 and 4-speed automatic transmission), power steering and brakes, AM/FM ST ET, tinted glass, cloth front seat with armrest, split folding second seat (wagon), rear defogger (wagon), outboard rear lap/shoulder belts, 185/75R14 SBR tires. **S/E adds:** 2.8-liter PFI V-6, 4-speed automatic transmission, tachometer, trip odometer, voltmeter, coolant temperature and oil pressure gauges, intermittent wipers, power windows and locks, AM/FM ST ET cassette, rear defogger, luggage rack (wagon), cloth reclining bucket seats, leather-wrapped steering wheel, sport suspension, Electronic Ride Control (wagon), 195/70R14 Goodyear Eagle GT + 4 tires on aluminum wheels. **STE All Wheel Drive** adds: 3.1-liter V-6, 3-speed automatic transmission, permanent, full-time 4-wheel drive, 4-wheel disc brakes, anti-lock braking system, accessory kit (flares, first aid kit, raincoat, gloves, spotlight), air conditioning, cruise control, power windows and locks, Driver Information Center, tire inflator, Lamp Group, intermittent wipers, Electronic Ride Control, AM/FM ET cassette with EQ and Touch Control, power antenna, power trunk release, rear defogger, electronic instruments, front reading lamps, lighted right visor mirror, fog lamps, 195/70R15 Goodyear Eagle GT + 4 tires on alloy wheels.

Optional Equipment:	Retail Price	Dealer Invoice	Low Price
2.8-liter PFI V-6, LE 4-door	610	519	561
4-speed automatic transmission, LE 4-door	175	149	161
Requires 2.8-liter V-6.			
Air conditioning	775	659	713
Option Pkg. 1SA, LE	605	514	557
Air conditioning, tilt steering column, intermittent wipers.			
Option Pkg. 1SB, LE 4-door	730	621	672
LE wagon	865	735	796
Pkg. 1SA plus cruise control.			
Option Pkg. 1SC, LE 4-door	1141	970	1050
LE wagon	1131	961	1041
Pkg. 1SB plus power windows and locks, Lamp Group, remote decklid release, power driver's seat, lighted right visor mirror, reading lamps.			
Option Pkg. 1SA, S/E 4-door	238	202	219
S/E wagon	188	160	173
Air conditioning, remote decklid release, power driver's seat, reading lamps.			
Value Option Pkg., LE 4-door	345	293	317
45/55 seat, AM/FM ST ET cassette, alloy wheels.			
Rear defogger, LE 4-door	145	123	133
California emissions pkg.	100	85	92
Two-tone paint	115	98	106
Power locks, LE 4-door	195	166	179
Power windows, LE 4-door	300	255	276
AM/FM ST ET cassette, LE	122	104	112
AM/FM ST ET cassette w/EQ, S/E 4-door	350	298	322
S/E wagon	315	268	290
AM/FM ST ET w/CD & EQ, S/E 4-door	554	471	510
S/E wagon	519	441	477
STE	204	173	188
45/55 seat, LE	133	113	122
Custom Interior Trim, LE	493	419	454
LE w/VOP	360	306	331
45/55 seat with recliners, gauges.			
WSW tires, LE	68	58	63
Alloy wheels, LE	215	183	198
Simulated woodgrain siding, LE wagon	295	251	271

Porsche 911 and 928 S4

Porsche 928 S4

What's New

Porsche's line of 6-cylinder, rear-engine sports cars, the 911 series, grabs the lion's share of news, with two new models for '89. On sale after January is the 911 Carrera 4, an all-wheel-drive version of the 911. It's the first street Porsche to borrow technology from the limited-edition 4WD 959 supercar. Its styling applies subtle aero shaping to the 25-year-old 911 body. The Carrera 4 uses a 3.6-liter version of the 911's naturally aspirated flat-6 engine producing 247 horsepower. The 4WD system employs a variable torque split with a rearward bias. Front wheel rims are six inches wide, rears are eight. A 5-speed manual transmission is used, and the Carrera 4 is the first 911 to have

Prices are accurate at time of printing; subject to manufacturer's change

either power steering or anti-lock brakes (ABS). Also due after January is the limited-edition 911 Speedster. Recalling Porsche's legendary 356 Speedster sports car of the 1950s, the new Speedster uses the base 911 Carrera running gear, but deletes the rear jump seats and adds a cutdown windshield to a convertible body. Its single-layer manual soft top is in contrast to the 911 Cabriolet's, which is power operated and uses a padded inner liner. The Speedster's lowered top also can be hidden beneath a tonneau cover, and a Turbo-look body style is available. The 911 Carrera, 911 Targa and 911 Cabriolet keep their 214-horsepower 3.2-liter engine but trade last year's 15-inch wheels for 16s and wider tires. The 911 Turbo, with its 282-horsepower turbocharged 3.3-liter flat-6, dumps its 4-speed manual transmission in favor of its first-ever 5-speed. The V8 flagship 928 S4 gets new ratios for its 4-speed overdrive automatic transmission and a new 10-speaker sound system. The 928 rear-drive 2+2 is otherwise unchanged.

For

- Performance • Handling • 4WD traction (911 Carrera 4)
- Anti-lock brakes (928 S4 and 911 Carrera 4)

Against

- Price • Ride (928 S4)

Summary

In the U.S., where Porsche sells about half its annual worldwide production, sales were down 31 percent in the first six months of 1988 compared to the same period a year earlier. This after a 22-percent drop from 1986 to 1987. Four-cylinder models fell most steeply—one factor behind Porsche's new emphasis on the upper ranges of its line, including the new 911 Carrera 4 and the Speedster. The high prices and ultra-high performance of these cars means few people can afford them, and even fewer can safely manage them. Porsche says the slowest has a top speed of 149 mph; the 928 S4 goes 168. Interested? If so, we hope you have the driving skill to match. If you don't, then shop elsewhere. These Porsches are not boulevard sports cars. They're race cars modified for street use.

Specifications

	911 2-door notchback	911 Turbo 2-door notchback	928 S4 3-door hatchback
Wheelbase, in.	89.5	89.5	98.4
Overall length, in.	168.9	168.9	178.1
Overall width, in.	65.0	69.9	72.3
Overall height, in.	52.0	51.6	50.5
Front track, in.	54.0	56.4	61.1
Rear track, in.	55.3	58.7	60.9
Turn diameter, ft.	35.9	35.9	37.7
Curb weight, lbs.	2756	2976	3505
Cargo vol., cu. ft.	NA	NA	NA
Fuel capacity, gal.	22.5	22.5	22.7
Seating capacity	4	4	4
Front headroom, in.	NA	NA	NA
Front shoulder room, in.	NA	NA	NA
Front legroom, max., in.	NA	NA	NA
Rear headroom, in.	NA	NA	NA
Rear shoulder room, in.	NA	NA	NA
Rear legroom, min., in.	NA	NA	NA

Powertrain layout 911: Longitudinal rear engine/rear-wheel drive or permanent 4-wheel drive. *928 S4:* Longitudinal front engine/rear-wheel drive.

Engines

	ohc flat-6	Turbo ohc flat-6	dohc V-8
Size, liters/cu. in.	3.2/193	3.3/201	5.0/302
Fuel delivery	PFI	PFI	PFI
Horsepower @ rpm	214 @ 5900	282 @ 5500	316 @ 6000
Torque (lbs./ft.) @ rpm	195 @ 4800	288 @ 4000	317 @ 3000
Availability	S	S[1]	S[2]

EPA city/highway mpg

5-speed OD manual	18/24	14/21	14/22
4-speed OD automatic			15/19

1. 911 Turbo   2. 928 S4

Prices

Porsche 911

	Retail Price	Dealer Invoice	Low Price
Carrera coupe	$51205	—	
Carrera 4 coupe	69500	—	—
Turbo coupe	70975	—	—

Dealer invoice, low price and Targa/Cabriolet/Speedster prices not available at time of publication. Retail price includes destination charge.

Standard Equipment:

Carrera: 3.2-liter PFI 6-cylinder engine, 5-speed manual transmission, 4-wheel disc brakes, automatic climate control, partial leather reclining front bucket seats with power height adjustment, leather-wrapped steering wheel, cruise control, power windows and locks, anti-theft alarm, heated power mirrors and windshield washer nozzles, fog lights, visor mirrors, power sunroof (coupe), AM/FM ST ET cassette, rear defogger (except Cabriolet), tinted glass, 205/55ZR16 front and 225/50ZR16 rear tires on forged alloy wheels; Carrera 4 has permanent 4-wheel drive. **Turbo** adds: 3.3-liter turbocharged, intercooled engine, limited-slip differential, power front seats, headlamp washers, power top (Cabriolet), rear wiper (except Cabriolet), 205/55ZR16 front and 245/45ZR16 rear tires. **Speedster** listings not available at time of publication.

OPTIONS prices not available at time of publication.

Porsche 928 S4

3-door hatchback	$74545	—	—

Dealer invoice and low price not available at time of publication. Retail price includes destination charge.

Standard Equipment:

5.0-liter DOHC 32-valve PFI V-8, 5-speed manual transmission, power steering, power 4-wheel disc brakes, anti-lock braking system, automatic climate control, leather power front seats, driver's seat and mirror position memory, driver information and diagnostic system, power windows and locks, anti-theft alarm, AM/FM ST ET cassette, fog lights, leather-wrapped steering wheel, tachometer, coolant temperature and oil pressure gauges, voltmeter, trip odometer, tilt steering column and instrument cluster, power liftgate release, rear wiper, cruise control, folding front armrests, cassette holder, heated power mirrors and washer nozzles, power sunroof, visor mirrors, 225/50ZR16 front and 245/45ZR16 rear tires on forged alloy wheels.

OPTIONS prices not available at time of publication.

> **KEY: ohv** = overhead valve; **ohc** = overhead cam; **dohc** = double overhead cam; **I** = inline cylinders; **V** = cylinders in V configuration; **flat** = horizontally opposed cylinders; **bbl.** = barrel (carburetor); **PFI** = port (multi-point) fuel injection; **TBI** = throttle-body (single-point) fuel injection; **rpm** = revolutions per minute; **OD** = overdrive transmission; **S** = standard; **O** = optional; **NA** = not available.

Prices are accurate at time of printing; subject to manufacturer's change

Porsche 944

Porsche 944 Turbo

What's New

A convertible will join the roster in the spring and the price-leader 924S has been dropped. Some engine displacements are increased and all 944s now have standard anti-lock brakes. This line of front-engine, rear-drive sports cars starts with the 944, which gets a 162-horsepower 2.7-liter 4-cylinder in place of the previous 158-horsepower 2.5-liter. It's the only 944 available with automatic transmission, a 3-speed automatic optional in place of the standard 5-speed manual. The new 944 S2 gets the bespoiled bodywork of the 944 Turbo, plus a 16-valve 3.0-liter four rated at 208 horsepower. This engine replaces the 188-horsepower 2.5 used in last year's 944 S. The 944 Cabriolet is based on the 944 S2. A manual folding top is standard, a power top optional. The convertible conversion is done at a plant set up in Germany by ASC, the Michigan-based specialty-car builder. The 944 Turbo replaces the 944 Turbo S. It retains the turbocharged and intercooled 8-valve 2.5-liter four, but now has 247 horsepower, a gain of 30 from 1988. It also gets the race-bred sport suspension, the 928 brakes, the alloy wheels and the driver's- and passenger-side air bags that were standard on last year's Turbo S. All Porsches get a new anti-theft system for '89.

For

• Performance • Handling • Braking • Air bags

Against

• Price • Noise • Interior room

Summary

Porsche's 4-cylinder models took the brunt of the significant sales decline affecting all of these high-priced, high-performance German sports cars in 1988. Porsche's response was to drop the "budget" 924S and to take the rest of the line even further upscale, where customer price-sensitivity is lower and corporate profits higher. Price increases, by the way, average around 20 percent for '89, though Porsche argues that higher levels of power and equipment account for much of the boost. It says 1989 cars actually cost only about two percent more than 1988 models similarly equipped, and that the adjusted cost of a turbo has ac-

tually dropped about five percent. Be that as it may, few cars offer the 944 range's blend of all-around performance, resale value, status, style and quality of materials and assembly. Japanese sports cars, such as the Mazda RX-7 Turbo and Toyota Supra Turbo, encroach on Porsche territory, and for less money. But to many serious drivers, there's no substitute for the serious sports cars from Porsche. And that kind of attitude costs serious money.

Specifications

	2-door convertible	3-door hatchback
Wheelbase, in.	94.5	94.5
Overall length, in.	168.9	168.9
Overall width, in.	68.3	68.3
Overall height, in.	50.2	50.2
Front track, in.	58.2	58.2
Rear track, in.	57.1	57.1
Turn diameter, ft.	35.2	35.2
Curb weight, lbs.	2932	2932
Cargo vol., cu. ft.	NA	18.3
Fuel capacity, gal.	21.1	21.1
Seating capacity	4	4
Front headroom, in.	NA	35.5
Front shoulder room, in.	NA	NA
Front legroom, max., in.	48.5	48.5
Rear headroom, in.	NA	32.0
Rear shoulder room, in.	NA	NA
Rear legroom, min., in.	NA	12.0

Powertrain layout: longitudinal front engine/rear-wheel drive.

Engines	ohc I-4	dohc I-4	Turbo ohc I-4
Size, liters/cu. in.	2.7/163	3.0/182	2.5/151
Fuel delivery	PFI	PFI	PFI
Horsepower @ rpm	162 @ 5800	208 @ 5800	247 @ 6000
Torque (lbs./ft.) @ rpm	166 @ 4200	207 @ 4100	258 @ 4000
Availability	S[1]	S[2]	S[3]

EPA city/highway mpg

5-speed OD manual	19/26	NA	19/27
3-speed automatic	18/22		

1. 944   2. 944 S2 and Cabriolet   3. 944 Turbo

Prices

Porsche 944	Retail Price	Dealer Invoice	Low Price
944 3-door hatchback	$35969	—	—
944 S2 3-door hatchback	45285	—	—
944 S2 Cabriolet	52650	—	—
944 Turbo 3-door hatchback	47600	—	—

Dealer invoice and low price not available at time of publication. Retail price includes destination charge.

Standard Equipment:

944: 2.7-liter PFI 4-cylinder engine, 5-speed manual transmission, power steering, power 4-wheel disc brakes, anti-lock braking system, air conditioning, power windows and locks, leatherette interior with cloth seat inlays, reclining front bucket seats with power height adjustment, anti-theft alarm, leather-wrapped steering wheel and shift boot, tachometer, coolant temperature and oil pressure gauges, voltmeter, trip odometer, digital clock, heated power mirrors and washer nozzles, power liftgate release, front center armrest with cassette storage, AM/FM ST ET cassette, tinted glass, visor mirrors, cruise control, power tilt/removable sunroof, fog lights, rear defogger, bodyside molding, 215/60ZR16 SBR tires on alloy wheels. **S2 adds:** 3.0-liter DOHC 16-valve engine, driver and passenger air bags, automatic climate control, 205/55ZR16 front and 225/50ZR16 rear tires on pressure-cast alloy

---

KEY: ohv = overhead valve; **ohc** = overhead cam; **dohc** = double overhead cam; **I** = inline cylinders; **V** = cylinders in V configuration; **flat** = horizontally opposed cylinders; **bbl.** = barrel (carburetor); **PFI** = port (multi-point) fuel injection; **TBI** = throttle-body (single-point) fuel injection; **rpm** = revolutions per minute; **OD** = overdrive transmission; **S** = standard; **O** = optional; **NA** = not available.

Prices are accurate at time of printing; subject to manufacturer's change

wheels. **Turbo** adds: 2.5-liter SOHC turbocharged, intercooled engine, limited-slip differential, adjustable Koni shock absorbers, upgraded stereo with 10 speakers, 225/50ZR16 front and 245/45ZR16 rear tires on forged alloy wheels.

OPTIONS prices not available at time of publication.

Range Rover

Range Rover

What's New

A larger, more powerful V-8 engine heads a list of 30 changes for Range Rover, the British-built luxury 4-wheel-drive vehicle. Rover's aluminum V-8 has grown from 3.5 liters to 3.9 through a larger cylinder bore. Horsepower has increased from 150 to 178 and torque from 195 pounds/feet to 220. Rover claims better low-end response and a 0-60 mph time of 10.9 seconds (last year's was clocked at 13.2 seconds). With the larger engine, the EPA's city fuel economy estimate has slipped by one mpg to 12; the EPA highway estimate remains 15 mpg. Rover's V-8 is adapted from an old General Motors engine that first appeared in the 1961 Buick Special. A new Borg Warner transfer case for the permanently engaged 4WD system eliminates the manual differential lock; a viscous coupling now automatically locks the center differential to compensate for wheel slip by dividing engine torque equally between the front and rear axles. An overdrive 4-speed automatic transmission remains standard. Among other changes announced by Rover are an electrically heated windshield and washer jets, a 20 percent increase in air conditioning capacity, improved fresh-air flow from new vents, a push button release for the fold-down rear seats, electronic locking of the rear hatch from the driver's seat and a new 6-speaker stereo system. Range Rover continues to be offered only as a 5-door wagon with leather upholstery and a power sunroof as the only options; 1989 prices hadn't been announced when this issue went to press.

For

- 4WD traction
- Ride
- Passenger room
- Cargo space
- Trailer towing ability

Against

- Fuel economy
- Noise

Summary

We haven't driven the 1989 model, so we can't say how much performance has improved with the new engine. The earlier version was burdened by too much weight, making

Range Rover sluggish. While performance is sure to improve this year, mileage should remain dismal; we averaged 14.8 mpg from an equal mix of city and highway driving. Rover has some strong credentials, including the ability to tow up to 7700 pounds for trailers with brakes (a trailer hitch is standard). The permanently engaged 4WD system eliminates the need to shift in or out of 4WD as road conditions change, an added convenience. Range Rover's high, square-cut build pays off in bountiful interior room. There's plenty of room for five adults, and generous cargo space to stow their gear. Range Rover irons out most surfaces with the supple absorbency of a passenger car. Noise levels, however, are more like a truck's. Cruising at 55 mph, the brick-like body produces too much wind noise. Off-road, the Rover is probably the equal of any competitor; on-road, it leads the field for ride comfort, snob appeal and price.

Specifications

	5-door wagon
Wheelbase, in.	100.0
Overall length, in.	175.0
Overall width, in.	71.4
Overall height, in.	70.8
Front track, in.	58.5
Rear track, in.	58.5
Turn diameter, ft.	39.4
Curb weight, lbs.	4372
Cargo vol., cu. ft.	82.8
Fuel capacity, gal.	20.0
Seating capacity	5
Front headroom, in.	38.4
Front shoulder room, in.	NA
Front legroom, max., in.	41.0
Rear headroom, in.	37.3
Rear shoulder room, in.	NA
Rear legroom, min., in.	32.7

Powertrain layout: longitudinal front engine/permanent 4-wheel drive.

Engines	ohv V-8
Size, liters/cu. in.	3.9/241
Fuel delivery	PFI
Horsepower @ rpm	178 @ 4750
Torque (lbs./ft.) @ rpm	220 @ 3250
Availability	S
EPA city/highway mpg	
4-speed OD automatic	12/15

Prices not available at time of publication.

Saab 900

What's New

The 900 Turbo 4-door sedan returns after a 3-year absence and the 8-valve, 110-horsepower naturally aspirated 2.0-liter engine has been dropped. The Turbo 4-door uses the same 160-horsepower, turbocharged and intercooled 2.0-liter four found in the 900 Turbo 3-door hatchback and the 2-door convertible. The 900 Turbo models otherwise are unchanged, but this year the optional SPG package for the 3-door, which adds a sport suspension, wider tires, aero body add-ons and five horsepower to the turbo engine, gets a leather-wrapped steering wheel and shift knob, and is now available in black as well as gray. On naturally aspi-

Saab

1988 Saab 900 3-door

rated 900s, the 16-valve, previously standard only on the 900S, is now standard on base models as well. It gets a horsepower boost from 125 to 128 and uses hydraulic instead of rubber engine mounts for 1989. The 900S still has more standard equipment than the 900, including cruise control, power windows and mirrors, a manual sunroof, and alloy wheels. Other changes: the base 900 gains stabilizer bars front and rear, and high-pressure gas shock absorbers in place of low-pressure shocks; the convertible gets the upgraded leather upholstery used in the SPG.

For

● Performance (Turbo)   ● Handling/roadholding
● Passenger room   ● Cargo space

Against

● Fuel economy (turbo)   ● Manual shift linkage

Summary

Saab's decision to scrap the 110-horsepower, 8-valve engine is a wise one, given the tepid performance it generated in the base 900, a $16,000 car. Though we haven't driven the 1989 version of the 16-valve engine, we doubt the additional three horsepower will mean much improvement. That engine feels adequate for these cars, but it hardly pins your ears back with neck-snapping acceleration. The turbo-charged engine furnishes exhilarating acceleration and excellent passing response. We prefer it with automatic transmission because the 5-speed manual has notchy shift linkage. With all models the handling is crisp, the steering sharp and precise, and the ride firm but absorbent. All 900s have capable 4-wheel disc brakes standard, but an anti-lock system is unavailable, an unfortunate omission. The 900 traces its lineage to the Saab 99, introduced in 1968, yet this car hides its age well. The interior is surprisingly roomy for the 99.1-inch wheelbase and 66.5 overall width. Saab uses clearly marked, well-lit analog gauges and mounts most controls high on the dashboard where they're easy to reach. Exceptions are that the power window switches are mounted on the floor between the seats, and on manual-transmission models, so is the ignition lock. The hatchback and sedan have generous cargo room, augmented by a folding rear seat. Though pricey, these cars have a lot of character and are enjoyable to drive.

KEY: ohv = overhead valve; **ohc** = overhead cam; **dohc** = double overhead cam; **I** = inline cylinders; **V** = cylinders in V configuration; **flat** = horizontally opposed cylinders; **bbl.** = barrel (carburetor); **PFI** = port (multi-point) fuel injection; **TBI** = throttle-body (single-point) fuel injection; **rpm** = revolutions per minute; **OD** = overdrive transmission; **S** = standard; **O** = optional; **NA** = not available.

1988 Specifications

	3-door hatchback	4-door notchback	2-door convertible
Wheelbase, in.	99.1	99.1	99.1
Overall length, in.	184.5	184.3	184.3
Overall width, in.	66.5	66.5	66.5
Overall height, in.	56.1	56.1	56.1
Front track, in.	56.4	56.4	56.3
Rear track, in.	56.8	56.8	56.7
Turn diameter, ft.	33.8	33.8	33.8
Curb weight, lbs.	2695	2735	2985
Cargo vol., cu. ft.	56.5	53.0	9.9
Fuel capacity, gal.	16.6	16.6	16.6
Seating capacity	5	5	4
Front headroom, in.	36.8	36.8	36.8
Front shoulder room, in.	52.2	53.0	52.2
Front legroom, max., in.	41.7	41.7	41.7
Rear headroom, in.	37.4	37.4	NA
Rear shoulder room, in.	53.5	54.5	NA
Rear legroom, min., in.	36.2	36.2	NA

Powertrain layout: longitudinal front engine/front-wheel drive.

Engines

	dohc I-4	Turbo dohc I-4
Size, liters/cu. in.	2.0/121	2.0/121
Fuel delivery	PFI	PFI
Horsepower @ rpm	125 @ 5500	160 @ 5500
Torque (lbs./ft.) @ rpm	125 @ 3000	188 @ 3000
Availability	S	S
EPA city/highway mpg		
5-speed OD manual	22/28	21/28
3-speed automatic	19/22	19/23

Prices

Saab 900	Retail Price	Dealer Invoice	Low Price
3-door hatchback	$16995	—	—
4-door notchback	17515	—	—
S 3-door hatchback	19695	—	—
S 4-door notchback	20245	—	—
Turbo 3-door hatchback	23795	—	—
Turbo 4-door notchback	24345	—	—
Turbo convertible	32095	—	—
Turbo SPG 3-door hatchback	26895	—	—
Destination charge	359	—	—

Dealer invoice and low price not available at time of publication.

Standard Equipment:

2.0-liter DOHC 16-valve PFI 4-cylinder engine, 5-speed manual or 3-speed automatic transmission, power steering, power 4-wheel disc brakes, air conditioning, tachometer, coolant temperature gauge, trip odometer, analog clock, rear defogger, intermittent wipers, power locks, tinted glass, driver's seat tilt/height adjustment, cloth heated reclining front bucket seats, folding rear seat, AM/FM ST ET, 185/65R15 SBR tires. **S** adds: cruise control, folding rear armrest, power windows and mirrors, AM/FM ST ET cassette, manual sunroof, alloy wheels. **Turbo** adds: turbocharged engine, sport seats, upgraded stereo with EQ; convertible has power top and leather interior. **Turbo SPG** has: higher-output engine, sport suspension, leather-wrapped steering wheel and shift knob, aero body addenda, wider tires.

Optional Equipment:

3-speed automatic transmission (NA SPG)	525	—	—
Metallic or special black paint	485	—	—
Leather Pkg., Turbos exc. conv.	1295	—	—

Prices are accurate at time of printing; subject to manufacturer's change

CONSUMER GUIDE®

Saab 9000

1988 Saab 9000S

What's New

Saab adds a trunk and a full complement of standard equipment to create a 4-door notchback version of its 9000 as the new flagship. Saab calls the new 4-door notchback the 9000CD. It joins the 9000S and 9000 Turbo 5-door hatchbacks that came to the U.S. for the 1986 model year. The CD is a regular sedan with a separate trunk, as opposed to the semi-station-wagon hatchback. Its body is about 6.5 inches longer than the hatchbacks, though curb weight is within a few pounds. Saab says the CD's trunk has 19.8 cubic feet of storage space, versus the 15.9 cubic feet of space available in the hatchback with the rear seatback upright. The CD's rear seatback does not fold for access to the trunk, though there is a small pass-through for skis and other long objects. Other features unique to the CD include additional backup lamps that illuminate a larger area than the standard backup lamps; a revised front end with a smoother bumper and grille than other 9000s; and a front spoiler and fog lamps. The bodyside moldings are wider and a rear accent panel bridges the taillamps, which are smoked in the CD. All 9000 CDs use the same 160-horsepower turbocharged and intercooled 2.0-liter 4-cylinder engine standard in the 9000 Turbo. CD suspension settings are softer than other 9000s, and it has more sound insulation. Among other changes, the 9000 Turbo 5-door gets standard power front seats—Saab's first power seats ever. The base 9000S gets a new option package that includes leather upholstery, which was previously a separate option.

For

- Performance (Turbo)
- Anti-lock brakes
- Handling/roadholding
- Passenger room
- Cargo space

Against

- Performance (9000S)
- Fuel economy (Turbo)
- Climate control

Summary

In turbocharged form, the 9000 delivers neck-snapping acceleration, plus it has fine handling, good traction in rain and snow, a capable anti-lock brake system, a roomy, well-furnished interior and ample cargo room. You get all of the above on the 9000S except the neck-snapping acceleration. The 9000S' engine has 128 horsepower for more than 3000 pounds of curb weight, enough to provide steady progress, but not brisk performance. What makes that hard to swallow is the 9000S' base price of close to $25,000; you can buy much better performance for a lot less in any number of 6-cylinder cars, though few match the quality and overall competence of the Saab. The Turbo hatchback runs about $5000 more, but it also runs a whole lot quicker to justify the higher price. Other complaints we have are that the turbo engine suffers too much lag time before boost is supplied, and the annoying automatic climate control system has a mind of its own and can't be shut off. Overall, we have high praise for these roadworthy, well-designed Swedish cars.

1988 Specifications

	5-door hatchback
Wheelbase, in.	105.2
Overall length, in.	181.9
Overall width, in.	69.4
Overall height, in.	55.9
Front track, in.	59.9
Rear track, in.	58.7
Turn diameter, ft.	35.8
Curb weight, lbs.	3022
Cargo vol., cu. ft.	56.5
Fuel capacity, gal.	17.9
Seating capacity	5
Front headroom, in.	38.5
Front shoulder room, in.	NA
Front legroom, max., in.	41.5
Rear headroom, in.	37.4
Rear shoulder room, in.	NA
Rear legroom, min., in.	38.7

Powertrain layout: transverse front engine/front-wheel drive.

Engines	dohc I-4	Turbo dohc I-4
Size, liters/cu. in.	2.0/121	2.0/121
Fuel delivery	PFI	PFI
Horsepower @ rpm	125 @ 5500	160 @ 5500
Torque (lbs./ft.) @ rpm	125 @ 3000	188 @ 3000
Availability	S	S
EPA city/highway mpg		
5-speed OD manual	22/28	22/28
4-speed OD automatic	18/24	19/26

Prices

Saab 9000	Retail Price	Dealer Invoice	Low Price
S 5-door hatchback	$24445	—	—
Turbo 5-door hatchback	30795	—	—
CD 4-door notchback	31995	—	—
Destination charge	359	—	—

Dealer invoice and low price not available at time of publication.

Standard Equipment:

2.0-liter DOHC 16-valve PFI 4-cylinder engine, 5-speed manual or 4-speed automatic transmission, power steering, anti-lock braking system, 4-wheel disc brakes, automatic air conditioning, AM/FM ST ET cassette, power antenna, trip computer, trip odometer, tachometer, coolant temperature gauge, front and rear reading lights, reclining front bucket seats, driver's seat height/tilt, lumbar and lateral support adjustments, velour upholstery, emergency tensioning front seatbelt retractors, rear shoulder belts, telescopic steering column, power tilt/slide steel sunroof, 185/65R15 SBR tires on alloy wheels. **Turbo and CD** add: turbocharged engine, leather upholstery, fog lights, power glass sunroof, upgraded stereo, 205/55VR15 tires.

Prices are accurate at time of printing; subject to manufacturer's change

Optional Equipment:

	Retail Price	Dealer Invoice	Low Price
4-speed automatic transmission	695	—	—
Leather Pkg., S	1595	—	—
Metallic or special black paint	485	—	—

Sterling 827

1988 Sterling 825SL

What's New

A 5-door hatchback body style joins the 4-door sedan and all versions of this front-drive luxury car from England get a larger, more powerful engine. Sterling's chassis and drive-train are shared with the Acura Legend Sedan in a joint venture between British manufacturer Austin Rover and Honda, which builds Acura. The Sterling 825 debuted in the U.S. in early 1987 as a 4-door notchback that used the Legend Sedan's 151-horsepower 2.5-liter V-6. The Legend Sedan got a 161-horsepower 2.7-liter V-6 for '88, and Sterling gets it for '89, becoming the 827. Two notchbacks are available. Standard on the flagship 827SL are anti-lock brakes, 4-speed overdrive automatic transmission and leather upholstery, all of which are optional on the base 827S. The new hatchback is called the 827SLi. Its rear suspension lacks the self-leveling feature of the sedan's, but Sterling says the ride is tuned for a sportier feel than the sedan's.

For

- Performance   • Anti-lock brakes
- Handling/roadholding

Against

- Driveability (automatic transmission)
- Instruments/controls

Summary

This year's larger engine should give Sterling a little more off-the-line pep and make it more compatible with auto-matic transmission. However, based on our experiences with that same engine in the Acura Legend, we still expect some driveability problems. This engine develops its power at higher speeds, which slows acceleration from a standstill, and the automatic tends to shift harshly and seems to be constantly changing gears in hilly terrain. Otherwise, the performance is quite satisfying, the handling is crisp, the ride controlled and, with the anti-lock brakes, stopping power is impressive. While the Sterling sedan is built from the same design as the Legend, it has its own interior de-sign, and we find it less appealing in some respects than Acura's. The gauges are right in front of the driver, yet at first glance the graphics are confusing. The power seat buttons

are hard to use, the temperature and sliding fan-speed con-trols work opposite of normal, and the headlamp switch is on the turn signal lever, where it's easy to hit accidentally. While all that may seem like nit-picking, the Legend doesn't suffer the same faults. The Legends we've tested appeared to have better overall assembly quality than the Sterlings we've tested, though a 1988 825SL was more impressive than previous models. In its favor, Sterling uses rich leather upholstery and real wood interior trim, giving the 827 models an added touch of class. Also in its favor, the 827 is priced lower than the Legend and other competitors. Slow sales make it likely you can buy a Sterling at a discount, making it even more tempting.

Specifications

	4-door notchback
Wheelbase, in.	108.6
Overall length, in.	188.8
Overall width, in.	76.8
Overall height, in.	54.8
Front track, in.	58.8
Rear track, in.	57.1
Turn diameter, ft.	36.5
Curb weight, lbs.	3164
Cargo vol., cu. ft.	12.1
Fuel capacity, gal.	17.0
Seating capacity	5
Front headroom, in.	37.8
Front shoulder room, in.	54.9
Front legroom, max., in.	41.2
Rear headroom, in.	36.3
Rear shoulder room, in.	54.3
Rear legroom, min., in.	36.4

Powertrain layout: transverse front engine/front-wheel drive.

Engines

	ohc V-6
Size, liters/cu. in.	2.7/163
Fuel delivery	PFI
Horsepower @ rpm	161 @ 5900
Torque (lbs./ft.) @ rpm	162 @ 4500
Availability	S

EPA city/highway mpg

5-speed OD manual	19/24
4-speed OD automatic	18/23

Prices

Sterling (1988 prices)	Retail Price	Dealer Invoice	Low Price
825S 4-door sedan	$20804	$17995	$19200
825SL 4-door sedan	25995	22356	23976
Destination charge	435	435	435

Standard Equipment:

825S: 2.5-liter PFI 24-valve V-6 engine, 5-speed manual transmission, power steering, power brakes, cruise control, intermittent wipers, remote fuel filler release, locking remote trunk release, gauges (tachometer, coolant temper-ature, oil pressure and voltage), Philips 6-speaker AM/FM ST ET cassette, courtesy delay interior lights, tilt steering column, reclining front bucket seats with adjustable lumbar support, burled walnut interior trim, coin tray, front and rear map/reading lights, composite halogen headlamps, heated power mirrors, central locking with infrared remote control, theft deterrent system, power moonroof with louvered blind, tinted glass, P195/65VR15 Goodyear tires. **825SL** adds: 4-speed automatic transmission (5-speed man-ual may be substituted at no charge), anti-lock braking system, Connolly leather upholstery, metallic paint, curbside illumination, power driver's seat

Prices are accurate at time of printing; subject to manufacturer's change

with 4-position memory (includes mirrors), power passenger seat, trip computer, graphic display for fluid levels, door ajar and outside air temperature, courtesy delay headlight switch, 8-speaker stereo with upgraded amplifier, alloy wheels.

Optional Equipment:	Retail Price	Dealer Invoice	Low Price
Connolly leather upholstery, 825S	1025	861	943
Anti-lock braking system, 825S	1150	966	1058
4-speed automatic transmission, 825S	625	525	575
Metallic paint, 825S	400	336	368
Power rear seat, 825SL	375	315	345

Subaru Hatchback

Subaru Hatchback

What's New

The oldest model in Subaru's lineup is back for another season with minor changes and new exterior colors. The Hatchback now comes only in GL trim, but with a choice of front-wheel drive or on-demand, part-time 4-wheel drive. Front-drive models come with either a 5-speed manual or 3-speed automatic, while the 4WD comes only with a Dual Range 4-speed manual transmission. The 4WD system is engaged by a button on the shift lever. A 73-horsepower 1.8-liter horizontally opposed engine with a 2-barrel carburetor is standard.

For
●4WD traction ●Fuel economy ●Maneuverability

Against
●Performance ●Passenger room ●Noise

Summary

Subaru's venerable Hatchback still makes sense as a compromise in size and price between the minicompact Justy and the larger Subaru Sedan/Wagon/3-Door Coupe trio. While the Hatchback is a foot longer than Justy on the outside, it's hardly any roomier on the inside, and in fact may have even less usable space for the front seats. The rear seat is too cramped for adults to fit for very long without discomfort. Acceleration is barely adequate with manual transmission in the front-drive model, and sluggish with the optional automatic. The optional 4WD system adds 120 pounds to the curb weight, which isn't much, though with only 73 horsepower, every little bit hurts. The engine has a coarse, guttural growl, yet runs fairly smoothly except when pushed to higher rpm. Subaru's convenient on-demand 4WD system allows shift-on-the-fly and gives the little Hatchback outstanding traction. Fuel economy

should be good and this is one of the models that established Subaru's reputation for building reliable, durable cars. It's short on space, frills and refinement, but is still one of the least expensive 4WD passenger cars available.

Specifications

	3-door hatchback
Wheelbase, in.	93.7
Overall length, in.	157.9
Overall width, in.	63.4
Overall height, in.	53.7
Front track, in.	52.8
Rear track, in.	53.0
Turn diameter, ft.	30.8
Curb weight, lbs.	2120[1]
Cargo vol., cu. ft.	33.9
Fuel capacity, gal.	13.2
Seating capacity	4
Front headroom, in.	38.2
Front shoulder room, in.	51.0
Front legroom, max., in.	39.3
Rear headroom, in.	36.6
Rear shoulder room, in.	52.4
Rear legroom, min., in.	30.2

1. 2240 lbs., 4WD

Powertrain layout: longitudinal front engine/front-wheel drive or on-demand 4WD.

Engines

	ohv flat-4
Size, liters/cu. in.	1.8/109
Fuel delivery	2 bbl.
Horsepower @ rpm	73 @ 4400
Torque (lbs./ft.) @ rpm	94 @ 2400
Availability	S

EPA city/highway mpg
4-speed OD manual	25/29
5-speed OD manual	26/31
3-speed automatic	23/25

Prices

Subaru Hatchback	Retail Price	Dealer Invoice	Low Price
GL 3-door hatchback	$8596	—	—

Dealer invoice and low price not available at time of publication. Destination charge varies by region.

Standard Equipment:

1.8-liter 2bbl. 4-cylinder engine, 5-speed manual transmission, power brakes, Hill Holder, cloth and vinyl reclining front bucket seats, driver's seat lumbar support adjustment, tachometer, coolant temperature and oil pressure gauges, trip odometer, tilt steering column, rear defogger, intermittent wipers, right visor mirror, 50/50 folding rear seat, AM/FM stereo, rear wiper, dual outside mirrors, 175/70SR13 tires.

Optional Equipment:

3-speed automatic transmission	560	—	—
On-Demand 4WD w/4-speed manual	700	—	—

KEY: ohv = overhead valve; **ohc** = overhead cam; **dohc** = double overhead cam; **I** = inline cylinders; **V** = cylinders in V configuration; **flat** = horizontally opposed cylinders; **bbl.** = barrel (carburetor); **PFI** = port (multi-point) fuel injection; **TBI** = throttle-body (single-point) fuel injection; **rpm** = revolutions per minute; **OD** = overdrive transmission; **S** = standard; **O** = optional; **NA** = not available.

Prices are accurate at time of printing; subject to manufacturer's change

Subaru Justy

Subaru Justy

pointed straight ahead. Justy's gutsy 3-cylinder engine eagerly produces decent acceleration and adequate passing power, and returns commendable fuel economy. The 90-inch wheelbase and firm suspension add up to a lively, even jumpy ride. You'll feel nearly every bump and be jarred by big ones. It's far from perfect, and so small that a strong wind gust can move you over a lane on the expressway. Justy is still a good buy among minicompacts, and your only buy if you want a 4WD mini.

What's New

An electronic continuously variable transmission (ECVT) will be optional on the front-drive Justy GL, and Subaru says it is nothing short of revolutionary. Like conventional automatics, it only has to be placed in "Drive" to operate automatically in the forward gears. However, it is essentially a "gearless" automatic transmission that uses pulleys to continuously vary the ratio of engine speed to drive-wheel speed, operating without any shifting or interruption of power flow. Conventional automatics have three or four pre-determined gear ratios that are engaged according to throttle position, engine load, vehicle speed and other factors. Subaru says the ECVT has a much wider spread of gear ratios and also provides livelier acceleration and lower engine speeds at cruising speed than conventional automatics. The ECVT has fewer moving parts than a conventional transmission, and is lighter and more compact. Justy also is restyled for 1989, gaining six inches in overall length to 145 and a more rounded appearance for its 3-door hatchback body. Subaru says the restyling gives its minicompact more head room front and rear, and additional luggage space behind the folding rear seat. Justy arrived in the U.S. during 1987 and gained optional 4-wheel drive last year. All models have new rear suspension geometry that's supposed to improve handling, while GL and RS models also get a new rear stabilizer bar and 13-inch wheels and tires (instead of 12-inchers). The base DL hatchback comes with front-wheel drive, and the mid-level GL model is available with either front-drive or on-demand 4WD that's engaged by a button on the shift lever. The sporty RS comes only with 4WD. A 66-horsepower 1.2-liter 3-cylinder engine and 5-speed manual transmission are standard on all versions.

For

●Fuel economy ●4WD traction ●Maneuverability

Against

●Ride ●Noise ●Weight/size

Summary

We haven't driven the new continuously variable transmission, but chalk up another technical innovation for Subaru. This is the same company that pioneered 4WD in small passenger cars back in 1975. Some other car companies are just getting into 4WD. Justy is the smallest passenger car sold with 4WD, and it couldn't be more convenient to use. The on-demand system is activated by a button on the shift lever and the only requirement is that the wheels be

Specifications

	3-door hatchback
Wheelbase, in.	90.0
Overall length, in.	145.5
Overall width, in.	60.4
Overall height, in.	55.9
Front track, in.	52.4
Rear track, in.	50.8
Turn diameter, ft.	32.2
Curb weight, lbs.	1745[1]
Cargo vol., cu. ft.	21.9
Fuel capacity, gal.	9.2
Seating capacity	4
Front headroom, in.	38.0
Front shoulder room, in.	51.9
Front legroom, max., in.	41.5
Rear headroom, in.	37.0
Rear shoulder room, in.	51.0
Rear legroom, min., in.	30.2

1. 1920 lbs., 4WD

Powertrain layout: transverse front engine/front-wheel drive or on-demand 4WD.

Engines

	ohc I-3
Size, liters/cu. in.	1.2/73
Fuel delivery	2 bbl.
Horsepower @ rpm	66 @ 5200
Torque (lbs./ft.) @ rpm	70 @ 3600
Availability	S

EPA city/highway mpg

5-speed OD manual	34/37
ECVT automatic	34/35

Prices

Subaru Justy (1988 prices)	Retail Price	Dealer Invoice	Low Price
DL 3-door hatchback	$5666	$5221	$5444
GL 3-door hatchback	6666	5987	6327
RS 4WD 3-door hatchback	7666	6868	7267

Destination charge varies by region.

Standard Equipment:

DL: 1.3-liter 2bbl. 3-cylinder engine, 5-speed manual transmission, power brakes, locking fuel door, reclining front bucket seats, one-piece folding rear seat, SBR tires. **GL** adds: rear defogger, tachometer, intermittent wipers, rear wiper, digital clock, AM/FM ST ET, 50/50 folding rear seat, luggage shelf, remote hatch release. **RS** adds: monochrome exterior, AM/FM ST ET, remote mirrors, carpet, digital clock, cloth and vinyl upholstery, graphic monitor, full wheel covers, all-season tires.

Optional Equipment:

On-Demand 4WD, GL	600	400	500
Air conditioning	685	NA	NA
Alloy wheels	550	NA	NA

Prices are accurate at time of printing; subject to manufacturer's change.

Subaru Sedan/Wagon/ 3-Door Coupe

Subaru RX 3-door Coupe

What's New

Subaru's high-volume line gains a Touring Wagon for 1989 with new rear styling that adds three inches of head room, more cargo room and larger rear and side rear windows. The Touring Wagon is available in GL and GL-10 trim levels with front-drive or 4-wheel-drive. The regular 5-door wagon returns as well. The sporty RX 3-Door Coupe this year gains as an option Subaru's Active 4WD system, a permanently engaged, full-time system with a 4-speed overdrive automatic transmission. With Active 4WD, the amount of power going to the front and rear wheels varies depending on road conditions. Active 4WD became optional last year on the 4-door sedan and wagon. Black has been added as a second color choice for the monochromatic RX, which last year came only in white. All three body styles are built from the same basic design, which debuted for 1985. All are available with front-wheel-drive, on-demand 4WD (engaged by a shift-lever button), Active 4WD, or Continuous 4WD, which comes with a 5-speed manual transmission and permanently splits power 50/50 between the front and rear wheels. Two engines are available: Base engine is a 90-horsepower 1.8-liter flat (horizontally opposed) 4-cylinder engine. Optional is a 115-horsepower turbocharged version of that engine. Rear shoulder belts are standard on all three body styles.

For

- 4WD traction   • Cargo space   • Maneuverability

Against

- Fuel economy (turbo)   • Performance (non-turbo)
- Passenger room (3-Door)

Summary

We've always found Subaru's on-demand 4WD system convenient, since it allows changing in or out of 4WD on the fly and requires a minimum of "dos and don'ts" from the driver. However, we've come to appreciate Subaru's permanently engaged Active 4WD system, which is installed on a 1988 turbocharged GL-10 wagon that's in our long-term test fleet. On smooth, dry surfaces, you hardly notice the 4WD. In rain or on loose gravel, you notice the additional traction immediately; there's simply no wheelspin. We're looking forward to winter (well, sort of) to test this 4WD system in the snow. With 4WD, automatic transmission and the other equipment standard on this model, base curb weight is 2770 pounds. That slows acceleration even with the turbocharged engine and takes a toll in fuel economy.

We've reached as high as 26 mpg in straight highway driving, unexceptional for a 4-cylinder car with an overdrive transmission, and dipped as low as 16.5 mpg in urban driving. This wagon has served our staff well over nearly 10,000 miles, though we still wince at the high price of this loaded version: nearly $10,000 at 1988 prices. For that reason we'd look first at the less expensive on-demand 4WD system or a front-drive model. If 4WD is a necessity, then this is the place to start your shopping.

Specifications

	3-door hatchback	4-door notchback	5-door wagon
Wheelbase, in.	97.2	97.2	97.2
Overall length, in.	174.6	174.6	176.8
Overall width, in.	65.4	65.4	65.4
Overall height, in.	51.8	52.5	53.0
Front track, in.	56.1	56.1	55.9
Rear track, in.	56.1	56.1	56.1
Turn diameter, ft.	34.8	34.8	34.8
Curb weight, lbs.	2280	2240	2370
Cargo vol., cu. ft.	39.8	14.9	70.3
Fuel capacity, gal.	15.9	15.9	15.9
Seating capacity	5	5	5
Front headroom, in.	37.6	37.6	37.6
Front shoulder room, in.	53.5	53.5	53.5
Front legroom, max., in.	42.2	41.7	41.7
Rear headroom, in.	35.8	36.5	37.7
Rear shoulder room, in.	52.8	53.5	53.5
Rear legroom, min., in.	32.6	35.2	35.2

Powertrain layout: longitudinal front engine/front-wheel drive, on-demand 4WD or permanent 4WD.

Engines

	ohc flat-4	Turbo ohc flat-4
Size, liters/cu. in.	1.8/109	1.8/109
Fuel delivery	TBI	PFI
Horsepower @ rpm	90 @ 5200	115 @ 5200
Torque (lbs./ft.) @ rpm	101 @ 2800	134 @ 2800
Availability	S	O

EPA city/highway mpg

5-speed OD manual	25/31	22/25
3-speed automatic	25/27	22/24
4-speed OD automatic		20/26

Prices

Subaru Sedan, Wagon & 3-door Coupe	Retail Price	Dealer Invoice	Low Price
DL 4-door notchback	$9731	—	—
GL 4-door notchback	11521	—	—
GL-10 Turbo 4-door notchback	16401	—	—
DL 3-door hatchback	10031	—	—
GL 3-door hatchback	11821	—	—
RX 3-door hatchback	16361	—	—
DL 5-door wagon	10181	—	—
GL 5-door wagon	11971	—	—
GL-10 Turbo 5-door wagon	16851	—	—

Dealer invoice and low price not available at time of publication. Destination charge varies by region.

> **KEY: ohv** = overhead valve; **ohc** = overhead cam; **dohc** = double overhead cam; **I** = inline cylinders; **V** = cylinders in V configuration; **flat** = horizontally opposed cylinders; **bbl.** = barrel (carburetor); **PFI** = port (multi-point) fuel injection; **TBI** = throttle-body (single-point) fuel injection; **rpm** = revolutions per minute; **OD** = overdrive transmission; **S** = standard; **O** = optional; **NA** = not available.

Prices are accurate at time of printing; subject to manufacturer's change.

Subaru

Standard Equipment:

DL: 1.8-liter TBI 4-cylinder engine, 5-speed manual transmission, power steering and brakes, reclining front bucket seats, cloth upholstery, tinted glass, digital clock, rear defogger, remote fuel door and hatch/trunk releases, bodyside moldings, trip odometer, 50/50 folding rear seat (wagon), rear wiper (wagon), 155SR13 all-season SBR tires. **GL** adds: power mirrors, tachometer, power windows and locks, AM/FM stereo, memory tilt steering column, 50/50 folding rear seat (except 4-door), driver's seat lumbar support adjustment, 175/70SR13 tires. **GL-10 Turbo** adds: PFI turbocharged engine, 4-wheel disc brakes, variable intermittent wipers, power sunroof, air conditioning, cruise control, digital instruments with trip computer, upgraded stereo. **RX** adds: Continuous 4WD, 5-speed dual-range manual transmission, upgraded suspension, white monochrome exterior treatment, analog instruments, performance tires.

Optional Equipment:	Retail Price	Dealer Invoice	Low Price
3-speed automatic transmission, DL & GL .	560	—	—
4-speed automatic transmission, RX	760	—	—
On-Demand 4WD, GL	700	—	—
Continuous 4WD/5-speed, GL-10 Turbo . .	1600	—	—
Active 4WD/4-speed automatic, GL-10 Turbo	2360	—	—
Continuous 4WD Turbo/5-speed pkg.,			
GL wagon	2475	—	—
Touring Wagon Pkg., GL	200	—	—
GL-10 Turbo	NC	—	—

Subaru XT

Subaru XT6

What's New

New high-performance, all-season tires are standard this year on XT6 4-wheel-drive models, and all models with rear seats also gain rear shoulder belts. The new tires for the XT6 are Goodyear Eagle GT+4s; size is the same as last year, 205/60HR14. Subaru's sporty coupe gained a 6-cylinder option last year, thus the XT6 badge. The six is a 2.7-liter horizontally opposed, or flat, engine making 145 horsepower. Base XT models use a 97-horsepower 1.8-liter flat 4-cylinder. The 4-cylinder engine is available in 2-seat DL trim and plusher 4-seat GL trim. The DL comes with front-drive only, the GL with front- or on-demand 4WD. The XT6 comes three ways: front-drive; Continuous 4WD with a 5-speed manual transmission; and Active 4WD with a 4-speed overdrive automatic transmission. Continuous and Active 4WD are permanently engaged systems, but the first splits power 50/50 between the front and rear wheels, while the second varies the power split according to road conditions.

> **KEY: ohv** = overhead valve; **ohc** = overhead cam; **dohc** = double overhead cam; **I** = inline cylinders; **V** = cylinders in V configuration; **flat** = horizontally opposed cylinders; **bbl.** = barrel (carburetor); **PFI** = port (multi-point) fuel injection; **TBI** = throttle-body (single-point) fuel injection; **rpm** = revolutions per minute; **OD** = overdrive transmission; **S** = standard; **O** = optional; **NA** = not available.

For

- 4WD traction • Maneuverability • Fuel economy
- Performance (6-cylinder) • Handling/roadholding

Against

- Performance (4-cylinder) • Passenger room
- Dashboard controls

Summary

The 6-cylinder engine that arrived last year gives the XT performance that comes much closer to matching its racy looks than previous engines. Smooth and responsive, the 2.7-liter six has enough torque at low speeds for quick getaways from stoplights and plenty of power at high speeds for brisk passing. We tested a 1988 front-drive XT6 with automatic transmission and found it quite enjoyable, though we were disappointed to average less than 20 mpg from mostly city driving. We strongly recommend the six over the 4-cylinder since it makes the XT coupe more refined as well as more fun to drive. The front-drive model handles capably, though with noticeable understeer (resistance to turning), while the XT's electronic power steering system is accurate and returns good road feel. Adding 4WD increases the price by several hundred dollars, but also increases the XT's usefulness in foul weather. On snow and ice, when many other sporty coupes are relegated to the slow lane, a 4WD XT continues to motor along with a glued-to-the-road feel. Inside, there's not much head or leg room even in front, and the tiny back seat is suitable for children only. An attractive analog gauge cluster is surrounded by a complicated control layout and a multitude of small buttons that are hard to find and operate while driving. While the interior still needs work, the XT offers respectable performance with the 6-cylinder and the advantages of 4WD traction.

Specifications

	2-door notchback
Wheelbase, in.	97.0
Overall length, in.	177.6
Overall width, in.	66.5
Overall height, in.	49.4
Front track, in.	56.5
Rear track, in.	56.1
Turn diameter, ft.	34.1
Curb weight, lbs.	2455
Cargo vol., cu. ft.	11.6
Fuel capacity, gal.	15.9
Seating capacity	4
Front headroom, in.	37.4
Front shoulder room, in.	52.8
Front legroom, max., in.	43.3
Rear headroom, in.	34.4
Rear shoulder room, in.	52.8
Rear legroom, min., in.	26.2

Powertrain layout: longitudinal front engine/front-wheel drive, on-demand 4WD or permanent 4WD.

Engines	ohc flat-4	ohc flat-6
Size, liters/cu. in.	1.8/109	2.7/163
Fuel delivery	PFI	PFI
Horsepower @ rpm	97 @ 5200	145 @ 5200
Torque (lbs./ft.) @ rpm	103 @ 3200	156 @ 4000
Availability	S	O

Prices are accurate at time of printing; subject to manufacturer's change.

EPA city/highway mpg	ohc flat-4	ohc flat-6
5-speed OD manual	24/28	18/25
4-speed OD automatic	23/30	20/28

Prices

Subaru XT	Retail Price	Dealer Invoice	Low Price
GL 2-door notchback	$13071	—	—
XT6 2-door notchback	17111	—	—

Dealer invoice and low price not available at time of publication. Destination charge varies by region.

Standard Equipment:

1.8-liter PFI 4-cylinder engine, 5-speed manual transmission, power steering and brakes, cloth reclining front bucket seats, passive restraint system (automatic front shoulder and manual lap belts), console, Hill Holder, tinted glass, remote fuel door and trunk releases, dual spot lamps, memory tilt steering column, coolant temperature gauge, trip odometer, tachometer, digital clock, rear defogger, telltale graphic monitor, variable intermittent wipers, AM/FM ST ET, power windows and locks, power mirrors, driver's seat lumbar and height adjustments, one-piece folding rear seat, oil pressure gauge, voltmeter, 185/70HR13 all-season tires. **XT6** adds: 2.7-liter PFI 6-cylinder engine, 4-wheel disc brakes, air conditioning, fog lamps, cruise control, upgraded stereo with cassette and EQ, headlamp washers, trip computer, 195/60HR14 all-season tires on alloy wheels.

Optional Equipment:

4-speed automatic transmission, GL	760	—	—
On-Demand 4WD, GL	700	—	—
Active 4WD/4-speed automatic	1660	—	—
Continuous 4WD/5-speed, XT6	840	—	—

Suzuki Samurai

1988 1/2 Suzuki Samurai

What's New

Suzuki insists the embattled Samurai will continue as part of the firm's long-term U.S. strategy, with minor year-to-year adjustments but less marketing emphasis now that the larger Sidekick is here. Samurai sales plunged to under 3000 a month following allegations that the vehicle is prone to rolling over in quick directional changes. Suzuki launched a direct, vigorous defense of the vehicle's dynamic capabilities and offered dealers a $2000 cash rebate for each vehicle sold. The result was record sales of more than 12,000 Samurais during August. For 1989, Samurai is carried over unchanged. Last spring the suspension was softened for improved ride comfort and the dashboard was

redesigned. Samurai, built from a design that is 19 years old, comes in a choice of convertible or fixed hardtop formats, both with a 63-horsepower 1.3-liter 4-cylinder engine and 5-speed manual transmission. Automatic transmission isn't offered. The standard 4WD system is an on-demand part-time system with manual locking front hubs. Automatic locking hubs, along with an extensive list of accessories and options, are available through dealers.

For

• 4WD traction • Price

Against

• Performance • Steering/handling • Ride • Noise

Summary

Is it or isn't it prone to rollovers? The federal government says not any more than other vehicles of this type, though the government also warns that 4WD vehicles are more prone to rollovers than passenger cars. Comparing Samurai's dimensions to that of passenger cars is revealing. On most cars, the track (width between the wheels on the same axle) is greater than the overall height, or at least roughly the same, for a low center of gravity. On the Samurai, the track is about 14 inches less than the overall height. This tall body on a narrow chassis results in ample body lean in turns, giving you plenty of warning that you should slow down. In addition, Samurai has enough understeer (resistance to turning) to reinforce the warning to take it easy around corners. On top of this, the steering is slow and vague; you turn the wheel and there's a pause before the front wheels respond. Suzuki did improve ride quality on the 1988½ models, but Samurai still bounces and bangs over rough pavement. The loud, coarse-sounding engine struggles to keep pace with traffic, even with the throttle mashed to the floor. This is supposed to be fun? Before the rollover controversy we warned that you get what you pay for in a Samurai, whose low price remains its main virtue. However, we're still not convinced Samurai is a good idea even as a weekend toy, let alone an everyday vehicle. Tinny body panels, flimsy interior materials, a general crudeness and the prospect of low resale value are more reasons to avoid it.

Specifications

	2-door wagon
Wheelbase, in.	79.9
Overall length, in.	135.4
Overall width, in.	60.2
Overall height, in.	65.6
Front track, in.	51.2
Rear track, in.	51.6
Turn diameter, ft.	33.4
Curb weight, lbs.	2094[1]
Cargo vol., cu. ft.	31.9
Fuel capacity, gal.	10.6
Seating capacity	4
Front headroom, in.	38.6
Front shoulder room, in.	48.5
Front legroom, max., in.	40.0
Rear headroom, in.	35.5
Rear shoulder room, in.	51.0
Rear legroom, min., in.	27.6

1. Convertible; 2127 lbs., hardtop

Powertrain layout: longitudinal front engine/rear-wheel drive or on-demand 4WD.

Prices are accurate at time of printing; subject to manufacturer's change.

Suzuki

Engines

	ohc I-4
Size, liters/cu. in.	1.3/81
Fuel delivery	2 bbl.
Horsepower @ rpm	63 @ 6000
Torque (lbs./ft.) @ rpm	74 @ 3500
Availability	S
EPA city/highway mpg	
5-speed OD manual	28/29

Prices

Suzuki Samurai	Retail Price	Dealer Invoice	Low Price
Standard soft top	$8495	—	—
Standard hardtop	8595	—	—
Deluxe soft top	9195	—	—
Deluxe hard top	9495	—	—

Dealer invoice, low price and destination charge not available at time of publication.

Standard Equipment:

Standard: 1.3-liter 2bbl. 4-cylinder engine, 5-speed manual transmission, power brakes, locking fuel cap, dual outside mirrors, vinyl reclining front bucket seats, trip odometer, intermittent wipers, 205/70R15 tires. **Deluxe** adds: front carpet, digital clock, rear defogger (hardtop), day/night mirror, right visor mirror, AM/FM cassette, cloth seat trim (hardtop), folding rear seat (hardtop), tachometer.

Optional Equipment:

AM/FM cassette	395	—	—
Air conditioning	595	—	—
Rear seat kit, Standard soft top	380	—	—
Standard hard top	395	—	—
Deluxe soft top	370	—	—

Suzuki Sidekick

Suzuki Sidekick hard top

What's New

Sidekick is Suzuki's new 4-wheel-drive vehicle that also's available through Chevrolet dealers as the Geo Tracker. The only differences involve badging and model mix. Suzuki offers five versions: convertible and fixed-roof hardtop in JX and JLX trim, plus a price-leader JA convertible with no direct Tracker counterpart. The JA employs a 64-horse-power 1.3-liter four with 2-barrel carburetor (not the same engine used in Samurai) that teams only with a 5-speed manual transmission; all others have the 80-horsepower 1.6-liter engine with single-point fuel injection as used in the Tracker and are available with optional 3-speed automatic. Sidekick and Tracker are otherwise peas in a pod, but Sidekick will be more readily available. Suzuki plans to bring in 30,000-35,000 Sidekicks through March 1989 versus Chevrolet's allotment of only some 10,000 Trackers. Base price of the Sidekick JA is $8998. In addition to having more power, Sidekick also is roomier inside than Samurai, thanks to a longer wheelbase (86.6 inches versus Samurai's 79.9 inches) and wider body (64.2 inches versus 60.2). An on-demand, part-time 4WD system (not for use on dry pavement) with a floor-mounted transfer case lever is standard. Sidekick comes with manual locking front hubs that have to be engaged by hand before 4WD can be engaged.

For

● 4WD traction ● Price ● Driveability

Against

● Inconvenient 4WD system ● Ride ● Rear seat room

Summary

Suzuki is pitching Sidekick at adults who'll "never grow up"—ironic considering that this new mini-Jeep is basically a grown-up Samurai. Though our exposure has so far been limited to a couple of hours on freeways and city streets, a few things stand out. Sidekick's 1.6-liter engine provides perceptibly stronger acceleration and more relaxed cruising than the 1.3-liter in the Samurai; the longer wheelbase and greater weight give the Sidekick a much smoother ride (though it still bounces a lot on washboard surfaces); and the new model feels more reassuringly stable at all times, thanks in part to its wider track. Braking seems better too, but there's still a lot of scrubby on-road understeer (resistance to turning) in tight turns, and the manual steering requires too much muscle when parking. Worse, the Sidekick reeks of the same tinny cheapness that are at least partly responsible for the Samurai's low resale value. It's evident most everywhere you look: thin doors, plasticky dashboard, blotchy paint, the convertible's convoluted snaps-and-Velcro top. True, you can't expect Mercedes quality with a starting price of $8995, but workmanship needs to be better than this. So though it's a better Samurai, the Sidekick is just as much a disposable 4-wheel drive vehicle.

Specifications

	2-door wagon
Wheelbase, in.	86.6
Overall length, in.	142.5
Overall width, in.	64.2
Overall height, in.	65.6
Front track, in.	54.9
Rear track, in.	55.1
Turn diameter, ft.	32.2
Curb weight, lbs.	2127[1]
Cargo vol., cu. ft.	NA
Fuel capacity, gal.	11.1
Seating capacity	4
Front headroom, in.	NA
Front shoulder room, in.	NA
Front legroom, max., in.	NA

Prices are accurate at time of printing; subject to manufacturer's change.

	2-door wagon
Rear headroom, in.	NA
Rear shoulder room, in.	NA
Rear legroom, min., in.	NA

1. Convertible; 2260 lbs., hardtop

Powertrain layout: longitudinal front engine/rear-wheel drive or on-demand 4WD.

Engines

	ohc I-4	ohc I-4
Size, liters/cu. in.	1.3/79	1.6/97
Fuel delivery	2 bbl.	TBI
Horsepower @ rpm	64 @ 5500	80 @ 5400
Torque (lbs./ft.) @ rpm	73 @ 3500	94 @ 3000
Availability	S[1]	S[2]

EPA city/highway mpg

5-speed OD manual	28/29	28/29
3-speed automatic		25/26

1. JA   2. JX, JLX

Prices

Suzuki Sidekick

	Retail Price	Dealer Invoice	Low Price
Standard JA soft top	$8998	—	—
Deluxe JX soft top	10995	—	—
Deluxe JX hard top	12495	—	—
Custom JLX soft top	12595	—	—
Custom JLX hard top	13595	—	—

Dealer invoice, low price and destination charge not available at time of publication.

Standard Equipment:

Standard JA: 1.3-liter 2bbl. 4-cylinder engine, 5-speed manual transmission, part-time 4WD with 2-speed transfer case, vinyl seats with head restraints, tinted glass, intermittent wipers, trip odometer, 205/75R15 tires. **Deluxe JX** adds: 1.6-liter TBI engine, AM/FM cassette, tachometer, half-type console, tricot upholstery, power mirrors, spare tire lock. **Custom JLX soft top** adds: 3-speed automatic transmission, spotlight, full console, cloth upholstery, tilt steering column, bronze tinted glass. **Custom JLX hardtop** adds: power steering, rear defogger and wiper/washer, power windows and locks, cruise control, remote rear door release.

OPTIONS prices not available at time of publication.

Suzuki Swift

What's New

Swift is Suzuki's version of the minicompact sold by Chevrolet dealers as the new Geo Metro, and the first passenger car sold in the U.S. by Suzuki. Both Swift and Metro share a front-drive design evolved from that of the first-generation Suzuki Cultus, which only GM sold here before as the Chevy Sprint (GM owns a small piece of Suzuki). Where Chevy pitches Metro in the economy-car sector, Suzuki is aiming Swift at two different niches. The Swift GLX comes only as a 5-door hatchback with a new 70-horsepower 1.3-liter 4-cylinder engine and a 3-speed automatic transmission. Standard equipment includes power door locks, rear wiper/washer, rear deroster, electric mirrors and a tachometer. The GLX is supposed to be on sale by the end of November with a base price of $7495. The Swift GTi, a 3-door hatchback priced at $8995 for its October introduction, sports racy rocker-panel extensions and a modest front spoiler holding a pair of foglamps. Under the hood, GTi

Suzuki Swift GTi

uses a 1.3-liter engine with multi-point injection, twin overhead camshafts and four valves per cylinder. Suzuki claims a healthy 100 horsepower with this engine and a 0-60 mph time of 8.2 seconds with the standard 5-speed overdrive manual transaxle. An electronically controlled 3-speed automatic is optional. Matching this extra performance are standard all-disc brakes (versus the GLX's front-disc/rear-drum system), a firmer suspension, and 14-inch wheels with 175/60R14 tires, instead of the GLX's 13-inchers and skinny 155/80R rubber. The GTi is also sportier than the GLX inside, with more heavily bolstered front bucket seats. Paint choices are limited to red, white and black on the GTi. We haven't driven either version of the Swift, so we cannot comment on the performance of these cars.

Specifications

	GTi 3-door hatchback	GLX 5-door hatchback
Wheelbase, in.	89.2	93.1
Overall length, in.	146.1	150.0
Overall width, in.	62.4	62.6
Overall height, in.	53.1	54.3
Front track, in.	53.7	53.7
Rear track, in.	52.8	52.8
Turn diameter, ft.	30.2	31.4
Curb weight, lbs.	1768	1741
Cargo vol., cu. ft.	NA	NA
Fuel capacity, gal.	10.6	10.6
Seating capacity	4	4
Front headroom, in.	NA	NA
Front shoulder room, in.	NA	NA
Front legroom, max., in.	NA	NA
Rear headroom, in.	NA	NA
Rear shoulder room, in.	NA	NA
Rear legroom, min., in.	NA	NA

Powertrain layout: transverse front engine/front-wheel drive.

Engines

	ohc I-4	dohc I-4
Size, liters/cu. in.	1.3/79	1.3/79
Fuel delivery	TBI	PFI
Horsepower @ rpm	70 @ 6000	100 @ 6500
Torque (lbs./ft.) @ rpm	74 @ 3500	83 @ 5000
Availability	S[1]	S[2]

> **KEY: ohv** = overhead valve; **ohc** = overhead cam; **dohc** = double overhead cam; **I** = inline cylinders; **V** = cylinders in V configuration; **flat** = horizontally opposed cylinders; **bbl.** = barrel (carburetor); **PFI** = port (multi-point) fuel injection; **TBI** = throttle-body (single-point) fuel injection; **rpm** = revolutions per minute; **OD** = overdrive transmission; **S** = standard; **O** = optional; **NA** = not available.

Prices are accurate at time of printing; subject to manufacturer's change.

city/highway mpg	ohc I-4	dohc I-4
5-speed OD manual		29/36
3-speed automatic	31/34	25/28

1. GLX   2. GTi

Prices not available at time of publication.

Toyota Camry

Toyota Camry V6

What's New

Automatic transmission is now available for the 4-wheel-drive All-Trac sedan as the major change for Camry, Toyota's best-selling line in the U.S. The 4-speed overdrive automatic is mandatory on the top-line LE version (the previous 5-speed manual LE All-Trac is dropped) and optional on the less costly Deluxe, which retains the 5-speed as standard. The All-Trac was new last fall and the Camry line was expanded again in the spring with the addition of a 2.5-liter V-6 engine. All-Trac models have a permanently engaged 4WD system; other Camrys have front-wheel drive. Redesigned engine mounts are standard on all 4-cylinder Camrys, whose 2.0-liter, 115-horsepower, twin-cam engine is unchanged. Likewise, the 24-valve, twin-cam V-6, rated at 153 horsepower, is unchanged. Anti-lock brakes (ABS), also added mid-1988, remain optional for the V-6 LE sedan and now are available on All-Trac models as well. ABS costs $1280 on the All-Trac and $1130 on the LE, down from $1250 last year. As before, 4-cylinder Camrys comprise standard, Deluxe and LE sedans, and Deluxe and LE wagons, plus Deluxe and LE All-Trac sedans. V-6 offerings consist of Deluxe and LE sedans and wagons. High price and the resulting prospect of limited sales have ruled out a V-6 All-Trac Camry, according to Toyota.

For

- Performance (V-6) • Anti-lock brakes • Ride
- 4WD traction (All-Trac) • Passenger room

Against

- Engine noise (4-cylinder)
- Driveability (automatic transmission)

Summary

Camry, Toyota's rival for the Honda Accord and other compacts, offers several features not available on Accord: anti-lock brakes, a V-6 engine, 4WD and a station wagon body style. The V-6 engine is smooth, flexible at low speeds, eager to work hard and strong enough for prompt highway passing. It works better with automatic transmission than the standard 4-cylinder engine, which doesn't have much low-end power and tends to sound coarse when worked hard. Even so, the 4-cylinder is adequate for Camry and returns good gas mileage. We recommend the anti-lock brakes for the improved stopping ability they provide. Note that they're available only on two of the more expensive models, which are quite pricey for the compact class. In fact, an LE 4-door with the V-6, ABS, air conditioning, leather upholstery and a little more will be close to $21,000 with the destination charge. A Deluxe 4-door with the V-6 and fewer options will be much more reasonable. A 4-cylinder Deluxe 4-door is price competitive with a Honda Accord LX at around $15,000 fully equipped. Camry is competitive with Accord in most other areas as well, including assembly quality, reliability and resale value.

Specifications

	4-door notchback	5-door wagon
Wheelbase, in.	102.4	102.4
Overall length, in.	182.1	183.1
Overall width, in.	67.4	67.4
Overall height, in.	54.1	54.5
Front track, in.	58.3	58.3
Rear track, in.	57.1	57.1
Turn diameter, ft.	34.8	34.8
Curb weight, lbs.	2690[1]	2855
Cargo vol., cu. ft.	15.4	65.1
Fuel capacity, gal.	15.9	15.9
Seating capacity	5	5
Front headroom, in.	37.9	38.2
Front shoulder room, in.	54.3	54.3
Front legroom, max., in.	42.9	42.9
Rear headroom, in.	36.6	37.7
Rear shoulder room, in.	53.7	53.7
Rear legroom, min., in.	34.4	34.4

1. 3086 lbs., All-Trac

Powertrain layout: transverse front engine/front-wheel drive or permanent 4WD (All-Trac).

Engines

	dohc I-4	dohc V-6
Size, liters/cu. in.	2.0/122	2.5/153
Fuel delivery	PFI	PFI
Horsepower @ rpm	115 @ 5200	153 @ 5600
Torque (lbs./ft.) @ rpm	124 @ 4400	155 @ 4400
Availability	S	O

EPA city/highway mpg

5-speed OD manual	26/32	19/24
4-speed OD automatic	24/30	19/24

Prices

Toyota Camry	Retail Price	Dealer Invoice	Low Price
4-door notchback, 5-speed	$11488	9880	10884
4-door notchback, automatic	12158	10456	11507
Deluxe 4-door notchback, 5-speed	12328	10503	11616
Deluxe 4-door notchback, automatic ...	13078	11142	12310
LE 4-door notchback, automatic	14658	12415	13737
Deluxe All-Trac 4-door notchback, 5-speed .	14108	12020	13264
Deluxe All-Trac 4-door notchback, automatic	15058	12829	14144
LE All-Trac 4-door notchback, automatic	16648	14101	15575

Prices are accurate at time of printing; subject to manufacturer's change.

	Retail Price	Dealer Invoice	Low Price
Deluxe 5-door wagon, 5-speed	13018	11091	12255
Deluxe 5-door wagon, automatic	13768	11730	12949
LE 5-door wagon, automatic	15438	13076	14457
Deluxe V6 4-door notchback, 5-speed . . .	13638	11620	12829
Deluxe V6 4-door notchback, automatic . .	14300	12260	13594
LE V6 4-door notchback, automatic	16428	13915	15372
Deluxe V6 5-door wagon, automatic	15078	12846	14162
LE V6 5-door wagon, automatic	17218	14584	16101

Dealer invoice and destination charge may vary by region.

Standard Equipment:

2.0-liter DOHC 16-valve PFI 4-cylinder engine, 5-speed manual or 4-speed automatic transmission, power steering, narrow bodyside moldings, coolant temperature gauge, trip odometer, tilt steering column, center console with storage bin, velour reclining front bucket seats with driver's seat height adjustment, passive restraint system (motorized front shoulder belts and manual lap belts), rear lap/shoulder belts, remote fuel door and trunk/liftgate releases, rear defogger, tinted glass, P185/70SR13 all-season SBR tires. **Deluxe** adds: 2.0-liter DOHC 16-valve PFI 4-cylinder or 2.5-liter DOHC 24-valve PFI V-6 engine, wide bodyside moldings, dual remote mirrors, tilt steering column, automatic-off headlamp feature, folding rear seatbacks (wagon), rear wiper/washer (wagon), digital clock, right visor mirror, cup holder. **LE** adds: air conditioning (V6), power mirrors, tachometer, console armrest, multi-adjustable driver's seat, folding rear armrest, cargo cover (wagon), illuminated entry with fadeout, upper windshield tint band, AM/FM ST ET with power antenna. **All-Trac** models have full-time 4-wheel drive.

Optional Equipment:

Anti-lock brakes, LE All-Trac	1280	1024	1152
LE V6 4-door	1130	904	1017
Air conditioning (std. LE V6)	795	636	716
All Weather Guard Pkg., base	55	46	51
California emissions pkg	70	59	65
Power sunroof (NA base)	700	560	630
Cruise control	210	168	189
Power Pkg., Deluxe	520	416	468
LE	565	452	509
Power windows and locks; LE includes lighted visor mirrors.			
Power Seat Pkg., LE	230	184	207
Power driver's seat, lighted visor mirrors; requires Power Value Pkg.			
Leather Pkg., LE 4-door	230	184	207
Requires Power Pkg. and Power Seat Pkg. or Power Seat Pkg. and Value Pkg.			
Alloy wheels, LE 4-cyl.	360	288	324
LE V6	380	304	342
AM/FM ST ET, base & Deluxe 4-doors . . .	330	247	289
Deluxe wagons	360	270	315
AM/FM ST ET cassette, base	480	360	420
Deluxe 4-doors	520	390	455
Deluxe wagons	550	412	481
LE	190	142	166
AM/FM cassette w/EQ & diversity antenna, LE	470	352	411

1988 Toyota Camry V6

	Retail Price	Dealer Invoice	Low Price
Speaker upgrade, base & Deluxe 4-doors .	140	112	126
Deluxe wagons	170	136	153
Cargo deck cover, Deluxe wagons	50	41	46
Two-Tone Paint Pkg., LE 4-doors	245	196	221
Full wheel covers, Deluxe 4-cyl. 4-doors .	75	60	68
Tachometer, Deluxe w/5-speed	60	48	54
CQ Convenience Pkg., base	195	156	176
Full-size spare tire, dual remote mirrors, digital clock.			
Tilt steering column, base	105	90	98
Mudguards	30	24	27
Fall 1989 Value Pkg., Deluxe 4-cyl. 4-doors .	1580	1346	1463
Deluxe 4-cyl. wagons	1600	1363	1482
Deluxe V6 4-doors	1530	1304	1417
LE 4-cyl. 4-doors	1190	1008	1099
LE V6 4-door exc. All-Trac	650	551	601
Air conditioning, cruise control, AM/FM ST ET, full wheel covers, Power Pkg.			

Toyota Celica

Toyota Celica All-Trac Turbo

What's New

Only detail changes attend Toyota's front-drive and All-Trac 4-wheel-drive sporty coupes. Full wheel covers are newly standard for the base ST and midrange GT models, and all but the ST have new interior fabric trim. A rear spoiler is a new option for GTs. Model selections reprise last year's: ST, GT and GT-S 2-door notchbacks; GT convertible; and GT, GT-S and All-Trac Turbo 3-door hatchbacks. The All-Trac Turbo, which has a permanently engaged 4WD system, comes only with a 5-speed manual transmission; optional for others is Toyota's electronically controlled 4-speed over-drive automatic. Anti-lock brakes remain an option for GT-S and All-Trac. Celica STs and GTs use a 115-horsepower 2.0-liter twin-cam 4-cylinder with four valves per cylinder. GT-S models get a 135-horsepower version of that engine, while the All-Trac Turbo has a turbocharged and inter-cooled engine that makes 190 horsepower.

For

- Performance (All-Trac, GT-S) •Handling/roadholding
- 4WD traction (All-Trac) •Anti-lock brakes
- Fuel economy

KEY: ohv = overhead valve; **ohc** = overhead cam; **dohc** = double overhead cam; **I** = inline cylinders; **V** = cylinders in V configuration; **flat** = horizontally opposed cylinders; **bbl.** = barrel (carburetor); **PFI** = port (multi-point) fuel injection; **TBI** = throttle-body (single-point) fuel injection; **rpm** = revolutions per minute; **OD** = overdrive transmission; **S** = standard; **O** = optional; **NA** = not available.

Prices are accurate at time of printing; subject to manufacturer's change.

Against

●Rear seat room   ●Engine noise   ●Entry/exit

Summary

Remember when you could buy a well-furnished Celica GT for around $10,000 about four or five years back? Now a similarly equipped Celica will run more than $15,000, which is more than a lot of younger drivers can afford. Since Celica appeals mainly to younger drivers, sales have dropped as its prices have increased. While the ST and GT have sporty looks, they lack sporty performance. Their 115-horsepower engine is economical, but hardly exciting, especially with automatic transmission. The GT-S and All-Trac models elevate performance considerably, but GT-S prices start at more than $15,000 and the All-Trac Turbo is more than $20,000, putting it in the same league as the Mazda RX-7. Still, the All-Trac is quite a package: a high-powered, responsive engine; pavement-hugging 4WD system; and optional anti-lock brakes. The GT-S is more reasonably priced and still provides brisk performance, plus you can order anti-lock brakes. Even so, you're probably looking at spending at least $17,000. At that price there are several alternatives, including a Ford Mustang GT or a low-line RX-7. For less than $17,000 you can get an Acura Integra, Chevrolet Beretta GT, Honda Prelude, Ford Probe or Mazda MX-6. Despite Celica's reputation for longevity, we find today's prices too high for what you get.

Specifications

	2-door notchback	3-door hatchback	2-door convertible
Wheelbase, in.	99.4	99.4	99.4
Overall length, in.	173.6	171.9	173.6
Overall width, in.	67.3	67.3	67.3
Overall height, in.	49.8	49.8	49.8
Front track, in.	57.9	57.9	57.9
Rear track, in.	56.5	56.5	56.5
Turn diameter, ft.	35.4	35.4	35.4
Curb weight, lbs.	2436	2524[1]	2680
Cargo vol., cu. ft.	NA	25.2	NA
Fuel capacity, gal.	15.9	15.9	15.9
Seating capacity	5	5	4
Front headroom, in.	37.8	37.8	38.4
Front shoulder room, in.	52.1	52.1	52.1
Front legroom, max., in.	44.4	44.4	44.4
Rear headroom, in.	33.9	33.9	35.8
Rear shoulder room, in.	50.9	50.9	36.6
Rear legroom, min., in.	27.9	27.9	27.9

1. 3197 lbs., All-Trac

Powertrain layout: front engine/front-wheel drive or permanent 4WD (All-Trac).

Engines

	dohc I-4	dohc I-4	Turbo dohc I-4
Size, liters/cu. in.	2.0/122	2.0/122	2.0/122
Fuel delivery	PFI	PFI	PFI
Horsepower @ rpm	115 @ 5200	135 @ 6000	190 @ 6000
Torque (lbs./ft.) @ rpm	124 @ 4400	125 @ 4800	190 @ 3200
Availability	S[1]	S[2]	S[3]

EPA city/highway mpg

5-speed OD manual	26/32	22/28	19/26
4-speed OD automatic	26/32	22/28	

1. ST, GT   2. GT-S   3. All-Trac

Prices

Toyota Celica	Retail Price	Dealer Invoice	Low Price
ST 2-door notchback, 5-speed	$11808	$10096	$10952
ST 2-door notchback, automatic	12478	10665	11572
GT 2-door notchback, 5-speed	13408	11397	12403
GT 2-door notchback, automatic	14078	11966	13022
GT 3-door hatchback, 5-speed	13658	11609	12634
GT 3-door hatchback, automatic	14328	12178	13253
GT 2-door convertible, 5-speed	18318	15570	16944
GT 2-door convertible, automatic	18988	16139	17564
GS-S 2-door notchback, 5-speed	15388	13003	14196
GT-S 2-door notchback, automatic	16138	13636	14887
GT-S 3-door hatchback, 5-speed	15738	13299	14519
All-Trac Turbo 3-door hatchback, 5-speed	20878	17642	19260

Dealer invoice and destination charge may vary by region.

Standard Equipment:

2.0iter DOHC 16-valve PFI engine, 5-speed manual or 4-speed automatic transmission, power steering, tinted glass, console, rear defogger, remote fuel door and decklid/liftgate releases, fabric reclining front bucket seats, one-piece folding rear seatback, automatic headlamps off, trip odometer, tachometer, coolant temperature and oil pressure gauges, voltmeter, digital clock, AM/FM ST ET, 165SR13 tires. **GT** adds: split folding rear seatback, right visor mirror, pushbutton heating/ventilation controls, power mirrors, memory tilt steering column, cargo cover (3-door), cargo area carpet, 185/70SR13 tires; convertible has power top and power rear windows. **GT** adds: higher-output engine, rear wiper/washer (3-door), variable intermittent wipers, sport seats with power lumbar and lateral supports, tilt/telescopic steering column with memory, 205/60HR14 tires on alloy wheels. **All-Trac Turbo** adds: turbocharged, intercooled engine, permanent 4-wheel drive, 205/60VR14 tires.

Optional Equipment:

Anti-lock brakes, GT-S, All-Trac	1130	904	1017
Air conditioning, ST & GT	795	636	716
w/auto temp, GT-S & All-Trac	960	768	864
Power Pkg., ST & GT exc. conv.	415	332	374
Power windows and locks, heated power mirrors.			
Power Pkg., GT conv.	490	392	441
Power windows, locks and mirrors.			
Power Pkg., GT-S	390	312	351
Power windows and locks.			
Power Pkg. 2, GT exc. conv.	390	312	351
Power windows and locks.			
AM/FM ST ET cassette (std. All-Trac)	190	142	166
w/EQ, GT & GT-S	430	322	376
w/EQ & diversity antenna, All-Trac	280	210	245
CD player, All-Trac	800	600	700
Alloy wheels, GT	340	272	306
GT conv.	370	296	333
Power sunroof	675	540	608
Cruise control	210	168	189
Color-Keyed Pkg., GT-S	50	40	45
Leather Sport Seat Pkg., GT-S	1550	1240	1395
All-Trac	1160	928	1044
Includes cruise control, power windows and locks.			
Two-tone paint	215	172	194
Rear wiper/washer, GT 3-door	135	111	123
Tilt/telescopic steering column, GT exc. conv.	70	60	65
Rear spoiler, GT	225	180	203
ST Value Pkg. 1 or 2	510	500	510
Pkg. 1: AM/FM ST ET cassette, air conditioner, pinstripe. Pkg. 2 deletes air conditioner and adds sunroof.			
GT Value Pkg.	620	610	620
AM/FM ST ET cassette, air conditioning, cruise control, rear spoiler, GT stripe.			
GT-S Value Pkg.	700	690	700
AM/FM ST ET cassette, automatic air conditioning, cruise control, Power Pkg.			

Prices are accurate at time of printing; subject to manufacturer's change.

Toyota Corolla

1988 Toyota Corolla All-Trac wagon

higher than most other small sedans, though a commendable reliability record and good resale value compensate in the long run. The Deluxe versions offer the most value for the money.

What's New

The subcompact Corollas are basically unchanged following last year's redesign, except that a new All-Trac 4-wheel-drive Deluxe sedan joins the All-Trac Deluxe and SR5 wagons that arrived during 1988 to replace the 4WD Tercel wagons. The All-Trac 4-door sedan uses the same permanently engaged 4WD system as the wagons. It includes a center differential that can be locked for maximum traction. With manual shift this is accomplished by a dashboard switch; with automatic, by electronic control. All-Trac Corollas are powered by a fuel-injected, 100-horsepower 1.6-liter four with dual overhead cams. A carbureted version of this engine developing 90 horsepower is used in the front-drive Deluxe and LE sedans, Deluxe wagon and SR5 coupe. The top-line GT-S coupe continues with a 115-horsepower, fuel-injected 1.6-liter engine. As before, a 5-speed manual transmission is standard across the board and your only choice on the GT-S and SR5 All-Trac wagon. A 3-speed automatic is optional for the front-drive Deluxe sedan and wagon; other models have a 4-speed overdrive automatic as optional.

For

●Fuel economy ●4WD traction (All-Trac) ●Ride

Against

●Engine noise ●Rear seat room

Summary

We haven't driven an All-Trac model or one of the Corolla coupes, so we're limiting our comments to the front-drive 4-door sedans, which we find competent and comfortable. The Deluxe sedan offers the same amount of interior and cargo room as the LE and similar performance since it uses the same 90-horsepower engine. One advantage the LE offers is a 4-speed automatic transmission with an overdrive top gear that provides quieter, more economical highway cruising than the 3-speed automatic available in the Deluxe. For best performance and maximum economy, stick with the standard 5-speed manual. LE models also have wider tires for a slightly larger footprint on the road, yet there isn't much difference in cornering or roadholding compared to the Deluxe. Interior room is comparable to other subcompact sedans, which means the front seats have to be moved forward some to generate enough leg room for adults to fit easily in the back. Corolla is priced

Specifications

	2-door notchback	4-door notchback	5-door wagon
Wheelbase, in.	95.7	95.7	95.7
Overall length, in.	172.3	170.3	171.5
Overall width, in.	65.6	65.2	65.2
Overall height, in.	49.6	52.4	54.5
Front track, in.	56.3	56.3	56.3
Rear track, in.	56.1	55.5	55.3
Turn diameter, ft.	NA	NA	NA
Curb weight, lbs.	2242	2207[1]	2282[2]
Cargo vol., cu. ft.	12.0	11.0	26.0
Fuel capacity, gal.	13.2	13.2	13.2
Seating capacity	5	5	5
Front headroom, in.	37.9	38.4	39.6
Front shoulder room, in.	51.0	53.2	53.2
Front legroom, max., in.	42.9	40.9	40.9
Rear headroom, in.	35.3	36.4	39.3
Rear shoulder room, in.	51.0	52.3	52.7
Rear legroom, min., in.	25.8	31.6	31.6

1. 2606 lbs., All-Trac   2. 2690 lbs., All-Trac

Powertrain layout: transverse front engine/front-wheel drive or permanent 4WD (All-Trac).

Engines

	dohc I-4	dohc I-4	dohc I-4
Size, liters/cu. in.	1.6/97	1.6/97	1.6/97
Fuel delivery	2 bbl.	PFI	PFI
Horsepower @ rpm	90 @ 6000	100 @ 5600	115 @ 6600
Torque (lbs./ft.) @ rpm	95 @ 3600	101 @ 4400	100 @ 4800
Availability	S[1]	S[2]	S[3]

EPA city/highway mpg

5-speed OD manual	30/35	23/28	26/31
3-speed automatic	27/30		
4-speed OD automatic	27/34	23/29	

1. Deluxe, LE, SR5   2. All-Trac   3. GT-S

Prices

Toyota Corolla	Retail Price	Dealer Invoice	Low Price
Deluxe 4-door notchback, 5-speed	$9198	$7909	$8554
Deluxe 4-door notchback, automatic	9668	8314	8991
LE 4-door notchback, 5-speed	10418	8928	9673
LE 4-door notchback, automatic	11088	9502	10295
Deluxe 5-door wagon, 5-speed	9788	8417	9103
Deluxe 5-door wagon, automatic	10258	8822	9540
Deluxe All-Trac 5-door wagon, 5-speed	11498	9887	10693
Deluxe All-Trac 5-door wagon, automatic	12268	10549	11409
SR5 All-Trac 5-door wagon, 5-speed	13088	11216	12152
Deluxe All-Trac 4-door notchback, 5-speed	10608	9122	9865
Deluxe All-Trac 4-door notchback, automatic	11378	9784	10581

Dealer invoice and destination charge may vary by region.

> **KEY: ohv** = overhead valve; **ohc** = overhead cam; **dohc** = double overhead cam; **I** = inline cylinders; **V** = cylinders in V configuration; **flat** = horizontally opposed cylinders; **bbl.** = barrel (carburetor); **PFI** = port (multi-point) fuel injection; **TBI** = throttle-body (single-point) fuel injection; **rpm** = revolutions per minute; **OD** = overdrive transmission; **S** = standard; **O** = optional; **NA** = not available.

Prices are accurate at time of printing; subject to manufacturer's change.

Toyota

Standard Equipment:

Deluxe: 1.6-liter DOHC 16-valve 2bbl. 4-cylinder engine, 5-speed manual or 3-speed automatic transmission, power brakes, cloth reclining front bucket seats (vinyl on wagons), split folding rear seatback (wagons), tinted glass, console with storage, door map pockets, cup holder, remote fuel door and trunk releases, trip odometer, coolant temperature gauge, 155SR13 tires. **LE** adds: 5-speed manual or 4-speed automatic transmission, tachometer, intermittent wipers, digital clock, 60/40 folding rear seatbacks, bodyside molding, remote mirrors, driver's seat height and lumbar support adjustments, upgraded trunk trim, 175/70SR13 all-season tires. **Deluxe All-Trac** models have: PFI engine, permanent 4-wheel drive, 5-speed manual or 4-speed automatic transmission. **SR5 All-Trac wagon** adds: power steering, cruise control, digital clock, AM/FM ST ET, cloth upholstery, tilt steering column, remote mirrors, intermittent wipers, rear wiper.

Optional Equipment:

	Retail Price	Dealer Invoice	Low Price
Air conditioning	745	596	671
All-Weather Guard Pkg. (std. All-Trac)	55	46	51
California emissions pkg.	70	59	65
CQ Convenience Pkg., Deluxe w/5-speed	195	156	176
Deluxe w/automatic	135	108	122
Intermittent wipers, dual remote mirrors, digital clock, tachometer (w/5-speed).			
Power steering (std. SR5)	250	214	232
Power sunroof	530	424	477
CL Convenience Pkg., Deluxe w/5-speed	385	308	347
Deluxe w/automatic	325	260	293
LE, SR5	210	168	189
Cruise control, intermittent wipers, digital clock, dual remote mirrors, tachometer (w/5-speed).			
Alloy wheels w/tire upgrade, LE 4-doors	370	296	333
SR5	415	332	374
AM/FM ST (4 speakers; std. LE & SR5; NA wagons)	330	247	289
w/2 speakers, Deluxe	210	157	184
AM/FM cassette, Deluxe 4-door & All-Trac wagon	480	360	420
Deluxe wagons	380	285	333
SR5	190	142	166
LE	150	112	131
Tilt steering column (std. SR5)	85	73	79
Power Pkg., LE 4-doors & SR5	570	456	513
Power windows, locks and mirrors.			
Two-Tone Paint Pkg., SR5	320	256	288
Cargo deck cover, wagons (std. SR5)	50	41	46
Fabric seats	70	60	65
RQ Convenience Pkg., Deluxe wagons w/5-speed	520	419	470
Deluxe wagons w/automatic	460	371	416
Rear wiper, cruise control, CQ Pkg.			
Rear wiper, wagons	155	127	141
Exterior Appearance Pkg., LE 4-doors	85	68	77
Speaker upgrade, Deluxe All-Trac wagons	115	92	104
Value Pkg., Deluxe	449	386	418
LE 4-doors	659	575	617
Deluxe 2WD wagons	599	515	557
Deluxe All-Trac wagons	539	464	502
Power steering, digital clock, dual remote mirrors, AM/FM ST, intermittent wipers, cargo cover (wagons), fabric seats (wagons), tilt steering column (LE), Power Pkg. (LE), trim rings (Deluxe), roof rack (2WD wagons).			

Toyota Corolla Sport

SR5 2-door notchback, 5-speed	$10628	$9108	$9868
SR5 2-door notchback, automatic	11298	9682	10490
GT-S 2-door notchback, 5-speed	12728	10882	11805

Dealer invoice and destination charge may vary by region.

Standard Equipment:

SR5: 1.6-liter DOHC 16-valve 2bbl. 4-cylinder engine, 5-speed manual or 4-speed automatic transmission, cloth reclining front bucket seats, trip odometer, coolant temperature gauge, tachometer, 175/70SR13 tires. **GT-S** adds: higher-output PFI engine, oil pressure gauge, voltmeter, tilt steering

---

column, automatic headlights-off system, power mirrors, leather steering wheel trim, intermittent wipers, 185/60R14 tires.

Optional Equipment:

	Retail Price	Dealer Invoice	Low Price
Air conditioning	596	745	671
Power steering	250	214	232
Power sunroof	530	424	477
CL Pkg., SR5 5-speed	365	292	329
SR5 automatic	305	244	275
GT-S	210	168	189
Cruise control, variable intermittent wipers, digital clock, dual mirrors.			
CQ Convenience Pkg., SR5	240	199	220
Intermittent wipers, digital clock, split folding rear seat, dual remote mirrors, driver's seat lumbar and height adjustments.			
Alloy wheels, SR5	445	356	401
GT-S	435	348	392
AM/FM ST ET, SR5	330	247	289
AM/FM ST ET cassette, SR5	520	390	455
GT-S	190	142	166
w/EQ, GT-S	430	322	376
Tilt steering column, SR5	85	73	79
Power Pkg.	390	312	351
Power windows and locks.			
Two-Tone Paint Pkg.	215	172	194
Sport seat, GT-S	180	144	162
KQ Pkg., SR5	415	339	377
Exterior Appearance Pkg., CQ Pkg., power mirrors.			
KL Pkg., SR5	480	384	432
Exterior Appearance Pkg., CL Pkg., power mirrors.			
All Weather Guard Pkg.	55	46	51
California emissions pkg.	70	59	65

Toyota Cressida

Toyota Cressida

What's New

Though familiar in format, Toyota's 6-cylinder rear-drive luxury flagship is all-new this year: slightly larger, with less straight-edged styling and a more powerful engine. As in '88, only a 4-door notchback sedan is offered. Replacing the 2.8-liter, twin-cam, inline 6-cylinder of previous models is a slightly detuned version of the 3.0-liter six from the Supra, also an inline design but with four valves per cylinder. At 190 horsepower and 185 pounds/feet torque, it delivers 22 percent more horsepower and 12 percent more torque than the previous 12-valve Cressida engine. It teams exclusively with Toyota's electronically controlled 4-speed overdrive automatic transmission, which now has a shift lock that prevents moving out of park unless the brakes are applied—a response to the current unintended-acceleration controversy; a similar device is used on Audi 100/200 and Nissan 300ZX models with automatic transmission. There's also an interlock that prevents removing the ignition key except in park. Compared to its 1985-88 predecessor, the new Cres-

---

Prices are accurate at time of printing; subject to manufacturer's change.

sida rides an inch-longer wheelbase, measures 1.8 inches longer overall and is 1.2 inches wider. Curb weight is up about 100 pounds. Toyota claims a 0-60-mph time of 9.3 seconds, a little quicker than the previous Cressida. Four-wheel disc brakes are standard and anti-lock control is a new option. As before, standard equipment is comprehensive: power windows, power mirrors, 2-stage central door locking, automatic climate control, tilt/telescope steering wheel, fold-down rear seatbacks and 6-speaker AM/FM/cassette stereo.

For

- Performance • Anti-lock brakes • Ride
- Handling/roadholding

Against

- Fuel economy • Climate system controls

Summary

The new Cressida is no big advance on the old one, which was a surprisingly capable sedan, staid looks notwithstanding. Cressida's new 3.0-liter twin-cam straight six delivers more than adequate punch, and the 4-speed overdrive automatic provides prompt, super-smooth shifts. The new standard shift-lock feature can be a nuisance until you get used to it, but we applaud this effort at reassuring buyers concerned about sudden acceleration. It prevents shifting out of park unless the brake pedal is depressed. Despite its slightly greater heft and bulk, the new Cressida feels more agile than the old, with less apparent understeer and body roll, and better rear-end adhesion, especially in tight corners. Braking is fine, again not obviously superior to that of most rivals but free from drama and, on our test car, aided by the reassuring presence of the optional anti-lock feature. The same is true of the ride, which is firm but supple and quite resistant to pitching. The inside story is that the new Cressida seems to have about the same total passenger room as the Acura Legend and slightly less than the Mazda 929. The driving position, seat comfort and general ergonomics give little cause for complaint—with two exceptions. The automatic-climate-system control panel slides out like a drawer to reveal manual fan-speed and mode buttons on its top face; the panel won't extend or retract without the ignition on. Equally silly is the bank of duplicate sound system controls just under the center dash vents. This is an idea inherited from the previous Cressida, where it made a certain amount of sense because the main audio unit was more remote from the satellite controls. Here they're separated by less distance and the main unit isn't mounted that low. We think Toyota ought to dump the duplicate controls, as they add six to 12 needless buttons to an already busy dashboard. In all, this new Cressida is just as much an enigma as the equally conservative 929: a wolverine in sheep's clothing—a smooth, competent and luxurious upscale sedan rendered almost invisible by unimaginative styling.

Specifications

	4-door notchback
Wheelbase, in.	105.5
Overall length, in.	189.6
Overall width, in.	67.3
Overall height, in.	53.7
Front track, in.	57.3
Rear track, in.	57.3
Turn diameter, ft.	NA

	4-door notchback
Curb weight, lbs.	3417
Cargo vol., cu. ft.	12.0
Fuel capacity, gal.	18.5
Seating capacity	5
Front headroom, in.	38.4
Front shoulder room, in.	51.6
Front legroom, max., in.	42.8
Rear headroom, in.	37.1
Rear shoulder room, in.	54.4
Rear legroom, min., in.	35.0

Powertrain layout: longitudinal front engine/rear-wheel drive.

Engines	dohc I-6
Size, liters/cu. in.	3.0/180
Fuel delivery	PFI
Horsepower @ rpm	190 @ 5600
Torque (lbs./ft.) @ rpm	185 @ 4400
Availability	S
EPA city/highway mpg	
4-speed OD automatic	19/24

Prices

Toyota Cressida	Retail Price	Dealer Invoice	Low Price
4-door notchback	$21498	$17628	$19563

Dealer invoice and destination charge may vary by region.

Standard Equipment:

3.0-liter DOHC 24-valve PFI 6-cylinder engine, 4-speed automatic transmission, power steering, power 4-wheel disc brakes, shift lock, automatic air conditioning, reclining front bucket seats, passive restraint system (motorized front shoulder belts, manual lap belts), rear lap/shoulder belts, cruise control, variable intermittent wipers, trip odometer, coolant temperature gauge, tachometer, AM/FM ST ET cassette with EQ, power antenna, power windows and locks, heated power mirrors, tilt/telescopic steering column, 205/60R15 tires.

Optional Equipment:

Anti-lock brakes	1130	904	1017
CD player	800	600	700
Power sunroof	810	648	729
Leather Pkg.	905	724	815
Power Seat Pkg.	540	432	486
Leather Power Seat Pkg.	1245	996	1121

Toyota Land Cruiser

What's New

This slow-selling 5-door wagon with part-time, on-demand 4-wheel drive is a competitor for the likes of the Jeep Cherokee/Wagoneer and Range Rover. It is virtually unchanged following its 1988 update, the most extensive since the model's last redesign for 1981. Last year's changes included a new 155-horsepower 4.0-liter straight six and 4-

> **KEY: ohv** = overhead valve; **ohc** = overhead cam; **dohc** = double overhead cam; **I** = inline cylinders; **V** = cylinders in V configuration; **flat** = horizontally opposed cylinders; **bbl.** = barrel (carburetor); **PFI** = port (multi-point) fuel injection; **TBI** = throttle-body (single-point) fuel injection; **rpm** = revolutions per minute; **OD** = overdrive transmission; **S** = standard; **O** = optional; **NA** = not available.

Prices are accurate at time of printing; subject to manufacturer's change.

Toyota

Toyota Land Cruiser

speed overdrive automatic transmission (replacing a 4.2-liter 6-cylinder and 4-speed manual), increased towing capacity from 2500 to 3500 pounds and a modified 4WD system. Land Cruiser has manually locking front hubs (they have to be locked or unlocked by hand), but an interior switch allows selecting 2- or 4-wheel operation once the hubs are locked. The standard velocity-sensitive shock absorbers automatically adjust suspension firmness to match driving conditions. On a smooth surface the shocks operate in a "soft" mode for ride comfort. On rough roads they stiffen automatically for firmer ride control. We haven't driven the Land Cruiser since it was revamped for 1988, so we cannot comment on its performance.

Specifications

	5-door wagon
Wheelbase, in.	107.5
Overall length, in.	184.0
Overall width, in.	70.9
Overall height, in.	68.9
Front track, in.	58.5
Rear track, in.	57.9
Turn diameter, ft.	40.7
Curb weight, lbs.	4480
Cargo vol., cu. ft.	99.3
Fuel capacity, gal.	23.8
Seating capacity	5
Front headroom, in.	40.0
Front shoulder room, in.	59.5
Front legroom, max., in.	39.2
Rear headroom, in.	40.4
Rear shoulder room, in.	59.6
Rear legroom, min., in.	34.6

Powertrain layout: longitudinal front engine/rear-wheel drive or on-demand 4WD.

Engines

	ohv I-6
Size, liters/cu. in.	4.0/241
Fuel delivery	PFI
Horsepower @ rpm	155 @ 4000
Torque (lbs./ft.) @ rpm	220 @ 3000
Availability	S

EPA city/highway mpg

4-speed OD automatic	12/14

Prices

Toyota Land Cruiser	Retail Price	Dealer Invoice	Low Price
5-door wagon	$20898	$17450	$19174

Dealer invoice and destination charge may vary by region.

Standard Equipment:

4.0-liter PFI 6-cylinder engine, 4-speed automatic transmission, freewheeling manual front hubs, power steering, power brakes, cloth reclining front seats, tachometer, trip odometer, coolant temperature and oil pressure gauges, digital clock, remote fuel door release, AM/FM ST ET, voltmeter, locking fuel filler door, tinted glass, dual outside mirrors, tilt steering column, front tow hook, intermittent wipers, 225/75R15 M&S tires.

Optional Equipment:

Air conditioner	890	712	801
AM/FM ST ET cassette	190	142	166
Power Pkg.	680	544	612
Power windows, locks and mirrors.			
Two-tone paint	245	196	221
California emissions pkg.	70	59	65

Toyota MR2

Toyota MR2

What's New

Despite fast-falling sales, caused partly by escalating insurance premiums for 2-seat cars, Toyota's mid-engine sports car receives only two changes for '89: a center high-mount stop lamp that employs LEDs instead of conventional bulbs (now included with the optional Aerodynamic Spoiler Package only) and a standard rear stabilizer bar for the Supercharged version introduced last year. Both "Mister Twos" continue with their familiar square-rigged styling, choice of 5-speed manual or 4-speed automatic transmissions, and all-disc brakes. Base engine is a naturally aspirated 1.6-liter, twin-cam 4-cylinder with 115 horsepower. The supercharged version of that engine has 145 horsepower.

For

●Performance ●Braking ●Fuel economy

Against

●Noise ●Cargo room ●Entry/exit

Summary

Toyota blames high insurance rates for declining MR2 sales, but high prices certainly aren't helping. Younger drivers, who want cars such as the MR2 the most, can least afford one that starts at $14,000 and climbs to more than

---

Prices are accurate at time of printing; subject to manufacturer's change.

$18,000 in supercharged trim. We prefer the MR2's supercharged engine to most turbocharged engines because it provides an instant, smooth rush of power, even at low speeds. By contrast, most turbo engines suffer a noticeable lag before delivering more power, which often comes in a big gob instead of a steady flow. That doesn't mean you have to order the supercharged model for good performance. On the contrary, the base engine delivers almost as much pleasure for substantially less money. Our staff is divided on the MR2's handling and roadholding; some rate both as excellent, others claim it feels loose in corners, lacks straight-line stability and has vague steering. Others apparently have voiced complaints since Toyota has added a rear stabilizer bar to the supercharged model (a rear stabilizer was standard on early MR2s, but was removed as a running change). Though the cockpit is tight, most drivers find a workable position. The gauges are clearly marked, all controls are within easy reach, except for the low-mounted stereo. A low build makes getting in or out a chore, and despite trunks in the front and rear, cargo space is minimal. However, MR2s are bought for driving pleasure, not practical considerations. This is a well-designed 2-seater that usually puts smiles on our faces, but we don't have to worry about the big dent it can put in your bank account.

Specifications

	2-door notchback
Wheelbase, in.	91.3
Overall length, in.	155.5
Overall width, in.	65.6
Overall height, in.	48.6
Front track, in.	56.7
Rear track, in.	56.7
Turn diameter, ft.	31.5
Curb weight, lbs.	2350[1]
Cargo vol., cu. ft.	7.8
Fuel capacity, gal.	10.8
Seating capacity	2
Front headroom, in.	37.4
Front shoulder room, in.	52.4
Front legroom, max., in.	43.0
Rear headroom, in.	—
Rear shoulder room, in.	—
Rear legroom, min., in.	—

1. 2493 lbs., Supercharged MR2

Powertrain layout: transverse mid-engine/rear-wheel drive.

Engines	dohc I-4	Supercharged dohc I-4
Size, liters/cu. in.	1.6/97	1.6/97
Fuel delivery	PFI	PFI
Horsepower @ rpm	115 @ 6600	145 @ 6400
Torque (lbs./ft.) @ rpm	100 @ 4800	140 @ 4000
Availability	S	S
EPA city/highway mpg		
5-speed OD manual	26/31	24/30
4-speed OD automatic	25/30	22/27

Prices

Toyota MR2	Retail Price	Dealer Invoice	Low Price
2-door, 5-speed	$13798	$11659	$12729
2-door, automatic	14548	12293	13421
2-door w/T-bar roof, 5-speed	15268	12901	14085
2-door w/T-bar roof, automatic	16018	13535	14777

	Retail Price	Dealer Invoice	Low Price
Supercharged w/T-bar roof, 5-speed	17628	14896	16262
Supercharged w/T-bar roof, automatic	18378	15529	16954

Dealer invoice and destination charge may vary by region.

Standard Equipment:

1.6-liter DOHC 16-valve PFI 4-cylinder engine, 5-speed manual or 4-speed automatic transmission, 4-wheel disc brakes, reclining bucket seats, tinted glass with shaded upper windshield band, remote fuel door and decklid releases, AM/FM ST ET, intermittent wipers, trip odometer, tachometer, coolant temperature and oil pressure gauges, voltmeter, 185/60HR14 tires. **T-bar roof models** add: Performance Interior Pkg. (sport seats, leather-wrapped steering wheel and shift knob, power mirrors, rear console, automatic headlamp off). **Supercharged** models have 1.6-liter supercharged and intercooled engine, Aerodynamic Spoiler Pkg., alloy wheels.

Optional Equipment:

Air conditioning	795	636	716
Moonroof w/sunshade, base	380	304	342
Power Pkg., base w/T-bar	390	312	351
Base w/o T-bar	920	736	828
T-bar: power windows, locks and mirrors. Standard roof adds leather-wrapped steering wheel, automatic headlamp off and rear console.			
Performance Interior Pkg., base w/o T-bar	530	424	477
Sport seats, leather-wrapped steering wheel, automatic headlamp off, power mirrors.			
Cruise control	225	180	203
Rear spoiler, T-bar models	285	228	257
Aerodynamic Spoiler Pkg., base	560	448	504
Rear spoiler, side skirts, rear mudguards, rear sunshade.			
Alloy wheels, base	435	348	392
Leather/Power/Performance Pkg., base w/o T-bar	1730	1384	1557
Base w/T-bar	1200	960	1080
Power windows and locks, power mirrors, leather-wrapped steering wheel, automatic headlamp off, rear console box, leather sport seat.			
Two-Tone Paint Pkg.	245	196	221
AM/FM ST ET cassette	280	210	245
w/EQ, base w/o T-bar	490	367	429
w/EQ, base w/T-bar	435	326	381
Alloy wheels & all-weather tires, base	435	348	392

Toyota Supra

What's New

Toyota's rear-drive grand touring car enters its fourth season with several touchups inside and out, a shift lock for the optional automatic transmission, a retuned suspension and slightly more power for the Turbo model. A modified wastegate actuator for the turbocharged engine results in a nominal two extra horsepower for a total of 232 and an additional eight pounds/feet torque for a total of 254. Supra's naturally aspirated 3.0-liter, 24-valve inline six is unchanged at 200 horsepower and 188 lbs/ft torque. Standard for the Turbo and optional for the base Supra is a new "progressive" speed-sensitive power-steering system that decreases assist as road speed increases (and vice-versa, of course). Rear suspension components have been revised for what

> **KEY: ohv** = overhead valve; **ohc** = overhead cam; **dohc** = double overhead cam; **I** = inline cylinders; **V** = cylinders in V configuration; **flat** = horizontally opposed cylinders; **bbl.** = barrel (carburetor); **PFI** = port (multi-point) fuel injection; **TBI** = throttle-body (single-point) fuel injection; **rpm** = revolutions per minute; **OD** = overdrive transmission; **S** = standard; **O** = optional; **NA** = not available.

Prices are accurate at time of printing; subject to manufacturer's change.

CONSUMER GUIDE®   227

Toyota

Toyota Supra Turbo

Toyota says is improved stability, crisper lateral response and greater roll rigidity. Other changes include a thicker steering wheel with spoke-mounted cruise control switches (previously divided between a column stalk and a main dash switch); black instrument-panel appliques; new gauge graphics, power-window switches and climate controls; the new automatic-transmission shift lock described under Cressida; and a revised 6-speaker audio system with newly optional compact-disc player. Outside are an enlarged air intake under the front bumper, a vertical grille piece that carries down the line of the prominent hood bulge, new taillamps and, on the Turbo, a restyled 3-piece spoiler with LED center high-mount stop lamp as on the MR2.

For

- Performance - Anti-lock brakes - Handling

Against

- Fuel economy   - Rear seat room   - Entry/exit
- Road noise

Summary

Supra succeeds despite its hefty curb weight, which slows acceleration. Even so, performance is still brisk with the base engine and even more satisfying with the turbocharged version. The weight also makes Supra feel ponderous and clumsy at lower speeds, but the wide performance tires give it impressive cornering ability and the suspension handles demanding roads with aplomb. The aggressive tires also generate a lot of thumping and road noise. It takes strong brakes to handle this much power and weight, and Supra has them. We recommend the optional anti-lock brakes for even better stopping ability. Inside, Supra has well-shaped, supportive front seats, a comfortable driving position and simple, convenient controls. The rear seat is tiny and the cargo area is only adequate for two people with the rear seats folded down, while entry/exit is complicated by the low build and wide doors that require a lot of space to be opened all the way. All of these complaints are common among sports cars, so Supra isn't worse than the norm. It's now quite expensive, but it should also be reliable and durable.

KEY: ohv = overhead valve; **ohc** = overhead cam; **dohc** = double overhead cam; **I** = inline cylinders; **V** = cylinders in V configuration; **flat** = horizontally opposed cylinders; **bbl.** = barrel (carburetor); **PFI** = port (multi-point) fuel injection; **TBI** = throttle-body (single-point) fuel injection; **rpm** = revolutions per minute; **OD** = overdrive transmission; **S** = standard; **O** = optional; **NA** = not available.

Specifications

	3-door hatchback
Wheelbase, in.	102.2
Overall length, in.	181.9
Overall width, in.	68.7
Overall height, in.	51.6
Front track, in.	58.5
Rear track, in.	58.5
Turn diameter, ft.	35.4
Curb weight, lbs.	3459
Cargo vol., cu. ft.	12.8
Fuel capacity, gal.	18.5
Seating capacity	4
Front headroom, in.	37.5
Front shoulder room, in.	52.5
Front legroom, max., in.	43.6
Rear headroom, in.	33.9
Rear shoulder room, in.	50.5
Rear legroom, min., in.	24.7

Powertrain layout: longitudinal front engine/rear-wheel drive.

Engines

	dohc I-6	Turbo dohc I-6
Size, liters/cu. in.	3.0/180	3.0/180
Fuel delivery	PFI	PFI
Horsepower @ rpm	200 @ 6000	232 @ 5600
Torque (lbs./ft.) @ rpm	188 @ 3600	254 @ 3200
Availability	S	S

EPA city/highway mpg

5-speed OD manual	18/23	17/22
4-speed OD automatic	18/23	18/23

Prices

Toyota Supra	Retail Price	Dealer Invoice	Low Price
3-door hatchback, 5-speed	$22360	$18335	$20348
3-door hatchback, automatic	23110	18950	21030
w/Sport Roof, 5-speed	23430	19213	21322
w/Sport Roof, automatic	24180	19828	22004
Turbo, 5-speed	24700	20254	22477
Turbo, automatic	25450	20869	23160
Turbo w/Sport Roof, 5-speed	25720	21090	23405
Turbo w/Sport Roof, automatic	26470	21705	24088

Dealer invoice and destination charge may vary by region.

Standard Equipment:

3.0-liter DOHC 24-valve PFI 6-cylinder engine, 5-speed manual or 4-speed automatic transmission, speed-sensitive power steering, power 4-wheel disc brakes, automatic air conditioning, power windows and locks, heated power mirrors, two-can cupholder, tilt/telescopic steering column, fog lights, theft deterrent system, bodyside moldings, tachometer, coolant temperature and oil pressure gauges, voltmeter, trip odometer, variable intermittent wipers, cruise control, console with storage area and padded armrest, cloth sport seats, driver's seat power lumbar and lateral support adjustments, folding rear seatbacks, lighted visor mirrors, remote fuel door and hatch releases, automatic-off headlight system, illuminated entry system, tinted glass with shaded upper windshield band, rear defogger, cargo cover, AM/FM ST ET cassette with EQ, diversity power antenna, 225/50VR16 Goodyear Eagle GT Gatorback tires on alloy wheels. **Turbo** adds: turbocharged, intercooled engine, oil cooler, turbo boost gauge, Sports Pkg. (Electronically Modulated Suspension, limited-slip differential, headlamp washers).

Optional Equipment:

Anti-lock brakes	1130	904	1017
Leather Seat Pkg.	1010	808	909

Prices are accurate at time of printing; subject to manufacturer's change.

	Retail Price	Dealer Invoice	Low Price
Power driver's seat	230	184	207
Sports Pkg., base	595	476	536
Turbo	360	288	324
Electronically Modulated Suspension, limited-slip differential.			
White Exterior Appearance Pkg.	40	32	36
CD player	800	600	700
Leather/Power Seat Pkg.	1240	992	1116
California emissions pkg.	70	59	65

Toyota Tercel

Toyota Tercel 2-door

What's New

Now bereft of wagons, Toyota's price-leader subcompact series gets only minor changes this year. Included are upgraded seat fabrics for Standard and Deluxe models, standard dual, manual remote-control mirrors, and newly optional full wheel covers for Deluxes. Model choices are down to a 2-door notchback coupe and 3- and 5-door hatchback sedans in standard and Deluxe trim, plus the bare-bones EZ 3-door, all with front-wheel drive. The EZ comes with a 4-speed manual transaxle instead of the others' 5-speed overdrive unit and isn't available with optional 3-speed automatic. Continuing as Tercel power is a carbureted 1.5-liter 4-cylinder engine with three valves per cylinder and 78 horsepower.

For

- Fuel economy ● Maneuverability
- Cargo space (hatchbacks)

Against

- Passenger room and cargo space (coupe)
- Performance (automatic transmission)

Summary

Tercel is smaller, a little more basic and less expensive than the Corolla, so this is the logical choice for those on a tight budget. Fuel economy is good with either transmission, but naturally better with the overdrive manual transmissions. Acceleration is surprisingly brisk with manual shift and barely adequate with automatic. However, we've encountered flat spots during acceleration that may be from the variable-venturi carburetor used on this engine. In addition, the air conditioner drains a lot of engine power. The coupe has minimal rear seat room, while the hatchbacks allow enough space for two adults to sit in somewhat cramped fashion.

Trunk space is modest on the coupe as well, while the hatchbacks have rear seatbacks that fold for more cargo room. While Tercel is a competent small car, neither its performance nor its price are exceptional for today. However, an admirable record for reliability and durability make it worth looking at.

Specifications

	2-door notchback	3-door hatchback	5-door hatchback
Wheelbase, in.	93.7	93.7	93.7
Overall length, in.	166.7	157.3	157.3
Overall width, in.	64.0	64.0	64.0
Overall height, in.	51.8	52.6	52.8
Front track, in.	54.5	54.5	54.5
Rear track, in.	54.1	54.1	54.1
Turn diameter, ft.	31.2	31.2	31.2
Curb weight, lbs.	2000	1970	2025
Cargo vol., cu. ft.	NA	36.2	37.8
Fuel capacity, gal.	11.9	11.9	11.9
Seating capacity	5	5	5
Front headroom, in.	37.8	38.4	38.56
Front shoulder room, in.	51.5	51.5	52.9
Front legroom, max., in.	40.2	40.2	40.2
Rear headroom, in.	35.9	36.6	36.3
Rear shoulder room, in.	50.7	51.5	52.4
Rear legroom, min., in.	30.8	32.1	32.1

Powertrain layout: transverse front engine/front-wheel drive.

Engines

	ohc I-4
Size, liters/cu. in.	1.5/89
Fuel delivery	1 bbl.
Horsepower @ rpm	78 @ 6000
Torque (lbs./ft.) @ rpm	87 @ 4000
Availability	S

EPA city/highway mpg

4-speed OD manual	35/41
5-speed OD manual	31/37
3-speed automatic	28/32

Prices

Toyota Tercel	Retail Price	Dealer Invoice	Low Price
Standard 2-door notchback, 5-speed	$7338	$6604	$6971
Standard 2-door notchback, automatic . . .	7798	7018	7408
Deluxe 2-door notchback, 5-speed	8398	7221	7810
Deluxe 2-door notchback, automatic	8868	7626	8247
EZ 3-door hatchback, 4-speed	6328	5789	6059
Standard 3-door hatchback, 5-speed	7178	6532	6855
Standard 3-door hatchback, automatic . . .	7648	6959	7304
Deluxe 3-door hatchback, 5-speed	8298	7136	7717
Deluxe 3-door hatchback, automatic . . .	8768	7540	8154
Deluxe 5-door hatchback, 5-speed	8538	7342	7940
Deluxe 5-door hatchback, automatic . . .	9008	7747	8378

Dealer invoice and destination charge may vary by region.

Standard Equipment:

EZ: 1.5-liter 12-valve 4-cylinder engine, 4-speed manual transmission, power brakes, locking fuel door, trip odometer, vinyl reclining front bucket seats, door pockets, fold-down rear seatback, 145/80SR13 tires on styled steel wheels. **Standard** adds: 5-speed manual or 3-speed automatic transmission, cloth seat inserts, cup holder, door-ajar light, bodyside moldings, rear ashtray. **Deluxe** adds: cloth seat trim, folding rear seatbacks, right visor mirror, rear defogger, tinted glass, 155SR13 all-season tires.

Prices are accurate at time of printing; subject to manufacturer's change.

Optional Equipment:

	Retail Price	Dealer Invoice	Low Price
Air conditioning (NA EZ)	735	588	662
Power steering (NA EZ)	250	214	232
Manual sunroof, Deluxe 3-door	370	296	333
Rear wiper/washer, Std & Deluxe hatchbacks	135	111	123
Full wheel covers, Deluxe	125	100	113
AM/FM ST ET	210	157	184
AM/FM ST ET cassette, Deluxe	480	360	420
Cargo cover, Std 3-door	50	41	46
CL Pkg., Deluxe hatchbacks	355	287	321
Deluxe 2-doors	385	311	348

Cruise control, variable intermittent wipers, tilt steering column, clock, lighted cigaret lighter.

	Retail Price	Dealer Invoice	Low Price
CQ Convenience Pkg., Standard	140	116	128
Deluxe hatchbacks	155	127	141
Deluxe 2-doors	185	151	168

Standard: remote left mirror, tilt steering column, remote decklid/liftgate release, full fabric seats, day/night mirror, trunk light. Deluxe adds: low-fuel warning lamp, rear console box, security latch (hatchbacks), clock, lighted cigaret lighter and trunk light.

Toyota Van

Toyota Van

What's New

Though a replacement for the current Toyota Van is definitely in the works, it likely won't appear before model-year 1991. Meantime, the familiar "forward-control" design, with its between-the-seats engine, carries into 1989 with only new colors, standard 3-point seat belts for all but middle passengers and revised model choices. A Deluxe-trim 4-wheel-drive passenger model is added and the previous manual-shift 4WD LE model is deleted. The 1989 passenger Van lineup goes like this: rear-drive Deluxe and LE models available with either a 5-speed manual or 4-speed automatic; 4WD Deluxe Van with the same choice of transmissions; and a 4WD LE with automatic only. All passenger models have seats for seven. Commercial offerings comprise windowless Panel Vans with 2- or 4-wheel drive and a 2WD Cargo Van with windows. All Toyota Vans use a 2.2-liter fuel-injected 4-cylinder, unchanged at 101 horsepower.

For

- Passenger room
- Cargo space
- 4WD traction

Against

- Handling/roadholding (except 4WD)
- Entry/exit
- Driving position

Summary

With a 4-cylinder engine mounted behind the front axle and between the front seats, Toyota's Van has the same mechanical layout as the Mitsubishi Wagon and Nissan Van. All three models suffer from that design. The Toyota's short wheelbase aids maneuverability, but around corners there's too much body lean and the inside rear wheel tends to lift and lose traction on slippery roads. On the open road, the tall Van is pushed around by crosswinds. The 4WD package greatly improves traction, but adds 400 power-robbing pounds to the Van's curb weight, which it can ill afford. The extra weight makes acceleration slow enough that merging into fast-moving expressway traffic becomes a risky venture. Toyota's compact van has ample room inside for passengers to move around, though the engine placement forces the driver and front passenger to get out of the vehicle to get to the rear. The driver gets little foot room and he's squeezed between the door and the engine compartment, leaving little room for his elbows. It's hard to squeeze through the front doorways while stepping up into the tall interior; the driver has an even harder time since he has to slip under the steering wheel. During the winter, it's easy to soil your clothes if there's salt and slush around the front doors. With all seats in place, there's modest cargo room behind the rear bench, but removing the rear and middle seats opens up a huge cargo area. Aside from the traction benefits of Toyota's 4WD model, we rate the front-drive Dodge Caravan/Plymouth Voyager as easier to drive and better for passenger use.

Specifications

	4-door van
Wheelbase, in.	88.0
Overall length, in.	175.8
Overall width, in.	66.3
Overall height, in.	70.3
Front track, in.	56.7
Rear track, in.	54.5
Turn diameter, ft.	30.2
Curb weight, lbs.	3038[1]
Cargo vol., cu. ft.	149.8
Fuel capacity, gal.	15.9
Seating capacity	7
Front headroom, in.	39.6
Front shoulder room, in.	54.9
Front legroom, max., in.	41.5
Rear headroom, in.	40.6
Rear shoulder room, in.	57.7
Rear legroom, min., in.	32.2

1. 3455 lbs., 4WD

Powertrain layout: longitudinal mid-engine/rear-wheel drive or on-demand 4WD.

Engines

	ohv I-4
Size, liters/cu. in.	2.2/137
Fuel delivery	PFI
Horsepower @ rpm	101 @ 4400
Torque (lbs./ft.) @ rpm	133 @ 3000
Availability	S

EPA city/highway mpg

5-speed OD manual	22/24
4-speed OD automatic	21/23

Prices are accurate at time of printing; subject to manufacturer's change.

Prices

Toyota Passenger Van

Toyota Passenger Van	Retail Price	Dealer Invoice	Low Price
Deluxe 4-door van, 5-speed	$13608	$11635	$12622
Deluxe 4-door van, automatic	14278	12208	10243
LE 4-door van, 5-speed	15538	13207	14373
LE 4-door van, automatic	16208	13777	14993
Deluxe 4WD, 5-speed	16068	13658	14863
Deluxe 4WD, automatic	16848	14321	15585
LE 4WD, automatic	10448	15681	17065

Dealer invoice and destination charge may vary by region.

Standard Equipment:

Deluxe: 2.2-liter TBI 4-cylinder engine, 5-speed manual or 4-speed automatic transmission, power brakes, removable rear seats, reclining middle seat, cloth upholstery, tinted glass, 195/75R14 SBR tires. **LE** adds: power steering, power door locks, rear defogger and wiper/washer, AM/FM ST ET, tilt steering column, 195/75R14 tires (2WD), 205/75R14 tires (4WD).

Optional Equipment:

Rear defogger & wiper/washer, Deluxe . . .	245	199	222
Power locks, Deluxe	225	180	203
Power Pkg., LE	440	352	396
Power windows and locks, remote mirrors, fog lamps.			
Two-tone paint, LE	245	196	221
Alloy wheels, LE 2WD	380	304	342
Dual Sunroof Pkg., LE 2WD	1030	824	927
AM/FM ST ET, Deluxe	360	270	315
w/cassette, Deluxe	550	412	481
w/cassette, LE	190	142	166
w/cassette & EQ, LE	505	379	442
Dual air conditioning w/rear heater	1420	1136	1278
w/cooler/icemaker, LE	1685	1348	1517
Cruise control, Deluxe	245	196	221
LE	220	176	198
Swivel Seat Pkg., LE	745	596	671
Front captain's chairs, mid-row swivel/reclining captain's chairs, rear bench seat.			
Privacy glass, LE	365	292	329
Chrome wheels, 4WD	220	176	198
Power steering, Deluxe 2WD	275	234	255
Tilt steering column, Deluxe	85	73	79
Tachometer & digital clock, Deluxe 5-speed .	80	64	72
California emissions pkg.	70	59	65

Toyota 4Runner

What's New

The 4Runner, a compact 4-wheel-drive sport utility vehicle, retains the same styling and suspension design as last year, but will soon be brought into line with Toyota's restyled '89 pickups, perhaps as early as mid-model year. For now, there are only new exterior graphics for the sporty SR5 models and a 5-horsepower increase for the optional 3.0-liter V-6 engine. As before, the V-6 is available with either 5-speed manual or 4-speed automatic transmission, as is the standard 116-horsepower 2.4-liter 4-cylinder. Models comprise 2- and 5-passenger 4-cylinder Deluxes, 2-seat V-6 Deluxe, and 5-passenger 4- and 6-cylinder SR5s. All have two front doors, 2-way rear tailgate, integral rear roll bar, removable rear roof and a part-time, on-demand 4WD system with manual locking front hubs.

For

● 4WD traction ● Passenger and cargo room
● Performance (V-6)

Toyota 4Runner V6

Against

● Fuel economy ● Performance (4-cylinder)
● Inconvenient 4WD system ● Ride

Summary

The optional V-6 engine, added last year, gives 4Runner a welcome performance boost over the standard 2.4-liter 4-cylinder, which is barely adequate for this heavy vehicle. The V-6 is smooth, responsive and powerful enough to perform well with automatic transmission. Unfortunately, fuel economy is low; we averaged only 14.4 mpg on a 1988 model from urban driving. Be advised that 4Runner rides and handles like a truck, not surprising since it's based on Toyota's 4x4 pickup, using the same chassis and powertrains. The ride is plagued by excessive bouncing and pitching, while the body leans heavily in turns. Off-road, the high ground clearance and sturdy suspension help 4Runner conquer rough terrain. The 4WD system lacks true shift-on-the-fly capability, discouraging intermittent use (you have to back up a few feet to disengage 4WD), and automatic locking front hubs aren't offered. Before you can engage 4WD, you have to lock the front hubs by hand. On the plus side, the removable rear roof is a feature not available elsewhere (though Jeep and Suzuki offer convertibles), there's plenty of passenger space and ample cargo room. Toyota offers many creature comforts as options on 4Runner, but the high cost of all those features makes the price of a loaded 4Runner comparable to a Jeep Cherokee, which has more power, greater payload and towing capacity, and a choice of two more convenient 4WD systems. In 4Runner's favor, it scored the highest among compact 4x4s in the most recent J.D. Power customer satisfaction survey.

Specifications

	3-door wagon
Wheelbase, in.	103.0
Overall length, in.	174.8
Overall width, in.	66.5
Overall height, in.	66.1
Front track, in.	56.3
Rear track, in.	55.5

KEY: ohv = overhead valve; **ohc** = overhead cam; **dohc** = double overhead cam; **I** = inline cylinders; **V** = cylinders in V configuration; **flat** = horizontally opposed cylinders; **bbl.** = barrel (carburetor); **PFI** = port (multi-point) fuel injection; **TBI** = throttle-body (single-point) fuel injection; **rpm** = revolutions per minute; **OD** = overdrive transmission; **S** = standard; **O** = optional; **NA** = not available.

Prices are accurate at time of printing; subject to manufacturer's change.

	3-door wagon
Turn diameter, ft.	37.4
Curb weight, lbs.	3605
Cargo vol., cu. ft.	80.9
Fuel capacity, gal.	17.2
Seating capacity	5
Front headroom, in.	39.2
Front shoulder room, in.	NA
Front legroom, max., in.	41.5
Rear headroom, in.	37.8
Rear shoulder room, in.	NA
Rear legroom, min., in.	34.3

Powertrain layout: longitudinal front engine/rear-wheel drive or on-demand 4WD.

Engines

	ohc I-4	ohc V-6
Size, liters/cu. in.	2.4/144	3.0/181
Fuel delivery	PFI	PFI
Horsepower @ rpm	116 @ 4800	150 @ 4800
Torque (lbs./ft.) @ rpm	140 @ 2800	180 @ 3400
Availability	S	O
EPA city/highway mpg		
5-speed OD manual	19/22	17/20
4-speed OD automatic	18/20	15/18

Prices

Toyota 4Runner	Retail Price	Dealer Invoice	Low Price
Deluxe 2-pass., 5-speed	$13988	$11890	$12939
Deluxe 2-pass., automatic	15158	12884	14021
Deluxe 5-pass., 5-speed	14668	12468	13568
Deluxe 5-pass., automatic	15838	13462	14650
Deluxe V6 2-pass., 5-speed	15078	12816	13947
Deluxe V6 2-pass., automatic	16248	13811	15030
SR5 V6 5-pass., 5-speed	17128	14473	15801
SR5 V6 5-pass., automatic	18248	15420	16834
SR5 5-pass., 5-speed	16318	13789	15054
SR5 5-pass., automatic	17438	14735	16087

Dealer invoice and destination charge may vary by region.

Standard Equipment:

Deluxe: 2.4-liter PFI 4-cylinder or 3.0-liter PFI V-6 engine, 5-speed manual or 4-speed automatic transmission, 2-speed transfer case, power steering (V-6 models), coolant temperature gauge, skid plates (front suspension, transfer case, fuel tank), cargo tiedown hooks, removable rear top, power rear windows, tinted side glass, 225/75R15 tires. **SR5** adds: 3.0-liter PFI V-6 engine, digital clock, rear defogger, coolant temperature and oil pressure gauges, voltmeter, tinted windshield, tilt steering column, variable intermittent wipers, rear wiper/washer, tinted glass, front vent windows, AM/FM ST ET.

Optional Equipment:

Air conditioning (NA Deluxe 2-pass.)	795	636	716
AM/FM ST ET, Deluxe	210	157	184
w/cassette, Deluxe 5-pass.	595	446	521
w/cassette, SR5	385	289	337
w/cassette & EQ, SR5	540	405	473
Power steering (std. SR5 & V6)	280	238	259

> **KEY: ohv** = overhead valve; **ohc** = overhead cam; **dohc** = double overhead cam; **I** = inline cylinders; **V** = cylinders in V configuration; **flat** = horizontally opposed cylinders; **bbl.** = barrel (carburetor); **PFI** = port (multi-point) fuel injection; **TBI** = throttle-body (single-point) fuel injection; **rpm** = revolutions per minute; **OD** = overdrive transmission; **S** = standard; **O** = optional; **NA** = not available.

	Retail Price	Dealer Invoice	Low Price
Moonroof w/sunshade, SR5	360	288	324
Fabric Sport Seats Pkg., SR5	290	232	261
Tilt wheel & variable int. wipers, Deluxe	120	102	111
Alloy wheels, SR5	480	384	432
Chrome wheels, SR5 w/Chrome Pkg.	220	176	198
Chrome Pkg., Deluxe 5-pass.	415	332	374
Deluxe 2-pass	330	264	297
SR5	220	176	198
Functional Upgrade Pkg., Deluxe	180	144	162
Tinted glass, front vent windows, digital clock.			
Cruise control/lights-on warning	245	196	221
Power Pkg., SR5	885	708	797
Power windows and locks, power mirrors, power antenna, fabric sunvisors; deletes vent windows.			
Rear heater, 5-pass. models	125	100	113
All Weather Guard Pkg., Deluxe	165	134	150

Volkswagen Cabriolet

1988 Volkswagen Cabriolet

What's New

Cabriolet, which received a facelift last year, returns unchanged this year. Sold in the U.S. since 1980, it is based on the Rabbit design that preceded the current Golf. Last year's facelift gave the Cabriolet new body-color bumpers, fender flares and rocker extensions, a new front spoiler and a new grille with round driving lamps mounted inboard of the headlamps. A manual folding top and integral roll bar are standard. The 4-seat Cabriolet is powered by a 90-horsepower 1.8-liter 4-cylinder engine and comes with a 5-speed manual or 3-speed automatic transmission.

For

- Performance
- Handling/roadholding
- Fuel economy

Against

- Rear seat room
- Cargo space
- Noise

Summary

Volkswagen's Rabbit-based Cabriolet is still one of the neatest convertibles around. It offers attractive styling, capable road manners and a sporting personality that makes it genuinely fun to drive. Its 1.8-liter engine lacks the sheer power of some newer multi-valve engines, yet it has a willing, free-revving nature and enough flexibility to pull well from low speeds. Fuel economy is generally in the 25-30 mpg range, which is quite acceptable given the sprightly performance, and better than just about all other convertibles.

The front-drive Cabriolet cuts through tight corners with confidence and has fine stability on the open road. Unlike most convertibles, the Cabriolet loses little body rigidity in the transformation from sedan to ragtop, partly due to the integral roll bar. The result is less body shake and fewer rattles than usual for a convertible. The convertible top installation takes up a lot of rear seat space, making it nearly impossible to fit adults in the back, and reduces the trunk to a modest rectangle with a small opening for loading cargo. Also, a deep exhaust is always noticeable and there's considerable wind noise even with the top up, making highway driving tiresome. One final caution: The Cabriolet has been highly popular among stereo thieves. They only have to cut through the top to get at the stereo, which is easy to remove from the dashboard. Check with your insurance agent about rates on a Cabriolet.

Specifications

	2-door conv.
Wheelbase, in.	94.5
Overall length, in.	153.1
Overall width, in.	64.6
Overall height, in.	55.6
Front track, in.	55.3
Rear track, in.	54.0
Turn diameter, ft.	31.2
Curb weight, lbs.	2274
Cargo vol., cu. ft.	6.5
Fuel capacity, gal.	13.8
Seating capacity	4
Front headroom, in.	37.4
Front shoulder room, in.	51.9
Front legroom, max., in.	39.4
Rear headroom, in.	35.6
Rear shoulder room, in.	50.4
Rear legroom, min., in.	31.0

Powertrain layout: transverse front engine/front-wheel drive.

Engines

	ohc I-4
Size, liters/cu. in.	1.8/109
Fuel delivery	PFI
Horsepower @ rpm	90 @ 5500
Torque (lbs./ft.) @ rpm	100 @ 3000
Availability	S

EPA city/highway mpg

5-speed OD manual	24/27
3-speed automatic	22/24

Prices

Volkswagen Cabriolet	Retail Price	Dealer Invoice	Low Price
2-door convertible, 5-speed	$15195	$13483	$14339
2-door convertible, automatic	15700	13948	14824
Bestseller 2-door convertible, 5-speed	15890	14142	15016
Bestseller 2-door convertible, automatic	16395	14607	15501
Boutique 2-door convertible, 5-speed	16740	14850	15795
Boutique 2-door convertible, automatic	17245	15315	16280
Destination charge	320	320	320

Standard Equipment:

1.8-liter PFI 4-cylinder engine, 5-speed manual or 3-speed automatic transmission, power brakes, cloth reclining front seats, cloth upholstery, folding rear seat, tachometer, coolant temperature gauge, trip odometer, rear defog-ger, tinted glass, intermittent wipers, front door pockets, remote mirrors, AM/FM ST ET cassette, 185/60HR14 tires. **Bestseller** adds: sport seats with driver's height adjustment, alloy wheels. **Boutique** adds: power steering, leather upholstery, leather-wrapped steering wheel, white alloy wheels.

Optional Equipment:	Retail Price	Dealer Invoice	Low Price
Power steering, base & Bestseller	290	250	270
Air conditioning	825	710	768
Cruise control	225	194	210
Metallic clearcoat paint, base & Bestseller	165	142	154

Volkswagen Fox

1988 Volkswagen Fox GL wagon

What's New

A 5-speed manual transmission arrived during the summer for Fox, Volkswagen's Brazilian-built, front-drive subcompact. Coming for 1989 are a new GL trim level for the 2-door sedan, formerly available only in base trim, and a removable glass sunroof as a new option. The 5-speed manual transmission is standard on the Fox GL Sport model that also has alloy wheels and a sport steering wheel. This model became available on the 4-door sedan as a 1988½ model, and will be available on the 2-door when 1989 models are introduced later this fall. The 5-speed will not be offered on the GL 3-door wagon. Fox debuted in 1987 and until now has been available only with a 4-speed overdrive manual transmission. The 4-speed remains standard on the base 2-door, GL 4-door and GL wagon. No word yet on when automatic transmission or power steering will be offered. All models use an 81-horsepower 1.8-liter 4-cylinder engine similar to the ones used in the Golf and Jetta, but mounted longitudinally instead of transversely.

For

- Fuel economy • Performance • Visibility

Against

- Shift linkage • Passenger room
- Cargo space (sedans)

Summary

We welcome the 5-speed manual since the widely-spaced gear ratios in the 4-speed manual leave huge gaps between the gears, so it's hard to find the right one for many occa-

Prices are accurate at time of printing; subject to manufacturer's change.

Volkswagen

sions. Even so, Fox still easily outruns most other subcompacts. This version of VW's 1.8-liter 4-cylinder lacks neck-snapping power, but compensates with a free-revving, hard-working nature. Fuel economy is impressive for the lively performance; we averaged nearly 28 mpg in urban driving with a 1988 wagon. On models we've tested, shift action was fine in the forward gears, but getting into reverse at times was a real wrestling match. Maybe that will improve with the 5-speed. GL models come with wider tires than base models for improved cornering grip. Even the base models feel sportier than most subcompacts and have a firm, well-controlled ride. In size, Fox falls between the Golf hatchbacks and the Jetta sedans in overall length, but has a shorter wheelbase, 92.8 inches versus 97.3 for Golf and Jetta, and is narrower overall. Head room is limited in front, so most drivers have to recline the seatback more than usual. Trunk space is skimpy on the sedans and just adequate on the wagon. It's nice to get a full-size spare tire, but it takes up a lot of cargo space in all three body styles. Fox is still VW's least expensive U.S. model, but the company has had to raise prices because of rising labor costs in Brazil, plus an unfavorable currency exchange rate. We're also convinced that assembly quality on the Brazilian-made Fox doesn't match VW's German-made cars. In its favor, Fox is more fun to drive than most other cars in its size and price range.

Specifications

	2-door notchback	4-door notchback	3-door wagon
Wheelbase, in.	92.8	92.8	92.8
Overall length, in.	163.4	163.4	163.4
Overall width, in.	63.0	63.0	63.9
Overall height, in.	53.7	53.7	54.5
Front track, in.	53.1	53.1	53.1
Rear track, in.	53.9	53.9	53.9
Turn diameter, ft.	31.5	31.5	31.5
Curb weight, lbs.	2126	2203	2203
Cargo vol., cu. ft.	9.9	9.9	61.8
Fuel capacity, gal.	12.4	12.4	12.4
Seating capacity	4	4	4
Front headroom, in.	36.6	36.6	36.6
Front shoulder room, in.	51.7	51.7	51.7
Front legroom, max., in.	41.1	41.1	41.1
Rear headroom, in.	35.4	35.4	35.8
Rear shoulder room, in.	52.1	51.1	51.5
Rear legroom, min., in.	30.2	30.2	30.2

Powertrain layout: longitudinal front engine/front-wheel drive.

Engines

	ohc I-4
Size, liters/cu. in.	1.8/109
Fuel delivery	PFI
Horsepower @ rpm	81 @ 5500
Torque (lbs./ft.) @ rpm	93 @ 3250
Availability	S

EPA city/highway mpg

4-speed OD manual	25/30
5-speed OD manual	24/29

KEY: ohv = overhead valve; **ohc** = overhead cam; **dohc** = double overhead cam; **I** = inline cylinders; **V** = cylinders in V configuration; **flat** = horizontally opposed cylinders; **bbl.** = barrel (carburetor); **PFI** = port (multi-point) fuel injection; **TBI** = throttle-body (single-point) fuel injection; **rpm** = revolutions per minute; **OD** = overdrive transmission; **S** = standard; **O** = optional; **NA** = not available.

Prices

Volkswagen Fox	Retail Price	Dealer Invoice	Low Price
2-door notchback	$6590	$5966	$6278
GL 4-door sedan	7640	6796	7218
GL 3-door wagon	7770	6911	7340
GL Sport 4-door notchback	8115	7217	7666
Destination charge	320	320	320

Dealer invoice and low price not available at time of pubication.

Standard Equipment:

1.8-liter PFI 4-cylinder engine, 4-speed manual transmission, power brakes, reclining front bucket seats, tweed upholstery, tinted glass, rear defogger, intermittent wipers, carpeting, padded steering wheel, trip odometer, coolant temperature gauge, clock, swing-out rear side windows, remote left mirror, front door storage pockets, bodyside moldings, 155/80SR13 tires on steel wheels with hubcaps and full-size spare. **GL** adds: tachometer (except wagon), LCD digital clock (wagon has analog clock), velour upholstery, three-point rear seatbelts, swivel map light, trunk carpeting, wider bodyside moldings, full wheel covers, 175/70SR13 tires. **GL Sport** adds: 5-speed manual transmission, alloy wheels.

Optional Equipment:

Heavy-Duty Cooling Pkg.	85	71	78
Air conditioning	685	575	630
AM/FM stereo cassette	415	349	382
4-speaker radio prep	110	92	101
Rear wiper/washer, wagon	145	122	134

Volkswagen Golf/Jetta

1988 Volkswagen Golf 3-door

What's New

Anti-lock brakes are a new option this year on two upper-level Jetta models, the Carat and GLI 16V. The $995 Teves-designed anti-lock system is scheduled to be on sale by December. Volkswagen expects to eventually offer it on the Golf-based GTI 16V, which, like the Carat and GLI, has 4-wheel disc brakes standard. With the Scirocco sport coupe discontinued for 1989, the GTI is VW's main performance model, using a 123-horsepower, dual-cam 1.8-liter engine. The top-line Jetta Carat becomes the stand-in for the Quantum, also dropped for '89, as a sport/luxury sedan. Carat equipment includes a 105-horsepower 1.8-liter engine, air conditioning and power locks, windows and mirrors. Golf and Jetta are built from the same front-drive design, but Golf comes as 3- and 5-door hatchbacks and Jetta as 2- and 4-door sedans. This year's Golf lineup consists of base and GL hatchbacks with a 100-horsepower 1.8-liter 4-cylinder engine, available with either a 5-speed manual or 3-speed automatic transmission, and the GTI 16V, avail-

Prices are accurate at time of printing; subject to manufacturer's change.

able only as a 3-door with manual transmission. Last year's Golf GT hatchbacks have been dropped. Jetta's roster includes base 2- and 4-door sedans with the 100-horsepower engine and either manual or automatic transmission; GL and Carat 4-door models with the 105-horsepower engine and same choice of transmissions; and sporty GLI 16V with the 123-horsepower, 16-valve engine from the GTI and 5-speed manual only.

For

- ● Performance ● Fuel economy ● Cargo space
- ● Anti-lock brakes option ● Passenger room
- ● Handling/roadholding

Against

- ● Noise ● Entry/exit

Summary

Brisk performance, good fuel economy, capable front-drive handling, a well-controlled ride, ample passenger room and generous cargo space are standard equipment on all Golfs and Jettas. The optional anti-lock brakes make Jettas even more tempting. There isn't a huge difference between the 100- and 105-horsepower engines, so don't think you have to avoid the base models for quick acceleration. Performance and fuel economy suffer noticeably with the optional automatic transmission, so we recommend the 5-speed manual. The Jetta Carat, Golf GTI and Jetta GLI ride on larger wheels and tires than other models and have firmer suspensions, so they have the best handling. Here, too, you won't lose that much by choosing a less expensive version. Road noise can be intrusive because of inadequate sound insulation. Golf and Jetta are compacts based on EPA interior volume, but really subcompacts if you look at the 97.3-inch wheelbase. The modest wheelbase makes for narrow rear doorways on the 5-door hatchbacks and 4-door sedans. Even so, all models offer good head room and at least adequate leg room front and rear. In addition, fold-down rear seats on Golfs increase cargo capacity, while Jettas have a cavernous trunk with a low liftover for easy loading. These cars are among the most frequently broken into because their stereos can be popped out of the dashboard in seconds. Volkswagen has been selling anti-theft stereos the past couple of years, but the word hasn't gotten around to everyone in the "smash-and-grab" crowd. Ask your insurance agent about rates on one of these; the lack of 5-mph bumpers doesn't help either.

Specifications

	Jetta 2-door notchback	Jetta 4-door notchback	Golf 3-door hatchback	Golf 5-door hatchback
Wheelbase, in.	97.3	97.3	97.3	97.3
Overall length, in.	171.7	171.7	158.0	158.0
Overall width, in.	65.5	65.5	65.5	65.5
Overall height, in.	55.7	55.7	55.7	55.7
Front track, in.	56.3	56.3	56.3	56.3
Rear track, in.	56.0	56.0	56.0	56.0
Turn diameter, ft.	34.4	34.4	34.4	34.4
Curb weight, lbs.	2305	2305	2194	2246
Cargo vol., cu. ft.	16.6	16.6	39.6	39.6
Fuel capacity, gal.	14.5	14.5	14.5	14.5
Seating capacity	5	5	5	5
Front headroom, in.	38.1	38.1	38.1	38.1
Front shoulder room, in.	53.3	53.3	53.3	53.3
Front legroom, max., in.	39.5	39.5	39.5	39.5
Rear headroom, in.	37.1	37.1	37.5	37.5

	Jetta 2-door notchback	Jetta 4-door notchback	Golf 3-door hatchback	Golf 5-door hatchback
Rear shoulder room, in.	53.3	53.3	54.3	54.3
Rear legroom, min., in.	35.1	35.1	34.4	34.4

Powertrain layout: transverse front engine/front-wheel drive.

Engines

	ohc I-4	ohc I-4	dohc I-4
Size, liters/cu. in.	1.8/109	1.8/109	1.8/109
Fuel delivery	PFI	PFI	PFI
Horsepower @ rpm	100 @ 5400	105 @ 5400	123 @ 5800
Torque (lbs./ft.) @ rpm	107 @ 3400	110 @ 3400	120 @ 4250
Availability	S[1]	S[2]	S[3]

EPA city/highway mpg

5-speed OD manual	25/34	25/34	22/29
3-speed automatic	23/28	23/28	

1. Base Golf and Jetta, Golf GL   2. Jetta GL and Carat   3. GTI and GLI

Prices

Volkswagen Golf	Retail Price	Dealer Invoice	Low Price
3-door hatchback, 5-speed	$8465	$7823	$8144
3-door hatchback, automatic	8970	8293	8632
GL 3-door hatchback, 5-speed	9170	8242	8706
GL 3-door hatchback, automatic	9675	8712	9194
GL 5-door hatchback, 5-speed	9380	8431	8906
GL 5-door hatchback, automatic	9885	8901	9393
GTI 16V 3-door hatchback, 5-speed	13650	11979	12815
Destination charge	320	320	320

Standard Equipment:

1.8-liter PFI 4-cylinder engine, 5-speed manual or 3-speed automatic transmission, power brakes, reclining front bucket seats, cloth upholstery, folding rear seat, tinted glass, intermittent wipers, rear defogger, rear wiper, analog clock, trip odometer, coolant temperature gauge, console, left remote mirror, wide bodyside moldings, 175/70SR13 tires. **GL** adds: digital clock, upgraded door trim with pockets, tachometer, dual remote mirrors, automatic seatbelts. **GTI 16V** adds: DOHC 16-valve engine, close-ratio 5-speed manual transmission, power steering, 4-wheel disc brakes, tachometer, trip computer, courtesy light with delay, lighted right visor mirror, sport seats with driver's height adjustment, 60/40 rear seat, leather-wrapped steering wheel, sport suspension, 205/55VR14 tires on alloy wheels.

Optional Equipment:

Air conditioning	805	692	749
Power steering, base & GL	275	231	253
Cruise control, GL	225	194	210
Floormats, GL, GTI 16V	50	43	47
Metallic clearcoat paint, GL	165	142	154
Mica paint, GTI 16V	210	180	195
AM/FM ST ET cassette, base	520	447	484
GL, GTI 16V	585	503	544
4-speaker radio prep	165	142	154
Manual sliding sunroof	395	340	368
Split rear seat, GL	150	129	140
Power Pkg., GTI 16V	605	552	579
Power windows, locks and mirrors.			

Volkswagen Jetta			
2-door notchback, 5-speed	$9690	$8950	$9320
2-door notchback, automatic	10195	9415	9805
4-door notchback, 5-speed	9910	9152	9531
4-door notchback, automatic	10415	9617	10016
GL 4-door notchback, 5-speed	11120	9820	10470
GL 4-door notchback, automatic	11625	10285	10955
Carat 4-door notchback, 5-speed	15140	13358	14249
Carat 4-door notchback, automatic	15545	13823	14684
GLI 16V 4-door notchback, 5-speed	14770	13032	13901
Destination charge	320	320	320

Prices are accurate at time of printing; subject to manufacturer's change.

Volkswagen

Standard Equipment:

1.8-liter PFI 4-cylinder engine, 5-speed manual or 3-speed automatic transmission, power brakes, reclining front bucket seats, cloth upholstery, rear defogger, tachometer, coolant temperature gauge, trip odometer, front door pockets, dual remote mirrors, wide bodyside moldings, intermittent wipers, 175/70SR13 SBR tires. **GL** adds: automatic front seatbelts, rear armrest, time-delay courtesy light, visor mirrors with lighted right, velour upholstery. **Carat** adds: 105-bhp engne, 4-wheel disc brakes, power steering, air conditioning, cruise control, AM/FM cassette, power windows and locks, power mirrors, metallic paint, manual seatbelts, sport seats with height adjustment, ski sack, leather-wrapped steering wheel, sport suspension, 185/60HR14 tires on alloy wheels. **GLI 16V** deletes air conditioning, cruise control, radio, power windows, locks and mirrors and adds: DOHC 16-valve engine, Recaro seats with height adjustment, trip computer.

Optional Equipment:	Retail Price	Dealer Invoice	Low Price
Anti-lock brakes, GLI & Carat	995	856	926
Air conditioning	805	692	749
Power Pkg.	705	644	675
Power windows, locks and mirrors.			
Power steering, base	280	241	261
Cruise control, GL & GLI 16V	225	194	210
Metallic clearcoat paint (std. Carat)	165	142	154
AM/FM ST ET cassette, base	520	447	484
GL & GTI 16V	585	503	544
4-speaker radio prep	215	185	200
Manual sliding sunroof	395	340	368
Cast alloy wheels, GL	460	377	419
Forged alloy wheels, Carat	210	180	195

Volkswagen Vanagon

1988 Volkswagen Vanagon GL

What's New

All models get new bumpers and a plush Carat model is supposed to be added for Vanagon, the pioneer of the compact-van market segment. Content of the new Carat will be similar to that of last year's Wolfsburg model, which included power locks and mirrors, two rear-facing seats and a rear bench seat that converts to a bed. The 1989 lineup looks like this: rear-drive, 7-passenger GL and Carat models; 4-wheel-drive, 7-passenger GL Syncro; rear-drive Camper; and 4WD Camper Syncro. Syncro's 4WD system sends 95 percent of engine power to the rear wheels on smooth, dry pavement, but in slippery conditions automatically delivers sufficient power to the front wheels to stabilize traction. Camper models seat four, plus they have a 2-place bunk that pops up from the roof, a refrigerator, a sink and built-in cabinets. All models use a 90-horsepower 2.1-liter flat 4-cylinder engine mounted at the rear. A 4-speed manual is standard on all versions and a 3-speed

automatic is optional on rear-drive Vanagons. A sliding side door and rear liftgate are standard.

For

- Passenger room
- Cargo space
- 4WD traction
- Ride

Against

- Fuel economy
- Performance
- Entry/exit
- Driving position

Summary

All Vanagons have bountiful passenger and cargo space, the 4WD Syncros provide impressive traction and the Camper models are uniquely outfitted as recreational vehicles, so Volkswagen's compact van lineup boasts an unusual array of features. The prices are steep on Syncros and Campers, but the only other compact van currently available with 4WD is the Toyota Van, and it's hardly cheap. Vanagon is one of the best riding vans on the market, with good bounce control and excellent highway stability. It's also the tallest compact van, making it prone to the wind on open roads. That can make this vehicle a handful on a snow-packed road, but the Syncro's 4WD system quickly solves traction problems in the nastiest conditions. Vanagon's biggest problem is lack of power, which is especially apparent on the Syncros because they weigh 320-330 pounds more than the rear-drive models. With only 95 horsepower for at least 3500 pounds of curb weight, there's too little engine for too much van. Mileage is no bargain either. The best we've seen was 19.2 mpg from mostly expressway driving; the worst was 14.2 mpg in the city. The tall styling makes it hard to get in or out of the front seats, since you have to go through narrow doorways that are a steep step off the ground. The driving position is bus-like behind the fixed steering wheel, forcing most drivers to sit too close to the pedals. Vanagon is still a great way to haul six or seven people; all will have plenty of head and leg room, and there's space to walk around inside. Even with all the seats in place cargo room is ample. With the rear and/or middle seats removed, Vanagon has a cavernous cargo hold that easily tops all competitors.

Specifications	4-door van	Syncro 4-door van
Wheelbase, in.	96.9	96.7
Overall length, in.	179.9	179.9
Overall width, in.	72.6	72.6
Overall height, in.	75.9	78.3
Front track, in.	62.4	61.7
Rear track, in.	61.8	61.4
Turn diameter, ft.	35.8	35.8
Curb weight, lbs.	3460[1]	3780[2]
Cargo vol., cu. ft.	201	201
Fuel capacity, gal.	15.9	18.4
Seating capacity	7	7
Front headroom, in.	NA	NA
Front shoulder room, in.	NA	NA
Front legroom, max., in.	NA	NA
Rear headroom, in.	NA	NA
Rear shoulder room, in.	NA	NA
Rear legroom, min., in.	NA	NA

1. 3622 lbs., Camper.   2. 3942 lbs., Camper.

Powertrain layout: longitudinal rear engine/rear-wheel drive or automatic 4WD.

Prices are accurate at time of printing; subject to manufacturer's change.

CONSUMER GUIDE®

Engines

	ohv flat-4
Size, liters/cu. in.	2.1/129
Fuel delivery	PFI
Horsepower @ rpm	90 @ 4800
Torque (lbs./ft.) @ rpm	117 @ 3200
Availability	S
EPA city/highway mpg	
4-speed OD manual	18/19
3-speed automatic	17/18

Prices

Volkswagen Vanagon	Retail Price	Dealer Invoice	Low Price
GL 4-speed	$17035	$15026	$16031
GL automatic	17550	15501	16526
GL Syncro 4-speed	20560	18128	19344
Carat 4-speed	19355	17067	18211
Carat automatic	19870	17542	18706
Camper GL, 4-speed	22235	19602	20919
Camper GL, automatic	22750	20077	21414
Camper GL Syncro, 4-speed . . .	25760	22704	24232
Destination charge	320	320	320

Standard Equipment:

GL: 2.1-liter PFI 4-cylinder engine, 4-speed manual transmission, power steering and brakes, 7-passenger seating with reclining front buckets, tachometer, coolant temperature gauge, trip odometer, digital clock, tinted glass, 185/80R14 tires; **Syncro** has 4-wheel drive and 205/70R14 tires. **Carat** adds: two rear-facing seats, rear seat that converts to bed, power locks and mirrors, clearcoat metallic paint, rear wiper/washer, front spoiler, folding table, 205/70R14 tires on alloy wheels. **Camper** adds to GL: 4-passenger seating, utensil and rear ceiling cabinets, removable refrigerator, rear insect screen, roof bunk (sleeps two), rear bench seat (converts to bed), folding side table, pop-up top, 110-volt electrical hookup, sink.

Optional Equipment:

Air conditioning	1030	991	1011
Cruise control	225	194	210
Power mirrors	190	164	177
Metallic clearcoat paint	390	336	363
AM/FM ST ET cassette	625	538	582
4-speaker radio prep	215	185	200
205/70R14 tires/alloy wheels, exc. Syncro .	460	377	419
Alloy wheels, Syncro	365	299	332
Power windows, Carat	235	202	219
Rear wiper/washer (std. Carat) . . .	190	164	177
Power Pkg., exc. Carat	470	404	437
Power windows and locks.			
Weekender Pkg., 7-seaters exc. Carat . . .	305	262	284

Volvo 240

What's New

Rear headrests are now standard on all 240 models, which comprise DL and GL price levels in 4-door sedan and 5-door wagon body styles. DL models also have new wheel covers and the DL wagon offers cloth upholstery as a no-cost option to the standard vinyl. DL models can now be ordered with optional factory-installed power windows. Previously, power windows were available only as a dealer-installed option. With the factory-installed option, the power window controls will be mounted on the armrests, rather than the dashboard. All models have rear-wheel drive and a 114-horsepower 2.3-liter 4-cylinder engine. The DLs are available with a 5-speed manual or 4-speed automatic transmission; the GLs come only with the automatic.

Volvo 240 DL

For

● Passenger room ● Handling ● Ride ● Visibility

Against

● Performance ● Price

Summary

At more than $17,000, the cheapest Volvo isn't cheap at all, though there are lesser cars that cost more. While there's nothing special about the 240's credentials, we applaud its overall competence. The 114-horsepower engine feels surprisingly adequate in the 3000-pound 240 sedan, so you can easily keep pace with traffic and get brisk acceleration by using a heavy throttle foot. Alas, there's a serious loss of vitality when the air conditioner is on, making the engine sluggish. On the plus side, it's an economical engine; we averaged 23 mpg with a 1988 DL 4-door/5-speed from mostly city and suburban driving. That's better than we've achieved with smaller, lighter cars. The base Volvo is competent and reassuring on the road: The responsive steering has good centering action while the firm suspension absorbs bumps pretty well without being floppy, so you get a stable ride over most surfaces. The 4-wheel disc brakes have good stopping power and control, though anti-lock brakes aren't offered on the 240s. Anti-lock brakes aren't the only thing missing; there should be more here for $17,000—a stronger engine, a more luxurious interior, an air bag—something to make this car stand above others in the same price range. However, when we consider that the 240 has a fine reputation for durability and Volvo backs its cars with generous warranties, and then we add in other virtues such as the roomy and comfortable interior, we conclude there's enough here to justify the high price.

Specifications

	4-door notchback	5-door wagon
Wheelbase, in.	104.3	104.3
Overall length, in.	189.9	190.7
Overall width, in.	67.3	67.7
Overall height, in.	56.3	57.1
Front track, in.	56.3	56.3
Rear track, in.	53.5	53.5
Turn diameter, ft.	32.2	32.2

KEY: ohv = overhead valve; **ohc** = overhead cam; **dohc** = double overhead cam; **I** = inline cylinders; **V** = cylinders in V configuration; **flat** = horizontally opposed cylinders; **bbl.** = barrel (carburetor); **PFI** = port (multi-point) fuel injection; **TBI** = throttle-body (single-point) fuel injection; **rpm** = revolutions per minute; **OD** = overdrive transmission; **S** = standard; **O** = optional; **NA** = not available.

Prices are accurate at time of printing; subject to manufacturer's change.

	4-door notchback	5-door wagon
Curb weight, lbs.	2919	3051
Cargo vol., cu. ft.	14.0	76.0
Fuel capacity, gal.	15.8	15.8
Seating capacity	5	5
Front headroom, in.	37.9	37.9
Front shoulder room, in.	NA	NA
Front legroom, max., in.	40.1	40.1
Rear headroom, in.	36.1	36.8
Rear shoulder room, in.	NA	NA
Rear legroom, min., in.	36.4	36.1

Powertrain layout: longitudinal front engine/rear-wheel drive.

Engines

	ohc I-4
Size, liters/cu. in.	2.3/140
Fuel delivery	PFI
Horsepower @ rpm	114 @ 5400
Torque (lbs./ft.) @ rpm	136 @ 2750
Availability	S

EPA city/highway mpg

5-speed OD manual	21/27
4-speed OD automatic	19/24

Prices

Volvo 240	Retail Price	Dealer Invoice	Low Price
DL 4-door notchback, 5-speed	$17250	—	—
DL 4-door notchback, automatic	17820	—	—
DL 5-door wagon, 5-speed	17740	—	—
DL 5-door wagon, automatic	18310	—	—
GL 4-door notchback, automatic	20035	—	—
GL 5-door wagon, automatic	20775	—	—
Destination charge	350	350	350

Dealer invoice and low price not available at time of publication.

Standard Equipment:

DL: 2.3-liter PFI 4-cylinder engine, 5-speed manual or 4-speed automatic transmission, power steering, power 4-wheel disc brakes, air conditioning, power door locks, analog clock, trip odometer, rear defogger, tinted glass, dual remote mirrors, driver's seat height adjustment, cloth reclining heated front bucket seats, rear head restraints, AM/FM ST ET cassette, 185/70R14 tires (4-door), 185R14 tires (wagon). **GL** adds: 4-speed automatic transmission, manual sunroof (4-door), leather upholstery (wagon), power windows, tachometer, alloy wheels.

Optional Equipment:

Power windows, DL	275	—	—
Metallic paint	380	—	—
Leather upholstery, GL 4-door	785	—	—

Volvo 740/760/780

What's New

A new 16-valve engine becomes standard on the 740 GLE sedan and wagon after January 1, while new 740 GL models have been slotted below the GLEs in equipment and price. The limited-production 780 coupe gets a stronger engine, a 175-horsepower turbocharged version of Volvo's venerable 2.3-liter 4-cylinder. The new engine for the 740

Volvo 740 GL

GLE uses the same block as last year's 2.3-liter 4-cylinder, but is topped by a dual-cam, 16-valve cylinder head that boosts horsepower from 114 to 153. In addition, the new engine has twin, counter-rotating balance shafts for smoother, quieter operation. The 16-valve engine is available with a 5-speed manual or a 4-speed automatic transmission, which gains a lockup torque converter this year. Other changes for the 740 GLE are that a driver's-side air bag is standard instead of optional and tire size has been increased from 185/70R14 to 185/65R15. The new base model in this range is the 740 GL, with a 114-horsepower, single-cam, 8-valve engine. The GL comes with either the 5-speed manual or 4-speed automatic, and has as standard air conditioning, power windows and a stereo with cassette player. To keep the price down, steel wheels replace alloy wheels, the outside mirrors have manual remote control instead of electric, and the sunroof is manual instead of power. Carryover models include the 740 Turbo sedan and wagon, 760 GLE sedan, and 760 Turbo sedan and wagon. Turbo models use a 160-horsepower turbocharged 2.3, while the 760 GLE uses a 144-horsepower 2.8-liter V-6. Volvo's highest-price model, the Bertone-built 780 2-door, is available this year with a turbocharged 2.3 that benefits from different electronic engine controls to boost horsepower to 175 and provide a flatter torque curve, though peak torque is unchanged at 187. The same revised engine management system is sold over-the-counter by Volvo dealers as the "Turbo +" kit. A 780 with the 144-horsepower V-6 engine returns from last year; only automatic transmission is available with either 780 engine. Also new this year for the 780 are spoke-type alloy wheels. All models are built off the same rear-drive chassis.

For

- Performance (turbos, V-6) • Anti-lock brakes
- Air bag • Handling • Passenger room • Cargo space

Against

- Performance (GL) • Price

Summary

Volvo's 700-Series cars now come in five performance levels, and since we haven't driven the new 16-valve engine or the 175-horsepower turbocharged engine, we can only comment on three of them. The 114-horsepower 740 GLs are underpowered for this price; nice cars, but we want more for our $20,000. The 160-horsepower turbocharged engine suffers turbo lag at midrange speeds, but is at home cruising in the high-speed European tradition. Passing power is sensational; at around 55 mph, the turbo kicks in at a touch of the accelerator. The V-6 isn't as muscular as the turbo engine,

Prices are accurate at time of printing; subject to manufacturer's change.

but it's smoother and quieter, and doesn't suffer any lag before delivering more power. All models are highly capable otherwise; their firm suspensions provide reassuring high-speed stability and crisp cornering ability, while the steering is precise and responsive. Anti-lock brakes are standard on all but the GL, where they're optional. The same is true of a driver's-side air bag, and even with it the steering column is height-adjustable to three positions (most cars with air bags have fixed steering columns). Overall, these high-priced Swedish cars are refined, roomy and well built, and they are backed by generous warranties.

Specifications

	780 2-door notchback	740/760 4-door notchback	740/760 5-door wagon
Wheelbase, in.	109.1	109.1	109.1
Overall length, in.	188.8	188.4	188.4
Overall width, in.	69.3	69.3	69.3
Overall height, in.	55.1	55.5	56.5
Front track, in.	57.9	57.9	57.9
Rear track, in.	57.5	57.5	57.5
Turn diameter, ft.	32.2	32.2	32.2
Curb weight, lbs.	3415	2954	3082
Cargo vol., cu. ft.	14.9	16.8	74.9
Fuel capacity, gal.	15.8	15.8[1]	15.8
Seating capacity	4	5	5
Front headroom, in.	37.2	38.6	38.6
Front shoulder room, in.	NA	NA	NA
Front legroom, max., in.	41.0	41.0	41.0
Rear headroom, in.	35.8	37.1	37.6
Rear shoulder room, in.	NA	NA	NA
Rear legroom, min., in.	34.7	34.7	34.7

1. 21.0 gals with turbo engine or V-6.

Powertrain layout: longitudinal front engine/rear-wheel drive.

Engines

	ohc I-4	dohc I-4	Turbo ohc I-4	ohc V-6
Size, liters/cu. in.	2.3/140	2.3/140	2.3/140	2.8/176
Fuel delivery	PFI	PFI	PFI	PFI
Horsepower @ rpm	114 @ 5400	153 @ 5700	160 @ 5300[1]	144 @ 5100
Torque (lbs./ft.) @ rpm	136 @ 2750	150 @ 4450	187 @ 2900	173 @ 3750
Availability	S[2]	S[3]	S[4]	S[5]

EPA city/highway mpg

4-speed manual + OD			20/25	
5-speed OD manual	21/27	NA/NA		
4-speed OD automatic	20/26	NA/NA	19/23	17/21

1. 175 horsepower in 780 2. 740 GL 3. 740 GLE 4. Turbo models
5. 760 GLE, 780

Prices

Volvo 740/760/780

	Retail Price	Dealer Invoice	Low Price
740 GL 4-door notchback, 5-speed	$19985	—	—
GL 4-door notchback, automatic	20555	—	—
740 GL 5-door wagon, 5-speed	20665	—	—
740 GL 5-door wagon, automatic	21235	—	—
740 GLE 4-door notchback, 5-speed	NA	—	—
740 GLE 4-door notchback, automatic	NA	—	—
740 GLE 5-door wagon, 5-speed	NA	—	—
740 GLE 5-door wagon, automatic	NA	—	—
740 Turbo 4-door notchback, 5-speed	24925	—	—
740 Turbo 4-door notchback, automatic	25405	—	—
740 Turbo 5-door wagon, 5-speed	25605	—	—
740 Turbo 5-door wagon, automatic	26085	—	—
760 GLE 4-door notchback	32155	—	—
760 GLE Turbo 4-door notchback	32940	—	—
760 GLE Turbo 5-door wagon	32940	—	—
780 2-door notchback	37790	—	—
780 Turbo 2-door notchback	38975	—	—
Destination charge	350	350	350
Gas Guzzler Tax, 760 & 780 V-6	500	500	500

Dealer invoice and low price not available at time of publication.

Standard Equipment:

740 GL: 2.3-liter PFI 4-cylinder engine, power steering, power 4-wheel disc brakes, air conditioning, cloth reclining front bucket seats, power windows and locks, tachometer, coolant temperature gauge, trip odometer, analog clock, AM/FM ST ET cassette, power antenna, 185/70R14 tires. **740 GLE** adds: DOHC 16-valve engine, anti-lock braking system, Supplemental Restraint System, cruise control, 185/65R15 tires on alloy wheels. **740 Turbo** adds: turbocharged, intercooled engine, 4-speed manual plus overdrive or 4-speed automatic transmission, power sunroof, power mirrors, velour and leather upholstery, 195/60R15 tires. **760** adds: 2.8-liter PFI V-6 or 2.3-liter turbocharged 4-cylinder engine, 4-speed automatic transmission, independent rear suspension, automatic air conditioning, leather upholstery, power seats, tilt steering column, upgraded stereo with EQ, front map lights, rear reading lights (Turbo). **780** adds: power moonroof, elm burl accents.

Optional Equipment:

Anti-lock brakes, 740 GL	1175	—	—
Supplemental Restraint System, 740 GL	850	—	—
Pearlescent paint, 780	320	—	—

Yugo

1988 Yugo GVX

What's New

Yugo plans to offer an automatic transmission and introduce a convertible during the 1989 model year. Eventually, Yugo plans to introduce a new 4-door sedan that's larger than the current models. Yugo is now part of GM—Global Motors, not General Motors—which also plans to import a Malaysian-built version of the Mitsubishi Mirage 4-door called the Proton Saga. The Proton Saga is in addition to the planned Yugo-badged sedan. Yugo's lineup at the end of 1988 included the base GV (the original Yugo), and more expensive GVL and GVS versions of the same car. All three use the same front-drive chassis, 3-door hatchback body, 52-horsepower 1.1-liter 4-cylinder engine and 4-speed manual transmission. New for a spring 1988 release was

KEY: ohv = overhead valve; **ohc** = overhead cam; **dohc** = double overhead cam; **I** = inline cylinders; **V** = cylinders in V configuration; **flat** = horizontally opposed cylinders; **bbl.** = barrel (carburetor); **PFI** = port (multi-point) fuel injection; **TBI** = throttle-body (single-point) fuel injection; **rpm** = revolutions per minute; **OD** = overdrive transmission; **S** = standard; **O** = optional; **NA** = not available.

Prices are accurate at time of printing; subject to manufacturer's change.

Yugo

the GVX, powered by a 64-horsepower 1.3-liter four with a 5-speed overdrive manual transmission. The GVX, the first Yugo priced at more than $5000, also has a front air dam, fog lamps, aero body trim and 13-inch tires instead of 12-inch. A "GVX Euro" package planned for next spring includes more prominent air dams and monochromatic white paint. A 3-speed automatic transmission is supposed to be introduced this year and eventually be available on all models, though different transmissions may be offered for the 1.1- and 1.3-liter engines. Also coming during the 1989 model year—maybe in the fall, maybe later—is the GVC, a convertible based on the GV design. Yugo is estimating the GVC's base price at $8300, which would make it the lowest-priced convertible car in the U.S. The GVC will use the 1.3-liter engine and 5-speed manual transmission, plus it will have a power folding top and a heated glass rear window.

For

- Price
- Fuel economy

Against

- Interior room
- Shift linkage
- Noise
- Ride
- Driving position

Summary

We tested a GVX this summer and we have to admit, it wasn't as bad as we expected. We were surprised by the 1.3-liter engine's adequate performance, commendable smoothness and good gas mileage (31.4 mpg overall). Now for the rest of the story: The shift linkage was vague and rubbery; the ride was stiff and the little Yugo could easily be jarred off course by sharp bumps; the steering wheel was angled so far forward to make normal steering awkward; the driver's seat had excessive play in the seatback; road, wind and engine noise were quite high; the cramped interior had thinly padded seats, cheap materials, confusing heat/vent system controls and no glovebox on the dashboard. Well, what can you expect for about $4000? This was not a $4000 Yugo, but a $7200 Yugo (including optional air conditioning and stereo, plus the destination charge), and that's not counting the dealer-installed sunroof and body stripes. Once you're over $7000, that's budget Toyota Tercel, Dodge Omni/Plymouth Horizon, Nissan Sentra territory. Any of those are a better bet for the long haul than a Yugo.

Specifications

	3-door hatchback
Wheelbase, in.	84.6
Overall length, in.	139.0
Overall width, in.	60.7
Overall height, in.	54.7
Front track, in.	51.5
Rear track, in.	51.7
Turn diameter, ft.	31.2
Curb weight, lbs.	1832
Cargo vol., cu. ft.	27.5
Fuel capacity, gal.	8.4

KEY: ohv = overhead valve; **ohc** = overhead cam; **dohc** = double overhead cam; **I** = inline cylinders; **V** = cylinders in V configuration; **flat** = horizontally opposed cylinders; **bbl.** = barrel (carburetor); **PFI** = port (multi-point) fuel injection; **TBI** = throttle-body (single-point) fuel injection; **rpm** = revolutions per minute; **OD** = overdrive transmission; **S** = standard; **O** = optional; **NA** = not available.

	3-door hatchback
Seating capacity	4
Front headroom, in.	37.0
Front shoulder room, in.	NA
Front legroom, max., in.	39.0
Rear headroom, in.	36.0
Rear shoulder room, in.	NA
Rear legroom, min., in.	NA

Powertrain layout: transverse front engine/front-wheel drive.

Engines

	ohc I-4	ohc I-4
Size, liters/cu. in.	1.1/68	1.3/79
Fuel delivery	2 bbl.	2 bbl.
Horsepower @ rpm	52 @ 5000	64 @ 5800
Torque (lbs./ft.) @ rpm	52 @ 4600	68 @ 4000
Availability	S[1]	S[2]

EPA city/highway mpg

4-speed manual	28/31	
5-speed OD manual		26/29

1. GV, GVL, GVS   2. GVX

Prices

Yugo (1988 prices)	Retail Price	Dealer Invoice	Low Price
GV 3-door hatchback	$4349	$3979	$4164
GVL 3-door hatchback	4599	4158	4379
GVS 3-door hatchback	4699	4230	4465
GVX 3-door hatchback	5699	5149	5424
Destination charge	400	400	400

Standard Equipment:

GV: 1.1-liter 2bbl. 4-cylinder engine, 4-speed manual transmission, power brakes, rear defogger, bodyside molding, reclining front seats, fabric upholstery, folding rear seats, console, right visor mirror, locking fuel cap, tool kit, 145SR13 all-season SBR tires with full-size spare. **GVL** adds: upgraded seats and upholstery. **GVS** adds: AM/FM ST cassette, velour upholstery, wheel covers. **GVX** adds: 1.3-liter engine, uprated suspension, front air dam, fog lamps, sill extensions, rear spoiler, 155/70SR13 tires.

Optional Equipment:

Comfort & Sound Pkg., GV	999	749	874
GVL, GVX	1099	825	962
Air conditioning, AM/FM stereo.			
Rear wiper/washer, GV, GVS	139	112	126
Air conditioning, GVS	799	639	719
AM/FM ST cassette, GV	239	160	200
GVL, GVX	379	254	317
Sport Pkg., GV	229	160	195
GVL	189	132	161
Right outside mirror, wheel covers, luggage rack, rear louvers.			
Appearance Pkg., GV	99	74	87
Wheel covers, right mirror.			
Touuister Pkg., GV	229	160	195
Right outside mirror, wheel covers, luggage rack.			
Alloy wheels, GVL	349	244	297
GVS	299	224	262
Metallic paint (NA GV)	149	125	137
Passive restraint, GVS & GVX	99	89	94
Right outside mirror, GVS	39	29	34
Luggage rack, GVS	119	89	104
Travel Pkg., GVL	189	132	161
Luggage rack and right outside mirror.			
Premium Travel Pkg., GVL	249	174	212
Travel Pkg. plus rear louvers.			
Catapult seats, GVL	149	112	131
Easy-entry feature with memory.			
Rear louver, GVX	149	89	119

Prices are accurate at time of printing; subject to manufacturer's change.

CONSUMER GUIDE®